D0929635

HANDBOOK OF COMPLEX ENVIRONMENTAL REMEDIATION PROBLEMS

170201

Jay Lehr

Marve Hyman

Tyler E. Gass

William Seevers

McGRAW-HILL

New York   Chicago   San Francisco   Lisbon   London   Madrid
Mexico City   Milan   New Delhi   San Juan   Seoul
Singapore   Sydney   Toronto

Library of Congress Cataloging-in-Publication Data

Handbook of complex environmental remediation problems / Jay Lehr . . . [et al.].
    p. cm.
    Includes bibliographical references and index.
    ISBN 0-07-027689-7
        1. Hazardous wastes—Purification—Handbooks, manuals, etc.   2. Hazardous waste
site remediation—Handbooks, manuals, etc.   3. Soil remediation—Handbooks, manuals, etc.
4. Groundwater—Purification—Handbooks, manuals, etc.   I. Lehr, Jay H., date–

TD1060.H34   2001

                                                                    2001044946

McGraw-Hill

A Division of The McGraw·Hill Companies

1 2 3 4 5 6 7 8 9 0   DOC/DOC   0 9 8 7 6 5 4 3 2 1

ISBN 0-07-027689-7

*The sponsoring editor for this book was Kenneth P. McCombs, the editing
supervisor was David E. Fogarty, and the production supervisor was
Pamela A. Pelton. It was set in Times Roman by North Market Street Graphics.*

Printed and bound by R. R. Donnelley & Sons Company.

This book was printed on recycled, acid-free paper containing
a minimum of 50% recycled, de-inked fiber.

McGraw-Hill books are available at special quantity discounts to
use as premiums and sales promotions, or for use in corporate training
programs. For more information, please write to the Director of Special Sales,
Professional Publishing, McGraw-Hill, Two Penn Plaza, New York, NY
10121-2298. Or contact your local bookstore.

CONTENTS

Chapter 3. Discharges of Hazardous Waste into the Atmosphere   3.1

Chapter 4. An Evaluation of Environmental Dredging for Remediation of Contaminated Sediment   4.1

Chapter 5. Hazardous Contaminants in Marine Sediments   5.1

Chapter 6. Management of Wastes from Nuclear Facilities　　　　6.1

Chapter 7. Innovative Strategies in Remediating Mining Wastes　　7.1

Chapter 8. The Remediation of Hazardous Wastes from Oil Well Drilling 8.1

Chapter 11. Natural Resource Damage from the Groundwater Perspective   11.1

Chapter 12. Pollution Prevention/Waste Minimization   12.1

CONTRIBUTORS

Olin Braids   *O. C. Braids & Associates, LLC, Tampa, Fla.* (CHAP. 2)

Randy Buffington   *Placer Dome, Bald Mountain Mine, Elko, Nev.* (CHAP. 7)

Mark Cal   *Department of Environmental Engineering, New Mexico Tech, Socorro, N.M.* (CHAP. 3)

Bradford S. Cushing   *Applied Environmental Management, Inc., Malvern, Pa.* (CHAP. 4)

Andrew Davis   *Geomega, Boulder, Colo.* (CHAP. 7)

J. Paul Doody   *Blasland, Bouck & Lee, Syracuse, N.Y.* (CHAP. 4)

Richard W. Dunford   *Triangle Economic Research, A BBL Company, Durham, N. Carolina* (CHAP. 11)

George Fennimore   *Geomega, Boulder, Colo.* (CHAP. 7)

Kim Fowler   *Battelle, Pacific Northwest Division, Richland, Wash.* (CHAP. 12)

Tyler E. Gass   *Blasland, Bouck & Lee, Golden, Colo.* (CHAPS. 4, 10, 11)

Ken Hladek   *Duratek Federal Services of Hanford, Inc., Richland, Wash.* (CHAP. 6)

Marve Hyman   *Bechtel National, Inc., Richland, Wash.* (CHAPS. 6, 12)

James A. Jacobs   *Fast-Tek Engineering Support Services, Redwood City, Calif.* (CHAPS. 8, 9)

David S. Lipson   *Blasland, Bouck & Lee, Inc., Golden, Colo.* (CHAP. 10)

Cynthia Moomaw   *Geomega, Boulder, Colo.* (CHAP. 7)

Kevin Phillips   *FPM-Group, Ronkonkoma, N.Y.* (CHAP. 1)

Stephen S. Testa   *Testa Engineering, Mokelumne Hill, Calif.* (CHAPS. 8, 9)

Jack Q. Word   *MEC Analytical Systems, Sequim, Wash.* (CHAPS. 5)

Lucinda S. Word   *MEC Analytical Systems, Sequim, Wash.* (CHAP. 5)

ABOUT THE EDITORS

JAY LEHR has a career in environmental sicence spanning 6 decades with 14 books in print and over 400 articles. He began his career with the U.S. Geological Survey and taught at the University of Arizona and at Ohio State. He presently teaches, lectures, and consults following 25 years at the helm of The Association of Ground Water Scientists and Engineers, where he also served as Editor of the *Journal of Groundwater.* His most recent book is McGraw-Hill's widely acclaimed *Standard Handbook of Environmental Science, Health, and Technology.*

MARVE HYMAN, P.E., CIH, has an M.S. in chemical engineering from the University of California, Berkeley, and a B.S. from Caltech. He has over 30 years experience in environmental engineering, alternative energy, and industrial safety, and since 1994 has been working with Bechtel on remediation of soil, groundwater, and tank wastes contaminated with organics, metals, and radioisotopes at Hanford, Washington. Mr. Hyman is the author of the remediation article in the Wiley *Encyclopedia of Environmental Analysis and Remediation* (1998), and principal author of the ASCE Press book *Groundwater and Soil Remediation: Process Design and Cost Estimating* (2001). He is the remediation instructor in the continuing education programs of the American Institute of Chemical Engineers, American Industrial Hygiene Association, U.C. Berkeley Extension, and Washington State University.

TYLER E. GASS has approximately 30 years of experience as a hydrogeologist focusing on a variety of environmental issues, as well as groundwater resource evaluation and development. Mr. Gass is a former Director of Research of the National Ground Water Association, and is a member of the U.S. Army Science Advisory Board, where he is the former Chairman of the Infrastructure and Environment Subcommittee. Mr. Gass has published numerous articles and has coauthored two textbooks on groundwater development, resource protection, and the fate and transport of chemicals in the subsurface environment. He has been an expert witness in a number of federal and state courts, and has testified on subjects related to groundwater flow, aquifer analysis, contaminant fate and transport, groundwater remediation and groundwater natural resource damages.

WILLIAM J. SEEVERS, P.H., is the cofounder and president of the Environmental Technology Group, Inc. (ETG) a Long Island, New York based engineering and environmental consulting firm. Mr. Seevers is an environmental hydrologist with 35 years experience specializing in groundwater technology. ETG is involved in projects in the United States and overseas. The staff and principals of the Environmental Technology Group played a significant role in the creation of this handbook. They organized the chapters and recruited many of the authors who created the final product.

PREFACE

Thirty-five years ago I stood before numerous committees of the U.S. House of Representatives and the U.S. Senate calling attention to the many crimes being perpetrated against our environment by thoughtless inadvertent acts of individuals as well as ignorant and malevolent acts of industry. In the following decade Congress became convinced of the problems at hand and passed adequate legislation to educate the public and regulate the nation so as to better protect our natural resources and community environments.

Today we are a nation of environmental advocates, and we are more than adequately preserving our resources for generations to come. If anything, we have become overzealous in our efforts to purify our air, water, and soil, often requiring unreasonable standards which produce excessive costs with no subsequent measurable advances in the protection of public health. The argument of how clean is clean has left the realm of common sense and entered the laboratory. There technological advances in identifying ever smaller concentrations of chemicals have stymied our efforts to convince the public that the existence of measurable chemicals in our environment do not necessarily pose a threat to either plants, animals, or us. The result of this ever more conservative approach to environmental protection is the potential desensitizing of the public to what remains as the real threats to our environment.

In this handbook we have attempted to catalog, define, and describe the more difficult environmental protection problems which continue to require complex remedial solutions. In a nation known for the world's most successful economy, we face a number of serious industrial waste problems which cannot be readily dismissed with simple protocols.

We have asked some of our nation's leading experts in waste remediation to focus upon our most difficult challenges. We believe that collectively, we have assembled the keys to eliminating some of the most complex environmental problems facing our industrial nation. We have addressed four general areas: soil, groundwater, pollution prevention, and natural resource damage; and eight specific problems the country faces: air emissions, oil drilling, oil spills, mining, manufactured gas, marine wastes, radioactive wastes, and stream deposition.

We are confident that we have created a unique information resource that will assist many consultants, regulators, industrialists, and citizen advocates to better understand how the most difficult environmental challenges can be met successfully.

Jay Lehr

CHAPTER 1
GROUNDWATER

Kevin John Phillips*

1.1 INTRODUCTION

"A yawn is a silent shout."
    GILBERT KEITH CHESTERTON, 1874–1936

1.1.1 Why Is Groundwater Contamination So Important?

Demand for groundwater as a resource has been increasing as population growth continues to build and opportunities to develop surface water supplies continue to diminish. Groundwater accounts for approximately two-thirds of all the freshwater resources of the world (Nace, 1971). If we subtract out the ice caps and glaciers, it accounts for over 99 percent of all the freshwater available to the planet (Nace, 1971). Clearly, with 99 percent of the available resources, it behooves environmental professionals to try and protect it and, should it become polluted, to treat it.

However, one aspect of its nature is its long residence time. While typical turnover times in river systems average around two weeks, groundwater systems move much slower. Indeed, groundwater in certain zones of the Lloyd Aquifer in Long Island, N.Y., has been around since the birth of Christ. Hence, in the past, the general viewpoint held by many groundwater professionals and policy makers was that once an aquifer had been polluted, its water usage must be curtailed or possibly eliminated because of the difficulty and time in cleaning up that aquifer. This viewpoint is changing, however, as a result of new methodologies for aquifer cleanup. However, as we enter a new century, aquifer cleanup is still a very difficult and a costly endeavor that takes a significant amount of time, often yields less than desirable results, and frequently relies more on risk assessments rather than groundwater standards for cleanup levels simply because it is not yet practical.

---

* Dedicated to Sue, Al, and Chris.

1.1.2   What Are the Sources of Pollution of Groundwater?

Pollution of groundwater can result from many activities, including leaching from municipal and chemical landfills, abandoned dumpsites, accidental spills of chemical or waste materials, improper underground injection of liquid wastes, surface impoundments, placement of septic tank systems in hydrological and geological unsuitable locations, and improper chemical application of fertilizers and pesticides for agricultural and domestic vegetative processes. The pollution from solid waste left on the ground surface needs to first be solubilized before it causes a problem. Rain or melting snow will solubilize some of the waste that has been disposed of on the land and then carry that dissolved constituent down through the unsaturated zone into the saturated groundwater.

Some wastes in liquid form are only slightly soluble in water. This class of compounds are called *nonaqueous-phase liquids* (NAPLs) and pose a significant threat to the groundwater system. Such waste becomes trapped in the pore spaces of the aquifer and remains there in groundwater, slowly dissolving and yielding a continuous source of pollution. There are two kinds of NAPLs—dense NAPLs (DNAPLs) and light NAPLs (LNAPLs). DNAPLs are compounds whose density exceeds that of water (e.g., chlorinated solvents), and LNAPLs are compounds whose density is less than that of water (e.g., oils and petroleum products).

1.1.3   What Is the Hydrology of Contamination?

Precipitation is the driving force that moves the groundwater system. The groundwater system moves slowly compared to surface water. Groundwater velocity is generally in the order of 1 foot per day to 1 foot per year throughout the United States, depending on the hydraulic conductivity and the gradient of the groundwater system. Groundwater movement is generated from precipitation that mounds up the freshwater resources in an aquifer, which begins to move toward a sink, usually a creek, river, or other surface body of water. These surface water bodies are lower in their energy state (elevation head), and hence the groundwater system flows from a higher energy head to that of a lower energy head and is frequently plotted and shown as water table contours or potentiometric surface maps. These water table contours or potentiometric surface maps show the energy level of the aquifer and in general determine the gradient by which the groundwater is moving. Flow lines are almost always drawn perpendicular to groundwater contours even though this only occurs in an isotropic homogeneous porous media (something the author has never seen).

1.1.4   What Aspects of Geochemistry Are Important in Understanding Groundwater Pollution?

As mentioned earlier, precipitation is a major factor in groundwater systems. Not only does it drive the groundwater system flow, but it also dissolves the contaminants that have been left on the surface of land, buried beneath land, or locked into the pore spaces. Hence, the *solubility* of these wastes becomes a significant factor in groundwater contamination. For example, road salt has almost unlimited solubility in water. Once a contaminant has solubilized, it will move downward by gravity in the unsaturated zone, enter the saturated zone, and move with the groundwater. However, certain contaminants absorb to and desorb from the organic material in the aquifer. This phenomenon, described as *retardation,* slows down the contaminant transport but does not affect the molecules themselves. Indeed, retardation coeffi-cients of 10 to 20 have been documented for some waste. Some of the inorganic com-pounds such as nitrates and chlorides show almost no retardation at all, moving with the speed of the groundwater.

Additional significant factors in contaminant transport include *biodegradation* and *biotransformation;* many compounds undergo biodegradation both aerobically and anaerobically. This process can account for significant amounts of destruction of toxic molecules. Indeed, biodegradation as recently as 10 years ago was considered as only a natural process, but today, biodegradation has been marketed by hundreds of companies for specific and nonspecific compounds where bacteria, fungi, and other micro-organisms have been grown to break down certain contaminants.

Chemical reactions occur in aquifers continually. Chemical transformations, in-cluding oxidation and reduction, can be major routes for destruction and transfor-mation of contaminants as they pass through the aquifer.

1.1.5   What Are the Effects of These Compounds on Human Health and the Environment?

The effects these contaminants have on human health and the environment are clearly demonstrated by the amount of concern that has been shown by the United States Congress since the 1970s when the first water pollution control act was passed. The threat from groundwater is one that is very real because 35 percent of the United States water supply comes from the ground. Outside the major cities, 95 percent of the water supply comes from the ground (Driscoll, 1983). Documentary movies and books, such as *A Civil Action,* have clearly demonstrated the effect of these chemicals, some of which are both toxic and carcinogenic and directly affect the human population.

1.1.6   What Are the Chemicals That Have the Greatest Impact on Groundwater Quality?

One of the first overview studies of aquifer cleanup that took place was written in 1977 by Lindorff and Cartwright (1977) when they surveyed the nation for case histories of aquifer cleanup. At that time, 116 cases of aquifer pollution were summarized, with most of the pollution caused by industrial waste or leaching from municipal landfills. In 1977, the most common groundwater pollutional sources were gasoline, cyanide, acrylonitrile, acetone, hydrochloric acid, solvents, acids, heavy metals, chlorides, alu-minum, fuel oils, insecticides, organic wastes, sulfite liquors, petrochemicals, zinc, lead, and cadmium.

Since 1977, when Lindorff and Cartwright did their survey, the most important new parameters to be recognized as a significant threat to our groundwater quality have been the chlorinated hydrocarbons. These contaminants have very low solubili-ties but very high toxicities and carcinogentic potential. In addition, they are denser then water and have been labeled as dense nonaqueous-phase liquids (DNAPLs). Their particular problematic attributes are that, even though they are very slow to dissolve and have low solubility, they are considered carcinogenic at extremely low concentrations and are denser than water, and hence sink through the saturated media, contaminating the deeper portion of the aquifer. These compounds are typi-cally not readily biodegradable, and if they do biodegrade, it is a slow process. These DNAPLs are nonwetting with respect to water and get trapped in the porous media for long periods of time slowly dissolving into the aquifer, causing significant groundwater contamination for very long periods of time. The other low-solubility contaminants that frequently show up are the petroleum hydrocarbons commonly called light nonaqueous-phase liquids (LNAPLs). These LNAPLs also have low-solubility characteristics and can exist in the subsurface environment as pure-phase liquids. However, they are lighter than water and will not sink through the aquifer but remain on the surface of the water table. In addition, another difference between the DNAPLs and the LNAPLs is that the LNAPLs in general are readily biodegrable, while the DNAPLs have slower rates of biodegradability.

1.1.7 Summary

As the cleanup of groundwater and groundwater remediation systems is extremely complicated, I have attempted to simplify it by using solubility as a organizer of the text in this section and throughout the chapter.

Section 1.2, "Investigative Methods," will report on investigative measures in three areas: (1) the aqueous groundwater contaminants that are dissolved and move in the groundwater, (2) DNAPLs, and (3) LNAPLs.

Section 1.3 deals with remediation methods, and again the section will be organized by: (1) the aqueous groundwater remediation methods that focus on either the in situ treatment or the removal and treatment of the groundwater, (2) DNAPLs, (3) LNAPLs. Section 1.3 will also compare treatment methodologies and include cost estimates for groundwater, DNAPLs, and LNAPLs cleanup. Section 1.4 will consist of case histories of aquifer restorations.

1.2 INVESTIGATIVE METHODS

"Every truth passes through three stages before it is recognized. In the first, it is ridiculed, in the second it is opposed, in the third it is regarded as self-evident." A. SCHOPENHAUER, 1788–1860

1.2.1 Introduction

This section will be dealing with investigative methods for aqueous groundwater, DNAPLs and LNAPLs. The aqueous groundwater portion will first discuss the investigative methods for the kinds of chemicals that are frequently targeted at contamination sites. The three major lists of compounds that are frequently investigated come from the three major pieces of legislation for the cleanup of water: the priority pollutant list (Clean Water Act), the target compound list (Comprehensive Environmental Response, Compensation, and Liability Act—CERCLA), and the SW-846 analyte list (Resource Conservation and Recovery Act—RCRA). The list most often used for screening groundwater at contaminated sites is the Target Compound List and the Target Analyte List, TCL and TAL, respectively, will be discussed in Sec. 1.2.2. Section 1.2.3, covering the DNAPLs investigative methods, will focus on the pure-phase DNAPLs. Finally, Sec. 1.2.4 will discuss investigative methods for LNAPLs.

1.2.2 Aqueous Groundwater

Prior to discussing investigative methods for groundwater, we first must define what kinds of compounds we are going to investigate. Table 1.1 shows the Target Compounds List (TCL/TAL), the priority pollutants, and SW-846 compounds. The most widely used list of investigative compounds today is the U.S. EPA TCL and TAL. The first list of compounds was the EPA Priority Pollutant List, which was established in 1974 and was the first comprehensive list of compounds identifying the most frequently used compounds in industry as well as the ones we had laboratory methods to test for. Since then, great accomplishments have been made in laboratory analysis, expanding this list. In addition, compounds that were toxic and persistent were included. Today the Target Compound List is usually the measure by which contaminated sites are characterized.

The TCL is broken up into several chemical categories. The first category is the volatile organic chemicals (VOCs), which have a vapor pressure greater than 1 mmHg. These chemicals are almost all organically based and present a class of compounds that can easily volatize in the environment. The number of compounds included in this category is 34.

The second group of compounds in the TCL is the semivolatiles made up of the base neutral and acid-extractable compounds. The base neutral compounds are so called because of the way they are extracted and analyzed in the laboratory. There are 49 of the base neutral compounds given in the TCL.

The acid-extractable compounds are so called because of the laboratory method of extraction. They are all organic. There are 15 acid-extractable compounds in the TCL.

The next groups of compounds are the pesticides and PCBs, and they comprise a total of 29 compounds in the TCL.

The final group are elements and are inorganic. This group has 23 metals associated with them.

The analyses of these compounds and elements are shown in Table 1.2 along with the recommended containers, preservation, holding time, and analytical methodology.

The More Important Chemicals. The more important chemicals are those that show up more frequently in the groundwater and are more toxic, thereby causing more problems for cleanup. The most frequently detected compounds in groundwater at the waste disposal sites in Germany and in the United States have been reported by Keeley (1999). Chlorinated hydrocarbons dominate the list of frequently detected compounds at these waste sites (Fig. 1.1). All of the top-ranked contaminants in the United States are chlorinated hydrocarbons. In the dissolved phase, most of these contaminants have drinking water standards in the low parts per billion range. In the pure phase they all would be classified as DNAPLs. Though EPA requires preliminary screening using the TCL and TAL, clearly some compounds are of more concern. Presently, the most important compounds are the chlorinated hydrocarbons in the pure phase (DNAPL) and in the dissolved phase. They can be carcinogenic at a very low level, they pose significant additional problems because of their ability to sink through the aquifer as a pure DNAPL, they are of low solubility so water cannot easily flush out the problem, their retardation is usually high so their movement is slow, and the compounds are usually resistant to biodegradation so their natural attenuation is low.

Monitoring Strategies. Prior to discussing monitoring strategies, a brief discussion of well drilling methods is needed. Table 1.3 is an adaptation of Cohen and Mercer's

TABLE 1.1   Comparison of Chemicals on Three Regulatory Lists

Compounds	TCL*	PPL†	SW-846‡
Volatiles and semivolatiles			
1. Acenaphthene	x	x	x
2. Acenaphthylene	x	x	x
3. Acetone	x		x
4. Acrolein		x	x
5. Acrylonitrile		x	x
6. Anthracene	x	x	x
7. Benzo (a) anthracene	x	x	x
8. Benzo (a) pyrene	x	x	x
9. Benzene	x	x	x
10. Benzidine		x	x
11. Benzo (b) flouranthene	x	x	x
12. Benzo (ghi) perylene	x	x	x
13. Benzo (k) flouranthene	x	x	x
14. Benzoic acid	x		x
15. Benzyl alcohol	x		x
16. Bis (2-chloroethoxy) methane	x	x	x
17. Bis (2-chloroethyl) ether	x	x	x
18. Bis (2-chloroisopropyl) ether	x	x	x
19. Bis (2-ethylhexyl) phthalate	x	x	x
20. Bromoform	x	x	x
21. Bromodichloromethane	x	x	x
22. Bromomethane	x	x	x
23. 4-Bromophenyl phenyl ether	x	x	x
24. 2-Butanone	x		x
25. Butyl benzyl phthalate	x	x	x
26. Carbon disulfide	x		x
27. Carbon tetrachloride	x	x	x
28. 4-Chloro-3-methylphenol (P-chloro-M-cresol)	x	x	x
29. 4-Chloroaniline	x		x
30. Chlorobenzene	x	x	x
31. Chloroethane	x	x	x
32. Chloromethane	x	x	x
33. Chlorodibromomethane	x	x	
34. 2-Chloroethyl vinyl ether		x	x
35. Chloroform	x	x	x
36. 2-Chloronaphthalene	x	x	x
37. 2-Chlorophenol	x	x	x
38. 4-Chlorophenyl phenyl ether	x	x	x

TABLE 1.1 Comparison of Chemicals on Three Regulatory Lists (*Continued*)

Compounds	TCL*	PPL†	SW-846‡
Volatiles and semivolatiles			
39. Chrysene	x	x	x
40. Di-n-butylphthalate	x	x	x
41. Di-n-octylphthalate	x	x	x
42. Dibenz (a,h) anthracene	x	x	x
43. Dibenzofuran	x		x
44. 1,2-Dichlorobenzene	x	x	x
45. 1,3-Dichlorobenzene	x	x	x
46. 1,4-Dichlorobenzene	x	x	x
47. 3,3-Dichlorobenzidine	x	x	x
48. 1,1-Dichloroethane	x	x	x
49. 1,2-Dichlorothane	x	x	x
50. 1,1-Dichloroethylene	x	x	x
51. 1,2-Dichloroethylene (total)	x		
52. Tran-1,2-dichloroethylene		x	x
53. 2,4-Dichlorophenol	x	x	x
54. 1,2-Dichloropropane	x	x	x
55. c-1,3-Dichloropropylene	x	x	x
56. t-1,3-Dichloropropylene	x	x	x
57. Diethyl phthalate	x	x	x
58. Dimethyl phthalate	x	x	x
59. 2,4-Dimethylphenol	x	x	x
60. 4,6-Dinitro-2-methylphenol	x	x	x
61. 2,4-Dinitrophenol	x	x	x
62. 2,4-Dinitrotoluene	x	x	x
63. 2,6-Dinitrotoluene	x	x	x
64. 1,2-Diphenylhydrazine		x	x
65. Ethylbenzene	x	x	x
66. Flouranthene	x	x	x
67. Flourene	x	x	x
68. Hexachlorobenzene	x	x	x
69. Hexachlorobutadiene	x	x	x
70. Hexachlorocyclopentadiene	x	x	x
71. Hexachloroethane	x	x	x
72. 2-Hexanone	x		x
73. Indeno (1,2,3,-cd) pyrene	x	x	x
74. Isophorone	x	x	x
75. Methylene chloride	x	x	x
76. 4-Methyl-2-pentanone	x		x

TABLE 1.1   Comparison of Chemicals on Three Regulatory Lists (*Continued*)

Compounds	TCL*	PPL[†]	SW-846[‡]
Volatiles and semivolatiles			
77. 2-Methylnaphthalene	x		x
78. 2-Methylphenol	x		
79. 4-Methylphenol	x		
80. N-Nitrosodipropylamine	x	x	x
81. N-Nitrosodimethylamine		x	x
82. N-Nitrosodiphenylamine	x	x	x
83. Naphthalene	x	x	x
84. 2-Nitroaniline	x		x
85. 3-Nitroaniline	x		x
86. 4-Nitroaniline	x		x
87. Nitrobenzene	x	x	x
88. 2-Nitrophenol	x	x	x
89. 4-Nitrophenol	x	x	x
90. Phentachlorophenol	x	x	x
91. Phenanthrene	x	x	x
92. Phenol	x	x	x
93. Pyrene	x	x	x
94. Styrene	x		x
95. 1,1,2,2-tetrachlorobenzene	x	x	x
96. Tetrachloroethane	x	x	x
97. Toluene	x	x	x
98. Total xylenes	x		x
99. 1,2,4-Trichlorobenzene	x	x	x
100. 1,1,1-Trichloroethane	x	x	x
101. 1,1,2-Trichloroethane	x	x	x
102. Trichloroethylene	x	x	x
103. 2,4,5-Trichlorophenol	x		x
104. 2,4,6-Trichlorophenol	x	x	x
105. Vinyl acetate	x		x
106. Vinyl chloride	x	x	x
Pesticides/PCBs			
107. Aldrin	x	x	x
108. Dieldrin	x	x	x
109. Chlordane		x	x
110. Alpha-chlordane	x		
111. Gamma-chlordane	x		
112. 4,4'-DDT	x	x	x

TABLE 1.1   Comparison of Chemicals on Three Regulatory Lists (*Continued*)

Compounds	TCL*	PPL†	SW-846‡
Pesticides/PCBs			
113. 4,4'-DDD	x	x	x
114. 4,4'-DDE	x		x
115. Endosulfan I	x	x	x
116. Endosulfan II	x	x	x
117. Endosulfan sulfate	x	x	x
118. Endrin	x	x	x
119. Endrin aldehyde		x	x
120. Heptachlor	x	x	x
121. Heptachlor epoxide	x	x	x
122. Methoxychlor	x		x
123. Endrin ketone	x		x
124. BHC (alpha)	x	x	x
125. BHC (beta)	x	x	x
126. BHC (gamma)	x	x	x
127. BHC (delta)	x	x	x
128. Toxaphene	x	x	x
129. PCB 1242	x	x	x
130. PCB 1254	x	x	x
131. PCB 1221	x	x	x
132. PCB 1232	x	x	x
133. PCB 1248	x	x	x
134. PCB 1260	x	x	x
135. PCB 1016	x	x	x
136. 2,3,7,8-TCDD	x	x	x
Metals			
137. Aluminum	x		x
138. Antimony	x	x	x
139. Arsenic	x	x	x
140. Barium	x		x
141. Beryllium	x	x	x
142. Cadmium	x	x	x
143. Calcium	x		x
144. Chromium	x	x	x
145. Cobalt	x		x
146. Copper	x	x	x
147. Iron	x		x
148. Lead	x	x	x

TABLE 1.1 Comparison of Chemicals on Three Regulatory Lists (*Continued*)

Compounds	TCL*	PPL†	SW-846‡
Metals			
149. Magnesium	x		x
150. Manganese	x		x
151. Mercury	x	x	x
152. Nickel	x	x	x
153. Potassium	x		x
154. Selenium	x	x	x
155. Silver	x	x	x
156. Sodium	x		x
157. Thallium	x	x	x
158. Vanadium	x		x
159. Zinc	x	x	x

* Targeted Compound List (TCL) and Target Analyte List (TAL) from the U.S. EPA Contract Laboratory Programs.
† Priority Pollutant List (PPL) from the Clean Water Act.
‡ SW-846 analyte list from the RCRA program.

work (1993). This table discusses the various methods for drilling and their applications, advantages, and limitations for each of the methods from hand augering through direct push methods. As one can see from the table, there are various methods for drilling and installing observation wells, or for taking soil samples. Although a myriad of methods exist, the hollow-stem auger is the most often used and preferred method for installing observation wells because of the lack of introduction of any foreign material such as bentonite clay, slurry, or artificial organic gum (Johnson Revert™). Hence, many states will only accept hollow-stem augered wells. Once the earth has been drilled, a monitoring well then must be set and gravel packed. Most states have specifications on installation of monitoring wells in unconsolidated and bedrock formation (see Figs. 1.2 and 1.3), double-cased wells, and deep aquifer wells. Selection of a screen length, diameter, and elevation for each observation well is a function of the groundwater contamination or plume one desires to identify.

Once a plume has been identified as a problem by a regulatory agency, establishing its nature and extent is usually mandatory. In order to accomplish this, the first thing that has to be identified is the conceptual geological model. The geological model must encompass both regional information from sources such as the U.S. Geological Survey, university geological reports, and local information from sources such as local borings for construction, water supply borings, or site borings. The objective is to develop an understanding of all the geological substrata that may channel the flow patterns below the surface by acting as barriers to or conductors of groundwater flow. Once this geological conceptual model is put together it must become a "living" model in that it needs to be updated and changed as frequently as necessary as more and more information becomes available at the site. Indeed, some of the monitoring wells that will be installed may have as a secondary objective verifying certain substrata or boundary conditions that the geological model has identified as significant.

TABLE 1.2   Analysis of Targeted Compound List/Targeted Analyte List (TCL/TAL)*

Parameter	Sample container	Container volume	Preservation	Maximum holding time[†]	Analytical methodology
Volatile organics	Aqueous glass, black phenolic plastic screw cap, Teflon-lined septum	Aqueous—40 mL	Cool, 4°C, dark, HCl to pH < 2	14 days	SW-846 8260
	Nonaqueous glass, polypropylene cap, white Teflon liner	Nonaqueous—100 g	Cool, 4°C, dark	14 days	SW-846 8260
Base neutral/acid-extractable organics	Aqueous amber glass, Teflon-lined cap	1000 mL	Cool, 4°C, dark, $Na_2S_2O_3$	Extraction/analysis 7/40 days	SW-846 8270
	Nonaqueous glass, amber	100 g	Cool, 4°C	Extraction/analysis 14/40 days	SW-846 8270
Pesticide/PCBs	Amber glass	1000 mL	Cool, 4°C, $Na_2S_2O_3$	Extraction/analysis 7/40 days	SW-846 8081 SW-846 8082
	Nonaqueous glass	100 g	Cool, 4°C	Extraction/analysis 14/40 days	SW-846 8081 SW-846 8082
2,3,7,8-TCDD	Glass	1000 mL	Cool, 4°C, $Na_2S_2O_3$	Extraction/analysis 7/40 days	EPA 625/8270
	Nonaqueous glass	100 g	Cool, 4°C	Extraction/analysis 14/40 days	EPA 625m/8270
Metals except Hg	Aqueous-plastic bottle, plastic cap, plastic liner	500 mL	HNO_3 to pH < 2, Cool, 4°C	180 days	SW-846 6010
	Nonaqueous flint glass bottle, black phenolic cap, polyethylene liner	Nonaqueous, 100 g	4°C until analysis	180 days	SW-846 6010
Cyanide	Aqueous plastic bottle	500 mL	0.6 g ascorbic acid if residual Cl, NaOH to pH > 12, Cool, 4°C until analyzed, $CaCo_3$ in presence of sulfide	14 days	SW-846 9012
	Nonaqueous glass	100 g	Cool, 4°C until analyzed	14 days	SW-846 9012

* Using U.S. EPA Contract Lab Program Methodologies for aqueous and nonaqueous samples.
[†] Verified time of sample receipt (at the laboratory).

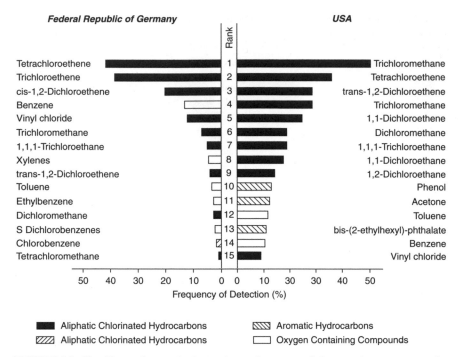

FIGURE 1.1 The 15 most frequently detected organic compounds in groundwater at waste disposal sites in Germany and the United States. (*Modified from Keeley, 1999.*)

Once the geological model has been conceptualized to fulfill the objective of establishing the vertical and horizontal extent of contamination, you must consider the nature of the plume you want to describe. Two aspects of a plume's vertical migration are its density and its regional hydrodynamics. When the plume's density exceeds 10,000 milligrams/liter (mg/L), it will have a tendency to sink in the aquifer. A common misconception among groundwater professionals is that DNAPL plumes sink. Almost no DNAPL plumes sink because their solubility is almost always less then the 10,000 mg/L. DNAPLs, in the pure phase, are indeed heavier than water and sink, but when they dissolve their solubility is so low that the resultant mixture usually cannot reach a density where it will sink in the aquifer.

The second aspect of vertical migration is the regional hydrodynamic flow pattern. Figure 1.4 demonstrates that vertically downward flow in an aquifer, and hence plume downward movement, is a reality in recharge areas, while vertically upward flow takes place in the discharge areas (Freeze and Cherry, 1979).

In addition to vertical movement, horizontal movement occurs. Very simple to very complex groundwater quality models have been used to describe both the contaminant transport and the potentiometric flow lines that represent the spread and transport of the plume.

Once this groundwater model is described, it gives us the predictive tool necessary to begin refining our estimates of the nature and extent of the plume by sampling at select locations for specific parameters if the source and the time when the initial contamination took place are known. Due to the high laboratory costs of the

TABLE 1.3 Drilling Methods, Application Advantages, and Limitations

Method	Applications/advantages	Limitations
Hand augers—A hand auger is advanced by turning it into the soil until the bucket or screw is filled. The auger is then removed from the hole. The sample is dislodged from the auger, and drilling continues. Motorized units are also available.	• Shallow soil investigations (0 to 15 ft) • Soil samples collected from the auger cutting edge • Water-bearing zone identification • Contamination presence examination; sample analysis • Shallow, small-diameter well installation • Experienced user can identify stratigraphic interfaces by penetration resistance differences as well as sample inspection • Highly mobile, and can be used in confined spaces • Various types (e.g., bucket, screw) and sizes (typically 1 to 9 in in diameter) • Inexpensive to purchase	• Limited to very shallow depths (typically < 15 ft) • Unable to penetrate extremely dense or rocky or gravelly soil • Borehole stability may be difficult to maintain, particularly beneath the water table • Potential for vertical cross-contamination • Labor intensive
Solid-flight augers—A cutter head (≥ 2-in diameter) is attached to multiple auger flights. As the augers are rotated by a rotary drive head and forced down by either a hydraulic pulldown or a feed device, cuttings are rotated up to ground surface by moving along the continuous flighting.	• Solid soils investigations (< 100 ft) • Soil samples are collected from the auger flights or by using split-spoon or thin-walled samplers if the hole will not cave upon retrieval of the augers • Vadose zone monitoring wells • Monitor wells in saturated, stable soils • Identification of depth to bedrock • Fast and mobile; can be used with small rigs • Holes up to 3 ft in diameter • No fluids required • Simple to decontaminate	• Low-quality soil samples unless split spoon or thin-wall samples are taken • Soil sample data limited to areas and depths where stable soils are predominant • Unable to install monitor wells in most unconsolidated aquifers because of borehole caving upon auger removal • Difficult penetration in loose boulder, cobbles, and other material that might lock up auger • Monitor well diameter limited by auger diameter • Cannot penetrate consolidated materials • Potential for vertical cross-contamination
Hollow-stem augers—Hollow-stem augering is done in a similar manner to solid-flight augering. Small-diameter drill rods and samplers can be lowered through the hollow augers for sam-	• All types of soil investigations to < 100 ft below ground • Permits high-quality soil sampling with split-spoon or thin-wall samplers • Water-quality sampling • Monitor well installation on all unconsolidated formation	• Difficulty in preserving sample integrity in heaving (running sand) formations • If water or drilling mud is used to control heaving, the mud will invade the formation

TABLE 1.3 Drilling Methods, Application Advantages, and Limitations (*Continued*)

Method	Applications/advantages	Limitations
pling. If necessary, sediment within the hollow stem can be cleaned out prior to inserting a sampler. Wells can be completed below the water table by using the augers as temporary casing.	• Can serve as a temporary casing for coring rock • Can be used in stable formations to set surface casing • Can be used with small rigs in confined spaces • Does not require drilling fluids	• Potential for cross-contamination of aquifers where annular space is not positively controlled by water or drilling mud or surface casing • Limited auger diameter limits casing size (typical augers are 6¼-in OD with 3¼-in ID, and 12-in OD with 6-in ID) • Smearing of clays may seal off interval to be monitored
Direct mud rotary—Drilling fluid is pumped down the drill rods and through a bit attached to the bottom of the rods. The fluid circulates up the annular space, bringing cuttings to the surface. At the surface, drilling fluid and cuttings are discharged into a baffled sedimentation tank, pond, or pit. The tank effluent overflows into a suction pit where drilling fluid is recirculated back through the drill rods. The drill stem is rotated at the surface by top head or rotary table drives and down pressure is provided by pulldown devices or drill collars.	• Rapid drilling of clay, silt, and reasonably compacted sand and gravel to great depth (> 700 ft) • Allows split-spoon and thin-wall sampling in unconsolidated materials • Allows drilling and core sampling in consolidated rock • Abundant and flexible range of tool size and depth capabilities • Sophisticated drilling and mud programs available • Geophysical borehole logs	• Difficult to remove drilling mud and wall cake from outer perimeter of filter pack during development • Bentonite or other drilling fluid additives may influence quality of groundwater samples • Potential for vertical cross-contamination • Circulated cutting samples are of poor quality; difficult to determine sample depth • Split-spoon and thin-wall samplers are expensive and of questionable cost effectiveness at depths > 150 ft • Wireline coring techniques for sampling both unconsolidated and consolidated formations often not available locally • Drilling fluid invasion of permeable zones may compromise integrity of subsequent monitor well samples • Difficult to decontaminate pumps
Air rotary—Air rotary drilling is similar to mud rotary drilling except	• Rapid drilling of semiconsolidated and consolidated rock to great depth (> 700 ft)	• Surface casing frequently required to protect top of hole from caving in

TABLE 1.3   Drilling Methods, Application Advantages, and Limitations (*Continued*)

Method	Applications/advantages	Limitations
that air is the circulation medium. Compressed air injected through the drill rods circulates cuttings and groundwater up the annulus to the surface. Typically, rotary drill bits are used in sedimentary rocks and downhole hammer bits are used in harder igneous and metamorphic rocks. Monitor wells can be completed as open hole intervals beneath telescoped casings.	• Good quality/reliable formation samples (particularly if small quantities of drilling fluid are used) because casing prevents mixture of cuttings from bottom of hole with collapsed material from above • Allows for core sampling of rock • Equipment generally available • Allows easy and quick identification of lithologic changes • Allows identification of most water-bearing zones • Allows estimation of yields in strong water-producing zones with short downtime	• Drilling restricted to semiconsolidated and consolidated formations • Samples reliable, but occur as small chips that may be difficult to interpret • Drying effect of air may mask lower-yield water-producing zones • Air stream requires contaminant filtration • Air may modify chemical or biological conditions; recovery time is uncertain • Potential for vertical cross-contamination • Potential exists for hydrocarbon contamination from air compressor or downhole hammer bit oils
Air rotary with casing driver—This method uses a casing driver to allow air rotary drilling through unstable unconsolidated materials. Typically, the drill bit is extended 6 to12 in ahead of the casing, the casing is driven down, and then the drill bit is used to clean material from within the casing.	• Rapid drilling of unconsolidated sands, silts, and clays • Drilling in alluvial material (including boulder formations) • Casing supports borehole integrity and reduces potential for vertical cross-contamination • Eliminates circulation problems common with direct mud rotary method • Good formation samples because casing (outer wall) prevents mixture of caving materials with cutting from bottom of hole • Minimal formation damage as casing is pulled back (smearing of silts and clays can be anticipated)	• Thin, low-pressure water-bearing zones easily overlooked if drilling is not stopped at appropriate places to observe whether water levels are recovering • Samples pulverized as in all rotary drilling • Air may modify chemical or biological conditions; recovery time is uncertain
Dual-wall reverse rotary—Circulating fluid (air or water) is injected through the outer casing and drill pipe, flows into the drill pipe through the bit, and carries cuttings to the surface through the drill pipe. As in rotary drilling	• Very rapid drilling through both unconsolidated and consolidated formations • Allows continuous sampling in all types of formations • Very good representative samples can be obtained with reduced risk of contamination of sample and/or water-bearing zone	• Limited borehole size that limits diameter of monitor wells • In unstable formations, well diameters are limited to approximately 4 in • Equipment available more commonly in the southwest United States than elsewhere

TABLE 1.3 Drilling Methods, Application Advantages, and Limitations (*Continued*)

Method	Applications/advantages	Limitations
with the casing driver, the outer pipe stabilizes the borehole and reduces cross-contamination of fluids and cuttings. Various bits can be used with this method.	• Allows for rock coring • In stable formations, wells with diameters as large as 6 in can be installed in open-hole completions	• Air may modify chemical or biological conditions; recovery time is uncertain • Unable to install filter pack unless completed open hole
Cable tool drilling—A drill bit is attached to the bottom of a weighted drill stem that is attached to a cable. The cable and drill stem are suspended from the drill rig mast. The bit is alternatively raised and lowered into the formation. Cuttings are periodically removed using a bailer. Casing must be added as drilling proceeds through unstable formations.	• Drilling in all types of geologic formations • Almost any depth and diameter range • Ease of monitor well installation • Ease and practicality of well development • Excellent samples of coarse-grained media can be obtained • Potential for vertical cross-contamination is reduced because casing is advanced with boring • Simple equipment and operation	• Drilling is slow, and frequently not cost-effective as a result • Heaving of unconsolidated materials must be controlled • Equipment availability more common in central, north central, and northeast sections of the United States
Rock coring—A carbide or diamond-tipped bit is attached to the bottom of a hollow core barrel. As the bit cuts deeper, the rock sample moves up into the core tube. With a double-wall core barrel, drilling fluid circulates between the two walls and does not contact the core, allowing better recovery. Clean water is usually the drilling fluid. Standard core tubes attached to the entire string of rods must be removed after each core barrel is withdrawn through the drill string by using an overshot device that is lowered on a wireline into the drill string.	• Provides high-quality, undisturbed core samples of stiff to hard clays and rock • Holes can be drilled at any angle • Can use core holes to run a complete suite of geophysical logs • Variety of core sizes available • Core holes can be utilized for hydraulic tests and monitor well completion • Can be adapted to a variety of drill rig types and operations	• Relatively expensive and slow rate of penetration • Can lose a large quantity of drilling water into permeable formations • Potential for vertical cross-contamination
Cone penetrometer—Hydraulic rams are used to push a narrow rod (e.g., 1.5-in diameter) with a	• Efficient tool for stratigraphic logging of soft soils • Measurement of some soil/fluid properties (e.g., tip pene-	• Unable to penetrate dense geologic conditions (i.e., hard clays, boulders, etc.) • Limited depth capability

TABLE 1.3 Drilling Methods, Application Advantages, and Limitations (*Continued*)

Method	Applications/advantages	Limitations
conical point into the ground at a steady rate. Electronic sensors attached to the test probe measure tip penetration resistance, probe side resistance, inclination and pore pressure. Sensors have also been developed to measure subsurface electrical conductivity, radioactivity, and optical properties (fluorescence and reflectance). Cone penetrometer tests (CPTs) are generally performed with a special rig and computerized data collection, analysis, and display system. To facilitate interpretation of CPT data from numerous tests, CPT data from at least one test per site should be compared to a log of continuously sampled soil at adjacent locations.	tration resistance, probe side fraction, pore pressure, electrical conductivity, radioactivity, fluorescence); with proper instrumentation, can be obtained continuously rather than at intervals, thus improving the detectability of thin layers (i.e., subtle DNAPL capillary barriers) and contaminants • There are virtually no cuttings brought to the ground surface, thus eliminating the need to handle cuttings • Process presents a reduced-potential for vertical cross-contamination if the openings are sealed with grout from the bottom up upon rod removal • Porous probe sampler can be used to collect groundwater samples with minimal loss of volatile compounds • Soil gas sampling can be conducted • Fluid sampling from discrete intervals can be conducted by using special tools (e.g., the Hydropouch™ manufactured by Q.E.D. Environmental Systems, Ann Arbor, Mich.)	• Soil samples cannot be collected for examination or chemical analyses, unless special equipment is utilized • Only very limited quantities of groundwater can be sampled • Limited well construction capability • Limited availability
Direct push methods— Hydraulic rams are used to push sampling devices into the ground. The sampling devices are affixed to the end of the rig rods and are typically 1 to 2 inches in diameter. Soil samples are collected in coring devices (macro- or large-bore corers), which may be either open ended (if the geologic materials are such that the hole stays open) or may be closed by a point that is either retracted or pushed out once the sampler reaches the target depth.	• Efficient, fast and inexpensive • Can sample groundwater and soil • Can be mounted on all-terrain vehicles or may be hand operated from a remote location allowing sampling in restricted access areas. • Except for the first few feet, no drill cuttings are produced. No costs associated with drill cutting disposal • Groundwater samples are obtained over a short interval (1 to 2 ft)	• One-time sampling only • Limited depth of sampling—100 ft and less • Limited amount of soil sample • Often the groundwater sample has high turbidity, necessitating samples of filtered and unfiltered groundwater • Not suitable for clay and silt • Vertical profiling should be performed from the top down to avoid cross-contamination.

TABLE 1.3 Drilling Methods, Application Advantages, and Limitations (*Continued*)

Method	Applications/advantages	Limitations
The opened coring device is lined with a dedicated disposable sleeve and is pushed by the hydraulic hammer through the interval to be sampled. The filled coring device is then brought to the ground surface by pulling up the rods. The sleeve is then removed from the coring device and sliced open for inspection and sampling.		
Groundwater samples are collected by using a closed screened rod or open		
Slotted rod which is affixed to the end of the rig rods. The sampling rod is driven to the desired depth and, if a closed screened rod is used, the screen is opened by pushing it out of the end of the sampling rod. Groundwater samples are obtained by using dedicated tubing inserted through the rig rods. The groundwater is brought to the surface either by using a peristaltic pump or by manually pumping the tubing if a downhole check valve is utilized. A vacuum pump may also be utilized if the samples will not be analyzed for VOCs. In cases where clays are present and groundwater does not flow readily into the sampling rod, a 1-in PVC well may be installed in the borehole created by the direct-push rods.		

Source: Modified from Cohen and Mercer (1993).

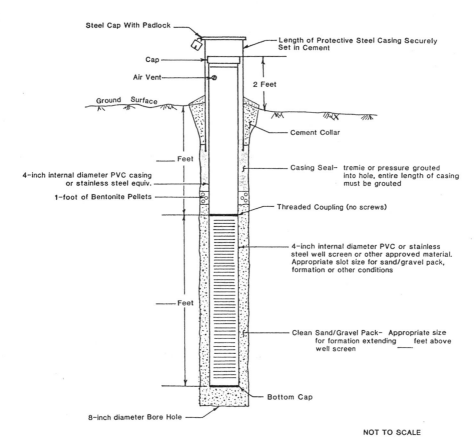

Steel Cap With Padlock

Length of Protective Steel Casing Securely Set in Cement

Cap

Air Vent

2 Feet

Ground Surface

Cement Collar

_____ Feet

4-inch internal diameter PVC casing or stainless steel equiv.

Casing Seal- tremie or pressure grouted into hole, entire length of casing must be grouted

1-foot of Bentonite Pellets

Threaded Coupling (no screws)

4-inch internal diameter PVC or stainless steel well screen or other approved material. Appropriate slot size for sand/gravel pack, formation or other conditions

_____ Feet

Clean Sand/Gravel Pack- Appropriate size for formation extending ____ feet above well screen

Bottom Cap

8-inch diameter Bore Hole

NOT TO SCALE

FIGURE 1.2 New Jersey Department of Environmental Protection monitor well specifications for unconsolidated formations, NJGS Revised 9-87. (*From N.J. DEP, 1988.*)

investigation, targeted compounds should be originally selected on the basis of use, solubility persistence in environment, sorption, biological degradation, toxicity, and expected breach of any standards or emerging regulatory concern (each site has its own special selection of targeted compounds to investigate). Additional parameters should also be investigated. Parameters such as pH, dissolved oxygen (DO), total organic carbon (TOC), chemical oxygen demand (COD), and reduction/oxidation potential all have significant value in interpreting the contamination patterns observed. When the plume enters the natural environment, certain things begin to happen. Pope and Jones (1999) consider the following processes as the most important: biodegration, absorption, dispersion and dilution, chemical reactions, and volatilization. Hence, in order to try to describe the plume as it migrates through the aquifer, we must also describe this natural or human-influenced attenuation process.

Biodegration is the ability of micro-organisms to break the chemical bonds of these compounds and transform them. Adsorption onto the soil refers to the physical phenomenon of the attraction of these compounds to the surface area of the solids in

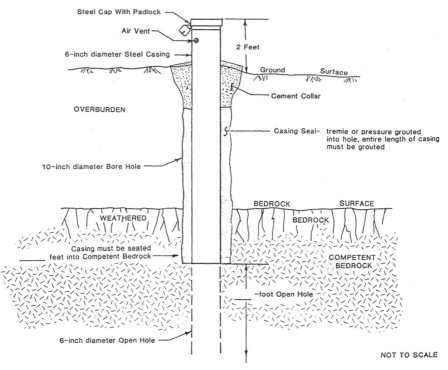

FIGURE 1.3 New Jersey Department of Environmental Protection monitor well specifications for bedrock formations, NJGS Revised 9-87. (*From N.J. DEP, 1988.*)

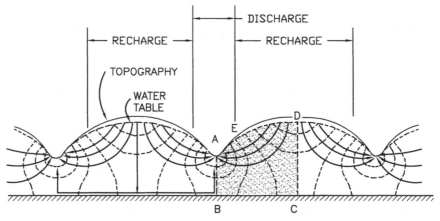

FIGURE 1.4 Recharge areas, discharge areas, and groundwater divides. Groundwater flow net in a two-dimensional vertical cross section through a homogeneous, isotropic system bounded on the bottom by an impermeable boundary. (*Modified from Freeze and Cherry, 1979.*)

the aquifer. This process is not destructive; however, it does retard the velocities at which contaminants travel through the aquifer. Dispersion and dilution can be described as the spreading out of the concentration distribution of compounds over time, both horizontally and vertically due to physical and hydrodynamic mechanisms. Chemical reaction occurs throughout the aquifer continuously breaking compounds down and forming new ones. Finally, volatilization is the ability of a compound to go from a liquid to a gaseous state.

Once the chemicals of concern have been identified, the natural attenuation process estimated, the geological hydrodynamic models developed, and the possible density model understood, then establishing the plume boundaries can begin. Our objective is to establish the vertical and horizontal boundary of this plume. The first step is to estimate the length of time the plume has migrated. Aquifer tests are conducted to determine the hydraulic conductivity, the potentiometric gradient, and dispersion characteristics and to compare these results to any studies in the area. The actual establishment of the plume boundaries is not an easy task. Indeed, plumes rarely are perpendicular to potentiometric gradients and have both vertical and horizontal heterogeneity that is difficult to predict. Hence the establishment of these plume boundaries is generally undertaken after several phases of explorations of the site model and making the necessary adjustments. Generally the phases of exploration begin with a series of wells or Geoprobes™, perpendicular to the centerline of the expected flow path that specifically explore one or more vertical zones. Then, on the basis of the results of this phase, the next set of wells also will be perpendicular to the centerline of the flow path, but slightly adjusted and always within the concept of the geologic and hydrodynamic models (Fig. 1.5).

The most important aspect of the plume is the source zone. The second most important is identifying the centerline of the plume and the third, the boundaries, (see Fig. 1.5). (Note different compounds may have different boundaries.) Hence, most monitoring strategies focus on the identification of the source zone, then identification of the centerline of the plume (intermediate zone), and finally the boundaries or fringe zone of the plume. Source wells seek to characterize the source and are shown in Fig. 1.5 as MW1 and MW2. Sometimes the source can be DNAPLs or LNAPLs. Identification of these are given in Secs. 1.3 and 1.4. In this section we are seeking only to characterize the aqueous portion of the source. MW1 and MW2 are examples of the delineation of the dissolved source; note that MW1 and MW2 are along the centerline of the plume. The objective of MW1 and MW2 is to verify the amount of source still available and impacting the aquifer and to try to assess a starting point for the plume to move downgradient.

Intermediate-zone wells along the centerline are shown in Fig. 1.5 as MW3 and MW4. Their objective is to further characterize the natural attenuation process. These wells should show steadily decreasing concentration if the source is continuous. This unfortunately is almost never the case, and interpretation of slugs of contamination down the centerline of the plume is of great importance and usually elusive. Indeed interpretation of variable source input is almost impossible without detailed knowledge of the source activity in time and a large number of intermediate wells frequently sampled over several years.

Boundary observation wells are required by most states to establish the boundary between the plume and the unaffected aquifer. These boundary wells are frequently used to determine whether a steady-state condition has been achieved, especially for the new "monitored natural attenuation" alternatives. These boundary wells are intended to describe the boundary of the plume and are shown in Fig. 1.5 as MW5, MW9, and MW10. Boundary wells, because they seek to identity the zero level of contamination, because they could be different for different contaminants, and also

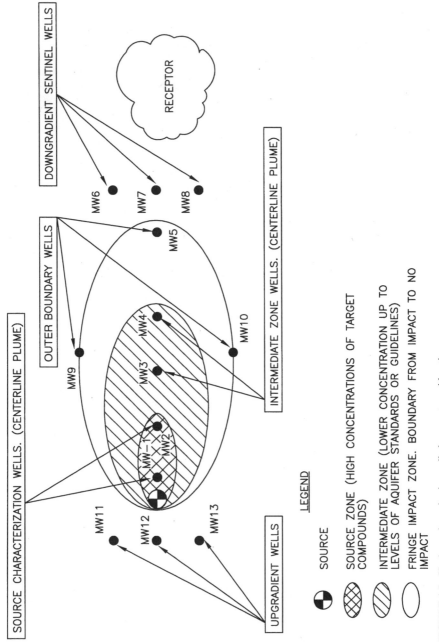

FIGURE 1.5 Horizontal monitoring well placement considerations.

1.22

because of "background" or other offsite sources, are one of the more elusive quarries of the groundwater professional.

Upgradient wells are necessary. Theoretically one upgradient well can describe the boundary conditions of the aquifer prior to any impact. Frequently multiple wells are used to establish boundary conditions for up gradient conditions for a particular plume. The location of these upgradient wells should cover the width of the downgradient plume as MW11, MW12, and MW13 do in Fig. 1.5.

Finally, downgradient sentinel wells quite often are used as an early warning system for a sensitive receptor. MW6, MW7, and MW8 are sentinel wells in Fig. 1.5. Sentinel wells can be utilized to establish a fail-safe or safety factor so as to identify a limit on contaminant transport at which action should take place. The sentinel wells must be placed with the realization that any action that needs to be taken requires time for construction and implementation. Hence, the time it takes for contaminants to travel from the sentinel wells to the sensitive receptor must be greater than or equal to the time to implement the remediation action. An example of a sensitive receptor could be a pumping water supply well or an ecologically sensitive marsh or preserve.

Figure 1.5 shows the placement of wells necessary to describe horizontal extent of a plume. In order to place the well screen properly, vertical profiling of the aquifer needs to take place as shown in Fig. 1.6. Typically this is done with a Geoprobe or

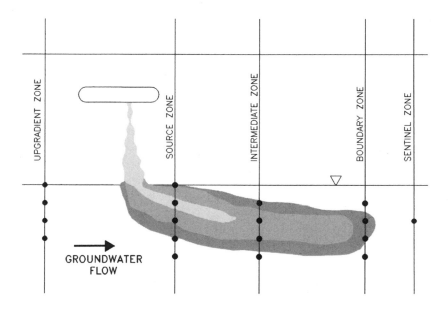

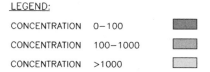

LEGEND:

CONCENTRATION   0–100

CONCENTRATION   100–1000

CONCENTRATION   >1000

FIGURE 1.6 Vertical profile monitoring well placement along the centerline of a plume.

other direct-push method where a small sample of water can be withdrawn at specific intervals in order to get a representative sample of the aquifer (usually on the way down), and thus describe the vertical distribution of contamination.

1.2.3 Dense Nonaqueous-Phase Liquids (DNAPLs)

For any investigative method to work for DNAPLs, it must recognize the three-phase system that exists in the saturated zone. Figure 1.7 shows a theoretical distribution of DNAPLs on a pore size scale between the three phases in the saturated zone (Huling and Weaver 1991): the water phase, the DNAPL phase, and the soil phase. The interaction mass transfer between the water phase and the DNAPL phase is described by the DNAPL water partition coefficient. The water-soil mass transfer between the two is governed by the soil water partition coefficient. Finally, where contaminants may adsorb or partition into the soil and back out is known as adsorption/desorption. This three-phase system makes it very difficult to sample for DNAPLs separately.

Many times DNAPLs are held in the soil matrix as part of the capillary forces, and hence, will not flow by itself (see Figs. 1.8 and 1.9). Therefore, if one were to put an observation well directly into an area where there were residual DNAPLs, one would not encounter any DNAPLs in the observation well. However, under these circum-

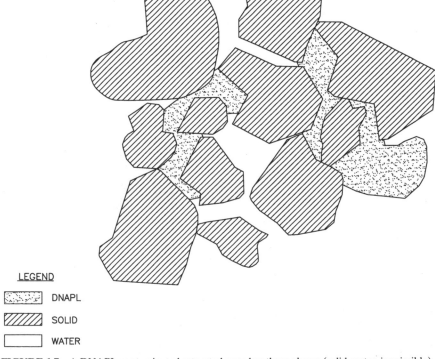

LEGEND

DNAPL

SOLID

WATER

FIGURE 1.7 A DNAPL-contaminated saturated zone has three phases (solid, water, immiscible).

FIGURE 1.8   Residual DNAPLs trapped by glass beads. (*From Schwille, 1988.*)

FIGURE 1.9   Residual DNAPLs trapped by glass beads. (*From Schwille, 1988.*)

stances, very high concentrations of DNAPLs in the groundwater at the observation well would be an indirect indicator of the DNAPL source being close by. Indeed, Cohen and Mercer (1993) suggest that one could infer DNAPL presence by interpreting concentrations of DNAPL chemicals in groundwater of greater than 1 percent of the pure-phase effective solubility. Effective solubility is defined as the mole fraction of a compound in the DNAPL mixture times the pure phase solubility of the compound. Note that this does not account for the phenomenon of cosolvency, where a mixture's solubility may increase in water (e.g., alcohol).

DNAPL Site Characterizations.   It is very difficult to find DNAPLs and characterize them properly. Historical site use is critical information to begin the process of the identification of DNAPLs. Indeed, careful examination of land use since the site was developed, including operations and processes and types and kinds of chemicals used, generated, stored, handled, and transported—both the chemical themselves and the operational residuals. The objective is to obtain a clear picture of the potential for DNAPL contamination at the site sliced in 5-year periods, or some suitable period that relates to the manufacturing or operating activities at the site. Next a clear understanding of the geological boundary conditions is essential for planning the scope of the investigation. The conceptual model of the geology at the site is extremely important, because DNAPLs migrate down because of gravity and choose the path of least resistance. Finer layers (such as a fine sand), with hydrologic conductivity as low as 10^{-2} cm/s, will inhibit the flow of DNAPLs downward and cause them to deflect (Schwille, 1988). So instead of moving vertically downward, DNAPLs can move sidewise, depending on the dip of low-permeability layers. The low-permeability layers, however, if flat or bowl shaped, will accumulate DNAPLs, and because of their fine particle size, the capillary forces will tend to hold on to them with greater tenacity (Schwille, 1988). Hence, pools of DNAPLs can develop in the unsaturated and saturated zones, as this material continues to cascade downward by gravity. Hence, in the investigation for DNAPLs, one must consider the possibility of pools of DNAPLs forming, perched in the unsaturated and saturated zones (see Fig. 1.10). Drilling through those finer layers may cause migration of the DNAPLs deeper into the aquifer. Hence, caution must be taken to first build a fairly accurate geologic model to understand and conceptualize where the DNAPLs may have gone to and to ensure that no further vertical migration occurs because of piercing the low-permeability layers that perch the DNAPL pools.

Noninvasive Characterization Methods.   Noninvasive methods can often be used early in field work to optimize cost-effectiveness of a DNAPL site characterization program. Typical methods such as geophysical surveys and soil gas analysis [organic vapor analyzer (OVA) and photoionizing detector (PID)] can facilitate the characterization of a contaminant source. These will all help in the conceptual geologic model refinement to reduce the risk of spreading any contaminants by piercing any low-permeable layers. However, surface geophysical techniques have been used with varied degrees of success to directly identity DNAPLs. The most common types of surface techniques include ground-penetrating radar, electromagnetic conductivity, electrical resistivity, seismic, and magnetic metal detection. All of these geophysical techniques have had less than stellar performances in trying to identify DNAPL presence. Their real worth is in identifying and confirming the geological conceptual model (Cohen and Mercer, 1993).

Another type of noninvasive technique is a soil gas analysis, which is a popular screening tool for detecting volatile organics in the vadose zone at contaminated sites (DeVitt et al., 1987; Marrin and Thompson, 1987). The American Society for Testing and Materials (ASTM) has developed a standard guide for soil gas monitoring in the

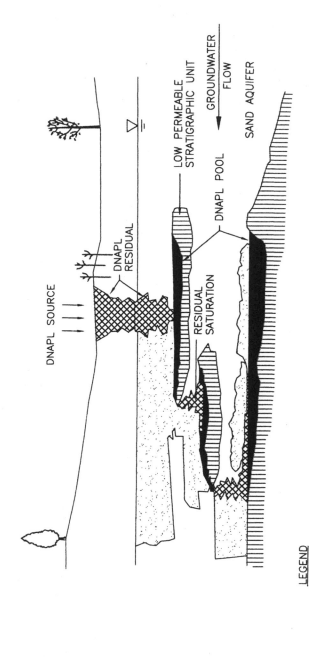

FIGURE 1.10 Perched residual and deep DNAPL reservoirs. *(Modified from Hulung, 1991.)*

LEGEND

	LOW PERMEABLE STRATIGRAPHIC UNIT
	DNAPL POOL
	DISSOLVED DNAPL
	RESIDUAL DNAPL

1.27

vadose zone. Soil gas surveys are relied upon to obtain extensive volatile organic gas information at a fraction of the cost of conventional methods and often with the benefit of real-time field data. However, because of the diffusion of the soil gas, pinpointing locations of source areas sometimes has been difficult, and these methods are best used as screening tools. Another reason why soil gas analysis is a good way to identify the source of DNAPLs is that experiments conducted at the Bordon, Ontario, DNAPL Research Site suggested the soil gas contamination usually is dominated by volatilization and vapor-phase transport from contaminated sources in the unsaturated zone, rather than from the groundwater. This implies that the upward transport of vapors from the dissolved groundwater to the unsaturated zone is very limited (Hughes et al., 1990). Therefore, soil gas site characterization is not a good indicator of distribution of DNAPLs in the saturated zone, but is an excellent characterization of the distribution of DNAPLs in the unsaturated zone and can therefore often be used to identify the source. It should be noted that the higher molecular weights and saturated vapor concentrations can engender density-driven gas migration in media with high gas-phase permeability. In density-driven gas flow, VOCs tend to sink and move outward and to some extent dissolve into the saturated zone. This phenomenon occurs only in and around high source concentrations.

Invasive Methods for Characterization of DNAPL. Invasive and soil sampling methods in the saturated zone generally involve a tradeoff between the advantages of the different techniques and the risks associated with drilling at DNAPL sites. Special consideration should be given to drilling methods that allow for: (1) continuous high-quality sampling to facilitate identification of DNAPL presence in low permeability barriers, (2) highly controlled well construction, and (3) well abandonment.

Drilling in unconsolidated media at DNAPL sites is most commonly done by using hollow-stem augers with either split-spoon samplers, Shelby tube open samplers, or thin-wall piston core samplers. These three methods are described next. Finally three additional methods of characterizing DNAPLs are presented.

Split-spoon sampling is part of a standard penetration test procedure. It involves driving a split-spoon sampler with a 140-lb hammer attached to a drill rig to obtain a representative soil sample. In addition, it measures soil penetration resistance. This sampling technique is described by ASTM test method D1586-84. The split-spoon sampler is either 18 or 30 in long with a 1½-in diameter, and made of steel. It is attached to the end of drill rods, lowered, and then hammered into the undisturbed soil by dropping a 140-lb weight a distance of 30 in onto an anvil that transmits the impact to the drill rods. The advantage of split-spoon sampling is that samples can be used to evaluate stratigraphy, and the physical and chemical properties can be tested. Steel, brass, or plastic liners can be used with split-spoon samplers so that samples can be sealed to minimize changes in samples' chemical and physical conditions prior to delivery to a laboratory. They are relatively inexpensive and widely available and frequently used. A limitation, however, is the stress created by hammering that can consolidate and disturb the sample. One has to remember that DNAPLs are held in the interstitial spaces of the aquifer by capillary action. That capillary action is determined by the size of the pore spaces; hence, when a split-spoon sample is being hammered into the aquifer, pore space can change radically. Hence the DNAPLs in the immediate area of the split-spoon may be altered.

Thin wall (Shelby) open-tube samplers consist of a connector head and a 30- or 36-in-long thin-wall steel, aluminum, brass, or stainless steel tube, which is sharpened at the cutting edge. The wall thickness should be less than 2½ percent of the tube outer diameter, which is commonly 2 or 3 in. A sampler is attached by its connector head to the end of the drill rod, lowered typically through a hollow stem auger to the bottom of the bore hole, which must be clean, then pushed down through the undisturbed soil by using the hydraulic or mechanical pulldown of the drilling rig. This procedure is described by ASTM method D1587-83. Advantages of the Shelby tube sampler are that it provides undisturbed samples in cohesive soils and representative samples in soft to medium cohesive soils. High-quality samples can be evaluated for mineralogy and stratigraphy and for physical and chemical properties. Samples can be preserved, stored within the sample tube, by sealing its ends (usually with wax), thereby minimizing disturbance prior to lab analysis. Shelby Tube Samplers are widely available and commonly used by geotechnical firms. Finally the cost of sampling is higher than with the split-spoon method. A disadvantage of this method is that the sampler should be at least 6 times the diameter of the largest particle size to minimize the disturbance of the sample. Large gravel or cobbles can disturb the finer-grain soils within and cause the deflection of the sampler. Because of the thin wall and limited structural strength, the sampler cannot easily be pushed into dense or consolidated soil. It's generally not very effective for sandy soils.

The *thin-wall piston core sampler produces* samples very similar to those of the Shelby tube sampler, except they have a piston in the tube that creates a vacuum as the sample is being pushed into the earth. An advantage of the thin-wall piston core sampler is that it provides an undisturbed sample of cohesive silts and sands above or below the water table. The vacuum enables recovery of the cohesionless soils (sands). High-quality samples can be evaluated for minerology and stratigraphy and for physical and chemical properties, and the samples can be preserved and stored within the sample tube, thereby minimizing the sample disturbance prior to lab analysis. A limitation, as with the Shelby tube sampler, is that large particles may disturb the sample. It is not as widely available as the split-spoon or open-tube samplers. It is relatively expensive compared to the other two types.

The *cone penetrometer* provides a new method for characterizing subsurface non-aqueous-phase liquids including chlorinated solvents and petroleum hydrocarbons. It uses a direct-push sensor probe, coupled with a laser-induced-florescence sensor with an in situ video imaging system. The laser-induced florescence (LIF) can cause florescence in polyclyclical aromatic hydrocarbons, which are compounds associated with most solvent extracted waste. These are not DNAPLs themselves but frequently are mixed with DNAPLs because DNAPLs are usually used as solvents or degreasers. These are commonly dissolved in solvents during the industrial process. The video imaging system is used to collect high-resolution images of the soil in contact with the probe. The video images provide direct visual evidence of the non-aqueous-phase liquid contaminants present in the soil. In a report by Lieberman (2000), the LIF imaging system was used on a site in Alameter Point (formerly NAS Alameter) that was contaminated with a TCE-rich petroleum product. The sensors were used to delineate the vertical and lateral extent of contaminant both before and after the site was remediated by steam-enhanced extraction. The initial sensor data showed that the DNAPL contamination occupied an area of about 2500 ft^2 that was limited to depths of 5 to 10 ft. Data collected showed that the distribution of observed microglobules and DNAPLs correlated closely with lithological changes estimated from cone and sleeve friction resistance measures by the cone penetrometer during the push. One great advantage of this system is that there are no waste cuttings to dispose of and that the cone penetrometer quickly advances through the formation. Another advantage is that there is no permanent pathway created that would allow DNAPLs to migrate. However, one of the disadvantages is that it can operate to depths of only 100 to 150 ft, depending upon the geology. Rocks and cobbles create significant problems for its penetration.

The *ribbon DNAPL sampler* (RDNS) is a direct-sampling device that can provide discrete sampling of nonaqueous-phase liquids in a borehole. The DNAPL identification technique uses a flexible liner underground technology (FLUTe) membrane to deploy hydrophobic absorbent ribbon into the subsurface. The system is pressurized against the wall of the borehole and the ribbon adsorbs DNAPLs that are in contact with it. A dye sensitive to the DNAPLs is impregnated into the ribbon and turns it bright red when the contaminants are contacted. The membrane is retrieved by the tether connected at the bottom of the membrane by turning the liner inside out. That surface liner is inverted and the ribbon is removed and examined. The presence in depth of DNAPLs is located and indicated by brilliant red marks on the ribbon (Riha et al., 2000). Riha described the ribbon NAPL sampler deployed at the DNAPL site at Savannah River (DOE), the Cape Canaveral Air Station, Paduca Gaseous Diffusion Plant, and a creosote-contaminated EPA Superfund site in both the vadose and saturated zones.

The *partitioning interwell-tracing test* (PITT) can be used not only to identify if DNAPLs are present, but also to identify how much mass is present. By injecting conservative and partitioning short-lived radioactive isotope traces into the subsurface and continually measuring their presence in monitoring wells with movable downhole sampling devices, the location and volume of DNAPLs can be measured to a much greater extent than currently can be achieved by any other method. Through this method, the DNAPLs can not only be identified but quantified as well. The method makes use of the fact that the partitioning compounds will partition at different times and rates, and hence will become separated in time, somewhat like a gas chromatograph separating gases through adsorption and desorption on the column. From time of travel in the downgradient well system and the sorption/desorption of the tracers, the DNAPL mass can be identified and quantified (Meinardus et al., 2000). Meinardus has applied PITT in a full-scale implementation program at Hill AFB Operable Unit II (OUII). After PITT, Meinardus performed a full-scale surfactant flood at OUII, followed by a second PITT to assess the performance of the surfactant flood. Meinardus reports that over 90 percent of DNAPLs have been removed by the surfactant flood, according to the results of before and after PITT.

1.2.4 Light Nonaqueous-Phase Liquids (LNAPL)

Light nonaqueous-phase liquids, like DNAPLs, get captured by soil matrices in similar ways. However, the significant differences between the two classes of compound is that the LNAPLs, being lighter than water, will float on top of the water table and therefore will not penetrate the water table. Hence, investigation need only take place at the top of the water table. Therefore, this poses much less of a problem than for DNAPLs. In addition, many LNAPLs are biodegradable, primarily because they have been around the earth as natural substances for millions of years, and bacteria have developed the necessary methods to break them down and use them as an energy source. Hence, because they pose less of a long-term problem and they are more biodegradable than many of the chlorinated solvents, they are considered less of a problem. One exception to this is methyl tert butyl ether (MTBE). MTBE, an oxygenative additive to gasoline, is very soluable in water and not particularly biodegradable. The investigative methods used to identify the pure-phase LNAPLs have been primarily focused on observation wells to identify "floating product" that is floating on top of the water table. This section will focus on the pure-phase product of the petroleum hydrocarbon, and not the dissolved phase. The dissolved phased was discussed in Sec. 1.2.1.

Although less of a problem than the DNAPLs from the standpoint of toxicity and persistence, the problem of LNAPLs is both diffuse and widespread For example, it has been estimated that over 75,000 underground storage tanks (USTs) alone annually release 11 million gallons of gasoline to the subsurface (Parker et al., 1994).

Hydrocarbons are fluids that are immiscible with water and are thus considered nonaqueous-phase liquids. In general, most hydrocarbon compounds are less dense than water and therefore termed LNAPLs. When released in the subsurface, LNAPLs remain as a distinct fluid separate from the water phase. The downgradient migration in the vadose zone is generally rapid and, depending upon the complexity and heterogeneity in the soil, may form an intricate network of pathways. Like DNAPLs, they have a residual level in the unsaturated zone, held there by capillary action. Once in the vicinity of the capillary fringe of the saturated zone, hydrocarbons will spread horizontally with no penetration below the water table, but some depression due to their weight. Contact with groundwater, as well as infiltrating precipitation resources, causes the chemical constituents of the LNAPLs to dissolve from the hydrocarbon phase into the groundwater, resulting in contamination of the aquifer. This aqueous-phase groundwater is dealt with in Sec. 1.2.1. This section will deal with the nonaqueous-phase portion, the LNAPLs.

The first step in assessing a hydrocarbon spill generally involves the delineation of the vertical and horizontal extent of the pure phase in the vadose zone and the smear zone on top of the water table. The smear zone and hydrocarbon pooling on the water table (floating product) will be discussed here.

The measurements of soil concentrations in the unsaturated zone (total petroleum hydrocarbons, or individual compounds) provide the most reliable quantitative information on the actual volume or mass of hydrocarbons. Estimation of hydrocarbon volume in the smear zone, or free-floating product, by observation well, is less straightforward. A general lack of understanding in this area compounded by promulgations of numerous methods of measurement, has resulted in widespread misunderstanding of the concept of apparent thickness and true thickness of the hydrocarbon in the well. Simplified practical theoretical approaches, such as that of dePastrovich et al. (1979), suggest that well product thickness will typically be about 4 times greater than the true free product thickness. Hall et al. (1984) investigated the relationship between soil product thickness and well product thickness in the laboratory and proposed a relationship to correct the discrepancies in the method of de Pastrovich.

Laboratory investigations by Hampton and Miller (1998) found the methods of both dePastrobich and Hall lacked accuracy. A theoretically based method for estimating oil specific volume from well product thickness was developed and reported independently by Lenard and Parker (1990) and Farr et al. (1990). The method is based on the assumption of vertical equilibrium pressure distributions near the water table, which can be inferred from well fluid levels and from the fluid pressure distribution. From the fluid pressure distribution and the general model for three-phase capillary pressure relations, vertical oil saturation distributions are computed and integrated to yield oil specific volume.

In addition to the free product that is sufficiently mobile to enter the monitoring well, a significant portion of the total spill volume may occur as the residual product confined in the interstitial spaces of the aquifer itself. As with DNAPL, these hydraulically isolated blobs, or ganglia, are effectively immobile because of capillary forces that hold them in place. Changes in water table elevations will generally result in increasing residual volume over time. As the water table rises, the free product will occupy the upper pore zones, and as the water table drops, the upper pore zones will then drain, resulting in a smear zone of hydrocarbons that can account for large

amounts of LNAPLs. These fluctuations can also occur from drawdown of overexu-
berant recovery projects, causing significant smearing of the aquifer. Hence, the key
to maximizing product recovery from spill sites involves minimizing the volume of
residual product that is induced as a result of recovery system operations. Parker et
al. (1994) provide a graphic illustration of the difference between apparent thickness
and true oil thickness, or free-oil specific volume. Figure 1.11 shows two theoretical
curves, one for silt and one for sand. These curves indicate the correlation between
the apparent oil thickness and the actual true oil thickness or the free-oil specific volume. Note that, in the case of silt, several feet of apparent oil thickness can be measured in observation wells, and free-oil specific volume is almost nothing. Indeed,
even in sand, a half-foot of apparent oil thickness implies that the true oil thickness is
almost zero.

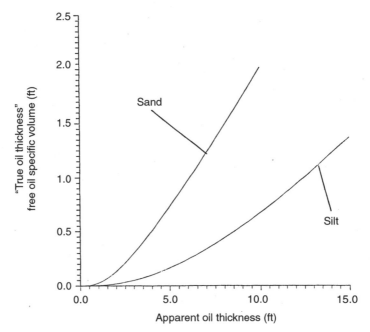

FIGURE 1.11   Free-oil specific volume versus well product thickness for gasoline in different soils. (*From Parker, 1994.*)

Observation wells that measure free product have been drilled by conventional
methods and generally have 10 ft of screen zone into the water table and 5 ft of
screen zone above the water table to measure the apparent thickness and relate the
apparent thickness to actual free-oil volume. The estimation of free-oil volume is
important because this is the volume that will actually continue to move and be
pumped out of the aquifer. Quantification of hydrocarbon volume in smear and
unsaturated zones requires that total petroleum hydrocarbons be sampled in these
zones and analytical quantification of hydrocarbons in terms of mass of hydrocarbons per mass of dry soil be performed.

1.3 REMEDIAL METHODS

*"To do easily what is difficult for others is the
mark of talent."*
HENRI FRÉDÉRIC AMIEL, 1821–1881

1.3.1 Introduction

Ask any groundwater professional in the field to list the most common perceptions
concerning aquifer restoration and you are likely to get this answer: (1) they are too
costly, (2) they are time-consuming, (3) they are not always effective, and (4) pertinent information is usually not available for the aquifer and the geological system.
These perceptions, having generally been accepted by groundwater professionals
over 2 decades, are slowly changing, however. An ever-increasing amount of information is becoming available about aquifer restorations and groundwater cleanup.
More and more information is being shared by aquifer professionals at conferences
and meetings. The federal government, however, has taken the lead and has established technology transfer programs and funded site demonstration programs.
Indeed, government sponsoring agencies such as the U.S. Air Force Center for Environmental Excellence, U.S. Army Environmental Center (ACE), Federal Remediation Technologies Roundtable, U.S. Environmental Protection Agency (EPA), U.S.
Navy (USN), and U.S. Department of Energy (DOE) have shed more light on remediation technologies in the last 8 years than all efforts in the preceding 50 years.
    This chapter summarizes the state of the art for aquifer restoration. As previously
stated, it focuses on the aqueous groundwater aquifer remedial methods, dense non-
aqueous-phased liquids (DNAPLs), and finally light nonaqueous-phase liquids
(LNAPLs). It is noted here, as earlier, that both the DNAPLs and LNAPLs pose the
most significant threat to groundwater systems, and this is the reason they are included
in this chapter.

1.3.2 Aqueous Groundwater Remediation Methods

Many different methods ranging from institutional mandates to physical, chemical,
and biological technologies have been proposed for the protection and/or cleanup of
groundwater. Institutional measures have reduced the risk of exposure to sensitive
receptors rather than reduce the contaminants themselves. These so-called risk-
based corrective actions have been slow to catch on, but, as we learn more and more
about exposure, their use will become more and more accepted.
    Federal guidelines associated with acceptable levels of contaminants in the envi-
ronment have come from several laws passed by Congress in the 1970s and 1980s.
These laws are the Comprehensive Environmental Response, Compensation, and
Liability Act (CERCLA, also known as Superfund), the Resource Conservation and
Recovery Act (RCRA), the Clean Air Act (CAA), and the Clean Water Act (CWA).
Different state programs have modeled themselves on each of these federal man-
dates. At the state level, property transfer has been the impetus for many cleanups.
Indeed, states like New Jersey and Massachusetts make it mandatory for sellers to
carry out groundwater cleanup prior to the transaction.
    Aquifer remedial methods can include hydrodynamic or physical containment of
the contaminated plume prior to extraction and treatment. Indeed, the hydrology of

pumping wells has long been known and applied for the development of groundwater. It is a small step to use this same technology for the removal and containment of plumes. This kind of technology became known as *pump-and-treat,* because it captures the plume by pumping and treating the contaminated liquids. Additional techniques from the construction industry such as grouting, slurry walls, and sheet piling are used to create impermeable barriers to constrain the plume and eliminate dispersion. At any one site, the remedial program employed to physically control the plume will usually consist of a combination of different technologies both hydrodynamic and physical. Each of these techniques will be discussed from the standpoint of construction, cost, advantages, and disadvantages.

Physical Methods of Controlling Groundwater

Sheet Piling. Sheet piling involves driving lengths of steel that are connected via a tongue-and-grove mechanism into the ground to form an impermeable barrier to flow. Sheet piling materials include steel and timber. However their application is primarily for the construction industry and not polluted groundwater. Sheet piling requires that the sections be assembled prior to being driven into the ground. The lengths of steel have connections on both edges so that the sheet piles actually connect to one another. Typical connections include slotted or ball and socket joints. The sections are then driven into the ground by a pile hammer. After the sheet piles have been driven to their desired depths they are cut off at the top. The problem with sheet piling is the permeability of the interconnections, and often a heavy grease is included to assure that the connections don't leak.

The cost of sheet piling for a 170-ft-long and 60-ft-deep lightweight steel cutoff wall is reported to range between $650,000 to $1 million (Tolman et al., 1978).

One of the advantages of sheet piling is that it is a simple technique known to every construction firm in the business. Another advantage of sheet piling is that no excavation is necessary, and no contaminated soils need disposal. Also, the kinds of equipment used are available throughout the United States. For small projects, construction can be economical and there is really no maintenance after construction, and the steel can be coated for corrosion protection to extend its service life.

A disadvantage is that the steel sheet piling initially is not watertight. In addition, diving piles through ground containing boulders is difficult and may result in the separation of sheets, thus causing large gaps in the impermeable wall. Finally, if certain chemicals are present, especially acids, they may attack the steel.

Grouting. Grouting is a technique that has been used in the construction industry for a long time. It is a process based on the injection of a stabilizing liquid slurry under pressure into the soil that can also be used to create an impermeable wall. The grout is injected into the soil until all spaces are completely filled. The grout will then set and solidify, thus resulting in a mass of solid material that will reduce the soil permeability to zero if properly constructed. Grouts are usually classified as particulate or chemical. The particulate (cementitious) grout solidifies within the soil matrix and the chemical grout consists usually of two or more liquids that, when mixed, create a gel of low permeability.

In the construction of grout curtains, the first consideration for design is the actual composition of the soil or geology. The success of the grouting will also be a function of several variables including soil temperature, the pollutant to be contained, and the time for installation. In general, chemical grouts are used for fine-grained soils. However, chemical grouts can be problematic because they are not suitable for highly acidic or alkaline environments and gel curing is generally an acid-base reaction. For gravel soil, cementitious particulate grouts are suitable. The amount of cement or bentonite in the particulate grout varies widely. A key design consideration is the pressure at which the grout is injected. The use of excess pressure may weaken the strata and increase the permeability. Highly permeable zones can take much more grout and reduce the necessary pressure for injection. Indeed, one of the problems of grouting is the variation of permeability with depth. The amount of grout injected can sometimes be very nonuniform. Orders of magnitude changes in permeability can cause gaps and poor seals in the adjacent grout column (Knox et al., 1986). It is extremely important that each grout column be keyed into the next grout column so that there are no gaps in the curtain. In some cases, a double- or triple-row curtain is used to ensure an impermeable wall. The cost of grout cutoff systems is quite high; hence, they will only be applicable to small, localized cases of groundwater pollution. Costs have been reported to range from $150 to $350 per installed cubic foot (Lu et al., 1981).

One of the advantages is that the technology of grouting has been used in the construction industry, and hence cut off walls have been installed successfully for many years for construction dewatering. Presently there are different kinds of grouts to suit a wide range of soil types and contaminant compatibilities. One of the disadvantages of grouting is that it is limited to granular types of soils, having pore space large enough to accept grout fluids under pressure. Grouting in a highly layered soil profile may result in an incomplete formation of grout columns such that the higher-permeability zones will accept more of the grout while the lower-permeability zones will accept little to none. The presence of rapidly flowing water will limit the groutability of a formation, while the presence of boulders may also limit the groutability of the soil. Some grouting techniques are proprietary, and final testing is a must for any cutoff wall. Finally, the interaction and compatibility of grouts with generic chemical classes is very important. This has been reviewed by Knox et al. (1986) and is reproduced here as Table 1.4.

Slurry Walls. Another method of impermeable wall formation is the use of slurry walls. Slurry walls represent a technology to prevent groundwater pollution or restrict the movement of previously contaminated groundwater. The technology is fairly simple; it involves digging a deep trench and concurrent in situ blending of a bentonite clay with the native soils (usually bentonite). Slurry walls are technically feasible up to approximately 100 ft in depth and can be very effective in cutting off all groundwater in flow. Like all of the other impermeable wall techniques, slurry walls have to be keyed into an impermeable or semipermeable bottom so that a bathtub effect is created.

The most common type of slurry wall construction is the trench method. In this method, a deep trench is excavated and a bentonite water slurry is added while the excavation is in progress. The original soil is then continually mixed with the bentonite or bentonite cement.

The New York City Transit Authority completed a bentonite slurry trench that was keyed into a rock formation 100 ft below the surface for its 63rd St. connection. The trench created a bathtub effect, reducing the likelihood that seepage from a nearby PCB-contaminated Superfund site would interfere with the construction. The construction of the slurry trench used a vibrating beam to guide the clamshell bucket used for digging down to a depth of 100 ft. The consistency of the bentonite slurry was controlled by continuously recirculating the slurry through a central mixing unit, which added bentonite cement to the excavated soils on an as-needed basis.

Critical design considerations for any slurry trench include the composition of slurry and the geology of the formation. The resulting permeability will depend upon the soil and the amount of bentonite that is blended in, but permeabilities of 10^{-7} cm/s are typically achieved in the field. The costs associated with slurry trench methods reported by Spooner (1982) are shown here in Table 1.5. The table shows costs per

TABLE 1.4   Predicted Grout Compatibilities

Chemical group	Silicate	Acrylamide	Phenolic	Urethane	Urea-formaldehye	Epoxy	Polyester
			Grout type				
			Polymers				
		Organic compounds					
Alcohols and glycols	1a	1—	3b	—	1	1—	1—
Aldehydes and ketones	1a	—	—	1a		1a	1a
Aliphatic and aromatic hydrocarbon	1d	1—	—	1—	—	1—	1—
Amides and amines	3a	3d	3b	—a	1a	1a	3a
Chlorinated hydrocarbons	1d	1—		1—	—	1	1—
Ethers and epoxides	1a	1—	1a	10		1a	1a
Heterocyclics	1d	1—	1a	1a	1a	1a	1a
Nitrites	1a	3—	1a	1a	1a	1a	1a
Organic acids and acid chlorides	1—	—	3—	2—	—	—	1—
Organometallics	1a	3a	—	—	1a	1a	3→?
Phenols	1a	1a	—	2—	1a	1a	1?
Organic esters	1—	?	?	?	?	?	1d
		Inorganic compounds					
Heavy metal salts and complexes	—a	—	—	—	—	3—	3?
Inorganic acids	—	—	2—	—	—	1—	1—
Inorganic bases	—	—	3—	—	—	—	1—
Inorganic salts	—d	—	—	—	—	—	3*—

Source: Knox et al. (1986).
Effect on set time
  1. No significant effect.
  2. Increase in set time (lengthen or prevent from setting).
  3. Decrease in set time.
Effect on durability
  a. No significant effect.
  b. Increase durability.
  c. Decrease durability (destructive action begins within a short time period).
  d. Decrease durability (destructive action occurs over a long time period).
* If metal salts are accelerators.
→ If metal is capable of acting as an accelerator.
? Data unavailable.

square foot based on the soil type and the depth of penetration for a soil bentonite wall. It should be noted that depths of greater than 150 ft have been accomplished.
    The advantages and disadvantages of the slurry trench are as follows:

• The construction methods are simple and widely used throughout the construction industry.

• The slurry wall method is essentially an excavation and soil-mixing process. The construction industry has vast experience with these activities.

• Bentonite minerals will not break down with age, and as long as the wall remains wet it will swell and maintain an excellent impermeable seal.

TABLE 1.5 Approximate Slurry Wall Costs as a Function of Medium and Depth

Soil type	Soil-bentonite wall*			Cement-bentonite wall*		
	Depth ≤30 ft	Depth 30–75 ft	Depth 75–120 ft	Depth ≤60 ft	Depth 60–150 ft	Depth > 150 ft
Soft to medium soil, $N \leq 40$	2–4	4–8	8–10	15–20	20–30	30–75
Hard soil, $N = 40$–200	4–7	5–10	10–20	25–30	30–40	40–95
Occasional boulders	4–8	5–8	8–25	20–30	30–40	40–85
Soft to medium rock, $N \leq 200$, sandstone, shale	6–12	10–20	20–50	50–60	60–85	85–175
Boulder strata	15–25	15–25	50–80	30–40	40–95	95–210
Hard rock granite, gneiss, schist	—	—	—	95–140	140–175	175–235

* Nominal penetration only, \$/ft^2.
Source: Spooner et al. (1982).
Note: For standard reinforcement in slurry walls, add \$8.99/ft^2. For construction in urban environments, add 25 to 50% of price.

- It is leachate resistant, and slurry walls are not attacked by any of the typical contaminants.
- They are low maintenance and have been used very successfully in the past.

  The main disadvantage is cost. Some other disadvantages include:

- Construction procedures are patented and may require a license.
- In rocky ground over excavation is necessary because of boulders, and sealing the wall becomes a problem.
- At the rock-soil interface, grout may have to added to ensure a complete seal.
- Finally, bentonite deterioration has occurred where there has been exposure to higher ionic strength leachate (Knox et al., 1986). It is also known that bentonite will dry out if any part of the wall is dewatered and exposed to the air.

Ex Situ Treatment Techniques. In addition to physically containing the groundwater contamination, groundwater can be treated by either ex situ or in situ methods. Ex situ treatment requires that the groundwater be removed and treated at the surface and then reinjected. The following treatment technologies will be considered in this section: air stripping, carbon adsorption, biological treatment, and chemical treatment.

Air Stripping. Air stripping is a process by which volatile compounds are removed from the aqueous waste stream. It is generally considered a mass transfer process in which a volatile compound in water is transformed into a vapor.

The driving force is actually the difference between the actual concentration in the air stripping unit and the conditions associated with equilibrium between the gas and the liquid phases. If equilibrium exists at the air-liquid interface, the liquid-phase concentration is related to the gas-phase concentration by Henry's law, which states that:

$$C_{iG} = H \times C_{iL}$$

where C_{iG} = equilibrium concentration in the gas phase
C_{iL} = equilibrium concentration in the liquid phase
H = Henry's law coefficient.

Henry's law coefficient is the ratio of the concentration of a compound in a gas relative to the solubility concentration of the same compound in water and is temperature sensitive. A compound with a high Henry's law concentration is more easily stripped from water than one with lower Henry's law constant. Figure 1.12 shows a graphical representation of the solubility versus vapor pressure of the Henry's law constant for selected DNAPLs at 25°C. This graph represents 60 compounds for strippability, without external thermal gradients. As a rule of thumb, a compound is considered strippable if its Henry's constant is above 10^{-4} atm-m³/mol. It should be noted that many of the chlorinated hydrocarbon compounds used as solvents are identified as the primary contaminants of concern in many of the Superfund sites (Fig. 1.1) and are almost all strippable, as shown in Fig. 1.12.

As shown in Fig. 1.13 there are several different types of equipment for air stripping, and each piece of equipment essentially accomplishes the same task, the transfer of the contamination from the water to the air. Figure 1.14 shows a series of stacked tray aerators at a major Superfund site in New York. The most widely used, the packed column, will be described here. The packing column material's function is to provide surface area for the countercurrent flow of water coming down and air being forced up. The turbulent action causes a great deal of air/water mixing to occur. Contaminated water is cascaded from the top of the column and splashes against the column packing while air is forced in from the bottom and out the top carrying with it the volatile material (see Fig. 1.15).

A stripping tower such as that in Fig. 1.15 is designed by selecting a combination of parameters. For example, the height of packing in the column is a function of several parameters:

- The water temperature
- The packing characteristics
- The liquid mass loading rate
- The required effluent concentration in the liquid
- The air-to-water ratio

If these parameters remain constant, then Henry's law is applicable. If Henry's law is applicable and the incoming air is contaminant-free, the following equation applies:

$$Z = \frac{L}{(K_L a)D_w} \frac{R}{R-1} \ln \left[\frac{(C_i/C_o)(R-1)+1}{R} \right]$$

where Z = packing height, m
$\quad\quad$ L = liquid loading rate, kg-mol/h per square meter of tower cross-sectional area
$\quad\quad$ $K_L a$ = liquid mass transfer rate times the interfacial areas of the packed column
$\quad\quad$ D_w = molar density of water = 55.6 kmol/m³ at 20°C
$\quad\quad$ R = stripping factor = $(H_A/P_t)(G/L)$
$\quad\quad$ C_i = influent concentration of water
$\quad\quad$ C_o = effluent concentration of water
$\quad\quad$ H_A = Henry's constant for compound A, atm
$\quad\quad$ P_t = total system pressure, atm
$\quad\quad$ G = air velocity, kg-mol/h per square meter of tower cross-sectional area

$K_L a$, the product of the overall mass transfer coefficient, is best estimated through pilot testing. An excellent description of a pilot plant design for air stripping is given by Boegel (1988).

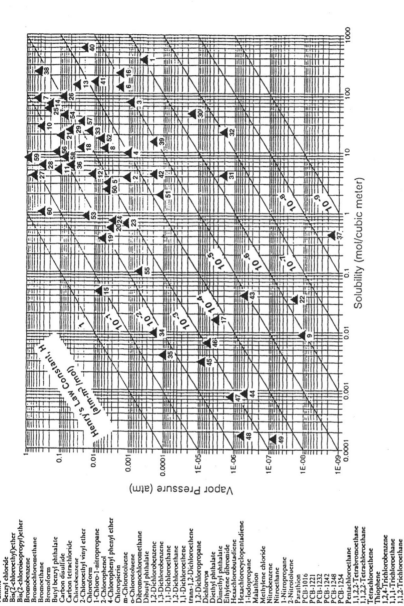

1 Aniline
2 Benzyl chloride
3 Bis(2-chloroethyl)ether
4 Bis(2-chloroisopropyl)ether
5 Bromobenzene
6 Bromochloromethane
7 Bromoethane
8 Bromoform
9 Butyl benzyl phthalate
10 Carbon disulfide
11 Carbon tetrachloride
12 Chlorobenzene
13 2-Chloroethyl vinyl ether
14 Chloroform
15 1-Chloro-1-nitropropane
16 2-Chlorophenol
17 4-Chlorophenyl phenyl ether
18 Chloropicrin
19 m-Chlorotoluene
20 o-Chlorotoluene
21 Dibromochloromethane
22 Dibutyl phthalate
23 1,2-Dichlorobenzene
24 1,3-Dichlorobenzene
25 1,1-Dichloroethane
26 1,2-Dichloroethane
27 1,1-Dichloroethene
28 trans-1,2-Dichloroethene
29 1,2-Dichloropropane
30 Dichlorvos
31 Diethyl phthalate
32 Dimethyl phthalate
33 Ethylene dibromide
34 Hexachlorobutadiene
35 Hexachlorocyclopentadiene
36 1-Iodopropane
37 Malathion
38 Methylene chloride
39 Nitrobenzene
40 Nitroethane
41 1-Nitropropane
42 2-Nitrotoluene
43 Parathion
44 PCB-101b
45 PCB-1221
46 PCB-1232
47 PCB-1242
48 PCB-1248
49 PCB-1254
50 Pentachloroethane
51 1,1,2,2-Tetrabromoethane
52 1,1,2,2-Tetrachloroethane
53 Tetrachloroethene
54 Thiophene
55 1,2,4-Trichlorobenzene
56 1,1,1-Trichloroethane
57 1,1,2-Trichloroethane
58 Trichloroethene
59 1,1,2-Trichlorofluoroethane
60 1,1,2-Trichlorotrifluoroethane

FIGURE 1.12 Solubility, vapor pressure, and Henry's law constants for selected DNAPLs. (*From Cohen and Mercer, 1993.*)

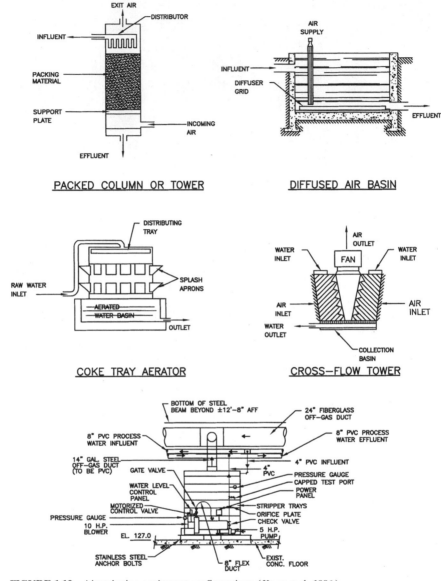

FIGURE 1.13 Air stripping equipment configurations. (*Knox et al., 1986.*)

The key design variables in air stripping design are the diameter of the columns, the liquid loading rate, the air-to-water ratio, the packing height, and the characteristics of the packing. Typical ranges of these variables are

- Diameter = 1 to 12 ft
- Liquid loading rates = 5 to 30 gal/min-ft^2

FIGURE 1.14 Tray aerators for polishing treated groundwater at a Superfund site, New York.

- Packing height = 5 to 50 ft
- Air-to-water ratio = 10 to 300

The groundwater flow rate is very often specified from the capture analysis of the plume. Knowing the flow rate, the diameter of the tower can be derived from the liquid loading rate. Knowing the tower diameter, the packing height is a trade-off with the air-water ratio. The higher the air-water ratio, the lower the packing

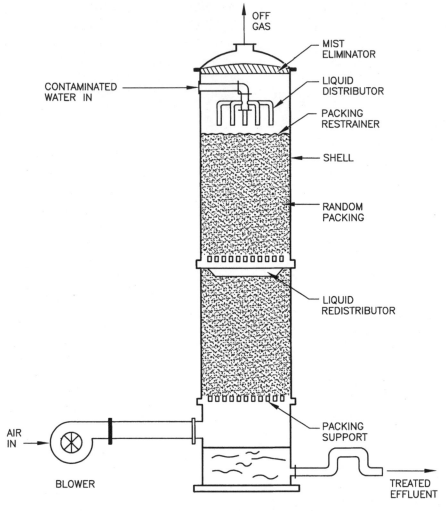

FIGURE 1.15 Packed-tower air stripper. (*Modified from Boege, 1988.*)

height required for a given contaminate removal efficiency. This trade-off is defined by the aforementioned equation for Z and can be readily depicted by plotting Z verses R for varying values of C_i/C_o.

The application of air stripping has been cost-effective for removing volatile organic contaminated groundwater. Volatile organics are probably the most troubling toxic compounds in the groundwater system. Table 1.6 shows the air-to-water ratio, influent, and effluent for several actual applications. Temperature plays a significant role in contaminant vapor pressures, and, consequently, soluble compounds that have low vapor pressure can be induced to vaporize when heated. Typical compounds of this type are the ketone and alcohol groups. These groups have higher solubility because they are slightly polar molecules. Indeed, methylethyl ketone has

TABLE 1.6 Packed Column Air Stripping of Volatile Organics

Organic contaminant	Air:water ratio	Influent, µg/L	Effluent, µg/L
1,1,2-Trichloroethylene	9.3	80	16
	96.3	80	3
	27.0	75	16
	156.0	813	52
	44.0	218	40
	75.0	204	36
	125.0	204	27
1,1,1-Trichloroethane	9.3	1200	460
	96.3	1200	49
	27.0	90	31
	156.0	1332	143
1,1-Dichlorethane	9.3	35	9
	96.3	35	1
1,2-Dichloropropane	27.0	50	<5
	146.0	70	<5
	156.0	377	52
Chloroform	27.0	50	<2
	146.0	57	<2
Diisopropyl ether	44.0	15	7
	75.0	14	6
	125.0	4	4

Source: Knox et al. (1986).

been removed with air stripping only after heating (to approximately 90°F). Knox reports that the removal rates of 92 percent occurred when methylethyl ketone was raised to a temperature of 90°F with an air-to-water ratio of 513 (Knox et al., 1986).

The capital costs of air stripping are relatively small; however, because of the length of time to treat a plume, annual costs can be substantial. An annual cost of per thousand gallons of water treatment has been presented by Knox et al. (1986), and Fig. 1.16 provides a summary. It should be noted that the costs given by Knox are in 1982 dollars.

Carbon Adsorptions. Adsorption is a physical process that occurs when a molecule is brought into contact with any surface and held there by physical forces. Which compounds can be adsorbed by activated carbon are determined by the balance between forces that keep a compound in solution and forces that attract the compound to the carbon surface. Factors that affect this balance include the following:

- *Solubility*—the higher the solubility of a compound, the less likely it will adsorb.
- The *pH* of the water can affect the adsorptive capacity: Organic acids adsorb better under acidic conditions and amino compounds favor alkaline conditions.

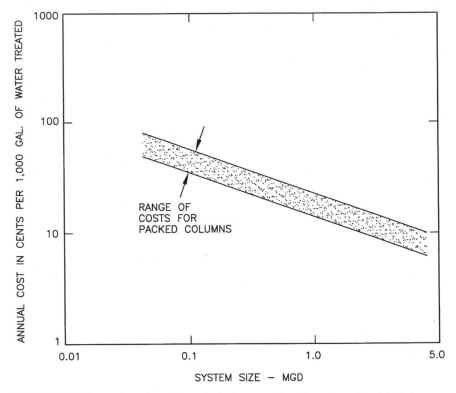

FIGURE 1.16 Comparison of costs for packed column aeration. (Notes: Annual costs include amortized capital costs and annual operating costs; system site represents average plant capacity.) (*Knox et al., 1986.*)

- *Aromatic and halogenated compounds* adsorb better then alphatic compounds.
- *Adsorption capacity* decreases with increasing temperature
- *Character* of the adsorbing surface
- Size of the molecule—the smaller the molecule the less likely it will adsorb

When activated carbon is placed in contact with water containing organic chemicals, the organic chemicals attach to the surface of the activated carbon. Hence the concentration in the water goes down. By conducting a series of adsorption tests, it is usually possible to obtain a relationship between equilibrium concentration and the amount of organics absorbed per unit mass of activated carbon. Such testing has been popularized by Freundlich and Langmuir and the results are often used to represent the adsorption equilibrium. One form of the Freundlich isotherm is:

$$\frac{X}{M} = KC_e^{1/n}$$

where X = mass of organic adsorbed, g
         M = mass of activated carbon, g
         C_e = equilibrium concentration of the organics in water, mole fraction
         n, K = experimental constants

Taking the natural logarithm of both sides and experimentally plotting ln (X/M) versus ln C_e yields a straight line with slope $1/n$ with an approximate y intercept of ln K. When n equals 1, the adsorption is said to be linear (see Fig. 1.17). In the activated carbon process, the contaminated liquid is passed through a column of activated carbon. The carbon is thought to have a set amount of receptors for the contaminant to cling to. Therefore, depending on the loading rate (mass/time) the column of carbon will reach the point where it needs to be either replaced or regenerated. Figure 1.18 shows a typical activated carbon column with a single column of water flowing down. Figure 1.19 is a carbon column in operation at a major Superfund site in New York.

The liquid loading needs to be stopped when the carbon column reaches breakthrough. Upflow systems have an advantage over downflow systems in that the fluidized carbon bath causes less plugging and less short circuiting. The hydraulic loading is usually from 2 to 5 gal/min per square foot. These rates have to be reasonably small because the higher the velocity, the more likely the physical attraction of the chemical to the surface area of the carbon will be overcome by friction and the chemical will be desorbed.

An active carbon column should be designed only after treatability studies of a particular wastewater. This is best done with a pilot column test at the site, using the contaminated groundwater. One of the reasons for field testing is that many contaminants will compete for the adsorbing space, causing breakthrough for the target compound earlier than expected. Indeed, carbon will indiscriminately adsorb chem-

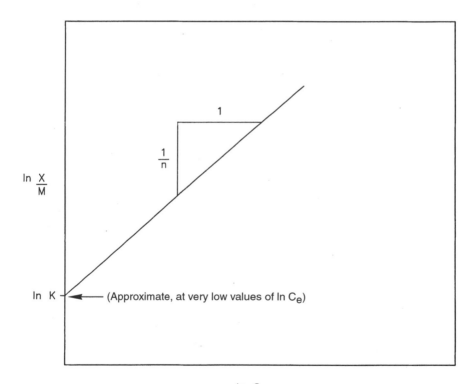

FIGURE 1.17 Absorption isotherm.

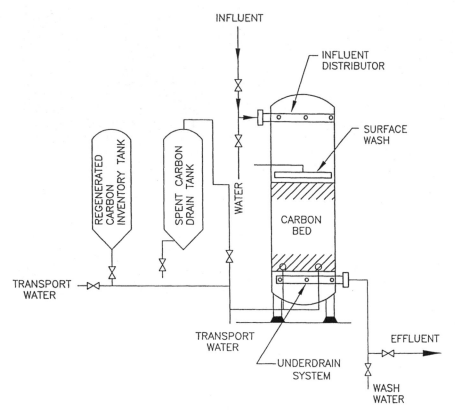

FIGURE 1.18 Fixed-bed adsorption system.

icals, and hence the receptor sites where chemicals can adsorb onto the surface will be occupied by other kinds of chemicals. Therefore, pilot field studies are a necessity in determining design parameters. These design pilot tests are designed to obtain the following information: contact time, bed depth, pretreatment requirements, breakthrough characteristics, head loss characteristics, and carbon dosage in pounds of pollutants removed per pound of carbon. The actual design of the activated carbon column can be accomplished by using a equation derived by Tomas in 1948 (Knox, 1986):

$$\frac{C_0}{C} = 1 + \exp\left[\frac{K_1}{Q}(A_0M - C_0V)\right]$$

where: C = effluent pollutant concentration, g/m^3
         C_0 = influent pollutant concentration, g/m^3
         K_1 = rate constant, $m^3/day\text{-}g$
         Q = flow rate, m^3/day
         A_0 = adsorption capacity, g/g carbon
         M = mass of carbon, g
         V = volume of water, m^3

FIGURE 1.19 Four large granular activated carbon vessels treating groundwater at a Superfund site, New York.

This equation can be rearranged and the algorithm of each side can be taken to yield an equation that can be represented as a straight line. Pilot column test provides all the parameters except K_1 and A_0, which can be determined by a graphical solution if the natural logarithm of $(C_0/C - 1)$ is plotted versus V. Hence, the mass of carbon needed for a select breakthrough concentration can be determined. Practical bed design frequently is a more conservative estimate of adsorption capacity (grams of contaminant versus grams of carbon) prior to breakthrough, the mass loading rate (groundwater concentration liquid loading rate), and the desired time between regenerations.

The application of activated carbon adsorption to groundwater contaminants has been successfully used for removing organics. As early as 1982, seventeen applications of carbon adsorption had been recorded (Kauffman, 1982). In each case, activated carbon successfully treated the contaminated aquifers. This nonselective technique will work on a number of different molecules. McDougal et al. (1980) reported successful treatment of the Love Canal landfill leachate by activated carbon absorption.

The cost of treating groundwater by activated carbon is dependent upon a number of factors. These factors include the contaminant concentration, molecular structures, flow rates, type of carbon, type of application, and site requirements. O'Brien and Fisher (1983) reported that the costs range from $0.48 to $2.52 per 1000 gallons treated. However, this is highly dependent upon the influent concentration. Indeed, if the influent contaminant concentrations are in the microgram per liter range, the costs varied from $0.22 to $0.55 per 1000 gallons treated.

Biological Treatment. Much of the biological treatment of groundwater contamination has been borrowed from the sanitary engineering profession, which has more than 50 years of experience treating sanitary sewage and industrial waste. Biological treatment, even though many of hazardous constituents are toxic, can still be used when the proper environment is provided. Heterotrophic micro-organisms are the most common group of micro-organisms providing the metabolic process for removing organic compounds from contaminated groundwater.

Heterotrophs use the contaminants as sources of carbon and energy. A portion of the organic material is oxidized to provide energy, while the remaining portions are used as building blocks for cellular synthesis. Three general methods exist by which heterotrophic micro-organisms can obtain energy. These are fermentation, aerobic respiration, and anaerobic respiration. For fermentation, the carbon and energy source are broken down by a series of enzyme-mediated reactions, which do not involve an electron transport chain. In aerobic respiration, the carbon and energy source are broken down by a series of enzyme-mediated reactions in which oxygen serves as an external electron acceptor. In anaerobic respiration, the carbon and energy source are broken down by a series of enzyme-mediated reactions in which sulfate, nitrates, and carbon dioxide usually serve as the external electron acceptors. The three processes of obtaining energy form the basis for the various biological waste treatment processes.

The biological treatment processes are typically divided into two categories: suspended growth systems and fixed-film systems. Suspended growth systems are more commonly referred to as *activated sludge processes,* of which several variations and modifications exist. The basic system consists of a large basin in which the contaminated water is introduced along with air or oxygen by either diffusers or mechanical aeration devices and possibly nitrogen or phosphorus as nutrients. The micro-organisms are present in the aeration as suspended material and grow on the waste that is being provided to them. The micro-organisms as they grow need to be separated from the liquid stream. This is accomplished by gravity settling. After the gravity settling, the biomass may be increased by sludge-thickening devices. The entire process's performance is dependent upon the recycling of sufficient biomass, termed *mixed liquor.* The process requires the skill of experienced operators, and can be upset by changes in influent or operating conditions.

Fixed-film biological processes differ from suspended growth in that micro-organisms attach themselves to media, which provide an environment for them. Biological towers and rotating biological contactors are the most common forms of fixed-film processes. Microbes form a slime layer on these fixed films and metabolize the organics, with oxygen being provided as air moves countercurrent to the water flow. Rotating biological contactors consist of a series of rotating disks connected by a shaft and set in a basin or trough. The contaminated water passes through the basins where the micro-organisms get a chance to feed on and metabolize the organic matter present in the water. At any one time, approximately 40 percent of the disk surface is submerged. After coming out of the water, contact with air provides oxygen to the microbes and the process starts all over again, as the disk continues to rotate back into the water. The removal efficiencies are approximately the same for the fixed-film and suspended growth processes; however, the fixed film has a tendency to be lower in cost because of the high energy requirement of providing oxygen to the large aeration tanks. The design of these biological treatment systems is usually based on a kinetic model. Most widely used models have been developed by Eckenfelder, McKenny, Lawrence and McCarthy, and Goudy (Kincannon and Stover, 1981). However, Kincannon and Stover (1981) have found that a great amount of variability exists in the biokinetic parameters for these models, and conclude that the models are not ideal for waters containing priority pollutants. Kincannon and Stover believe that the biokinetic parameters should be determined by conducting laboratory or pilot plant studies. After the biokinetic constants are determined, the required volume, aeration tank, surface area, biological tower, and rotating biological contactor can be determined for any flow rate.

The application of these aboveground biological treatment systems to groundwater have been sparse. These biological systems generally require an operator and can be slightly more costly than other types of physical or chemical treatment. Moreover, groundwater contamination is often dilute and made up of recalcitrant organics. Hence the buildup of enough micro-organisms to effectuate a rapid destruction of the contamination is sometimes problematic.

Chemical Precipitation. Chemical precipitation for the removal of inorganic compounds is a well-established technology. There are three common chemical addition systems that will precipitate inorganics at a specific pH: (1) carbonate system, (2) hydroxide system, and (3) sulfide system. The sulfide system removes the most inorganics, with the exception of arsenic, because of the low solubility of arsenic sulfide compounds. However, the difficulty in handling the chemicals after the sulfide has precipitated can sometimes lead to the resolubilization of the metal. The carbonate system is a method that relies on the use of soda ash and pH adjustment between 8.2 and 8.5. The carbonate system, although workable in theory, is difficult to control. The hydroxide system is most widely used for inorganic metal removal. The system responds directly to pH adjustments and usually uses either lime (calcium hydroxide) or sodium hydroxide as the chemical to adjust the pH upward. Sodium hydroxide has the advantage of ease and chemical handling and low volume of sludge. However, sodium hydroxide sludge is often gelatinous and difficult to dewater.

The hydroxide process can be described very simply. Water is treated either by batch tank or continuous flow. The contaminant concentration of influent water is measured and the amount of hydroxide is fed into the solution with mixing to produce the desired precipitate. A settling tank is used to separate the treated liquid from the settled solids. When the flow exceeds 30,000 gal/day, batch treatment is usually not feasible because of the large tankage required. Continuous treatment may require a preliminary tank for pH adjustment. A reaction tank may also be required before the treated waste is transferred. A polyelectrolyte may be added to assist the solids to settle faster, produce a higher-quality liquid effluent, and improve sludge dewatering procedures.

DESIGN PARAMETERS AND PROCEDURES: The important design factors that must be determined for a particular water treatability study include:

1. Best chemical addition system
2. Optimum chemical dose
3. Optimal pH control system
4. Rapid mix requirements
5. Flocculation requirements
6. Sludge production/holding tank
7. Sludge flocculation settling and dewatering characteristics
8. Chemical storage/reaction tanks
9. Space requirements
10. Effluent discharge system

Laboratory-scale test procedures consisting of jar test studies have been used for years and are the norm for establishing design parameters. For large systems, especially continuous flow-through systems, a small pilot plant treatability study may be required.

APPLICATION TO GROUNDWATER: Chemical precipitation has been successfully used for removing heavy metals from various waters. Kincannon and Stover (1981) reported the results of treating contaminated groundwater by various processes.

Table 1.7 shows the removal of metals via lime treatment at different pHs. All these metals showed remarkably good treatment levels for the identified precipitation techniques. Brantner and Cichon (1981) also compared the three precipitation processes, and the results are shown in Table 1.8. Their work shows that chemical precipitation is an effective way of removing metals in groundwater.

TABLE 1.7   Removal Chemical Precipitation Data for Metals

| | Concentrations in groundwater, mg/L | | | |
| | | Lime-treated water (hydroxide) | | |
Compound	Raw water	pH 9.1	pH 9.9	pH 11.3
Arsenic	0.12	0.03	0.03	0.03
Barium	0.24	0.17	0.15	0.19
Cadmium	0.003	<0.001	<0.001	<0.001
Chromium (total)	0.09	0.006	0.006	0.006
Lead	0.03	0.006	0.006	0.006
Mercury	<0.001	<0.001	<0.001	<0.001
Selenium	<0.001	<0.001	<0.001	<0.001
Silver	<0.001	<0.001	<0.001	<0.001
Copper	0.1	<0.001	<0.001	<0.001
Iron	352	0.07	0.07	1.05
Manganese	90	0	0	0
Nickel	1.95	0.05	0.3	0.45
Zinc	0.69	0.36	0.09	0.61

Note: For chromate removal with hydroxide, the hexavalent form must first be reduced to trivalent at a low pH.

Other Treatment Techniques for Inorganics.   A wide variety of other techniques also exists for the treatment of inorganic contaminants in groundwater. Table 1.9 was compiled from Paterson (1978) and represents viable techniques for each of the individual compounds. It should be noted that even though these treatment methods have been used, they have primarily been used for industrial waste treatment where concentrations have been higher than those of groundwater. So these treatment methods, although successful in other areas, do not have a proved track record in groundwater.

In Situ Technologies.   This section addresses aquifer restoration by treatment in place (in situ). A major problem with in situ technologies has been not so much the treatment technologies, which have been proved in the past, but delivering the necessary ingredients to the right location at the right concentration and the right time. A major problem to overcome is mixing of the feed material into the aquifer, which is a very slow process because of slow movement of groundwater, which creates laminar conditions. In situ physical/chemical treatment will be discussed first and followed by in situ biological treatment.

In Situ Physical/Chemical Technologies.   In situ physical/chemical treatment generally involves the installation of a bank of injection wells at the head of a plume

TABLE 1.8 Comparison of Precipitation Treatment Processes

Parameter	Influent, ppm		Clarifier effluent, ppm		Filtered effluent, ppm	
	Mean	Range	Mean	Range	Mean	Range
Hydroxide precipitation data summary						
Suspended solids	42	20–63	22	9–33	9	4–14
pH	7.5	7.0–8.1	9.9	9.7–10.2	9.8	9.5–10.4
Total cadmium	1.66	0.13–4.3	0.05	0.03–0.1	0.04	0.02–0.06
Total chromium	1.11	0.07–2.9	1.04	0.07–2.8	0.97	0.06–2.9
Total copper	0.29	0.12–1.5	0.03	0.02–0.03	0.03	0.02–03
Total lead	1.7	0.8–2.6	0.2	0.1–0.3	0.2	0.1–0.3
Total zinc	31	6–91	0.4	0.23–0.75	0.28	0.1–0.66
Carbonate precipitation data summary						
Suspended solids	43	16–75	27	14–52	6	2–10
pH	7.1	6.7–7.7	8.3	8.1–8.4	8.1	7.8–8.5
Total cadmium	1.37	0.26–2.9	0.14	0.02–0.27	0.04	0.02–0.06
Total chromium	0.67	0.23–1.8	0.62	0.17–1.8	0.6	0.14–2.00
Total copper	0.18	0.06–0.27	0.04	0.03–0.06	<0.03	<0.02–0.04
Total lead	1.4	0.7–2.1	0.2	0.2–0.4	<0.1	<0.1–0.2
Total zinc	26	1–67	3.2	0.37–5.0	1.18	0.19–5.00
Sulfide precipitation data summary						
Suspended solids	90	30–210	37	10–63	5	2–8
pH	6.8	6.4–7.7	8.2	8.0–8.4	8.1	7.8–8.5
Total cadmium	3.3	0.65–5.4	0.18	0.09–0.29	0.06	0.02–0.12
Total chromium	0.52	0.02–1.9	0.12	0.08–0.2	<0.05	<0.05–0.05
Total copper	0.35	0.19–0.96	<0.04	<0.04–0.05	<0.03	<0.02–0.03
Total lead	4.5	2.3–8.1	0.4	0.2–0.5	<0.1	<0.1–0.2
Total zinc	93	5.8–220	3.1	2.0–6.2	0.68	0.11–1.8

of contaminated groundwater. A treatment agent is then pumped into the aquifer. The selected treatment agents are specific for each type of contamination. For example, heavy metals may be made insoluble (precipitates) and immobile by sulfides. Cyanides can be destroyed with oxidizing agents, cations may be precipitated with various anions or by aeration, and hexavalent chromium could be made insoluble with reducing agents.

In situ physical/chemical technologies include:

- Air sparging
- Hot water or steam flushing/stripping
- In-well air stripping
- Circulating wells
- Passive/reactive treatment walls

TABLE 1.9   Treatment Alternatives for Inorganics

Inorganic	Treatment method
Arsenic	Charcoal filtration
	Lime softening
	Precipitation with lime + iron
	Precipitation with alum
	Precipitation with ferric sulfate
	Precipitation with ferric chloride
	Precipitation with ferric hydroxide
	Precipitation with sulfide
	Ferric sulfide filter bed
	Iron or lime coagulation + settling + dual media filtration + carbon absorption
Barium	Iron or lime coagulation + settling + dual media filtration + carbon absorption
	Precipitation as sulfate
	Precipitation as carbonate
	Precipitation as hydroxide
	Ion exchange
Boron	Evaporation
	Reverse osmosis
	Ion exchange
Cadmium	Precipitation as hydroxide
	Precipitation as hydroxide + filtration
	Precipitation as sulfide
	Coprecipitation with ferrous hydroxide
	Reverse osmosis
	Freeze concentration
Chloride	Ion exchange
	Electrodialysis
	Reverse osmosis
	Other (holding basins, evaporative ponds, deep well injection)

Engineers have a remarkable way of looking at successful treatment technologies and modifying them. Such is the case with *air sparging*. Air sparging, borrowed from ex situ air stripping, is an in situ technology in which air is injected into a contaminated aquifer. Injected air traverses horizontally and vertically in channels through the soil column, creating an underground stripper that removes contaminants by volatilization. The injected air helps to create mixing zones to further enhance removal rates. Volatile contamination rises to the unsaturated zone, where a vapor extraction system is usually implemented to remove the generated vapor-phase contamination. This technology is designed to operate at high gas-flow rates.

Oxygen added to contaminated groundwater and vadose zone soils can also enhance biodegradation of contaminants below and above the water table. Typical design and operating parameters have been presented by Marley et al. (2000) (see Table 1.10). After a review of 37 systems, Marley concluded that the sphere of influence of air sparging wells generally ranges between 10 and 26 ft. A typical well is 2 in in diam-

TABLE 1.10   Typical Design and Operating Parameters for in Situ Air Sparging Wells

Parameter and range	Most often used value (no. of sites)	Second most often used value (no. of sites)	Third most often used value (no. of sites)	Total number of sites
Screen length 0.15–3.05 m (0.5–10 ft)	0.61 m (2 ft) 16 sites	0.91 m (3 ft) 8 sites	1.52 m (5 ft) 7 sites	40
Well diameter 2.54–10.16 cm (1–4 in)	5.08 cm (2 in) 17 sites	10.16 cm (4 in) 7 sites)	2.54 cm (1 in) 5 sites	37
Overpressure* 2.41–125.67 kPa (0.35–18.2 psi)	2.41–34.54 kPa (5–10 ft) 10 sites	34.45–68.90 kPa (5–10 psi) 9 sites	68.90–103.35 kPa (10–15 psi) 5 sites	31
Well screen depth below water table 0.61–8.08 m (2–26.5 ft)	1.52–3.05 m (5–10 ft) 10 sites	3.05–4.57 m (10–15 ft) 8 sites	0.61–1.52 m (2–5 ft) 6 sites	31
In situ sparging flow rate 2.21–67.96 m³/h (1.3–40 ft³/min)	2.21–8.50 m³/h (1.3–5 ft³/min) 16 sites	8.50–16.99 m³/h (5–10 ft³/min) 9 sites	25.48–33.98 m³/h (15–20 ft³/min) 5 sites	39
In situ sparging pressure 24.11–172.25 kPa (3.5–25 psi)	34.45–68.90 kPa (5–10 psi) 17 sites	68.90–103.35 kPa (10–15 psi) 8 sites	137.80–172.25 kPa (20–25 psi) 6 sites	40
(SVE ROI)/(IAS ROI)[†] ratio 0.16–7.42	1–2 12 sites	0.16–1 6 sites	3–4 3 sites	26

   * Defined as the excess pressure delivered to the aquifer above that which is necessary to overcome the static head of the water (depth below water table to the screen).
   [†] SVE = soil vapor extraction, ROI = radius of influence, IAS = in situ air sparging
   Source: From Marley et al. (2000).

eter with a 2-ft screen positioned 5 to 10 ft below the water table, with an overpressure of 5 psi and a flow rate of 5 ft³/min.

Air sparging systems may last from a few months to a few years. Limitations to this technology are as follows:

- Airflow through the saturated zone may *not* be uniform, which implies that not all the plume is being treated.
- Depth of contaminants and specific site geology must be considered.
- Air injection wells must be designed for site-specific conditions.
- Soil heterogeneity and low-permeability zones may cause some zones to be relatively unaffected.
- Some contaminant spreading may result from the mixing of the aquifer (U.S. EPA, 1999) especially in the case where a floating product of petroleum is encountered.
- The influence from the well may be variable (Acomb et al., 2000).

Acomb et al. discuss the amount of oxygen as percent saturation. Their experiments show that the air saturation distribution changed with time, initially expanding outward then contracting from the air sparging well. They have divided up the aquifer zone into two areas: the cone area where remedial rates are vastly increased, defined by a 20 percent saturation value, and the outer area where remedial processes will be limited by diffusion (see Fig. 1.20).

In *hot water or steam flushing/stripping,* hot water or steam is forced into an aquifer through injection wells to vaporize volatile organic compounds (VOC) and semi-VOC (SVOC) contaminants. Vaporized components rise through the saturated zone to the unsaturated (vadose) zone where they are removed by vacuum extraction and then treated. Hot water or steam-based techniques include contained recovery of oily waste (CROW), steam injection and vacuum extraction (SIVE), in situ steam-enhanced extraction (ISEE), and the steam-enhanced recovery process (SERP). Hot water or steam flushing/stripping is a pilot-scale technology. A bonus of this technology is that in situ biological growth and treatment may follow the application. Biological processes flourish in warmer environments, and growth rates are geometrically increased as a result of temperature increases.

Steam injection is applicable to both shallow and deep contaminated areas and readily available by using mobile equipment. Hot water/steam injection is typically of short to medium duration, lasting a few weeks to several months. Because of the high temperatures, the applicability of this technology is for VOCs and SVOCs, and, although VOC compounds can be treated with this method, there are more cost-effective ways to treat VOCs.

Limitations for this technology are as follows:

- Low-permeability zones and heterogeneity may affect heating time.
- Mobilization of compounds can sometimes be uncontrolled, especially DNAPLs, since their viscosity and hence surface tension is a direct function of temperature (U.S. EPA, 1999).

In *in-well air stripping,* air is injected into a double-screened well, lifting the water in the well and forcing it out the upper screen. Simultaneously, additional water is drawn in the lower screen. Once in the well, some of the VOCs in the contaminated groundwater are transferred from the dissolved phase to the vapor phase by air bubbles. The contaminated air rises in the well to the water surface where vapors are drawn off and treated by a soil vapor extraction (SVE) system. This SVE system, in addition to collecting the vapors from within the well, can be designed to collect vapors from the surrounding vadose zone. The partially treated groundwater is never brought to the surface. It is forced into the unsaturated zone and the process is repeated as water follows a hydraulic circulation pattern, or cell, that allows continuous cycling of groundwater. As groundwater circulates through the treatment system in situ, contaminant concentrations are gradually reduced.

Modifications to the basic in-well stripping process may involve additives injected into the stripping well to enhance biodegradation (e.g., nutrients, air/oxygen, electron acceptors). In addition, the area around the well affected by the circulation cell (sphere of influence) can be modified through the addition of certain chemicals to allow in situ stabilization of metals originally dissolved in ground water.

The duration of in-well air stripping is short to long term, depending on contaminant concentrations (Henry's law constants of the contaminants), the sphere of influence, and site hydrogeology.

Circulating wells (CWs), a variation on the in-well air stripping, provide a technique for subsurface remediation by creating a three-dimensional circulation pattern of the groundwater. Groundwater is drawn into a well through one screened section and reintroduced to the aquifer at another location through a second screen.

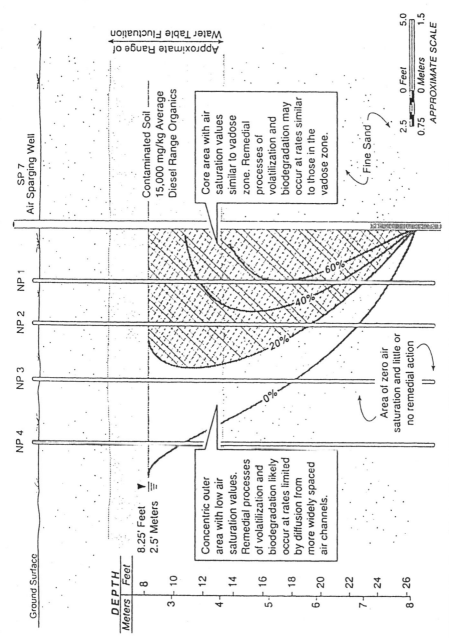

FIGURE 1.20 Remedial processes may occur at different rates in concentric areas around the air sparging well. (*From Acomb et al., 2000.*)

1.55

The flow direction through the well can be specified as either upward or downward to accommodate site-specific conditions. Because groundwater is not pumped above ground, pumping costs are reduced, permitting issues are eliminated, and problems associated with storage and discharge are removed. In addition to groundwater treatment, CW systems can provide simultaneous vadose zone treatment in the form of bioventing or SVE. This technique seeks to eliminate the problem of the lack of mixing in the groundwater system, but it requires capture and treatment of the off-gas. A series of adjacent CWs can create a treatment wall that a contaminant plume must pass through, thereby resulting in remediation.

CW systems can provide treatment inside the well, in the aquifer, or a combination of both. For effective in-well treatment, the contaminants must be adequately soluble and mobile so they can be transported by the circulating groundwater. Because CW systems provide a wide range of treatment options, they provide some degree of flexibility to a remediation effort. Target contaminants are VOCs.

Limitations of this technology and of in-well air stripping include:

- In general, in-well air strippers are more effective at sites containing high concentrations of dissolved contaminants with high Henry's law constants.
- Fouling of the system may occur by infiltrating precipitates containing oxidized constituents and may require periodic screen cleaning.
- Shallow aquifers may limit process effectiveness.
- Effective CW installations require a well-defined contaminant plume to prevent the spreading or smearing of the contamination. They should not be applied to sites containing NAPLs, to prevent the possibility of smearing the contaminants.
- CWs are limited to sites with horizontal hydraulic conductivities greater that 10^{-5} cm/s and should not be utilized at sites that have lenses of low-conductivity deposits.
- In-well air stripping may not be efficient in sites with strong natural flow patterns.
- The ratio of horizontal/vertical hydraulic conductivity should be between 3 and 10 for ideal performance (U.S. EPA, 1999).

Passive/reactive treatment walls involve the installation of a permeable reaction wall across the flow path of a contaminant plume, allowing the water portion of the plume to passively move through the wall. These barriers allow the passage of water while prohibiting the movement of contaminants by employing such agents as zero-valent metals, chelators (ligands selected for their specificity for a given metal), sorbents, and microbes.

The contaminants will either be degraded or retained in a concentrated form by the barrier material. The wall could provide permanent containment for relatively benign residuals or provide a decreased volume of the more toxic contaminants for subsequent treatment.

Modifications to the basic passive treatment walls may involve a *funnel-and-gate system*. The funnel-and-gate system for in situ treatment of contaminated plumes consists of low hydraulic conductivity (i.e., 10^{-6} cm/s) cutoff walls (the funnel) with a gate that contains in situ reaction zones. Groundwater primarily flows through high-conductivity gaps (the gates). The type of cutoff walls most likely to be used in the current practice are slurry walls or sheet piles. Innovative methods such as deep soil mixing and jet grouting are also being considered for funnel walls.

An *iron treatment wall* consists of iron granules or other iron-bearing minerals for the treatment of chlorinated contaminants such as TCE, DCE, and VC. As the iron is oxidized, a chlorine atom is removed from the compound by one or more reductive dechlorination mechanisms, using electrons supplied by the oxidation of iron. The iron granules are dissolved by the process, but the metal disappears so slowly that the remediation barriers can be expected to remain effective for many years, possibly even decades. One possible problem is the formation of an iron precipitate that could limit the rate transfer and even clog the permeable barriers; however, this process is not fast. Indeed, iron permeable walls have already been in place for 5 to 10 years.

Barrier and postclosure monitoring tests are being conducted by the USAF, U.S. Navy, and Department of Energy (DOE) in field-scale demonstration plots and are being designed for actual contaminated sites.

The range of materials available for augmenting existing barrier practice is broad. Two types of barriers have been the focus of initial efforts of this program, i.e., permeable reactive barriers and in-place bioreactors.

Passive treatment walls are generally intended for long-term operation to control migration of contaminants in groundwater. The target contaminants for this technology are VOCs, SVOCs and inorganics. The limitations for this technology are as follows:

- Passive treatment walls may lose their reactive capacity, requiring replacement of the reactive medium.

- Passive treatment wall permeability may decrease because of precipitation of metal salts.

- Depth and width of barrier are sometimes too small for a large plume.

- The treatment is limited to a subsurface lithology that has a continuous aquitard at a depth that is within the vertical limits of trenching equipment.

- Biological activity or chemical precipitation may limit the permeability of the passive treatment wall.

- Long-term monitoring is mandatory (U.S. EPA, 1999).

In Situ Biological Technologies. Cometabolism is one form of secondary substrate transformation in which enzymes produced for primary substrate oxidation are capable of degrading the secondary substrate (contaminant). This is fortuitous, as the secondary substrates usually do not afford sufficient energy to sustain the microbial population. An emerging application involves the injection of water containing dissolved primary substrate (e.g., methane, toluene) and oxygen into groundwater to support the cometabolic breakdown of targeted chlorinated organic contaminants.

The addition of methane or methanol supports methanotrophic activity, which has been demonstrated effective to degrade chlorinated solvents, such as vinyl chloride and TCE, by cometabolism. Although toluene, propane, and butane are used to stimulate a different class of micro-organisms, not methanotrophs, they have been used successfully for supporting cometabolism of TCE.

Cometabolic technologies may be classified as long-term technologies, which may take several years to decades to clean up a plume.

The primary targeted contaminants for this technology are the chlorinated hydrocarbons. Limitations of this technology are as follows:

- This technology is still under development

- Regulatory approval for use of specific cometabolites may be required and are themselves contaminants.

- Where the subsurface is heterogeneous, it is very difficult to circulate the methane solution throughout every portion of the contaminated zone. Higher-permeability zones are cleaned up much faster because groundwater flow rates are greater.

- Safety precautions (such as removing all ignition sources in the area) must be used when handling methane.
- A surface treatment system, such as air stripping or carbon adsorption, may be required to treat extracted groundwater prior to reinjection or disposal.
- High copper concentrations affect methanotrophic cometabolism.
- Predation affects methanotrophic cometabolism (U.S. EPA, 1991).

Bioremediation is a process in which indigenous or inoculated micro-organisms (i.e., fungi, bacteria, and other microbes) degrade (metabolize) organic contaminants found in soil and/or groundwater.

Bioremediation attempts to accelerate the natural biodegradation process by providing nutrients, electron acceptors, and competent degrading microorganisms that may otherwise be lethargic and hence limit the rapid conversion of contamination organics to innocuous end products.

Oxygen enhancement can be achieved by either sparging air below the water table or by circulating hydrogen peroxide (H_2O_2) throughout the contaminated groundwater zone. Under anaerobic conditions, nitrate is circulated throughout the groundwater contamination zone to enhance bioremediation. Additionally, solid-phase peroxide products [i.e., oxygen-releasing compounds (ORCs)] can also be used for oxygen enhancement and to increase the rate of biodegradation.

Oxygen enhancement with *air sparging* below the water table increases groundwater oxygen concentration and enhances the rate of biological degradation of organic contaminants by naturally occurring microbes. Air sparging also increases mixing in the saturated zone, which increases the contact between groundwater and soil. The ease and low cost of installing small-diameter air sparging systems have led to their extensive use as a remediation system. Oxygen enhancement with air sparging is typically used in conjunction with SVE or bioventing to enhance removal of the volatile component under consideration. Offgases from these systems are collected and treated prior to release to the atmosphere. If VOCs are not the target, then air sparging needs to be designed to supply oxygen to the biological community and thereby enhance biodegradation.

In *oxygen enhancement with hydrogen peroxide,* a dilute solution of hydrogen peroxide is circulated through the contaminated groundwater zone to increase the oxygen content of groundwater and enhance the rate of aerobic biodegradation of organic contaminants by naturally occurring microbes.

In *nitrate enhancement,* solubilized nitrate is circulated throughout groundwater contamination zones to provide an alternative electron acceptor for biological activity and enhance the rate of degradation of organic contaminants. Development of nitrate enhancement is still at the pilot scale. This technology enhances the anaerobic biodegradation through the addition of nitrate.

Fuel spills have been shown to degrade rapidly under aerobic conditions, but success often is limited by the inability to provide sufficient oxygen to the contaminated zones as a result of the low oxygen solubility in water and because oxygen is rapidly consumed by aerobic microbes. Nitrate also can serve as an electron acceptor and is more soluble in water than oxygen. The addition of nitrate to an aquifer results in the anaerobic biodegradation of toluene, ethylbenzene, and xylenes. The benzene component of fuel has been found to biodegrade slower under strictly anaerobic conditions. A mixed oxygen/nitrate system would prove advantageous in that the addition of nitrate would supplement the demand for oxygen rather than replace it, allowing for benzene to be biodegraded under varying aerobic conditions.

These technologies may be classified as long term, and may take several years to several decades for plume cleanup. Targeted contaminants for enhanced bioremediation are nonhalogenated hydrocarbons, SVOC, and fuels. Limitations for this technology are as follows:

* Where the subsurface is heterogeneous, it is very difficult to deliver the nitrate or hydrogen peroxide solution throughout every portion of the contaminated zone. Higher-permeability zones will be cleaned up much faster because groundwater flow rates are greater.
* Safety precautions must be used when handling hydrogen peroxide.
* Concentrations of hydrogen peroxide greater than 100 to 200 ppm in groundwater are inhibiting to microorganisms.
* Microbial enzymes and high iron content of subsurface materials can rapidly reduce concentrations of hydrogen peroxide and reduce zones of influence.
* A groundwater circulation system must be created so that contaminants do not escape from zones of active biodegradation.
* Because air sparging increases pressure in the vadose zone, vapors can build up in building basements, which are generally low-pressure areas.
* Many states prohibit nitrate injection into groundwater because nitrate is regulated through drinking water standards.
* A surface treatment system, such as air stripping or carbon adsorption, may be required to treat extracted groundwater prior to reinjection or disposal (U.S. EPA, 1999).

1.3.3   Dense Nonaqueous-Phase Liquids (DNAPLs)

Introduction.   In this section, in situ DNAPL technologies will be reviewed. The focus will be on the pure-phase contaminant rather than the aqueous-phase contaminant. Thermal desorption, chemical oxidation, and surfactant flushing will be reviewed.

Thermal Desorption.   Thermal desorption is a thermally induced physical separation process. Contaminants generally have low boiling points, and heating the soil will transform them from the liquid phase to the vapor phase, in both saturated and unsaturated soil. When the vapors are transferred into a gas they can easily be managed by a vapor extraction system. Options of management may include condensation, collection, combustion, or adsorption to media such as granular activated carbon. Of the most often utilized technologies for thermal adsorption, three are low temperature and one is high temperature. The three low-temperature technologies are steam stripping, six-phase heating, and electrical heating, while the high-temperature one is vitrification.
    Steam stripping has been used for the unsaturated zone as well as the saturated zone. In the saturated zone, steam is directly injected into the water-bearing zone and heats up the water and soil. The key to any of these heating technologies is heating the subsurface to steam temperatures. Once reaching these temperatures, complete DNAPL vaporization and removal from the subsurface can occur. Udell (1966) has shown this with experiments. The fundamentals of the steam injection and the extraction technology were borrowed from the enhanced oil recovery industry and renamed as steam-enhanced extraction at the University of California Berkley. Steam-enhanced extraction was later combined with electrical heating in a process called dynamic underground stripping by Lawrence Livermore National Laboratory (Newmark, 1994).

The Henry's law coefficient for TCE increases tenfold in heating from ambient to steam temperatures. Bench-scale studies have shown removals on the order of 99 percent from soils during heating, using both direct stream injection and electrical heating to produce steam within the soil. As a bonus to the development of this thermal technology, hydropyrolysis/oxidation can take place. When heated in an oxygenated zone, contaminants can be oxidized and degraded to benign products. This process can be stimulated by injection of atmospheric air with the steam and the oxygen to fuel the reaction.

Six-Phase Heating. Six-phase heating was developed by Battelle Memorial Institute for the U.S. Department of Energy. It involves conducting electricity through the subsurface. The resistance to the flow of electrical currents results in the generation of heat. Electrodes are installed in a pattern where a central neutral electrode is surrounded by six charged electrodes. The electrodes are sequentially charged 60° out of phase from each other. The result of this method is an even distribution of heat throughout the treatment zone, creating an in situ source of steam to strip the volatile organic compounds from the soil and groundwater.

Electric Heating. Electric heating technology has been used sparingly in the past. In typical applications of this process, electrodes are strategically placed in a contaminated zone. The pattern of electrodes is designed so that the conventional three-phase power can be used to heat the soil. Also, the distance between electrodes and their locations are determined from the heat transfer mechanisms associated with vapor extraction. Electrical heating and fluid movement in the contaminated zone without consideration of all the heat transfer mechanisms will result in a less than efficient process. To determine the ideal pattern of electrodes in extraction wells, a multiphased, multicomponent three-dimensional thermal model is used to simulate the process.

Electric heating increases the temperature of the soil by conducting current through the resistive water that fills the pore spaces of the soil matrix. Maximum temperatures are limited to the boiling temperatures of the water; otherwise, the electrical path is boiled off. The increase in temperature increases the vapor pressure of volatile and semivolatile contaminants, thus increasing the volitalization and removal of the soil by vapor extraction. Typical temperatures in the soil are somewhere around 80°C as a result of electrical heating, and the average pressure can be reduced by vapor extraction to approximately one-third atmospheric pressure. The average temperature, for example, need only exceed 50°C for benzene to change from a liquid to a gaseous phase during soil vapor extraction operations.

In Situ Vitrification. In situ vitrification is a similar technology, but is a high-temperature technology, also using electrical current. In this technology the soil is brought to temperatures that can melt the rock and turn it into a single mass unit. This technology cannot be operated below the water table. Some of the advantages of thermal treatment are that they readily create conditions for mass transfer for the chlorinated hydrocarbons from a DNAPL state directly to a gaseous state. Thermal treatment techniques have the added benefit of not being dependent on the heterogeneity of the formation. One of the disadvantages is that it takes time to heat up the soil, sometimes as much as 3 months before temperatures are high enough to begin to boil off the contaminants. If the system is carelessly planned, the contaminants may adsorb onto the soil matrix and go from the saturated zone to the unsaturated soil matrix and condense there. Another disadvantage is that these technologies tend to be a lot more costly then some of the chemical oxidation or surfactant methods.

Chemical Oxidation Techniques. Chemical oxidation techniques involve the oxidation of the chlorinated hydrocarbon compounds by another active chemical agent. There are many oxidants that are currently being used for the treatment of

DNAPLs; these include ozone, hydrogen peroxide, Fenton's reagent, sodium permanganate, potassium permanganate, and oxygen. Potassium permanganate is the most common oxidant used because of its relative stability and ease of handling, relatively low cost (about $3 per kilogram), and the ability to visually see the results of the application (the soluble permanganate ion is purple). One of the disadvantages of its use is that it forms a precipitate (manganese dioxide) and can limit the oxidation of the DNAPLs (Conrad, 2001). Potassium permanganate has been widely used for over 40 years by the drinking water and wastewater industries. The reaction kinetics are well understood and they work over a very wide range of pH values. It reacts quickly without the generation of heat and completely oxidizes a wide range of common recalcitrant organic contaminants including TCE, PCE, DCE, and phenols. Groundwater subject to permanganate treatment can be visually observed to change color from purple to brown to clear upon complete oxidation. One of the disadvantages, especially with potassium permanganate, is that the manganese dioxide precipitate may partially plug the aquifer.

Surfactants. The use of surfactants has been the subject of many experiments and some full-scale characterization studies. Their use increases the solubility of the DNAPLs by 10 to 1000 times the normal levels and greatly reduces the time necessary to flush the contaminants out. Often electrolyte control and cosolvents are used in this process. The cosolvents typically used are alcohols and the electrolyte is typically calcium chloride.

One of the advantages of surfactants is that, with the partitioning interwell tracer test (PITT), they provide a means for identifying not only where the solvent is in the aquifer but also its volume. One of the disadvantages is that if the solvent mixture is not perfectly fine tuned, the DNAPLs can cascade into the lower portions of the aquifer, making it much more difficult to remediate.

Physical Techniques. Air sparging for dissolved contaminants is very similar to air sparging for DNAPL contaminants. The design concepts and the resulting remediation are very much the same.

1.3.4 Light Nonaqueous-Phase Liquids (LNAPLs)

In Situ LNAPLs. Treatment technologies for LNAPLs have focused on separation because its density is less than that of water. This section will review the physical techniques that have been used to remove LNAPLs.

Dual Pumping and Skimming Pump. The first technique is a purely physical activity called *dual pumping.* The water level is lowered by pumping. The cone of depression creates a steep gradient that results in the movement and accumulations of the free product at the center of the pumping cone. The hydrocarbons that are mobile will concentrate and build to a thicker layer of oil that can be skimmed off. The first recovery of LNAPLs used dual pumping systems, but it was later found out that this process created a smear zone just above the water table that was very persistent and difficult to remove. Petroleum hydrocarbons locked into the smear zone were immobile and created more problems. Presently very few LNAPL recoveries utilize dual pumping.

Many of the new techniques rely on the ability to simply skim oil off of the surface of the groundwater. In this technique, a hydrophobic pump floats in a relatively large diameter well and just pumps oil from the surface. This kind of skimming apparatus works very well and does not increase the smear zone. However, it has a low output of oil. Other kinds of hydrophilic hydrocarbon techniques include a moving belt system that dips into a layer of hydrocarbons and selectively picks up and removes the oil. The oil is then squeegeed off the belt which in turn is returned to the oil layer to pick up another batch.

Bioslurping. LNAPL techniques can also combine physical and biological processes. The first technique in this section is bioslurping. Bioslurping is the adaptation of a vacuum-enhanced dewatering technology to remediate hydrocarbon-contaminated sites. Bioslurping utilizes the elements of both bioventing and free-product recovery to address two separate contaminant targets: the smear zone and the free-floating product. The free product is removed by simply applying a high vacuum to the very top of the oil-water interface that draws up an oil, air, and water mixture, which is subsequently separated at the surface. Bioslurping can improve free-product recovery efficiency without extracting large quantities of groundwater. Vacuum-enhanced pumping allows LNAPLs to be lifted off the water table and released from the capillary fringe. This minimizes the change in the water table elevation, which minimizes the creation of a smear zone.

Bioventing of vadose zone soils is achieved by drawing large volumes of air through the soils. The system is designed to minimize environmental discharge of groundwater while enhancing free-product removal. Some of the disadvantages of this technique are:

- Bioslurping is less effective in tight, low-permeability soils.

- Low soil moisture content may limit biodegradation and the effectiveness of bioventing, which tends to dry out the soils.

- Aerobic biodegradation of a mixture of chlorinated compounds may not be effective.

- Frequently the offgas from the bioslurping system requires treatment before discharge; however, treatment of the offgas may be required only for a short period after the start-up of the system, as the contaminant concentration rapidly decreases.

- At some sites, bioslurping systems can extract large volumes of water that may need to be treated prior to discharge, depending on the concentration of contaminants of the process water.

- Since petroleum, water, and air are removed from the subsurface in one stream, mixing of the other phases may require special oil-water separators, or treatment before the process water can be discharged.

- Many times emulsions develop because of the high rate of extraction mixes air, water, and oil.

- Typical costs for bioslurping pilot-scale system installation are $50,000, while full-scale bioslurping systems cost approximately $100,000 to $125,000.

Dual-phase extraction, also known as *multiphase extraction* or *vacuum-enhanced extraction,* is a technology that uses a high-vacuum system to remove various combinations of contaminated groundwater, separate-phase petroleum products, and hydrocarbon vapor from the subsurface. Extracted liquids and vapors are treated and collected for disposal and sometimes reinjected into the subsurface where state laws permit this to be done. In the dual-phase extraction system for vapor and liquid treatment, a high-vacuum system is utilized to remove petroleum, water, and gas from low-permeability formations. The vacuum extraction well includes a screen section in the zone of contaminated soil and groundwater. It removes contaminants from above and below the water table. The system lowers the water table around the well, exposing more of the formation contaminants in the newly exposed vadose zone.

Dual-phase extractions are frequently combined with bioremediation, air sparging, and bioventing for target compounds including the volatile and biodegradable hydrocarbons. The dual-phase extraction process is used primarily in cases where fuel hydrocarbon lenses are more than 8 in thick. The free product is brought up to the surface by a pumping system. After recovery, it can be disposed of or reused directly in an operation not requiring a high-purity petroleum product. Systems may be designed to recover only product, mixed product and water, or separate streams of petroleum and water. Some of the limitations of dual-phase extraction fall into the same category as bioslurping. For example, the site geology and contaminant characteristics' distribution are important in dual-phase extraction. Finally, dual-phase extraction requires both water treatment and vapor treatment.

1.4 CASE HISTORIES

*"It is one thing to make experiments to
determine, as Darcy did, the amount of water that
can pass through a sand filter. It is quite another
to apply the results to the geologic materials of an
aquifer."*   CHAPMAN, 1981.

1.4.1 Introduction

Six case histories are briefly reviewed here, covering the history of the site, site characterization, treatment remedy, and comments and discussion of results. Two of the cases involve aqueous groundwater cleanups, three describe DNAPL cleanups, and one describes an LNAPL cleanup. The aqueous groundwater cleanups are for the Lockheed Martin site, Lake Success, New York, and the IBM site in South Brunswick, New Jersey. The DNAPLs cases are for a surfactant flooding site at Camp Lejeune, North Carolina; six-phase heating for an electronic manufacturing site in Illinois, and chemical oxidation at Cape Canaveral. The LNAPLs case is for a diesel fuel problem in Pontiac, Michigan.

1.4.2 Aqueous Groundwater

Lockheed Martin
    History. The site, which is located in Nassau County, New York, adjacent to New York City, is approximately 94 acres in size, and is above the largest sole-source aquifer in the United States. All of the water supply for both Nassau and Suffolk Counties, with a total population of over 3 million people, is supplied by this aquifer. There are seven buildings on the site—the main manufacturing building and six smaller buildings (south of the main building), with a total floor area of about 1.5 million square feet. The site was an active manufacturing facility from its start in 1941 until 1995. The facility was originally designed and built by the U.S. government and operated under a contract with the Sperry Gyroscope Corporation from 1941 to 1951. The property was sold to Sperry, which merged with Burroughs in 1986 to form Unisys Corporation. In 1995, Loral Corporation acquired the assets of Unisys Defense Systems, which was purchased by Lockheed Martin in 1996, current owners of

the property. Past manufacturing processes included casting, foundry, etching, degreasing, plating, painting, machining, and assembly. Chemicals used in the manufacturing at the time included halogenated and nonhalogenated hyrodocarbons, solvents, cutting oils, paints, and fuel oils, as well as inorganic plating compounds. Groundwater has been used for noncontact cooling purposes since the facility was constructed.

The noncontact cooling water system consists of three extraction wells and four diffusion wells, which are located to the north and south of the main building (see Fig. 1.21 for a site location map). The center of attention is the five dry wells located

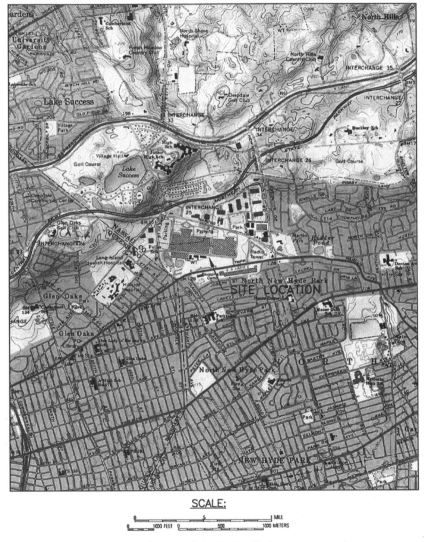

SCALE:

FIGURE 1.21 Site detail map. Lockheed Martin, New York.

at the southeast corner of the site, which were reported to have received waste solvents and oils from 1941 to 1978. In January of 1978, the Nassau County Department of Health inspected the facilities, and required the plant to begin an investigation of the contamination situation. At that time, tetrachloroethlyene, trichloroethylene, and 111-trichlorethane were identified as contaminants in the soil samples. During the field investigation, it was found that wastes from various points in the manufacturing facility were piped directly to the southeast area, and then disposed of in the dry wells. Since a limited amount of work on the contamination problem was carried out between 1978 to 1988, the Nassau Department of Health required additional tests to be carried out in 1988, which showed the need for installation of 29 monitoring wells, 32 borings, two recovery wells, and taking 65 soil samples. In addition, a vapor extraction pilot test was performed in the solvent room (dry well area).

On the basis of the results of soil and groundwater tests, DEC and Unisys entered into an administrative consent order in 1991 for remediation of the site. By 1993, a pump-and-treat remediation system with activated carbon was installed. Subsequently, in January 1994, a soil vapor extraction system with catalytic incineration was installed for the unsaturated zone of the dry well area. When it was found that the activated carbon system was not operating properly, it was supplemented with an air stripping polishing unit in February 1995 (N.Y.S. DEC, 1997).

Site Characterization.   The contaminated site was divided into two operable units—on site (OU1) and off site (OU2). This case study focuses on the OU1 on-site contamination problem. As part of the remedial investigation that took place from October 1993 to March 1995, additional monitoring wells were required in November 1996, bringing the present number of on-site monitoring wells to 47. The monitoring wells were screened at four depths in the Glacial and Magothy aquifers directly below the manufacturing facility—in the Glacial aquifer at 90 to 115 ft below grade in the Upper Glacial section and 125 to 185 ft in the Lower Glacial section, and in the Magothy aquifer at 200 to 250 ft below grade in the Upper Magothy and 300 to 400 ft in the Lower Magothy. It should be noted that the Glacial and the Magothy aquifers are hydraulically connected.

Screens were set at four different depths because DNAPLs in their pure forms can migrate vertically and cause deeper plumes. Also, because of the large amount of cooling water used, the three on-site extraction wells would discharge any generated plume back into the four diffusion wells, and create more of a mixture than a plume. New York State DEC maintains that there is only one source, in the southeast corner of the building. However, the concentration pattern that is a result of the sampling is a very diffused plume covering a much larger area than the one-source theory suggests. Cooling water extraction and reinjection is thought to be causing contamination spreading. Figure 1.22 shows the contamination in all four of the aquifer segments—some of the samples that were taken show very high levels of contamination. Several wells on the northern portion of the site (over 1500 ft from the source) show chlorinated hydrocarbon levels in the parts per million range. In fact, one well (Number 28 in the Lower Glacial) shows concentration of 1,2 DCE in the 11,000 parts per billion range.

Chemicals of concern found in the wells at the Lockheed Martin facility are 1,2, dichloroethylene (total) (2 to 11,000 ppb) and trichlorethylene (nondetectable to 320 ppb), tetrachloroethylene (nondetectable to 350 ppb). Clearly, there is a transition from PCE to TCE to DCE, which has probably occurred biologically. These solvents, which were used to degrease parts, resulted in both chlorinated and nonchlorinated solvents being deposited into the groundwater system.

Treatment.   As indicated above, the site was divided into two operable units— OU1 (on site), and OU2 (everything off site). For the on-site solution, a pump-and-

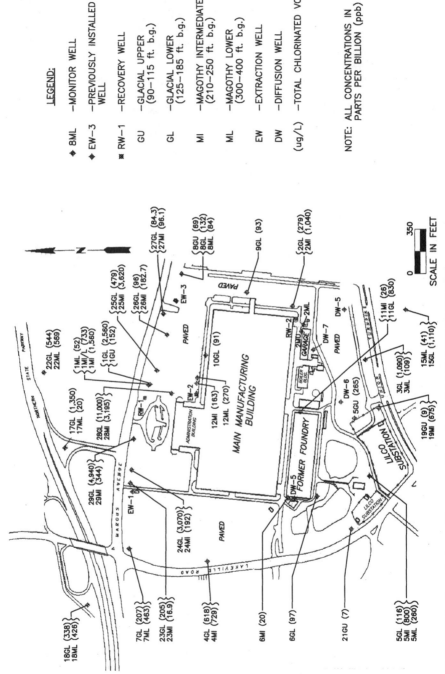

FIGURE 1.22 Groundwater quality summary, November 1994, Lockheed Martin, New York.

treat system was installed that is designed to capture all of the contaminants in the groundwater moving northward from the site. An interim remedial measure, consisting of three-extraction wells with carbon adsorption, was started in April 1993. Two years later, an air stripping polishing step was needed since the activated carbon was not removing contaminants to drinking water standards. As of March 1997, this system had removed over 8000 lb of VOCs from groundwater (N.Y.S. DEC, 1997).

Groundwater remedial alternatives that were investigated were all based upon a pump-and-treat scenario. The treatment for each of the alternatives was as follows:

1. Carbon adsorption

2. Air stripping

3. Air stripping with vapor carbon adsorption

4. Air stripping with catalytic incineration

5. UV oxidation

Table 1.11 shows the comparisons between each alternative, including present worth costs, capital costs, and annual operation and maintenance cost. The costs for each alternative were estimated for a 30-year period. The selected alternative, air stripping with vapor carbon adsorption, had the lowest cost for attaining the following: reaching the applicable standards, criteria, and guidance; protecting human health and the environment; short-term effectiveness; long-term goals; reducing toxicity, mobility, or volume; and implementability. The costs associated for these treatment technologies were based upon a groundwater flow model and a capture zone analysis for five extraction wells operating across the existing site, with a total flow of approximately 1800 gal/min.

Results and Discussion. The interim remedial action, which was started in 1993, operated with carbon adsorption only, and has since been supplemented with air stripping. As of March of 1997, 8000 lb of chlorinated hydrocarbons had been removed from groundwater. The final system, which is being modified to include air stripping and vapor phase carbon adsorption, has been evaluated with a better groundwater model, and a corresponding reduction in flow from 1800 to 740 gal/min has been proposed.

The influent concentrations into the recovery system since 1973 are plotted in Fig. 1.23. The combination of extraction well 1 and recovery well shows a leveling off of chlorinated hydrocarbons of concentrations to about 1700 μg/L. That seems to

TABLE 1.11 Present Worth, Cost, Capital Costs, and Annual Operation Cost at the Lockheed Martin Site, Great Neck, New York

Candidate technology	Present worth	Capital costs	Annual operation maintenance costs
1. Carbon adsorption	$30,570,000	$2,289,640	$1,079,300
2. Air stripping	$15,800,000	$2,297,640	$515,300
3. Air stripping with vapor carbon adsorption	$18,641,000	$2.5 million	$615,000 per year
4. Air stripping/cadlic incineration	$19,800,000	$3.1 million	$639,000
5. UV oxidation	$28.8 million	$3 million	$1 million

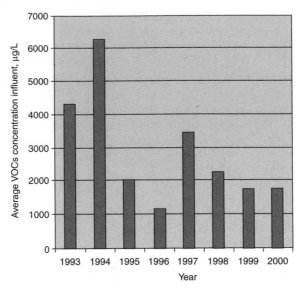

FIGURE 1.23   Influent total VOCs data for Lockheed Martin site.

imply that there may be an additional source under the building that has yet to be explored, as well as the possibility that DNAPLs are bound into the soil matrix. It should be pointed out that contamination in this aquifer is at the thousands of parts per billion level to depths of greater then 100 ft. This may indicate that DNAPLs have sunk into the aquifer up to levels of approximately 100 ft, resulting in a source of groundwater contamination for many decades, until the entire mass is dissolved by groundwater. At the southeast corner of the building, where the disposal of the chlorinated hydrocarbon occurred, the vapor extraction system has removed over 50,000 lb of chlorinated hydrocarbons from the unsaturated zone. With this large amount of chlorinated hydrocarbons, is it possible that none of it entered the saturated zone? Indeed, it appears likely that at least some DNAPLs have migrated into the saturated zone, where they will probably be a continuous source of chlorinated hydrocarbon contamination for many decades to come.

IBM Dayton Site, South Brunswick, New Jersey   The IBM Dayton site was first featured in an article in the *Wall Street Journal* (Stipp, 1991) describing the limitations of aquifer restoration when DNAPLs are present. This case history was selected to illustrate the difficulty of remediating when undetected DNAPLs are the source of contamination constantly feeding a groundwater plume.

   History.   This site is located in the township of South Brunswick, New Jersey. Organic contaminants, consisting primarily of 111-trichlorethane (TCA) and tri-chloroethylene (PCE), were discovered in a public water supply well in December 1977 (Fig. 1.24 shows the distribution of TCA in the groundwater). Plants A and B, both of which are the sources of contamination, are located approximately at the center of Fig. 1.24 and labeled as well SB 11. This is a good example of an early pump-and-treat system for chlorinated hydrocarbons, and was one of the first large studies for remediation of a chlorinated hydrocarbon problem. In January 1978, SB 11 was shut down, and IBM began a site assessment. During 1978, more than 60 monitoring wells and 10 on-site recovery wells were installed. The first groundwater

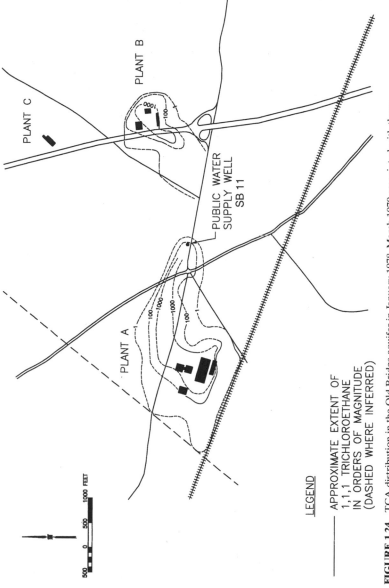

FIGURE 1.24 TCA distribution in the Old Bridge aquifer in January 1978–March 1979 associated with three facilities near SB 11. (*From Cohen et al. 1988.*)

extraction began in March of 1978 (Cohen, 1993). There was no preferred method of treatment—groundwater was discharged directly into a sanitary sewer system. The suspected contamination sources were various solvent storage tanks, which were removed in 1978. Eleven more extraction wells were installed by 1981. In 1981, nine additional injection wells were installed along the northeast boundary to contain the plume. The pump-and-treat system continued until September 9, 1984, at which time there were only six extraction wells and seven off-site injection wells. All parties (including EPA and IBM) agreed that further reductions in contaminant concentrations could not be achieved by the continued operation of the six on-site extraction wells. All the parties agreed that if the off-site TCA concentration increased above 100 μg/L, the extraction would be restarted.

Site Characterization. The water table aquifer under the site is known as the Old Bridge aquifer. This was thought to be a confining bed known as the Woodbridge Clay, below which was an aquifer known as the Farrington aquifer, believed to be a confined aquifer. This Farrington aquifer is highly productive, and is tapped by many large-capacity wells such as SB 11. The Woodbridge Clay was later found to be discontinuous (especially in the SB 11 area), resulting in the conclusion that the Farrington aquifer is a semiconfined aquifer. In order to determine this, 104 monitoring wells and borings had to be installed through both aquifers to develop a geological model that would make sense. The logs showed that DNAPL could migrate into the lower aquifer. Although DNAPLs were never directly found, TCA concentrations in the Old Bridge aquifer were as high as 12,000 μg/L, and PCE concentrations were as high as 8000 μg/L. These concentrations are 0.8 and 5.37 percent of aqueous solubilities for TCA and PCE, respectively. Although DNAPLs were never investigated, this suggests that there is a strong possibility that DNAPLs are present. The suspected contamination source is the chemical storage tanks that were removed in 1978, but no soil samples were collected at that time. Later on, in 1985, soil samples were taken from boreholes that showed a maximum concentration of 13,255 μg per kilogram of total VOCs at a depth of 22.5 ft (Cohen, 1993).

Treatment. There was no treatment; groundwater was pumped and discharge to a sanitary sewer.

Result and Discussion. This remediation, one of the earliest in the nation, is a milestone because of the very large pump-and-discharge system. At one time, there were over 10 extraction wells that did nothing more then extract the groundwater for discharge into a sanitary sewer system. Figure 1.25 shows concentrations of TCA and PCE over time. Each point represents a 6-month average, which shows a classic pattern of reduction of TCA and PCE over a period of time. It can be seen that, in 1978, both of the chlorinated hydrocarbons were fairly high, and dropped down rapidly by 1980 (with all the extraction wells working). In 1984, on-site pumping was stopped and extraction wells turned off, resulting in a upward rebound of concentrations of both PCE and TCA. TCA from January to June of 1988 reached high concentrations in the thousand of parts per billion. This is shown graphically in Figs. 1.26 and 1.27, where contaminants were fairly low in 1985, but increased 4 years later. At this particular site, DNAPLs are suspected to have been locked in the pore spaces and caused the increase in groundwater concentration due to the increase in contact time resulting in a VOC rebound. The only way that such a rebound could occur was to have DNAPLs trapped in the saturated zone.

1.4.3 DNAPL Cleanup

Marine Corps Base at Camp Lejeune, North Carolina. The first case history for DNAPL remediation is a study that was carried out at the Marine Corps Base at Camp Lejeune, North Carolina.

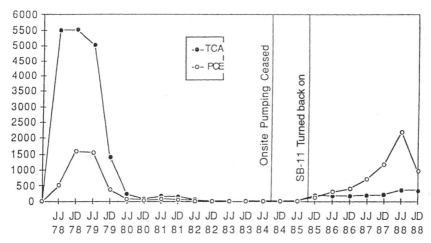

FIGURE 1.25 History of TCA and PCE variations in extraction well GW 168B, 6-month average concentration in parts per billion. (*From Cohen, 1993.*)

History. Camp Lejeune is the Marine Corps center for training recruits. Site 88, which was contaminated with DNAPLs, was selected to test whether surfactant-enhanced aquifer remediation (SEAR) was a cost-effective means for cleaning up an aquifer. This particular shallow aquifer was contaminated with PCE and DNAPLs, located at a depth of approximately 5 to 7 m beneath ground surface at the base dry cleaning facility (Holzmer et al., 2000).

Site Characterization. The initial remediation began with a partitioning inter-well tracer test (PITT) and soil core sampling to characterize the DNAPL zone. The DNAPL zone at Site 88 in Camp LeJuenne is located beneath Building 25. The depths, which ranged from 16 to 20 ft, included an area that extended about 20 ft north of the building. The DNAPLs occur immediately above and within a relatively low permeability layer of clayey silt sediments. The site conceptual model (or geosystem) is shown in cross section in Fig. 1.28. The hydraulic conductivities are estimated as follows: 5×10^{-4} cm/s for the upper zone and 1×10^{-4} cm/s for the middle zone (or about 5 times less permeability than the upper zone). The lower zone is predominantly composed of clayey silt with a hydraulic conductivity that is believed to be approximately 5×10^{-5} cm/s.

Treatment. The selected remedy is treatment by surfactant flooding. The concept is to increase the solubility of the DNAPLs in water by adding a surfactant. Since the surfactant was intended to be recycled, the design process included an extensive laboratory testing phase to select a surfactant not only with excellent subsurface performance, but also with characteristics that would enable recovery of the surfactant from the effluent. The selected surfactant was a propoxylated alcohol ether sulfate surfactive (Alfoterra 145-4-po sulfate™). Isopropanol was used as a cosolvent, and calcium chloride was also added as the electrolyte to control the micelle phase behavior. The design process utilized a series of simulations with a mathematical model to optimize well configurations and flow rates. The hydrodynamic design consists of three central injection wells and six extraction wells arranged in a line configuration, as shown in Fig. 1.29. The entire test area measured 20 ft wide by 30 ft long. Cross sections of extraction injection wells are shown in Fig. 1.28. A partitioning interwell tracer test (PITT) was used in 1998 to measure the vol-

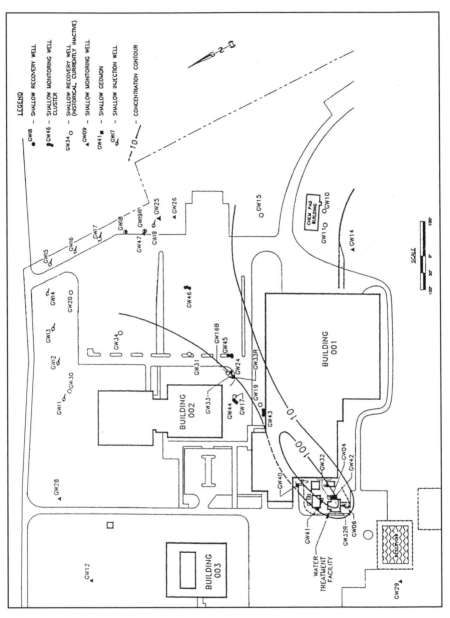

FIGURE 1.26 TCA Distribution in the Old Bridge aquifer in January 1985 associated with IBM facilities. (*From Cohen, 1993.*)

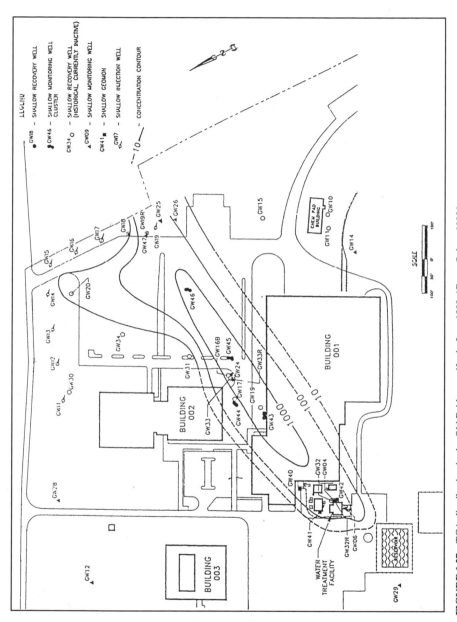

FIGURE 1.27 TCA distribution in the Old Bridge aquifer in June 1989. *(From Cohen, 1993.)*

1.73

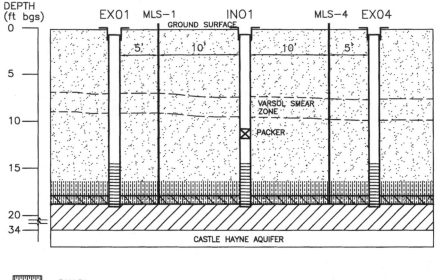

DNAPL	
CLAYEY SILT	
CLAY AQUITARD	
FINE SAND & SILT	

FIGURE 1.28 Generalized geosystem cross section of DNAPL zone at Site 88. (*From Holzmer, et al., 2000.*)

umes and relative distribution of the DNAPLs present in the test zone, before surfactant flooding was initiated. The results indicated that approximately 74 to 88 gal of DNAPLs were present in the test zone. The injection formulation consisted of 4 percent Alfoterra 145-4PO sulfate surfactant (16 percent by weight), isopropanol, and 0.16 to 0.19 percent by weight of calcium chloride mixed with the water source. The duration of the surfactant flooding was 58 days. Approximately 30,000 gal of surfactant mixture was injected into the test zone (equivalent to approximately 5 pore volumes). The total mass injected was 9718 lb of surfactant, 38,600 lb of isopropanol, and 427 lb of calcium chloride. The total discharge of surfactant injection was approximately 0.4 gal/min (because of the low soil permeability), and the extraction rate was about 1 gal/min (higher to maintain hydrodynamic control).

Results and Discussion. A total of 76 gal of PCE was recovered during the surfactant flood and subsequent water floods. Out of the 76 gal, approximately 32 gal of PCE was recovered as soluble DNAPLs, and 44 gal as free-phase DNAPLs. The recovery of free-phase DNAPLs is due to the increased mobilization of the DNAPLs sinking through the aquifer and pooling on the low permeability zone. This very often is not the case at other sites—mobilization of DNAPLs is one of the biggest fears with this technique.

A post-test was conducted by sampling the soil. Each soil sample was collected for analysis, and measured for total volatile organic compounds. Soil sampling was generally carried out for the purpose of defining the three-dimensional distribution

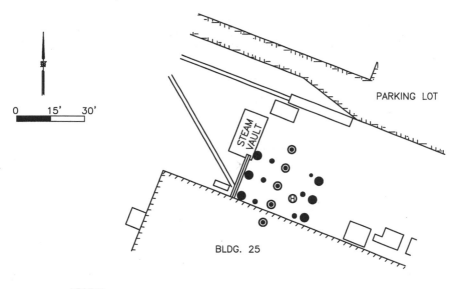

PARKING LOT

0   15'   30'

LEGEND
● EXTRACTION WELL
◎ INJECTION WELL
Ⓗ HYDRAULIC CONTROL WELL
• MULTILEVEL SAMPLER

FIGURE 1.29 Well array and multilevel sampling points for surfactant flush. (*Modified from Holzmer et al., 2000.*)

of DNAPLs remaining in the test zone. The vertical distribution indicates that the DNAPLs were effectively removed from the more permeable sediments, generally above 17.5 ft, and that DNAPLs still remain in the lower, less permeable clayey silt layer. This points to another limitation in the use of surfactants, which is that surfactants should not be used in low-permeability material. Based on the volume of DNAPLs distributed in the upper permeable zones, it can be demonstrated that the surfactant flood recovered between 92 and 96 percent of DNAPLs that were initially present. However, considering the lower, less permeable material, approximately 72 percent of the DNAPLs from the entire site were removed.

Six-Phase Heating at Electronics Manufacturing Site in Chicago

History. This work took place at a former electronics manufacturing facility where various electronic machining, plating, soldering, semiconductor manufacturing, and silicon chip production operations were carried out. In 1987, underground storage tanks were removed from the site, and one of the tanks originally believed to be empty was found to contain a mixture of trichloroethylene and 111-trichloethane.

Pools of these TCE and TCA mixtures were discovered in eight areas, resulting in 2.4-m-thick layers of these mixtures in monitoring wells. The groundwater-impacted area represented approximately 1.3 hectares. The facility is located on a lacustrine stream sequence of sediments, consisting of fine sands with some clay lenses, to depths of approximately 17 to 20 ft. The lacustrine sediments, which have hydraulic conductivities ranging from 1×10^{-5} to 1.2×10^{-4} cm/s, are underlain by a relatively dense, impermeable, and continuous glacial till throughout the site area. The water table is relatively high, at approximately 6 to 7 ft below the plant floor. This site was selected because it was the location of an underground remediation system using steam and enhanced biodegradation for approximately 7 years. Man-made subsurface features were limiting the effectiveness of the steam injection. Six-phase heating, which relies on electrical conductance through soil moisture, was selected since it is not influenced by physical obstructions or subsurface conductivity variations.

Six-Phase Heating. Six-phase heating was developed by the Battelle Memorial Institute for the U.S. Department of Energy, and involves the conducting of electricity through the subsurface via electrodes. The natural resistance of the soil to the flow of electrical current results in heat buildup in the aquifer. Electrodes are installed in a pattern where a central neutral electrode is surrounded by six charged electrodes. The surrounding electrodes are sequentially charged 60° out of phase with each other, resulting in an even distribution of heat throughout the treatment zone. This creates enough heat quickly to create steam to strip the volatile organic compounds from the soil and the groundwater and/or directly vaporize the compound.

The six-phase heating was performed in two phases. The first phase began on March 30, 1998, and lasted until October 1, 1998. The second phase started in October 1998 and continued to December 1998. The system was then expanded. Treatment continued until April 1999. On April 23, 1999, sampling confirmed that all wells showed concentrations below the target concentrations, with many well below the Illinois Class 2 Standards.

After six-phase heating began on June 4, 1998, temperatures throughout the entire 24,000 yd^3 treatment volume reached the boiling point of water within 60 days. Table 1.12 shows the groundwater concentration changes during the six-phase operation at various observation wells. All of the percent reductions were 90 percent or higher for the three most common compounds—12 dichloromethane, 111 trichlorethane, and trichlorethylene. With an average of 98 to 99.9 percent reduction in overall concentrations for this particular cell, approximately 4 months after start-up, the cleanup criteria had been achieved.

Treatment. Six-phase heating relies on resistance to electricity to heat up volumes of soil. High temperatures are achieved, which results in VOCs changing from the pure phase to the gas phase. This technique must be used with a vapor extraction system, because large amounts of VOC gas are stripped from the saturated zone.

Results and Discussions. Six-phase heating provided very rapid heating of the subsurface environment to the boiling point of water within a 60-day period. At the boiling point of water, many of the volatile organic DNAPLs will volatilize and be captured at the surface by a vapor extraction system. Although this technique was successful here, it could have the undesirable result of DNAPL mobilization, like surfactant flooding. By heating the DNAPL, the interfacial surface tension is changed, which may result in DNAPLs sinking deeper into the aquifer. Fortunately, at this location there is a lower layer that precludes vertical migration of the DNAPLs.

It is believed that, as the concentration of the volatile organic compounds reaches the regulatory criteria, the vapor pressure of the solution moves closer to that of water. From a cost-effectiveness standpoint, there appears to be a limit as to how far one can proceed to meet regulatory criteria with six-phase heating, which may require boiling away a significant portion of the aquifer. This would not be the best use of this technique, which could be more effective with pure-phase DNAPLs (Beyle et al., 2000).

Chemical Oxidation at NASA Cape Canaveral

History. The LC-34 site was used as a launch site at Cape Canaveral for the Apollo rocket from 1960 to 1968. TCE was historically used for rocket engine flush-

TABLE 1.12 Groundwater Concentrations during Six-Phase Heating

Well	Compound	March 1998, pg/L	October 1998, pg/L	November 1998, pg/L	Reduction
B3	cis 1,2-DCE	49,000	780	140	99.7%
	1,1,1-TCA	82,000	<500	31	99.96%
	TCE	34,000	790	120	99.6%
Ba6	cis 1,2-DCE	9,800	200	1,200	87.8%
	1,1,1-TCA	66,000	<50	<100	>99.8%
	TCE	7,000	510	470	93.3%
C4	cis 1,2-DCE	43,000	1,300	450	99.0%
	1,1,1-TCA	11,000	<100	15	99.9%
	TCE	76,000	1,600	100	99.9%
Cab	cis 1,2-DCE	15,000	4,100	250	98.3%
	1,1,1-TCA	16,000	14	<20	>99.9%
	TCE	63,000	81,000	1,600	97.5%
Da2	cis 1,2-DCE	18,000	120	51	99.7%
	1,1,1-TCA	28,000	290	<100	>99.9%
	TCE	47,000	8,800	320	99.3%
F13	cis 1,2-DCE	510	480	38	92.5%
	1,1,1-TCA	18,000	<250	<10	>99.9%
	TCE	800	260	12	98.5%
Fa2	cis 1,2-DCE	3,900	470	180	95.4%
	1,1,1-TCA	24,000	<50	24	99.9%
	TCE	22,000	1,200	70	99.7%
Average	cis 1,2-DCE	19,900	1,060	330	98.3%
	1,1,1-TCA	35,000	110	26	99.9%
	TCE	35,700	13,500	380	98.9%

ing, metal cleaning, and equipment degreasing. A cross section through the geology is shown on Fig. 1.30. There are four geologic units: the upper sand unit, the middle fine-grain unit, the lower sand unit, and a lower confining unit. Mass distribution of the TCE was defined by baseline sampling in 1999—a TCE mass of 6000 kg was estimated by the demonstration project. The vertical profiling used to estimate TCE contamination showed that the TCE is highly influenced by layered lithology. The aerial distribution is variable across the cell, with the highest mass adjacent to suspected historical TCE sources.

Treatment. The selection of chemical oxidation for remediation of DNAPLs was chosen because of the following advantages over other treatment technologies: It is a destructive technology, it has a short treatment time frame (hours to months), it requires minimal energy and equipment, it does not require vapor phase treatment, and it generates minimal wastes. Many oxidants were examined for treatment of DNAPLs, including ozone, hydrogen peroxide, Fenton's reagent, sodium permanganate, potassium permanganate, and oxygen. Potassium permanganate was selected because of its relative stability, ease of handling, relatively low cost ($3 per kilogram), and the ability to see the results of the application (groundwater subject to permanganate treatment changes color from purple to brown to clear upon complete oxidation). Potassium permanganate reacts relatively quickly without the generation of heat, and completely oxidizes a wide range of common organic contaminants includ-

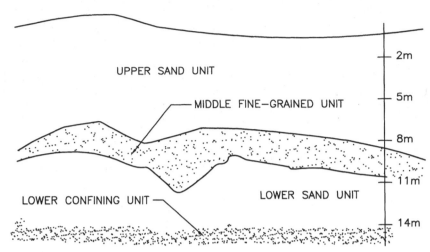

FIGURE 1.30 Lithology of Cape Canaveral. (*From Smith, 2000.*)

ing TCE, PCE, and DCE isomers; vinyl chloride; and phenols. With chlorinated ethenes, such as TCE, it reacts with the double bond, and ultimately breaks down into carbon dioxide, magnesium dioxide, potassium ions, chloride ions, and hydrogen ions. Some of the opponents for use of potassium permanganate have pointed out that it develops a precipitate of magnesium dioxide that can literally crust the DNAPLs and shield the reaction from going to completion.

The overall effectiveness of in situ permanganate treatment is a function of the reaction kinetics, the transport and contact of potassium permanganate with DNAPLs, the specific DNAPLs present, and the competitive reaction with other oxidizable compounds such as iron and natural organics.

The design of the treatment system was based on direct-push pressure injection. Since the vertical and horizontal distribution of TCE varies significantly, the system was designed to deliver permanganate at small vertical intervals of approximately 0.5 m. The permanganate dose, application rate, and duration were adjusted at each point to reflect the corresponding levels of contaminant mass at that location. Figure 1.31 shows the injection network. The injection tip consists of a 0.6-m-long perforated drive stem with 0.25-in-diameter holes and a 0.01-in-slot continuous wire-wound stainless steel screen. The permanganate was delivered to the site by bulk shipments of approximately 20,000 kg (free-flowing grade) and transferred into a bulk feed silo. Flow rates range from 0.1 to 6.1 gal/min at wellhead pressures of 15 to 55 psig.

In order to determine the effectiveness of the approach, initial treatment steps consisted of a tracer test performed at injection point, to arrive at in situ permanganate transport and reaction rates. A sodium fluoride tracer (2 percent solution) and permanganate solution (1.5 to 2 percent solution) were injected at 3-m intervals for a 3-day period. Tracer test results showed that the potassium permanganate could be injected, did show significant consumptive use by the DNAPL phase, and allowed engineers to modify initial conceptual designs to change such things as injection flow rate and pressure. The first full cell injection, which was initiated on September 8 and ended on October 29, extended across the entire treatment zone that was about 15 to 45 ft below grade at 11 locations.

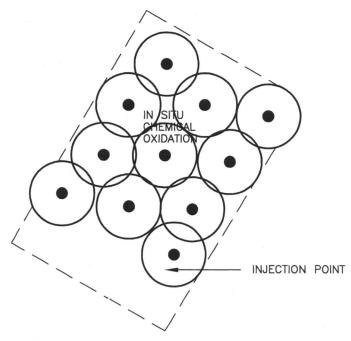

INJECTION POINT

FIGURE 1.31   Injection layout of Cape Canaveral. (*From Smith et al., 2000.*)

Results and Discussion.   The monitoring results showed that the upper sand unit showed reductions of about 99 percent removal in seven of the eleven wells sampled for TCE (see Table 1.13). The purple color was observed in most of the cell wells, indicating that sufficient permanganate was being distributed in the aquifer. Soil samples in the upper sand unit showed greater than 90.5 percent reduction of VOCs. Slightly less remediation occurred in the middle fine-grain unit—seven of the fifteen wells showed greater then 90 percent TCE decrease. Other reductions ranged from a low of 4.3 percent to a high of 88.6 percent. However, some wells in this unit dramatically increased in concentration. One well was reported to increase from 120,000 ppb to 15.3 million ppb. Such an increase is hard to explain. Out of the six wells tested, the lower sand unit had fairly good success—five of the six wells showed 88 to 99 percent reduction, while one of the wells showed only an 8.6 percent reduction.

One of the conclusions that can be drawn from the oxidation of these DNAPLs is that there was significant TCE mass reduction achieved. However, in some cases, groundwater concentrations increased dramatically. TCE concentrations in groundwater are a good indication of treatment effectiveness, but are very dynamic and sometimes difficult to interpret. TCE soil data is really the most reliable way to document mass destruction, but is very difficult to interpret because of the extreme variability. No hazardous chemical by-products were generated by the reaction of the potassium permanganate. However, the evaluation of the impact on secondary drinking water standards (such as manganese and chloride) could be a potential problem. Significant formation plugging by a manganese dioxide precipitate was not evident, and injection flow rates remained relativity constant. No DNAPLs were

TABLE 1.13 Soil Analytical Results

Sample	Depth, ft	Baseline TCE, ppb (μg/kg)	TCE, ppb (μg/kg)	Percent reduction TCE	Color present
B1	15–19	400	<20	95.00%	Purple
B1*	17–19	540	<20	96.30	Purple
B1	20–21	940	419	55.43%	None
B2	20–21	19,150	5,870	69.35%	Brown
B3	20–21	30,550	24	99.92%	Purple
B4	20–21	48,000	42	99.91%	Purple
B5	20–21	10,430	138	98.68%	Brown
B6	19–20	9,000	<20	99.78%	Purple
B6	22–23	15,000	<20	99.87%	Brown
B1*	24–27	110,000	22,000	80.00%	Brown
B3	25–26	104,500	43,400	58.7%	Brown
B6	28–29	170,000	126	99.93%	No
B1	29–30	180,000	242,500	−34.72%	No
B2	30–31	120,000	15,300,000	−12650%	No
B3	30–31	160,000	1,010	99.35%	No
B4	30–31	315,000	57,100	81.87%	Brown
B5	30–31	11,000,000	1,810,000	83.55%	Brown
B6	32–33	200,000	<20	99.99%	Purple
B3	34.5–.5	120,000	80	99.93%	Brown
B5	34–35	350,000	236,000	32.57%	Brown
B1*	35–39	119,000	14,000	88.24%	Brown
B5*	38–39	10,000,000	15,400	99.85%	Brown
B5*	40–41	6,800,000	10,000	99.85%	Brown/gray
B5*	41–43	1,800,000	25,200	98.60%	Brown
B1	43–45	86,000	78,600	8.60%	Brown
B5*	44–45	2,300,000	16,600	99.28%	Brown/gray

Note: Calculated TCE reductions are based on soil borings collected at close locations from pre-demonstration soil boring locations.
* Semiquantitative results (exceeded sample hold times; samples frozen).

reported to fall into the lower confining unit. Hence, unlike thermal and surfactant techniques, oxidation does not seem to mobilize the DNAPLs.

1.4.4 LNAPLs

Diesel Fuel, Pontiac, Michigan
History.   Diesel fuel that was released over several decades infiltrated into the water table at a railroad locomotive refueling operation site in Pontiac, Michigan.

An interceptor trench was installed in late 1996 to limit off-site migration, and a feasibility study was performed in early 1997 to evaluate conventional technology, multiphase extraction, and bioslurping. Bioslurping was determined to be aggressive and cost-effective on a net present value basis. A 2-week bioslurping pilot study was performed on three wells in June 1997 to evaluate oil recovery rates in response to various vacuum levels and flow rates. The system began operating in March 1998. Project objectives are to prevent further oil migration and to remove all recoverable oil in 4 years or less.

Site Characterization.   The diesel fuel that was released over several decades passed through sand to a depth of approximately 20 ft to a clay aquitard where the average water table is about 10 ft below grade. The LNAPLs covered about 2 acres, with an average observed thickness of 1 ft, and a maximum observed thickness of 4 ft. The volume of recoverable oil was estimated at 21,500 gal and groundwater has not been impacted by any contaminants at this location. Hence, the treatment of groundwater is not an issue here.

Treatment.   Bioslurping was selected after being compared with other skimming systems. Bioslurping is the adaptation of vacuum-enhanced dewatering technologies, utilizes elements of both bioventing and free product recovery, and is sometimes referred to as *dual-phase* or *multiphase extraction*. Bioslurping combines elements of both technologies to simultaneously recover free product and bioremediate vadose soils. Bioremediation occurs because of the high amount of oxygen that is brought into the vadose zone soils. This degrades many of the compounds that are trapped in the vadose and smear zones. Bioslurping vacuum-enhanced pumping allows the LNAPLs to be lifted from the water table and released from the capillary fringe. This decreases changes to the water table, and minimizes the creation of a smear zone. Conventional pumping, where a cone of depression is formed so that oil can be skimmed from the surface of the cone, creates an unusually large smear zone, increasing the difficulty of removing all of the LNAPLs.

Figure 1.32 shows the bioslurping layout and site map. The locations of the bioslurping tubes, which are approximately 50 to 100 ft apart, are determined by the amount of floating product measured in the observation wells. Figure 1.33 shows the full system installed, including the following: twenty-seven 6-in bioslurping wells, a 400-gal air-liquid separator, two 1800-lb vapor-phase carbon vessels, a 75-hp blower, heat exchangers, a 7.5-hp rotary lobe air injection blower, two air injection wells, a 3000-gal oil-water separator, a 50-gal/min inclined plate oil-water separator, a 100-gal oil transfer tank with a 1-hp discharge pump, six 6-in water injection wells, a programmable logic controller, an autodialer, and a 22 × 30-ft building.

Oil, water, and soil vapors are sucked into the air-liquid separator through the bioslurping tubes. Soil vapor at the top of the oil-water separator passes through the granular activated carbon (GAC) and is ultimately reinjected. Water and oil are pumped through and separated by the oil-water separator system, oil is skimmed off and disposed of off site, and the groundwater is reinjected into the capture zone to reduce water treatment costs. The groundwater that is in intimate contact with the oil is clean enough so that the reinjection of this water in not an environmental problem. This is frequently the case with older diesel and oil spills (fuel No. 4 or greater), where the oils have been weathered so that the lighter fractions have either biodegraded or volatilized. In this system, two groundwater recovery wells (with submersible pumps) maintain hydrodynamic control over the oil pool, and prevent its expansion by discharging to a storm sewer. These wells, which are located at the center of the system, pump groundwater from the bottom of the water table through a 100-lb activated carbon system prior to discharge to a storm sewer.

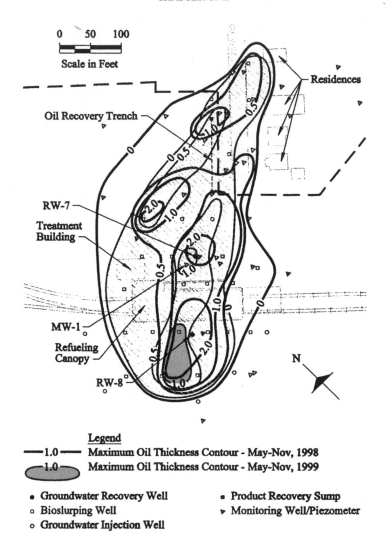

0   50   100
Scale in Feet

Oil Recovery Trench

Residences

RW-7
Treatment
Building

MW-1
Refueling
Canopy

RW-8

N

Legend

——1.0—— Maximum Oil Thickness Contour - May-Nov, 1998

‾1.0‾ Maximum Oil Thickness Contour - May-Nov, 1999

● Groundwater Recovery Well   ■ Product Recovery Sump
▫ Bioslurping Well   ▾ Monitoring Well/Piezometer
○ Groundwater Injection Well

FIGURE 1.32 Bioslurping layout and 1998 and 1999 oil thickness contour site map. (*Modified from Christian et al., 2000.*)

The system requires treatment of the air in order to reinject it—the recovered oil flash point is about 130°F. There is no possibility for the oil to reach this temperature since the lower explosive limit (LEL) of 0.7 percent (7000 ppm) should never be reached. Potential for the gas stream to be exceeded at the blowers is eliminated by the GAC. Cutoff was set to trigger system shutdown when LEL reached 1000 ppm to provide added safety.

Results and Discussion. In 22 months of operation beginning in March of 1998, 21,700 gal of oil was recovered. Figure 1.34 shows that oil recovery, even after 2 years

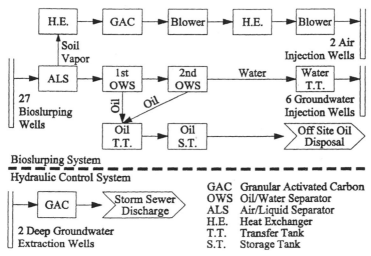

FIGURE 1.33 Schematic system diagram. (*From Christian et al., 2000.*)

of operation, seems to have remained on a very steady pace since the first year of operation. About the same amount of oil is being recovered on a weekly basis. From April 1, 1998, to February 7, 2000, the system had operated for 12,627 hours and removed the following: 21,700 gal of oil (at an average rate of 41 gal/day); 2,404,000 gal of groundwater, which is discharged back into the aquifer (at an average rate of 3.2 gal/min), and 126,000,000 ft^3 of extracted air from the vadose zone (at an average rate of 66 standard ft^3/min). The ratio of water to oil to air by volume is 1:116:45,000. Vacuum levels at the blower inlet may range from 8 to 10 inHg, with occasional extremes (low of 7 in and high of 12 in). Vacuum levels at the wellhead range from 0 to 50 inH$_2$O, while air velocities in the horizontal transmission pipe (1.5- to 4-in PVC) range from 600 to 1200 ft/min.

As indicated above, expansion of the oil pool was avoided by extracting groundwater from the hydraulic control wells, which were pumped at an average rate of 3.3 and 0.5 gal/min, respectively. These small extraction quantities produced a small depression of 1.5 to 2 ft in the water table below the plume, which hydraulically prevented the plume from migrating.

The oil thickness has changed considerably since the initiation of the remediation program. Present estimates of oil quantities are 5 times the original estimates. This exemplifies the problem of estimating free-flowing oil in an LNAPL pool using theoretical calculations. The engineering remediation program that is developed from these calculations should have the flexibility to increase the remediation parameters whenever necessary.

This particular site had its own set of physical difficulties. It is an active train yard with very aggressive project schedules. The site is intersected by numerous active roadways and railroad tracks that are closely spaced, allowing little room for construction. Wintertime excavation and backfilling of trenches under railroad tracks required extra effort in order to prevent eventual sloping of tracks under the heavy weight of locomotives. In addition to the numerous underground obstructions (resulting from

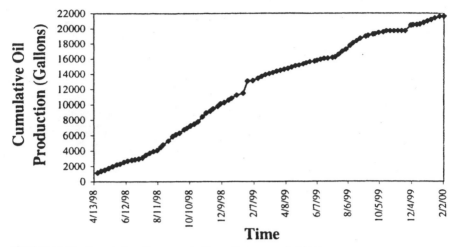

FIGURE 1.34 Cumulative oil recovered. (*From Christian et al., 2000.*)

about 100 years of railroad use), no walls, control equipment, or wells could extend above ground. Also, because of the water-table fluctuation, the drop had to be adjusted fairly frequently. Pipes connecting to manholes and sewers had to be located below the frost line and had to maintain the proper slope. Because of the high water table, manhole bottoms were quite close to the relatively shallow capillary fringe where the oil was located.

REFERENCES

Acomb, L. J., McKay, D., Currier, P., Berglund, S. T., Sherhort, T. V., and Bendiktssnon, C. V., 1995. "Neutron Probe Measurements of Air Saturation Near an Air Sparging Well," *In Situ Aeration: Air Sparging, Bioventing, and Related Remediation Processes,* R. E. Hinchee (ed.), Battelle Press, Columbus, Ohio, p. 47.

Ball, W. P., 1984. "Mass Transfer of Volatile Organic Compounds in Packed Towers Aeration," *JWPCF,* vol. 56, no. 2, pp. 127–136, February.

Beyle, G., Smith, G., and Jurka, V., 2000. "DNAPL remediation closure with six phase heating," *Physical and Chemical Technologies: Remediation of Chlorinated and Recalcitrant Compounds,* Battelle Press.

Boegel, J. V., 1997. "Air Stripping and Steam Stripping" *Standard Handbook of Hazardous Waste Treatment and Disposal,* H. Freeman (ed.), 2d ed., McGraw-Hill, N.Y.

Branter, K. A., and Cichon, E. J., 1981. "Heavy Metals Removal: Comparison of Alternative Precipitation Processes," *Industrial Waste—Proc. 13th Mid-Atlantic Conference.*

Christian, B., Clark, B., Hindi, N., Wolf, M., and Gibson, R., 2000. "Diesel Fuel Recovery Using Bioslurping—A Case Study," *Physical and Thermal Technologies: Remediation of Chlorinated and Recalcitrant Compounds,* Battelle Press, pp. 85–92.

Cohen, R. M., and J. W. Mercer, 1993. *DNAPL Site Evaluation,* Robert S. Kerr Environment Res. Lab., CRC Press.

Conrad, S. H., Glass, R. J., and Peplinski, W. J. 2001. "Bench Scale Visualization of DNAPL Remediation Processes in Analog Heterogeneous Aquifer: Surfactant Floods and In situ Oxidation Using Permanganate," draft submitted to *Journal of Contaminate Hydrology,* Feb.

de Pastrovich, T. L., Baradat, Y., Barthel, R., Chiarelli, A., and Fussell, D. R., 1979. "Protection of Groundwater from Oil Pollution," CONCAWE, Report 3/79, The Hague, p. 61.

DeVitt, D. A., Evans, R. B., Jury, W. A., Starks, T. H., Eklund, B., and Gholson, A., 1987. "Soil Gas Sensing for Detection and Mapping of Volatile Organics," EPA/600/8-87/036, U.S. EPA Environmental Monitoring Systems Lab, Las Vegas, Nev.

Driscoll, F. G., 1983. *Groundwater and Wells*, Johnson Division, St. Paul, Minn.

Farr, A. M., Houghtalen, R. J., and McWorter, D. B., 1990. "Volume Estimation of LNAPL in Porous Media," *Groundwater*, vol. 28, pp. 48–56.

Freeze, R. A., and Cherry, J. A., 1979. *Ground Water*, Prentice Hall, Upper Saddle River, N.J.

Hall, R. A., Blatie, S. B., and Champlin, Jr., S. C., 1984. "Determination of Hydrocarbon Thickness in Sediments Using Borehole Data," *Proc. 4th National Symposium on Aquifer Restoration and Groundwater Monitoring*, NWWA, pp. 300–304.

Hallberg, R. O., and Martinelli, R., 1976. "Vyredox—In-Situ Purification of Ground Water," *Ground Water*, vol. 14, pp. 88–93.

Hampton, D. R., and Miller, P. D. G., 1988. "Laboratory Investigation of the Relationship Between Actual and Apparent Thickness in Sand," *Proc. Petroleum Hydrocarbon and Organic Chemicals in Groundwater*, NWWA.

Holzmer, F. J., Hope, G. A., and Yeh, L., 2000. "Surfactant-enhanced aquifer remediation of PCE-DNAPL in low permeability sand," *Treating Dense Non-Aqueous–Phase Liquids (DNAPLs): Remediation of Chlorinated and Recalcitrant Compounds*, 2d International Conference on Remediation and Chlorinated Recalcitrant Compounds, Monterey, Calif., May 22–25, Battelle Press.

Hughes, B. M., Gillham, R. W., and Mendoza, C. A., 1990. "Transport of Trichloroethylene Vapors in the Unsaturated Zone: A Field Experiment," *International Association of Hydrogeologists Conference on Subsurface Contamination by Immissable Fluids*, Calgary, pp. 18–20.

Huling, S. G., and Weaver, J. W., 1991. "Dense Nonaqueous Phase Liquids," EPA/540/4-91-002, March.

Hunt, G. E., et al., no date. "Collection of Information on the Compatibility of Grouts with Hazardous Wastes," U.S. Environmental Protection Agency, Cincinnati.

Kaufmann, H. G., 1982. "Granular Carbon Treatment of Contaminated Supplies," *Proceedings of the 2d National Symposium on Aquifer Restoration and Ground Water Monitoring*, May 26–28, Fawcett Center, Columbus, Ohio.

Keeley, A. A., Russell, H. H., and Sewell, G. W., 1999. "Microbial Processes Affecting Monitored Natural Attenuation of Contaminants in the Subsurface." Sept. USEPA/540/S-99/001.

Kincannon, D. F., and Stover, E. L., 1981. "Stripping Characteristics of Priority Pollutants During Biological Treatment," *74th Annual AlChE Meeting*, New Orleans, November.

Knox, R. C., Conter, L. W., Kinconnon, D. F., Stover, E. L., and Ward, C. H., 1986. *Aquifer Restoration*, Noyes Publication.

Lieberman, S. H., Boss, P., and Anderson, G. W., 2000. "Characterization of NAPL Distributions Using in situ Imaging and Laser Induced Flurescence (LIF)," SPAWAR System Center San Diego Remediation of Chlorinated and Recalcitrant Compounds, Monterey, Calif., May 22–25.

Lenhard, R. J., and Parker, J. C., 1990. "Estimation of Free Hydrocarbon Volume from Fluid Levels in Observation Wells," *Groundwater*, vol. 28, pp. 57–67.

Lindorff, D. E., and Cartwright, K., 1977. "Ground Water Contamination: Problems and Remedial Actions," Report No. 81, May, Illinois State Geological Survey; Urbana, Ill.

Lu, J. C. S., Morrison, R. D., and Sterns, R. J., 1981. "Leachate Production and Management from Municipal Landfills: Summary and Assessment," *Land Disposal: Municipal Solid Waste-Proceedings of the Seventh Annual Research Symposium USEPA*, pp. 1–17.

Marley, M. C., Bruell, C. J., and Hopkins, H. H., 1995. "Air Sparging Technology: A Practical Update," *In Situ Aeration: Air Sparging, Bioventing, and Related Remediation Processes*, Hinchee, R. E., (ed.), Battelle Press, Columbus, Ohio, p. 31.

Marrin, D. L., and Thompson, G. M., 1984. "Remote Detection of Volatile Organic Contaminants in Groundwater via Shallow Soil Gas Sampling," *Proc. Petrol., Hydro., and Org.*

Chemical in Groundwater Prevention, Detection, and Restoration, NWWA—API, Houston, pp. 172–187.

Marrin, D. L., and Kerfoot, H. B., 1988. "Soil Gas Surveying Techniques" *Environmental Science and Technology,* vol. 22, no. 7, pp. 740–745, ACS.

McDougall, W. J., Fusco, R. A., and O'Brien, R. P., 1980. "Containment and Treatment of the Love Canal Landfill Leachate," *Journal WPCF,* vol. 52, no. 12, December.

Meinardus, H. W., Dwarakanath, V., Jackson, R. E., Jim, M., Ginn, J. S., and Stotler, G. C., 2000. "Full Scale Characterization of a DNAPL Source Zone Using PITTs," *Treating Dense Non Aqueous-Phase Liquids (DNAPL),* Battelle Press, Columbus, Ohio, pp. 17–24.

Nace, R. L., ed., 1971. "Scientific framework of world water balance," UNESCO Tech. Papers Hydrol., pp. 7, 27.

Newmark, R. L., 1994. "Demonstration of Dynamic Underground Stripping at the LLNL Gasoline Spill Site," Final UCR-ID-116964, vols. 1–4. Lawrence Livermore National Laboratory, Livermore, Calif.

NJDEP, 1988. *Field Sampling Procedures Manual.* Feb.

N.Y.S. DEC, 1997. Record of Decision Lockheed Martin Tactical Systems Inc. Site Operable Unit 1, Site No. 1-30-045, March.

O'Brien, R. P., and Fisher, J. L., 1983. "There Is an Answer to Ground-water Contamination," *Water Engineering and Management,* vol. 130, no. 5, May.

Parker, J. C., Waddill, D. W., and Johnson, J., 1994. "UST Corrective Action Technologies: Engineering Design of Free Product Recovery Systems," Superfund Technology Demonstration Division, Risk Reduction Engineering Laboratory, Edison, N.J.

Patterson, J. W., 1978. *Wastewater Treatment Technology,* 3d ed., Ann Arbor Science.

Phillips, K. J., 1995. Design drawings "Lockheed Martin Stripping Tower."

Phillips, K. J., Costanino, C., Atik, G. A., and Doriski, T., 1993. "Prediction of Surface Substance and PCB Oil Migration Resulting from Dewatering and Mitigation Measure at the 63rd Street/Queens Blvd. Line Construction Area," ASCE Geotechnical Group, "Control of Groundwater," New York Metropolitan Section ASCE, Nov. 9 and 10.

Pope, D. F., and Jones, J. N., 1999. "Monitored Natural Attenuation of Chlorinated Solvents," EPA/600/F-98/022, May.

Riha, B., Rossabi, J., Dilek, C. E., and Jackson, D., 2000. "DNAPL Characterization Using the Ribbon Sampler: Methods and Results Poster Session; Remediation of Chlorinated and Recalcitrant Compounds," Monterey, Calif., May 22–25.

Schwille, F., 1988. *Dense Chlorinated Solvents in Porous and Fractured Media,* Lewis Publishers, Boca Raton, Fla.

Smith, E. M., Leonard, W. C., Lewis, R., Clayton, W. S., Ramirez, J., and Brown, R., 2000. "In-site Oxidation of DNAPL using permanganate: IDC Cape Canaveral demonstration," *Chemical Oxidations and Reactive Barriers: Remediation of Chlorinated and Recalcitrant Compounds,* Battelle Press.

Spooner, P. A., Wetzel, R. S., and Grube, W. E., 1982. "Pollution Migration Cut-off Using Slurry Trench Construction," *Management of Uncontrolled Hazardous Waste Sites,* Hazardous Materials Control Research Institute, pp. 191–197.

Stipp, D., 1991. "Super Waste? Throwing Good Money at Bad Water Yields Scant Improvements," *The Wall Street Journal,* May 15, p. 1.

Tolman, A. L., et al., 1978. "Guidance Manual for Minimizing Pollution from Waste Disposal Sites," EPA-600/2-78-142, August, U.S. Environmental Protection Agency, Cincinnati.

Udell, K. S., 1966, "Heat and Mass Transfer in Cleanup in Underground Toxic Waste," *Annual Reviews of Heat Transfer,* vol. 7, Begell House, New York, pp. 333–405.

U.S. EPA, 1999. "Remediation Technologies Screening Matrix and Reference Guide." http://www.frtr.gov/matrix2/section1/toc.html.

CHAPTER 2
SOIL

Olin C. Braids
O. C. Braids & Associates, LLC

2.1 INTRODUCTION

Soil is the term used to describe the geologic mantle that covers most of the terrestrial earth's surface. Soils vary in development from old, strongly weathered types to young types that have only recently been exposed to weathering as they were deposited from active geological processes such as volcanism or floods and erosional processes. Except where bare stone or water forms the earth's surface, we all live on top of a soil layer. This situation results in soil contributing to beneficial activities of humans, such as growing row and horticultural crops. On the other hand, soil is subjected to deleterious activities such as disposing of waste, spilling of petroleum products and chemicals, discharging of air contaminants that fall to the ground, and subsistence farming that drains the soil of nutrients.

Pedology, the science of soil formation, development, and classification, defines soil as the surficial few inches to feet of the earth's crust that is altered over time in a way that differentiates it from the geologic parent material. The processes acting on specific geologic minerals alter them and develop a vertical profile of textural and color differences that are the bases for soil classification. Such influences as temperature range, native plant populations, precipitation amount and timing, and indigenous fauna produce a vertical soil profile that is characteristic of the influences and their effect on the geologic parent material that develops into the soil profile.

As will be discussed and developed in this chapter, many of the processes that lead to natural soil development are processes that assist in the dissipation of chemical contaminants that find their way into soil. On the other hand, there are also processes that tend to preserve chemical contaminants over extended periods of time. In either case, the combination of contaminants and soil is an important one because soil is the medium in contact with humans' activities on land and is the depository for a significant amount of waste.

2.2 SOIL CHARACTERISTICS

2.2.1 Mineralogy

Metamorphic, igneous, and sedimentary rocks are the parent materials from which soils are derived. Weathering processes, including leaching by percolating water, freezing and thawing cycles, and the influence of macro- and micro-organisms slowly change the characteristics of the parent material and develop a vertical profile of leached and redeposited zones of secondary minerals. Soil profile development produces unique characteristics as a function of the lithology and climatic environment of the soil. Soil, by definition, is a natural body of mineral and organic materials that harbors living organisms and supports plant growth. Soils are classified according to these characteristics.

The moisture regime in a soil determines how the soil will develop. Soils that are slow to drain or are inundated for extended periods of time develop characteristics that differ from soils formed from the same parent material that are well drained. When soils are examined in a landscape with vertical relief, the soil at higher elevation will differ from the soil at lower elevation where water accumulates even though the parent geology is the same. These differences are categorized and reflected in the soil series name. The name is derived from a geographical location where the soil was first described (e.g., Orlando series) and the characteristics of the soil are described in terms of color, moisture regime, temperature, organic matter content, and other aspects that give clues to its conditions of development (USDA, 1989). Although much of the classification information may not influence the soil/waste relationship, the information does provide textural and prevailing moisture conditions that do influence the way waste chemicals will behave.

Soil mineral particles are generally a mixture of the original parent mineral particles that have not weathered and particles of secondary minerals formed from parent mineral weathering. The weathered minerals are predominantly clay minerals that are very fine particles in the range of 1 nm (10^{-9} m) to 1 μm (10^{-6} m) in size. Particles of this size range are colloidal and have a large surface area per unit weight. The clays have surfaces that are chemically active and attract ions from soil solution that act as plant nutrients. The net result is that ions such as calcium, magnesium, potassium, and sodium are held in the soil where plant roots can contact them and absorb them into the plant. Clays are highly influential in determining soil chemical characteristics. Their interaction with inorganic ions in soil solution and the concept of cation exchange capacity will be discussed in more detail below.

Other soil mineral particles are classified as silt size, sand size, and gravel. The relative proportion of each of the particle sizes determines how the texture of the soil will be described. From soil of low permeability and small particle size (a silty clay), soil ranges to high permeability and coarse grain (a gravelly sand). Generally, the more homogeneous, or better sorted, the soil, the higher will be its permeability compared to similar soil with heterogeneous intrusions.

Clay and silt-sized particles also hold a greater amount of soil moisture than coarser sand and gravel. Field capacity in a soil is defined as the moisture percentage at which free water ceases draining from a soil column. A sandy, coarse-textured soil will retain 12 to 15 percent moisture at this point, whereas a silt loam (sand with 70 to 90 percent silt) will retain about 30 percent moisture (Buckman and Brady, 1960). This implies that less energy is required to remove water from coarse-textured soil than from fine-textured soil. This is an important relationship in either flushing soil for the purpose of removing contaminants or performing vapor extraction for contaminant removal. The efficiencies quickly decrease with increasing silt and clay in the soil. Reducing the soil moisture film to the wilting point of plants requires 15 atmospheres of tension. The moisture content at this same tension in a sandy soil will be about 5 or 6 percent, whereas a silt soil will retain about 20 percent (Buckman and Brady, 1960).

Another component of many mineral soils that is not always recognized for the amount of chemical influence it exerts is hydrous oxide. Hydrous oxide is formed from iron or manganese as an iron or manganese oxide/hydroxide polymer. Iron hydrous oxide is more common, as iron is usually more abundant in soil than manganese. Hydrous oxide coats soil particles and exhibits similar properties of ion adsorption as do clays, but with less activity per unit weight. When iron hydrous oxide becomes high in concentration, it may also act as a cementing agent for soil particles.

2.2.2 Organic Matter

Residues of plants including aerial portions of leaves and stems, as well as roots, accumulate in soil during and after the plant growth cycle. Micro- and macro-organisms utilize this organic matter as a food supply and partially digest it. Other organisms ingest these organisms' excreta and tissue after their life cycle. Over the course of months to years, this recycling of fresh organic matter turns it into a product that is composted and stabilized. This reworked, structureless organic matter is referred to as *humic material*. As it becomes more weathered, it becomes more resistant to further decomposition. Depending upon climate, humic material may accumulate and represent organic matter from decades to centuries old.

The molecular structure of humic material has been a matter of investigation for at least a century. As instrumental methods of chemical analysis have developed over the past 40 years, infrared spectrophotometry, nuclear magnetic resonance, differential thermal techniques, and other ways of probing have elucidated some of the humic material structure. However, humic material is composed of amorphous, high-molecular-weight polymers that do not yield easily to structural analysis. The humic material has a variety of functional groups such as carboxylic acid, phenol, phenoxy, and alcohol that produce the high chemical activity per unit weight that was previously referred to. The net functional group influence in humic material is acidic. Therefore, humic acid is a term used to describe the stabilized organic matter.

Organic matter in a generic way influences the environmental behavior of organic soil contaminants. Because organic matter and organic compounds have compatible characteristics, the organic compounds, including organic contaminant compounds, interact both physically and chemically with organic matter. This interaction includes both dissolution of organic compounds into the organic matter matrix and chemical reaction between the two substances. Whether in soil or deeper geologic materials, organic compounds, because of this interaction, are retarded while being transported by soil water or groundwater. In calculating or modeling organic compound transport in a soil/groundwater system, it is important to directly measure or accurately estimate the organic matter content of the geologic matrix.

2.3 SOIL MICROFLORA/FAUNA

Soil harbors a large population of plant and animal organisms. Filamentous and single-celled bacteria, fungi, actinomycetes, algae, and small animals from protozoa to worms, insects, and mammals are native to productive soils. Photosynthetic higher plants produce plant tissue and chemical exudates that become nourishment for the micro-organisms. Bacteria and fungi are decomposers of organic matter. By-products that include high-molecular-weight, resistant, complex molecules become humic materials that accumulate in soil.

Most mineral soils include a large population of bacteria having aerobic, facultative, and anaerobic metabolic capacities. These diverse bacteria are able to biodegrade a wide variety of organic compounds when sufficient time is provided for their acclimation. Numerous compounds that fall into the hazardous waste category have been shown to be accessible to soil micro-organisms, and with time, become biodegraded. In addition to micro-organisms, algae, fungi, and actinomycetes are indigenous to soil. These organisms, in the plant family, are also effective in reacting with and degrading certain types of organic compounds. Fungi, particularly, excrete enzymes that are effective in hydrolyzing polysaccharides and cellulose. As these organisms die, their tissue enters the sink of organic matter in the soil and becomes humified.

Most bacteria in soil are heterotrophic, meaning that they require an external source of energy and nutrient elements for metabolism and population growth. The energy source is organic matter produced by the plant community growing in the soil. Heterotrophs are the destroyers of organic matter, releasing mineral nutrients as the carbon-based tissue is degraded. This process allows the mineral elements to become recycled. Mineral nutrients are also derived from the natural dissolution of soil minerals and may be supplemented with nutrients added as part of agricultural fertilization. The energy in organic food sources is released by respiration under aerobic conditions or fermentation under anaerobic conditions. Normally oxygen acts an electron acceptor in oxidation reactions, but when molecular oxygen becomes low or absent, sulfate and nitrate can also act as electron acceptors.

Heterotrophic bacteria may biodegrade contaminant chemicals directly as in the case of petroleum hydrocarbon degradation. In this case, the contaminants are acting as energy sources to the bacteria. Alternatively, they may biodegrade chemicals by means of cometabolism. The cometabolic process enzymatically attacks and degrades compounds such as the chlorinated hydrocarbons in the process of utilizing another substrate for energy. The degradation of the cometabolized compound does not yield energy to the organism. The utility of this metabolic pathway is that it biodegrades compounds that are bioresistant and would not be directly metabolized as an energy source.

There are bacteria that are autotrophic; that is, they can manufacture tissue from energy they receive as light (photosynthesis) or from inorganic oxidation reactions. A nonsoil example is the bacterial colonization of deep oceanic trench fumaroles where the energy cycle is completely chemical, as there is no light penetration. In a soil ecosystem, the photosynthetic bacteria would have to be at the soil surface where light is available to drive the photosynthetic process. Because of this limitation, autotrophic bacteria have limited significance in interacting with contaminant chemicals.

2.4 CHARACTERISTICS OF CHEMICALLY HAZARDOUS WASTES

2.4.1 Chemical Speciation

Waste in the context of this chapter is going to be defined as chemical waste containing one or more hazardous or toxic chemical components. According to the Resource

Conservation and Recovery Act (RCRA), corrosivity and flammability are characteristics imparting a hazardous definition to waste, but they do not necessarily carry hazardous characteristics to soil into which they are disposed. However, transport of hazardous chemical substances into soil with waste disposal renders the soil hazardous if the substances exceed regulatory tolerances. This discussion of the interaction of soil and hazardous wastes will be limited to the hazardous chemicals contained in waste and will not cover biological hazards. The term *contaminant* will be assumed to refer to chemical and not biological substances.

Because waste as a term covers the spectrum of inorganic, organic, and biological substances, waste's relationship to soil cannot be described in generic terms. Each of these waste classifications and their subsets must be defined and described individually, as they react differently with soil and soil components. Yet, most hazardous waste sites have a mixture of these waste components, frequently intimately blended, with synergistic effects upon one another. Because of these complexities, it is important to understand how the characteristics of soil and of the contaminants interact on a fundamental basis.

One of the greatest difficulties in dealing with waste chemistry is to determine the actual speciation of the chemicals involved. This issue is significant from two points of view. One is that the chemical behavior of an element in soil and water is a function of its chemical speciation. The other is that the toxicity of an element is a function of its chemical speciation. The Environmental Protection Agency (EPA) Hazardous Substance List is a compilation of chemical elements and substances that pose a toxic threat to the environment and human health. However, the list does not specify which chemical form an element such as lead or copper is in. Therefore, metallic chromium present in a sample that was dissolved during the sample digestion and reported as a sample component might classify the sample hazardous, when the chromium was actually present in an inert form. For the purposes of this discussion, the treatment of speciation will be kept as simple as possible and limited to the most common ionic configurations.

Metal ions are the most common inorganic ions that cause waste to be classified as hazardous. Metals are used industrially as catalysts and pigments, and for electrolytic plating as well as in the structure of goods constructed of metal. Ag, As, Cd, Cr, Cu, Ni, Pb, Ti, V, Zn, and other less common metals find their way into industrial wastes as process by-products or products added directly in waste treatment. The metals in wastes that would be classified as hazardous are seldom in elemental form. They are usually present as dissolved or precipitated salts. In that form, the metals are subject to leaching and can enter into the soil and groundwater as they move with percolating water that has passed through the waste materials.

As salts, the metals may be highly water soluble or may be extremely insoluble, depending upon the salt species that is formed. When metal salts dissolve in water they do so by forming ions that separate and disperse in the water. The metallic elements, with certain exceptions that will be discussed later, form positively charged *cations*. The companion ions, usually nonmetallic elements and oxygen complexes, form negatively charged ions called *anions*. This distinction is important in soil chemistry because the two classes of ions react differently with the soil matrix.

Cations interact with the clay fraction of soil because of the clay's attractive forces that lead to adsorption of cations. This phenomenon is the subject of the next section in this chapter. Anions also have limited interaction with soil, as there are positively charged locations on clay and hydrous oxides that hold ions such as phosphate and arsenate. A number of metals in the inner transition series and those known as quasimetals form *oxo anions*. Common oxo anions found naturally or in hazardous waste are shown in Table 2.1.

TABLE 2.1 Names and Formulas of Important Oxo Anions
(the Ions Are in Periodic Table Order)

Borate BO_3^{3-}	Carbonate CO_3^{2-}	Nitrate NO_3^- Nitrite NO_2^-		
Aluminate AlO_4^{5-}	Silicate SiO_4^{4-}	Phosphate PO_4^{3-}	Sulfate SO_4^{2-} Sulfite SO_3^{2-}	Perchlorate ClO^- Chlorate ClO_3^-
		Arsenate AsO_4^{3-} Arsenite AsO_3^{3-}	Selenate SeO_4^{2-} Selenite SeO_3^{2-}	
	Stanate SnO_6^{8-}	Antimonate SbO_6^{7-}		Periodate IO_6^{5-}
	Plumbate PbO_6^{8-}	Vanadate VO_4^{3-}	Chromate CrO_4^{2-} Molybdate MoO_4^{2-} Tungstate WO_4^{2-}	Permanganate MnO_4^-

Source: Wulfsberg, 1991.

Phosphate and arsenate to a lesser degree interact with iron hydrous oxide and become adsorbed to its surface. Phosphate appears to chemically react over time to become incorporated as part of the solid structure, so becomes "fixed" in the sense of losing its mobility and availability to plant uptake. The other common oxo anions generally behave as conservative ions; that is, they tend to move with water and interact minimally with soil and aquifer components. For example, chromate ion, in which chromium is hexavalent is mobile in soil and groundwater. Only when chromium is reduced to trivalent does it become a cation and either sorb to geologic material or react with hydroxide and precipitate.

One of the difficulties posed by the variances in speciation is that normal analytical methods such as atomic absorption spectroscopy (AAS) and inductively coupled plasma spectroscopy (ICP) provide metal concentration values for all forms of the metal in the sample.

2.4.2 Cation Exchange Capacity

The component with the most influence on inorganic chemical species in soil is the clay fraction. This mineral fraction interacts with cations in a way that the cations are both removed from and released to soil solution. To understand this, some explanation of the clay structure is necessary. Clay minerals are aluminosilicates having the major elements of oxygen, silicon, and aluminum. The crystal structure is in the form of plates or layers. The silicate layer has a tetrahedral configuration and the aluminum layer has an octahedral configuration. The tetrahedron has four sides, whereas the octahedron is eight sided.

Silicon tetrahedra have four oxygen atoms associated with the silicon atom. Aluminum octahedra have six oxygen atoms associated with the aluminum atom. These crystal configurations are the building blocks of clay minerals. The minimum unit that repeats itself in three dimensions to form the clay crystal structure is the unit cell. This is analogous to the monomer unit that repeats to form an organic polymer such as polyethylene. Both silicon tetrahedra and aluminum octahedra form sheets by shared bonding with oxygen atoms. These sheets layer with one another to form the different clay mineral structures. The kandite mineral group has one silicon sheet and one aluminum sheet. The smectites and vermiculites have 2 silicon sheets to 1 aluminum sheet, but differ in their properties. Chlorites also have a 2:1 silicon to aluminum sheet ratio, but have a sheet of magnesium atoms between the silicon sheets.

The simplest clay mineral group is the kandite, to which the clay mineral kaolinite belongs. The mineral structure is made of two sheets, one of silicon/oxygen tetrahedra and one of aluminum/oxygen octahedra with an equal number of silicon and aluminum atoms. Some of the oxygen atoms on the surface of the mineral are actually hydroxyls (—OH functional groups). It is the hydrogen that provides a hydrogen bond that singly is very weak, but in aggregate holds the crystal sheets together and prevents the interlayer entry of water molecules or cations in solution. This structural characteristic presents only the outside surfaces of the clay crystal to the soil solution. Therefore, cation exchange is limited to only the outer surfaces of the kandite group clay minerals.

However, some cations that are present in soil solution can also react with the elements in the clay lattice and substitute for them. Aluminum can substitute for silicon and iron or magnesium can substitute for aluminum. This substitution is called *isomorphous* substitution, and it creates electrical imbalances within the clay crystal. The electrical imbalances are usually translated into a net negative charge on the crystal. The crystal structure is unchanged with substitution because the atoms that substitute are similar in size and do not change the tetrahedral or octahedral configuration. The net negative charge is also supplemented by broken bonds where clay crystals fracture. It is this net negative charge that attracts cations to the clay surface.

Smectites are 2:1 layer aluminosilicates. Two silicon tetrahedral layers sandwich an aluminum octahedral layer between them, thus resulting in the 2:1 layer configuration. Montmorillonite is the most abundant member of the smectite family. There are many clay types in this group because isomorphous substitution is prevalent. The crystal layers are not tightly fixed with hydrogen or other bonds, so water molecules can enter the interlayer. The water molecules take on an orderly configuration, thereby expanding the interlayer spacing. This brings about significant swelling of the clay upon hydration.

Aluminum substituting for silicon, and ferrous and ferric iron and magnesium substituting for aluminum, results in a wide variety of smectite minerals with differing characteristics. The ability for water carrying dissolved cations to enter the interlayer of the mineral lattice gives the mineral a variable thickness and an increased cation exchange capacity. On a gross scale, montmorillonite exhibits a large degree of swelling when it becomes hydrated. It also presents an inside active surface as well as the outside surface to the soil solution.

Illite, another 2:1 layer aluminosilicate, is stabilized by interlayer presence of potassium atoms. The interlayer bonding resulting from the potassium prevents entry of water between the layers, so the mineral is nonswelling. Soils rich in illite tend to "fix" postassium, that is, remove it from soil solution. For this reason, excess potassium or potash is necessary for fertilizing crops grown in soils containing this type of clay. Unlike montmorillonite, illite's interlayer is not accessible to other cations, so its cation exchange sites are limited to the outside surfaces. This reduces its cation exchange capacity relative to montmorillonite.

Cation exchange capacity (CEC) is the measure of a clay mineral's or a soil's capacity to attract and hold cations from soil solution. CEC is measured in milliequivalents of cation per 100 grams of soil. The CEC term also indicates that the cations are not held permanently, but may be held and released, depending on conditions. Cations in soil solution are attracted and held with weak quasi-bonding forces. These include electrostatic and van der Waals forces. Monovalent cations are held less tightly than divalent cations, which in turn are held less tightly than trivalent cations. This correctly implies that cations of higher valence may displace and replace cations of lower valence. This phenomenon occurs when soils are limed with calcium carbonate and the calcium displaces hydrogen and sodium ions that are adsorbed on the clay mineral surfaces. It also occurs with metals such as lead and chromium, which are held much more strongly on the cation exchange sites than are sodium, potassium, and some other divalent cations.

Metallic cations that are more closely associated with hazardous waste than with plant nutrients undergo the same CEC reactions in soils as the plant nutrient ions. Landfill leachate collected within a landfill frequently has a different complement of cations than the leachate collected after it has percolated through the unsaturated soil and been transported some distance in the aquifer. This change in composition is a result of cation exchange with the natural ion complement present in the soil and aquifer geologic matrix. The exchangeable ions may be displaced by higher valence ions displacing lower valence ions or by a significantly higher concentration of an ion displacing ions held on exchange sites. This concentration effect is observed with the leachate compositional change, as leachate usually has a large concentration of sodium, potassium, and calcium in relation to natural concentrations in percolating soil water.

2.4.3 Analytical Challenges

Soil chemists have been developing analytical techniques to characterize soil for more than a century. Three major objectives have motivated soil chemists to develop the science of soil chemistry:

1. To characterize soil in terms of its potential as a medium for plant growth
2. To characterize the soil in terms of its mineral composition
3. To elucidate the fate and behavior of chemicals within the soil

Plant mineral nutrition is supplied by inorganic ions that are produced and held in soil in a range of concentrations and attractive strengths. Separating the ions available to plants from the total ion complement has been a continuing challenge. This is important in the establishment of the fertility level in soil. Nitrogen, phosphorus, and potassium are the primary elements that determine the potential for crop yields.

Disposal of chemical wastes in soil adds a waste component to the soil's mineral and natural organic components. The waste will become part of the soil matrix that is sampled and analyzed. The waste will also change the chemistry of the soil to the extent that waste chemicals dissolve or otherwise commingle with the soil. Inorganic ions released by the waste will interact with soil clay minerals and organic matter to become part of the soil matrix. After the chemicals become a soil component instead of a waste component, they will exhibit different chemical characteristics than were exhibited when the ions were part of the waste.

The CEC characteristic of soil is significant in dealing with soil contamination because it prevents a simple flushing of the soil from removing undesirable metals and to a certain extent, organics. An ionic complement, or an extreme pH level, in the flushing fluid is necessary to displace the metals held by the soil. To effectively strip the contaminating ions, an excess of replacement ions over the stoichiometric equivalent is necessary. This produces a solution that retains much of its original composition, with addition of the contaminants stripped from the soil.

Frequently, organic wastes contain a large amount of amorphous, undefined organic matter in the form of sludge or tars. When this is added to the humified organic matter in soil, the resulting organic matter poses an interference problem with the analysis of specific compounds. When the soil is extracted, the tars and humic materials accompany the target compounds and must be removed through a preliminary cleanup step.

Although it isn't strictly an analytical technique challenge, the addition of contaminant metals to soil requires a careful selection of background soil samples. Most soils contain a natural complement of so-called trace metals that are the same metals that may be viewed as contaminants in waste products. Zinc, chromium, nickel, lead, manganese, copper, and arsenic are some of the more common ones. To establish that the metals are derived from waste requires the establishment of natural background concentrations. This not only includes reference soils for the near surface, but also soil samples from the subsurface. Increasing clay content in soil will generally increase the trace element content in soil as well.

2.5 METHODS FOR REMEDIATING SOIL CONTAMINATION

The most direct remedial measure for soil is excavation and transport to an approved disposal site. Remedial methods that allow soil to remain on site fall into three broad categories: chemical and biological reaction; mobilization, separation, and extraction; and solidification, stabilization, and containment. Table 2.2 lists some of the contaminant classes and remedial techniques that may be effective in controlling them.

2.5.1 Phytoremediation

Phytoremediation refers to the use of trees and other plants established in locations where uptake, stabilization, or biodegradation of soil and groundwater contaminants by the trees can take place. The term *phytoremediation* is also applied to applications of growing trees to control water. For example, trees may be planted on a landfill cap to reduce or eliminate rainfall infiltration. Trees may also be planted over a groundwater contaminant plume in shallow groundwater so that transpiration of shallow groundwater will remove and contain the contaminants. Planting tree groves also has the advantage of providing air filtration, where the trees absorb volatile chemical compounds. Trees also act as sound barriers and windbreaks. There is also a desirable aesthetic quality in establishing a grove of trees on what had been a site that was visually damaged by chemical contaminants or is part of a barren industrial landscape.

There are numerous plants that may be chosen for phytoremediation application, although trees are generally preferred because of their deep roots, rhizosphere development, and volume of water transpired. When plant uptake of nutrient elements or metals is desired from the point of view of harvesting and disposing of the crop containing the contaminants, grasses or row crops would be chosen. When "mining" soil by plants is the objective, legumes such as clover or alfalfa should be considered. Legumes grow with a symbiotic relationship with *Rhizobium* bacteria that live in nodules on the plant roots and fix nitrogen from the air so that it is available as a nutrient to the plant. This growth pattern helps to establish a self-sustaining crop.

Selection of trees for phytoremediation should reflect the climate of the location and the fact that the trees will be planted in closer proximity than they would in a reforestation program, for example. Hybrid poplars are frequently chosen because they do not reproduce, so do not introduce a nuisance species. Poplars also grow quickly, tolerate a wide range of climatic conditions, and respond to fertilization. Transpiration by poplars is so great that it can depress the water table.

TABLE 2.2 Contaminant Classes and Appropriate Treatment Technologies*

	Petroleum hydrocarbons	Chlorinated solvents	Semivolatiles	PCBs	Inorganic chemicals	Pesticides and explosives
	Solidification, stabilization, and containment					
Asphalt batching	X	NA	X		X	
Biostabilization	X	NA	?	?	NA	
Excavation	X	X	X	X	X	X
Lime addition	X(h)	NA			X	
Pozzolanic agents	X(h)	NA	?	?	X	
Vitrification	NA	NA	NA	?	X	
Barrier walls			X	X	X	
	Separation, mobilization, and extraction					
Dual-phase extraction	X(l)					
Soil washing	X					
Sparging (air/steam)	X(l)					
Thermally enhanced soil vapor extraction (SVE)	X(h) `					
Solvent extraction	X					
SVE	X(l)	X	NA	NA	NA	
	Chemical and biological reaction					
Biopiles	X	NA	X	?	NA	X
Biosparging/venting	X(l)	?	?		NA	NA
Chemical oxidation	?	?				
Chemical reduction		X			X	X
Incineration	X	X	X	X	NA	X
Intrinsic bioremediation	X	X	?	?	NA	
Land farming	X	NA	X	?	NA	X
Phytoremediation	X	X	X	?	X	X
Thermal destruction/ reduction	X	?	X		NA	
Passive/reactive barriers	X	X			X	

 * Modified from MacDonald and Rao, 1997.
 NA = not applicable.
 (l) = light, (h) = heavy.
 Blank = not enough information or experience.
 ? = information being developed.

Plant root systems, especially most tree roots, naturally establish a symbiotic relationship with fungi, actinomycetes, and bacteria in the immediate vicinity of the root.
That zone is termed the *rhizosphere*. The roots produce chemical exudates that are
food for the micro-organisms, while the micro-organisms produce metabolites that
can benefit the tree's mineral nutrition. Some metabolites are chemicals that can
complex with, or chelate, metals. The metal chelates become water soluble, thus are
available for the tree root to absorb. Forests receive virtually no nutrient input ex-

cept for nitrogen and sulfur from atmospheric fallout. The nutrient requirements for tree growth are met by recycling of plant tissue and minute amounts of dissolution of soil minerals. The ability of soil micro-organisms to enhance the mineral dissolution by chelation and complexation significantly enhances mineral nutrition of the growing plants.

This ability to enhance the solubility of metals from minerals also enhances the solubility of metals contained in metal wastes. Plants will absorb the metals and fix them in the plant tissue. If the plants are harvested and disposed of, metals will be "mined" and removed from the system. The University of Pennsylvania's Plant Science Institute is presently conducting genetic research on *Arabidopsis,* a plant that produces natural chelating chemicals, to enhance its effectiveness in absorbing heavy metals (*Orlando Sentinal,* 1999). *Arabidopsis* is particularly sensitive to cadmium. A recent note in *Nature* reported on a fern that hyperaccumulates arsenic (Ma et al., 2001).

The interrelationship between soil microbes and plant roots leads to a diversification of the soil microbes. The complexity of organic substances produced and the level of microbial activity relative to surrounding soil provide a range of metabolic capabilities. Fungi, actinomycetes, and bacteria are the primary players, with algae and protozoa also taking part in the synthesis and degradation of organic substances. This diversity in capabilities and species of organisms provides the opportunity for synthetic organic compounds, regarded as recalcitrant to biodegradation, to become biodegraded. Presence of a metabolizable substrate provided by root exudates provides the opportunity for organisms to cometabolize otherwise resistant compounds.

Recalcitrant compounds such as polycyclic aromatic hydrocarbons (PAHs) and pentachlorophenol (PCP), a wood preservative, as well as those more biodegradable such as alcohols, aliphatic hydrocarbons, and monocyclic aromatics, have all been effectively treated with the application of phytoremediation. Decomposition is not the only treatment mechanism. The relative abundance of organic matter, living and residual, in the rhizosphere allows sorption and fixation of organic contaminants, arresting their mobility.

Phytoremediation has proved effective with manufactured-gas plant sites where coal tars, PAHs, and simpler hydrocarbons are present as contaminants. This hydrocarbon assemblage is resistant to soil micro-organisms under normal conditions and will persist in the subsurface for decades. However, when trees have established rhizosphere colonization, the hydrocarbons become amenable to biodegradation.

Metabolism isn't limited to the rhizosphere, as the trees also metabolize chemicals that are absorbed by their roots in addition to synthesizing chemicals from the photosynthetic process. Depending upon the chemical in question, the tree may chemically transform it, store it unchanged in its tissue, or transpire it with water. Conifer trees growing in locations where chemicals have been land disposed may show the influence of the chemicals in their tissue. When a pine tree growing in a disposal area was cut, wood synthesized during the time that chemicals were present, as exhibited by the growth rings, was nearly black in color. Earlier growth at the core of the tree trunk was a light natural light tan color (personal observation). This showed that the tree absorbed the chemicals and provided an accurate timing of the disposal events.

There is also research on genetically engineered plant species that are designed for particular contaminant interactions. One of these has a bacterial gene, mercuric ion reductase, inserted in the plant DNA (Boyajian and Devedjian, 1997). This modification allows the plants to germinate in mercury-contaminated soil. After they are established, they volatilize mercury into the atmosphere. Mercury vapor concentrations in test trials have not proved to pose a risk.

Another company is experimenting with an Indian mustard plant, *Brassica,* that has been bred to hyperaccumulate lead (Boyajian and Devedjian, 1997). A chelating agent such as EDTA is applied to the soil to solubilize lead so it is available to the plant roots. When the plant matures, it is harvested and landfilled. Several generations of plants will result in lowered lead concentrations in the soil.

Trees transpire large quantities of water. For this reason, they can be used to draw down aquifers in order to contain and transpire contaminated groundwater. Soluble volatile compounds in soil water or groundwater are taken up into the tree where they may be metabolized by the tree or may be passed into the atmosphere with transpired water. Control of water is also a function of phytoremediation. For example, a 17-acre land parcel was planted with 35,000 trees designed to treat 7 million gallons of leachate per year at the Riverbend Landfill in McMinnville, Oregon (Licht, 1998).

Phytoremediation has the advantages of relatively low operation and maintenance costs and effective treatment for numerous classes of contaminant compounds. Its disadvantage is that it takes one to two growing seasons for trees to achieve a size where their influence is significant. It then may take several years for trees to nurture the reactions that dissipate contaminant chemicals. Maintaining sufficient moisture and nutrients for the trees and an appropriate undercover is important in providing continuous treatment.

2.5.2 Thermal Applications

Organic and inorganic contaminants, as the preceding discussion has illustrated, exhibit a range of bonding strength with the soil matrix. The energy thresholds required to release the contaminants range from simply physically releasing waste substances from interstitial entrainment to desorbing chemicals held with quasi-chemical bonding. Chemical release agents include acids, bases, salt solutions, surfactants, and organic solvents. Adding heat with release agents, or adding heat alone, provides a higher vapor pressure for organic compounds and added molecular agitation, weakening the bonds that hold the contaminants. One drawback of thermal application is that there are usually large masses of soil to be treated. The soil is moist, and water has a high specific heat capacity that requires a large Btu input for a given heat gain. Soil is also a relatively poor conductor of heat, so heat additions in situ must be on a closely spaced grid. This poor efficiency of heating makes thermal application relatively expensive.

Cost/benefit ratios may be tolerable when high contaminant concentrations occur in a confined, definable body of soil, or when contaminant low volatility or high viscosity prevent other remedial methods from being effective. It could also be worth the added cost of incorporating heat in the remedial method if cleanup time is critical. Heat will speed up any soil/contaminant separation process.

In Situ Thermal Treatment. Heat injection into soil can be accomplished through several routes. Perhaps the most common is injection of steam. Vertical pipes with well screens, or perforated sections like well screens, carry pressurized steam into zones where contaminants are located. The efficacy of this method, as with almost all remedial soil methods, is a function of the permeability of the soil. Fine-textured soils such as silts and clays will not transmit steam readily or as far from the injection point as coarser-textured soils. If the soils have voids or fractures, there may be preferential dispersion with less efficient heating of the soil mass. Contaminant chemical vapors and contaminated steam condensate must be accommodated when this technique is applied.

Hot water injection is another common method of heating the soil matrix. This method has advantages with some contaminants as it performs a flushing function as well as heating. It is analogous to hot-water injection for secondary oil recovery. This operation generates a volume of contaminated water that must be collected, treated, and disposed of in an approved manner. For viscous contaminants such as coal tars and high-boiling bunker-type fuel oils, steam or hot water may be the only methods to reduce the viscosity and mobilize the product for recovery.

A manufactured-gas plant (MGP) site in Pennsylvania was remediated by means of hot water injection through a recovery technique called Contained Recovery of Oily Wastes (CROW™) (Leuschner et al., 1997). Two areas of free coal tar on the site showed coal tar accumulation on a fine sand layer 6 to 9 meters below land surface. An estimated 23,000 to 24,000 liters of coal tar was estimated to be present. Hot water was injected around the perimeter of the coal tar accumulations and recovered with production wells from the center of the free product area. The coal tar was mobilized by the hot water and was recovered from the production wells. The heat reduced viscosity and density so that the coal tar, which was a dense nonaqueous phase liquid (DNAPL) when cold, became less dense than water and floated on the recovered water.

Recovered water was passed through an initial step of oil/water separation. The next step oxidized the water and adjusted pH to precipitate iron and manganese recovered in solution. Following treatment, the water was reheated and reinjected. A greater volume of water was recovered than injected to keep a hydraulic isolation of the treatment area. The excess water was treated and discharged into an adjacent creek.

Direct application of heat to soil to effect thermal desorption has been demonstrated by using thermal blankets (Anonymous, 1997). This application is appropriate for shallow contaminants in the first two or three feet of soil. The thermal blanket containing electrical heating elements is spread over the contaminated soil and heated to 800 to 1000°C. As the heating front moves downward into the soil, contaminant chemicals are thermally destroyed and volatilized. A vacuum system scavenges the vapors that are not destroyed, and they go to a treatment combustion system. This system was effective in destroying and volatilizing PCBs where the average concentration exceeded 500 ppm. A cleanup standard of <2 ppm was achieved while meeting air quality standards and worker exposure standards. A variant of this remediation technique useful for deep contaminants is electrical heating in wells. An emerging technology is radiofrequency heating.*

Ex Situ Thermal Treatment. High- and low-temperature heat stripping of soil is performed on soil that has been excavated and brought to a stationary facility. The principal of low-temperature operation is that heating and mixing burns combustible compounds and volatilizes compounds with higher flash points. For example, one low-temperature thermal desorption (LTTD) unit feeds soil into the primary treatment unit that operates at 480°C. This thermal desorber was designed to accept soil contaminated with manufactured-gas plant wastes and other petroleum hydrocarbons. The gas stream is directed to a primary dust collector and the soil is directed to a cooling unit. The gas stream leaves the dust collector and enters the secondary treatment unit where a burner raises the temperature to 980°C, incinerating any remaining organic compounds for which the system is certified (D'Angelo and Chiesa, 1998). Gas from the secondary treatment unit then goes through a heat exchanger to a bag house and is exhausted with gas analyzer monitoring. Table 2.3 lists the process conditions for this system and Table 2.4 lists the efficacy of PAH removal.

---

* KAI Technologies, Inc., Woburn, Mass.

TABLE 2.3 Low-Temperature Thermal
Desorber Operating Parameters

Soil feed rate	36 to 45 tonnes/h
Soil moisture	15 to 20 percent
MGP/TPH feed (max)	3%
Primary treatment unit	480°C
Fuel input	37.72 million Btu/h
Secondary treatment unit	980°C
Fuel input	33.84 million Btu/h
Stack discharge	190°C
VOC removal	>99%
Particulate removal	>99.5%

MGP facilities used several feedstocks as hydrocarbon sources for their gas manufacture, but coal was the most common. Light and medium coal tars are byproducts that were produced in large quantity and sold as a commercial product. The coal tars also were disposed of on site, spilled, and otherwise distributed, leaving substantial soil contamination and potential for groundwater contamination. Coal tars contain a large percentage of polycyclic aromatic hydrocarbons (PAHs) that are only slightly water soluble and are resistant to biodegradation. These compounds are classified as semivolatile compounds for analytical purposes. PAHs in coal tars have boiling points ranging from 159 to 448°C (O'Shaughnessy and Nardini, 1998). These coal tar characteristics are the reason that MGP sites are still contaminated decades after the gasification processes have ceased. Although the compounds are not volatile, they vaporize sufficiently under the conditions of thermal desorption to be effectively removed as shown in Table 2.4. O'Shaughnessy and Nardini (1998) reported that thermal desorption operating at temperatures between 427 and 482°C was effective in reducing TPH and PAH concentrations to below detection limits and below prevailing standards.

A variation on the thermal treatment described so far is a high-vacuum, low-temperature, ex situ approach described by Dagdigian, Findley, et al. (1997). Soil contaminated with highly chlorinated insecticides such as DDT, chlordane, and methoxychlor was placed in a steel sealed chamber capable of accommodating ap-

TABLE 2.4 PAH Removal in the Thermal Desorption Unit

Compound	Initial compound concentration, ppb	Treated soil concentration, ppb	Cleanup maximum contaminant level (N.I.), ppb
Anthracene	4000	41	100,000
Benzo(a) anthracene	3000	55	900
Benzo(a) pyrene	4000	43	660
Benzo(k) fluoranthene	3000	49	900
Chrysene	5000	41	9,000
Pyrene	5000	80	100,000

proximately 3.8 m^3 of soil on a tray with infrared heat sources above it. The infrared sources generate about 137,500 Btu/h, raising the top few centimeters of soil to a 175 to 260°C temperature. A vacuum pump produces a vacuum of 635 mmHg. While under vacuum, a centrifugal blower circulates air at about 6000 absolute cubic feet per minute (acfm). The hot air exiting the chamber carries moisture and pesticide vapor that is condensed in a two-stage condenser system. In this application, 10,400 m^3 of contaminated soil were treated at an average production rate of 5.4 metric tons per hour.

Dagdigian, Czernec, et al. (1997) report that the same low-temperature, high-vacuum treatment was successful in remediating soil at the Rocky Flats nuclear facility in Colorado. Organic solvents were mixed with radionuclides in wastes disposed of at the facility. The goal of removing chemical waste from the low-level radioactive waste in the soil was met through low-temperature vacuum desorption. Once the chemical component had been removed, the radioactive component was low enough in activity to permit replacement of the treated soil.

2.5.3 Chemical and Biological Reaction

Chemical Reaction. Chemical reactions that destroy contaminant compounds come under the classifications of oxidative and catalytic reactions. The type of reaction chosen is based on the chemical characteristics of the target contaminant. Some contaminants such as hydrocarbons, phenols, and other nonhalogenated solvents are amenable to oxidative destruction. Other contaminants such as chlorinated ethanes, ethenes, and aromatics are effectively destroyed with catalytic reactions, especially with elemental iron and palladium.

Chemical oxidants such as permanganate, peroxide, peroxy acid, perchlorate, chromate, and Fenton's reagent can be effective in the oxidative destruction of organic contaminants. However, there are obstacles in managing and controlling oxidizing reactions under field conditions. Soil usually contains a complement of natural organic matter in the form of humic substances, with the possibility of larger amounts of less humified organic residues. Since addition of a chemical oxidant induces a chemical reaction in which any oxidizable substrate is reacted upon, the natural organic residues may consume a significant amount of the oxidant before the target organic contaminants are oxidized. Moreover, reduced forms of iron, manganese, and sulfur may also consume oxidant as they are oxidized. The total oxidation demand is termed the *poise* of the system. Several times as much oxidant as the calculated stoichiometric amount necessary to destroy the target contaminant must be utilized in order to satisfy the poise of the system being remediated.

The use of oxidants also carries a threat that the exothermic oxidation reactions may become so vigorous that excess gases and heat are generated. This could go to the extent that the ground surface is disturbed by expanding gases. An outright explosion may be theoretically possible, but that would require a large organic mass capable of reaction.

Biological Reaction. The predominant biological reactions that influence contaminant chemicals are those performed by bacteria. The heterotrophs are the destroyers of organic substances, as they are required for cellular energy. Bacteria are present in agriculturally productive soil in a population of about a billion cells per cubic inch of soil (Buckman and Brady, 1960), but until relatively recently, bacteria deeper in the earth than the tilled zone were assumed to be scarce to absent. Water associated with oil recovery containing carbonates and hydrogen sulfide posed the first suggestion of deep subsurface microbial activity. A geologist, Edson S. Bastin, studied samples in the 1920s and concluded that bacteria were present several hundred meters below the surface in the oil formation. It wasn't certain that the samples hadn't been biologically contaminated, so the discovery was not generally accepted (Fredrickson and Onstott, 2000). By the late 1970s and early 1980s, deep groundwater contamination at the Savannah River Department of Energy site prompted investigation of subsurface microbiology. Bacteria were recovered from a few hundred feet to depths extending to 2.6 km (1.7 mi) below the surface (Fredrickson and Onstott, 2000). This discovery brought with it the potential for the deep bacteria to act as biodegraders of deep organic contaminants.

Soil contaminants seldom extend beyond a few tens of feet before they encounter the saturated zone and become groundwater contaminants. The presence of bacteria in deep aquifers and even deeper zones is encouraging for in situ biodegradation of groundwater contaminants, a subject covered in Chap. 1. The biological population in soil generally exceeds that in aquifers, but it may be inhibited by a lack of nutrients, slow oxygen flux, or smaller numbers of micro-organisms at depths below a few feet.

Application of nitrogen and phosphorus to soil provides nutrients that stimulate micro-organism activity. An adjustment in pH may also be necessary to raise or lower the ambient pH to an optimum range for bacterial activity. The application may be as topical fertilizer application, or it may be mixed into the soil by tillage. One effective method of dealing with petroleum hydrocarbon contamination in soil is to strip the contaminated soil and stockpile it. The soil is then spread as a layer a few inches deep on uncontaminated ground and fertilized, limed, watered, and periodically turned with a plow or special implement that turns over a windrow of soil if the layer is thicker than about 8 in. This treatment allows aeration, which is important, as aerobic decomposition is more rapid and efficient than anaerobic decomposition.

A similar biodegradation scheme for soil contaminants is to place a 2-ft or deeper layer of the fertilizer- and lime-amended soil over a network of perforated pipe, where the soil remains undisturbed, but compressed air is passed through the soil to aerate it. In cold climates, either the air can be heated or the system can be under cover where heat can be provided. Generally, soil micro-organisms are active at temperatures that exceed 10°C (50°F). This system requires monitoring for moisture level, as the air effects rapid drying.

Stimulating the micro-organisms present where recalcitrant compounds have been disposed of can accelerate the biodegradation process. Bacteria require a period of acclimation to conditions and the chemical substrate. Presumably there has been acclimation in a disposal area, but there may be limitations on activity from lack of nutrients, extreme pH, or even the concentration of the contaminants themselves. Adding nutrients, adjusting pH, and perhaps adding clean soil to dilute the contaminant will optimize the system to act as a more efficient bioreactor. Compounds such as chlorinated benzenes and pentachlorophenol have been biodegraded when indigenous organisms were appropriately stimulated.

Composting soil can sometimes make biodegradation more effective, as it provides an organic substrate that is easily decomposed, thus providing a high bacterial population and the potential for cometabolism. An example is provided by ammonium picrate $C_6H_6N_4O_7$, an ingredient in explosives and rocket propellants. Ammonium picrate contaminated soil at the Hawthorne Army Depot in Nevada as a result of handling explosive ordnance since 1928 (Potter et al., 1999). A compost mixture of wood chips, steer manure, potato waste, hay, and 30 percent contaminated soil was placed into windrows and watered. The average ammonium picrate concentration was 3500 ppm. A windrow turner moved the windrows each day for aeration and more homogeneous mixing. After 16 days of treatment, the ammonium picrate concentration had been reduced below the detection limit.

Techniques incorporating biological and physical mechanisms to dissipate organic contaminants are bioventing and biosparging. Bioventing, as the name implies, injects air into the soil to introduce oxygen that will stimulate biological activity. Air injection and extraction wells are used to move air through the soil matrix. This method will remove some volatile compounds physically while enhancing the biodegradation of soil contaminants. This is particularly applicable to removal of hydrocarbon products.

Biosparging adds air injection into the top of the saturated zone to air-strip volatiles from groundwater. The air also performs the bioventing function as it passes through unsaturated soil to air extraction wells. Volatile compounds purged from groundwater are likely to be biodegraded with soil contaminants as the vapor passes through soil in contact with micro-organisms. Aeration of both groundwater and soil stimulates indigenous bacteria to consume the organic contaminants resident in soil and groundwater.

2.5.4 Separation, Mobilization, and Extraction

Soil Vapor Extraction. Volatile solvents and petroleum products are frequently inadvertently released from aboveground or underground storage tanks. The solvents may exist as soil contaminants for an extended period of time when recharge of rainwater is low, soils have low permeability, or the ground surface is paved or beneath a structure. Soil vapor extraction (SVE) is a technique whereby a vacuum created in the soil pulls the vapors into a collection system at the surface. There the vapor is generally treated before being released to the atmosphere. The effectiveness of SVE is a function of the soil permeability, volatility of the volatile compound(s), and their presence in the unsaturated zone.

The efficacy of SVE is enhanced by heating the soil while operating the SVE system. This raises the vapor pressure of the target chemicals, increasing their removal rate. Most SVE systems are operated at ambient temperatures because the scope of the removal is relatively small or the expenses of adding heat are not warranted. SVE is frequently used at service stations to remove gasoline in the unsaturated zone where it acts as a source for groundwater contamination. Heat addition is possible with injected hot air, steam, or electrical resistance heating elements inserted into the ground. None of these methods is particularly efficient because soil has low heat conductivity. Moreover, adding hot fluids to a low-permeability formation to enhance recovery is inhibited by the formation just as the removal of vapor was inhibited.

Soil can be heated resistively by passing electric current through it. An application of this technique using six-phase electric current has been implemented at several locations where contaminants are present deep in the ground or where they are in high concentration in a limited area (Bergsman and Trowbridge, 1997). Use of six-phase current creates a more uniform heating field where six electrodes are placed in a circle. Each electrode is fed with a separate phase so each conducts to all of the others and to a central neutral electrode.

The heated soil produces steam by vaporizing the natural soil moisture. This assists in vaporizing and releasing the volatile chemical contaminants. Steam distillation is an effective method of vaporizing compounds with lower vapor pressures that require a high degree of heat to directly vaporize. The six-phase soil heating method has been demonstrated at sites such as the Department of Energy's Savannah River facility, Dover Air Force Base, and a fire-training pit in Niagara Falls.

Soil Washing and Solvent Extraction. Soil washing is an ex situ technology that is self-explanatory. It is applied for removing certain organic contaminants as well as metals from coarse-grained soils. A liquid extractant composed of an aqueous solution of surfactant or other substance is mixed with excavated soil, thereby removing the contaminants. The solution effects a physical removal of the contaminant from the soil matrix. This treatment results in a liquid product with a high concentration of the soil contaminant and soil that has been brought to a cleanup-level concentration. The soil is then eligible for replacement. The cleaning solution becomes a hazardous waste that requires appropriate disposal. Whatever the ultimate fate, the liquid will be relatively expensive to dispose of.

The soil washing machinery is relatively large and expensive to transport and set up. For this reason, soil washing has been applied to larger sites where an on-site treatment system can be economically justified. Many contaminants, for the reasons given in the previous discussion, are not readily released from their attachment to soil. Thus, the washing solution proposed for any given site should be pilot tested to assure that contaminant removal efficiency will be equal to the regulatory cleanup requirements.

The technique is also used to separate coarse-grained soil fractions from fine-grained by using wet screens, flotation cells, spiral classifiers, and/or hydraulic classifiers. Contaminants tend to stay adsorbed on fine soil particles, which need further treatment such as stabilization/solidification (fixation). The coarse fractions often comprise a very high proportion of the soil and do not need further cleanup.

Solvent extraction uses nontoxic hydrocarbons for removing nonpolar organic contaminants or liquefied carbon dioxide for polar organics. The scheme is directly applicable to fine-grained as well as coarse soils.

Summary

Cleanup technology has come a long way in the past couple of decades. When dirt became recognized as dirty, the application of biotechnology, chemical technology, and physical principles to cleanup gained momentum and attracted research and development. There are numerous variations on the themes that have been covered in this chapter. Unique contaminants that are resistant to conventional treatments, or unusual site conditions, prompt modifications to fundamental remedial techniques that are aimed at making the removal or decomposition of the target contaminants more efficient. For example, bacteria have been cultured to be effective in decomposing specific compounds. These then have been used as inoculum in soil bioremediation. More recently, genetic engineering has been used to effect bacterial transformations that were not possible with natural organisms.

Chemical innovations have also been developed that release oxygen or effect oxidative reactions to chemically oxidize organic contaminants to the end products of carbon dioxide and water. Whatever the attempted solution, the removal or decomposition of chemical contaminants requires energy, time, and generally a substantial capital investment. With experience and experimentation, the remedial process should become less difficult and more effective.

REFERENCES

Anonymous, 1997. "Throwing a blanket on the problem," *Soil & Groundwater Cleanup,* May.

Bergsman, Theresa, and Bretton Trowbridge, 1997. "Soil vapor extraction to the sixth degree," *Soil & Groundwater Cleanup,* July, pp. 6–11.

Boyajian, George E., and Deborah L. Devedjian, 1997. "Phytoremediation: It grows on you," *Soil & Groundwater Cleanup,* February/March.

Buckman, Harry O., and Nyle C. Brady, 1960. *The Nature and Properties of Soils,* Macmillan, New York.

Dagdigian, Jeffrey, Janice Czernec, Steve Corney, and Ronald Hill, 1997. "Mixed waste without mixed results," *Soil & Groundwater Cleanup,* July, pp. 30–32.

Dagdigian, Jeffrey, Nancy Findley, Jeffrey O'Ham, and Edward Walsh, 1997. "Ex situ thermal technology exterminates pesticides," *Soil & Groundwater Cleanup,* June.

D'Angelo, Chip, and Anthony Chiesa, 1998. "Take it to the MART," *Soil & Groundwater Cleanup,* June.

Fredrickson, James K., and Tullis C. Onstott, 2000. "Microbes Deep Inside the Earth," *Scientific American, Earth from the Inside Out,* pp. 10–15.

Leuschner, Alfred P., Mark W. Moeller, Jason A. Gerrish, and Lyle A. Johnson, 1997. "MGP in hot water," *Soil & Groundwater Cleanup,* October.

Licht, Louis, 1998. "Phytoremediation Installation Experience Control of Water and Waterborne Pollutants by Plants," *Advances in Innovative Ground-Water Remediation Technologies,* Atlanta, December 15.

Ma, Lena Q., Kenneth M. Komar, Cong Tu, Weihua Zhang, Yong Cai, and Elizabeth D. Kennelley, 2001. "A fern that hyperaccumulates arsenic," *Nature* vol. 409, February 1, p. 579.

MacDonald, Jacqueline A., and P. Suresh Rao, 1997. "Shift needed to improve market for innovative technologies," *Soil & Groundwater Cleanup,* August/September.

Orlando Sentinel, 1999. "Handy plant could clean toxic waste," July 4, p. A-20.

O'Shaughnessy, James C., and Christine Nardini, 1997. "Heating the tar out," *Soil & Groundwater Cleanup,* June.

Potter, Dennis L., Judy Soutiere, Dick Brunner, Kathleen Bishop, and Andrea Hatch, 1999. "Hawthorne Army Depot, Nevada, Ammonium Picrate Contaminated Soil," *Soil & Groundwater,* August/September, pp. 12–14.

Singer, Michael J., and Donald N. Munns, 1996. *Soils—An Introduction,* Prentice Hall.

Sawhney, B. L., and K. Brown, eds., 1989. *Reactions and Movement of Organic Chemicals in Soils,* Soil Science Society of America, Madison, Wis.

USDA, 1989. "Soil Survey of Hillsborough County, Florida," U.S. Department of Agriculture, Soil Conservation Service, May.

Wulfsberg, Gary, 1991. *Principles of Descriptive Inorganic Chemistry,* University Science Books, Sausalito, Calif.

CHAPTER 3
DISCHARGES OF HAZARDOUS WASTE INTO THE ATMOSPHERE

Mark P. Cal

Department of Environmental Engineering
New Mexico Tech
Socorro, New Mexico

3.1 SOURCES OF HAZARDOUS CHEMICALS EMITTED INTO THE ATMOSPHERE

Hazardous or toxic air pollutants are those pollutants that are known or suspected to cause cancer or other serious health effects, such as reproductive or birth defects, or to cause adverse environmental effects. The degree to which a toxic air pollutant affects a person's health depends on many factors, including the quantity of pollutant and length of exposure, the toxicity of the chemical, and the person's current state of health and susceptibility. The 1990 Clean Air Act Amendments (CAAA) listed 188 hazardous air pollutants that the U.S. Environmental Protection Agency (EPA) is required to control. The list contains mainly organic pollutants, such as benzene, toluene, and dioxins, but some inorganic pollutants, such as mercury, cadmium, and lead compounds are also included.

Each year, millions of tons of hazardous air pollutants (HAPs) are emitted from hundreds of source categories, and include emissions from stationary (e.g., factories, refineries, power plants), mobile (e.g., cars, buses, trucks), fugitive (e.g., leaking equipment and containers), and area (e.g., multiple gas stations in a neighborhood area) sources. Under Sec. 112 of the 1990 Clean Air Act Amendments, the U.S. EPA developed a promulgation schedule for categories of HAPs by industry group. This list is published periodically and now includes about 200 source categories. Updated source category lists contain revisions, additions, and deletions for the National Emission Standards for Hazardous Air Pollutants (NESHAP). Updates can be found in the *Federal Register* (vol. 64, no. 222, pp. 63025-63035) and on the U.S. EPA Web site (*http://www.epa.gov*). A recent list of NESHAP sources is presented in Table 3.1, and a list of hazardous air pollutants is presented in Table 3.2.

TABLE 3.1   Categories of Sources of Hazardous Air Pollutants

Industry group source category (Revision date: November 18, 1999)	

Fuel Combustion:
  Combustion Turbines
  Engine Test Facilities
  Industrial Boilers
  Institutional/Commercial Boilers
 	
Process Heaters:
  Reciprocating Internal Combustion
    Engines
  Rocket Testing Facilities
  Stationary Internal Combustion Engines
  Stationary Turbines
 	
Nonferrous Metals Processing:
  Primary Aluminum Production
  Primary Copper Smelting
  Primary Lead Smelting
  Primary Magnesium Refining
  Secondary Aluminum Production
  Secondary Lead Smelting
 	
Ferrous Metals Processing:
  Coke By-Product Plants
  Coke Ovens: Charging, Top Side, and
    Door Leaks
  Coke Ovens: Pushing, Quenching, and
    Battery Stacks
  Ferroalloys Production
  Ferroalloys Production: Silicomanganese
    and Ferromanganese
  Integrated Iron and Steel Manufacturing
  Iron Foundries
  Steel Foundries
  Steel Pickling—HCl Process
  Steel Pickling—HCl Process Facilities and
    Hydrochloric Acid Re-generation
    Plants
 	
Mineral Products Processing:
  Alumina Processing
  Asphalt Concrete Manufacturing
  Asphalt Processing
  Asphalt Roofing Manufacturing
  Asphalt/Coal Tar Application—Metal
    Pipes
  Chromium Refractories Production
  Clay Products Manufacturing
  Lime Manufacturing
  Mineral Wool Production
  Portland Cement Manufacturing
  Refractories Manufacturing
  Taconite Iron Ore Processing
  Wool Fiberglass Manufacturing
	Petroleum and Natural Gas Production and
	  Refining:
	  Oil and Natural Gas Production
	  Natural Gas Transmission and Storage
	  Petroleum Refineries—Catalytic
	    Cracking (Fluid and other) Units,
	    Catalytic Reforming Units, and Sulfur
	    Plant Units
	  Petroleum Refineries—Other Sources
	    Not Distinctly Listed
	
	Liquids Distribution:
	  Gasoline Distribution (Stage 1)
	  Marine Vessel Loading Operations
	  Organic Liquids Distribution
	    (Nongasoline)
	
	Surface Coating Processes:
	  Aerospace Industries
	  Auto and Light Duty Truck (Surface
	    Coating)
	  Flat Wood Paneling (Surface Coating)
	  Large Appliance (Surface Coating)
	  Magnetic Tapes (Surface Coating)
	  Manufacture of Paints, Coatings, and
	    Adhesives
	  Metal Can (Surface Coating)
	  Metal Coil (Surface Coating)
	  Metal Furniture (Surface Coating)
	  Miscellaneous Metal Parts and Products
	    (Surface Coating)
	  Paper and Other Webs (Surface Coating)
	  Plastic Parts and Products (Surface
	    Coating)
	  Printing, Coating, and Dyeing of Fabrics
	  Printing/Publishing (Surface Coating)
	  Shipbuilding and Ship Repair (Surface
	    Coating)
	  Wood Building Products (Surface
	    Coating)
	  Wood Furniture (Surface Coating)
	
	Waste Treatment and Disposal:
	  Hazardous Waste Incineration
	  Municipal Landfills
	  Off-Site Waste and Recovery
	    Operations
	  Publicly Owned Treatment Works
	    (POTW) Emissions
	  Sewage Sludge Incineration
	  Site Remediation
	  Solid Waste Treatment, Storage and
	    Disposal Facilities (TSDF)

TABLE 3.1   Categories of Sources of Hazardous Air Pollutants (*Continued*)

Industry group source category (Revision date: November 18, 1999) (*Continued*)

Agricultural Chemicals Production:
  Pesticide Active Ingredient Production
  4-Chloro-2-Methylphenoxyacetic Acid
    Production
  2,4-D Salts and Esters Production
  4,6-Dinitro-o-cresol Production
  Butadiene-Furfural Cotrimer (R-11)
    Production
  Captafol Production
  Captan Production
  Chloroneb Production
  Chlorothalonil Production
  Dacthal™ Production
  Sodium Pentachlorophenate Production
  Tordon™ Acid Production

Fibers Production Processes:
  Acrylic Fibers/Modacrylic Fibers
    Production
  Rayon Production
  Spandex Production

Food and Agriculture Processes:
  Baker's Yeast Manufacturing
  Manufacturing of Nutritional Yeast
  Cellulose Food Casing Manufacturing
  Vegetable Oil Production

Pharmaceutical Production Processes:
  Pharmaceuticals Production

Polymers and Resins Production:
  Acetal Resins Production
  Acrylonitrile-Butadiene-Styrene
    Production
  Alkyd Resins Production
  Amino Resins Production
  Boat Manufacturing
  Butyl Rubber Production
  Carboxymethylcellulose Production
  Cellophane Production
  Cellulose Ethers Production
  Epichlorohydrin Elastomers Production
  Epoxy Resins Production
  Ethylene-Propylene Rubber Production
  Flexible Polyurethane Foam Production
  Hypalon™ Production
  Maleic Anhydride Copolymers
    Production
  Methylcellulose Production
  Methyl Methacrylate-Acrylonitrile-
    Butadiene-Styrene Production
  Methyl Methacrylate-Butadiene-Styrene
    Terpolymers Production

  Neoprene Production
  Nitrile Butadiene Rubber Production
  Nitrile Resins Production
  Non-Nylon Polyamides Production
  Phenolic Resins Production
  Polybutadiene Rubber Production
  Polycarbonates Production
  Polyester Resins Production
  Polyether Polyols Production
  Polyethylene Terephthalate Production
  Polymethyl Methacrylate Resins
    Production
  Polystyrene Production
  Polysulfide Rubber Production
  Polyvinyl Acetate Emulsions
    Production
  Polyvinyl Alcohol Production
  Polyvinyl Butyral Production
  Polyvinyl Chloride and Copolymers
    Production
  Reinforced Plastic Composites
    Production
  Styrene-Acrylonitrile Production
  Styrene-Butadiene Rubber and Latex
    Production

Production of Inorganic Chemicals:
  Ammonium Sulfate Production—
    Caprolactam By-Product Plants
  Antimony Oxides Manufacturing
  Carbon Black Production
  Chlorine Production
  Cyanide Chemicals Manufacturing
  Cyanuric Chloride Production
  Fumed Silica Production
  Hydrochloric Acid Production
  Hydrogen Cyanide Production
  Hydrogen Fluoride Production
  Phosphate Fertilizers Production
  Phosphoric Acid Manufacturing
  Quaternary Ammonium Compounds
    Production
  Sodium Cyanide Production
  Uranium Hexafluoride Production

Production of Organic Chemicals:
  Ethylene Processes
  Quaternary Ammonium Compounds
    Production
  Synthetic Organic Chemical
    Manufacturing
  Tetrahydrobenzaldehyde Production

TABLE 3.1   Categories of Sources of Hazardous Air Pollutants (*Continued*)

Industry group source category (Revision date: November 18, 1999) (*Continued*)	
Miscellaneous Processes:	Leather Tanning and Finishing Operations
Aerosol Can-Filling Facilities	OBPA/1,3-Diisocyanate Production
Benzyltrimethylammonium Chloride	Paint Stripper Users
Production	Paint Stripping Operations
Butadiene Dimers Production	Photographic Chemicals Production
Carbonyl Sulfide Production	Phthalate Plasticizers Production
Cellulosic Sponge Manufacturing	Plywood and Composite Wood Products
Chelating Agents Production	Plywood/Particle Board Manufacturing
Chlorinated Paraffins Production	Polyether Polyols Production
Chromic Acid Anodizing	Pulp and Paper Production
Commercial Dry Cleaning	Rocket Engine Test Firing
(Perchloroethylene)—Transfer	Rubber Chemicals Manufacturing
Machines	Rubber Tire Manufacturing
Commercial Sterilization Facilities	Semiconductor Manufacturing
Decorative Chromium Electroplating	Symmetrical Tetrachloropyridine
Dodecanedioic Acid Production	Production
Dry Cleaning (Petroleum Solvent)	Tetrahydrobenzaldehyde Production
Ethylidene Norbornene Production	Tire Production
Explosives Production	
Flexible Polyurethane Foam Fabrication	*Categories of Area Sources:*
Operations	Chromic Acid Anodizing
Friction Products Manufacturing	Commercial Dry Cleaning
Halogenated Solvent Cleaners	(Perchloroethylene)—Dry-to-Dry
Hard Chromium Electroplating	Machines
Hydrazine Production	Commercial Dry Cleaning
Industrial Cleaning	(Perchloroethylene)—Transfer
(Perchloroethylene)—Dry-to-Dry	Machines
Machines	Commercial Sterilization Facilities
Industrial Dry Cleaning	Decorative Chromium Electroplating
(Perchloroethylene)—Transfer	Halogenated Solvent Cleaners
Machines	Hard Chromium Electroplating
Industrial Process Cooling Towers	Secondary Lead Smelting

There are two types of stationary sources that generate routine emissions of air toxics: *major* sources and *area* sources. *Major* sources are defined as sources that emit 10 tons per year of any of the listed toxic air pollutants, or 25 tons per year of a mixture of air toxics. Examples of major sources include chemical plants, pharmaceutical plants, petroleum refineries, hazardous waste incinerators, and steel mills (Table 3.1). Major sources may discharge HAPs from stacks, vents, or leaking equipment, or during materials handling. *Area* sources consist of smaller sources, each releasing smaller amounts of HAPs into the atmosphere. Area sources are defined as sources that emit less than 10 tons per year of a single air toxic, or less than 25 tons per year of a combination of air toxics. Examples include neighborhood dry cleaners and gas stations. Though emissions from individual sources within an area might be small, when there are a large number of sources in an area, they may pose a collective health concern.

TABLE 3.2 List of 188 HAPs by CAS Number

Chemical name	CAS no.	Chemical name	CAS no.
Acetaldehyde	75-07-0	Chlorine	7782-50-5
Acetamide	60-35-5	Chloroacetic acid	79-11-8
Acetonitrile	75-05-8	Chloroacetophenone (2-)	532-27-4
Acetophenone	98-86-2	Chlorobenzene	108-90-7
Acetylaminofluorene (2-)	53-96-3	Chlorobenzilate	510-15-6
Acrolein	107-02-8	Chloroform	67-66-3
Acrylamide	79-06-1	Chloromethyl methyl ether	107-30-2
Acrylic acid	79-10-7	Chloroprene	126-99-8
Acrylonitrile	107-13-1	Chromium compounds	0
Allyl chloride	107-05-1	Cobalt compounds	0
Aminobiphenyl (4-)	92-67-1	Coke oven emissions	0
Aniline	62-53-3	Cresol (m-)	108-39-4
Anisidine (o-)	90-04-0	Cresol (o-)	95-48-7
Antimony compounds	0	Cresol (p-)	106-44-5
Arsenic compounds (inorganic including arsine)	0	Cresols/cresylic acid (isomers and mixture)	1319-77-3
Asbestos	1332-21-4	Cumene	98-82-8
Benzene (including benzene from gasoline)	71-43-2	Cyanide compounds	0
		D (2,4-), salts and esters	94-75-7
Benzidine	92-87-5	DDE	3547-04-4
Benzotrichloride	98-07-7	Diazomethane	334-88-3
Benzyl chloride	100-44-7	Dibenzofurans	132-64-9
Beryllium compounds	0	Dibromo-3-chloropropane (1,2-)	96-12-8
beta-Propiolactone	57-57-8	Dibutylphthalate	84-74-2
Biphenyl	92-52-4	Dichlorobenzene(p) (1,4-)	106-46-7
Bis(2-ethylhexyl)phthalate (DEHP)	117-81-7	Dichlorobenzidene (3,3-)	91-94-1
Bis(chloromethyl)ether	542-88-1	Dichloroethyl ether (bis(2-chloroethyl)ether)	111-44-4
Bromoform	75-25-2	Dichloropropene (1,3-)	542-75-6
Butadiene (1,3-)	106-99-0	Dichlorvos	62-73-7
Cadmium compounds	0	Diethanolamine	111-42-2
Calcium cyanamide	156-62-7	Diethyl sulfate	64-67-5
Captan	133-06-2	Dimethoxybenzidine (3,3-)	119-90-4
Carbaryl	63-25-2	Dimethyl aminoazobenzene	60-11-7
Carbon disulfide	75-15-0	Dimethyl benzidine (3,3-)	119-93-7
Carbon tetrachloride	56-23-5	Dimethyl carbamoyl chloride	79-44-7
Carbonyl sulfide	463-58-1	Dimethyl formamide	68-12-2
Catechol	120-80-9	Dimethyl hydrazine (1,1-)	57-14-7
Chloramben	133-90-4	Dimethyl phthalate	131-11-3
Chlordane	57-74-9	Dimethyl sulfate	77-78-1

TABLE 3.2   List of 188 HAPs by CAS Number (*Continued*)

Chemical name	CAS no.	Chemical name	CAS no.
Dinitro-o-cresol (4,6-), and salts	534-52-1	Lead compounds	0
Dinitrophenol (2,4-)	51-28-5	Lindane (all isomers)	58-89-9
Dinitrotoluene (2,4-)	121-14-2	Maleic anhydride	108-31-6
Dioxane		Manganese compounds	0
(1,4-)(1,4-diethyleneoxide)	123-91-1	Mercury compounds	0
Diphenylhydrazine (1,2-)	122-66-7	Methanol	67-56-1
Epichlorohydrin		Methoxychlor	72-43-5
(1-chloro-2,3-epoxypropane)	106-89-8	Methyl bromide (bromomethane)	74-83-9
Epoxybutane (1,2-)	106-88-7	Methyl chloride (chloromethane)	74-87-3
Ethyl acrylate	140-88-5	Methyl chloroform	
Ethyl benzene	100-41-4	(1,1,1-trichloroethane)	71-55-6
Ethyl carbamate (urethane)	51-79-6	Methyl ethyl ketone	
Ethyl chloride (chloroethane)	75-00-3	(2-butanone)	78-93-3
Ethylene dibromide		Methyl hydrazine	60-34-4
(dibromoethane)	106-93-4	Methyl iodide (iodomethane)	74-88-4
Ethylene dichloride		Methyl isobutyl ketone (hexone)	108-10-1
(1,2-dichloroethane)	107-06-2	Methyl isocyanate	624-83-9
Ethylene glycol	107-21-1	Methyl methacrylate	80-62-6
Ethylene imine (aziridine)	151-56-4	Methyl tert butyl ether	1634-04-4
Ethylene oxide	75-21-8	Methylene bis (4,4-)	
Ethylene thiourea	96-45-7	(2-chloroaniline)	101-14-4
Ethylidene dichloride		Methylene chloride	
(1,1-dichloroethane)	75-34-3	(dichloromethane)	75-09-2
Fine mineral fibers	0	Methylene diphenyl diisocyanate	
Formaldehyde	50-00-0	(MDI)	101-68-8
Glycol ethers	0	Methylenedianiline (4,4-)	101-77-9
Heptachlor	76-44-8	N,N-diethyl aniline	
Hexachlorobenzene	118-74-1	(N,N-dimethylaniline)	121-69-7
Hexachlorobutadiene	87-68-3	N-nitroso-N-methylurea	684-93-5
Hexachlorocyclopentadiene	77-47-4	N-nitrosodimethylamine	62-75-9
Hexachloroethane	67-72-1	N-nitrosomorpholine	59-89-2
Hexamethylene-1,6-diisocyanate	822-06-0	Naphthalene	91-20-3
Hexamethylphosphoramide	680-31-9	Nickel compounds	0
Hexane	110-54-3	Nitrobenzene	98-95-3
Hydrazine	302-01-2	Nitrobiphenyl (4-)	92-93-3
Hydrochloric acid	7647-01-0	Nitrophenol (4-)	100-02-7
Hydrogen fluoride (hydrofluoric		Nitropropane (2-)	79-46-9
acid)	7664-39-3	Parathion	56-38-2
Hydroquinone	123-31-9	Pentachloronitrobenzene	
Isophorone	78-59-1	(quintobenzene)	82-68-8

TABLE 3.2 List of 188 HAPs by CAS Number (*Continued*)

Chemical name	CAS no.	Chemical name	CAS no.
Pentachlorophenol	87-86-5	Tetrachloroethylene (perchloroethylene)	127-18-4
Phenol	108-95-2	Titanium tetrachloride	7550-45-0
Phenylenediamine (p-)	106-50-3	Toluene	108-88-3
Phosgene	75-44-5	Toluene diamine (2,4-)	95-80-7
Phosphine	7803-51-2	Toluene diisocyanate (2,4-)	584-84-9
Phosphorus	7723-14-0	Toluidine (o-)	95-53-4
Phthalic anhydride	85-44-9	Toxaphene (chlorinated camphene)	8001-35-2
Polychlorinated biphenyls (aroclors)	1336-36-3	Trichlorobenzene (1,2,4-)	120-82-1
Polycylic organic matter	0	Trichloroethane (1,1,2-)	79-00-5
Propane sultone (1,3-)	1120-71-4	Trichloroethylene	79-01-6
Propionaldehyde	123-38-6	Trichlorophenol (2,4,5-)	95-95-4
Propoxur (baygon)	114-26-1	Trichlorophenol (2,4,6-)	88-06-2
Propylene dichloride (1,2-dichloropropane)	78-87-5	Triethylamine	121-44-8
Propylene oxide	75-56-9	Trifluralin	1582-09-8
Propylenimine (2-methyl aziridine) (1,2-)	75-55-8	Trimethylpentane (2,2,4-)	540-84-1
Quinoline	91-22-5	Vinyl acetate	108-05-4
Quinone	106-51-4	Vinyl bromide	593-60-2
Radionuclides (including radon)	0	Vinyl chloride	75-01-4
Selenium compounds	0	Vinylidene chloride (1,1-dichloroethylene)	75-35-4
Styrene	100-42-5	Xylenes (isomers and mixture)	1330-20-7
Styrene oxide	96-09-3	Xylenes (m-)	108-38-3
Tetrachlorodibenzo-p-dioxin (2,3,7,8-)	1746-01-6	Xylenes (o-)	95-47-6
Tetrachloroethane (1,1,2,2-)	79-34-5	Xylenes (p-)	106-42-3

3.2 ESTIMATING EMISSIONS OF TOXIC AIR POLLUTANTS

Emissions estimates are important in the determination of applicable regulations and requirements that must be satisfied. If emissions are too high, project construction may not be allowed, and typically, the higher the emissions, the more extensive the requirements that must be satisfied. Emissions estimates of hazardous air pollutants can vary greatly in accuracy, depending on the methods used. Estimates are usually performed in sequential steps of increasing accuracy, with the first step being only a crude and conservative estimate of emissions. If regulatory requirements based on a crude estimate are acceptable, more accurate and more costly estimates are usually not required. If a first estimate results in costly control measures or other unacceptable requirements, a more accurate estimate is justified.

There are four accepted methods for estimating air pollutant emissions: (1) emissions factors, (2) engineering calculations, (3) material balances, and (4) emissions monitoring. Each emissions estimation technique has a range of accuracy and cost, but in general, the more accurate the estimate, the higher the cost (Fig. 3.1).

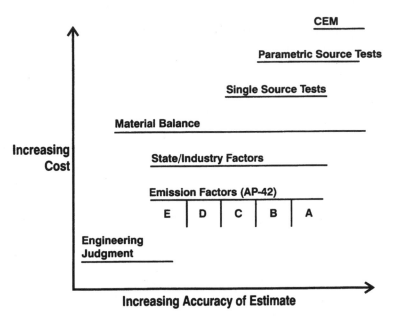

FIGURE 3.1 Accuracy and cost of different approaches to emission estimation. *[Source: U.S. EPA, Compilation of Air Pollutant Emission Factors (AP-42), 5th ed., PB 86-142906, 1995.]*

Emission factors for various pollutants and processes can be found in the U.S. EPA document *Compilation of Air Pollutant Emission Factors (AP-42)*, reference books, and journal articles. Emission factors are easy to use and are available for a wide variety of pollutants and processes. They also provide a quick, although often crude, estimate of air pollution emissions from a particular source. Often emission factors provide an adequate estimation of emissions for permitting purposes. Emission factors relate mass emission rate to production rate or use rate:

$$\text{Emission rate} = \text{emission factor} \times \text{activity data} \qquad (3.1)$$

As an example, AP-42 emission factors for charcoal manufacturing are presented in Table 3.3.

If process conditions vary widely from those used to obtain the emission factor, emission estimates using that factor may have a high degree of uncertainty. Emission factors obtained from AP-42 are averages of all data available of acceptable quality for a particular source type. They have been attached ratings from A to E, with A being the most reliable estimate and E being the least reliable. Emission factors obtained from AP-42 are generally considered to be conservative estimates for con-

TABLE 3.3 Uncontrolled Emission Factors for Charcoal Manufacturing*

Pollutant	Average emission factor	
	kg/Mg	lb/ton
Total PM[†]	160	310
CO	140	290
NO_x	12	24
CO_2	560	1100
VOC[‡]	140	270
Methane	54	110
Ethane	26	52
Methanol	76	150
POM	0.0047	0.0095

* Factor units are kg/Mg and lb/ton of product. All factors are for uncontrolled emissions. Manufacturing process used a charcoal kiln. The data have been given an EPA rating of E (poor).
† Includes condensibles and consists primarily of tars and oils; approximately 74% consists of organic material
‡ Consists primarily of methanol, acetic acid, formaldehyde, pyroacids, and unsaturated hydrocarbons.

trolled air pollution sources, meaning that they will probably overestimate the emissions from a particular source. This is probably because AP-42 emission factors are usually several years old and may not take into account advances in technology that have been used to lower emissions.

EMISSION RATE EXAMPLE *Estimate the uncontrolled volatile organic compound (VOC) emissions from a charcoal kiln producing 2.0 Mg (metric tons) of product per day.*

SOLUTION   From Table 3.3, the uncontrolled emission factor for VOCs from a charcoal kiln is 140 kg/Mg, so

$$E_{VOC} = (140 \text{ kg/Mg})(2.0 \text{ Mg/day}) = 280 \text{ kg VOC/day}$$

Note that the emission factors in Table 3.3 have a rating of E (poor). Since emission factors tend to provide conservative estimates of emissions, VOC emissions in this example have probably been overestimated.

3.3   OVERVIEW OF AIR QUALITY REGULATIONS

The most recent comprehensive revision to U.S. air quality regulations came with the 1990 Clean Air Act Amendments (CAAA) (Public Law 101-549, November 15, 1990). The 1990 CAAA contains 11 major divisions or titles, numbered I to XI (Table 3.4). The CAAA titles most applicable to toxic gas emissions are Titles I, III,

TABLE 3.4 Summary of Titles I to XI for CAAA of 1990

Title I	Provisions for Attainment and Maintenance of NAAQSs
Title II	Provisions Relating to Mobile Sources
Title III	Hazardous Air Pollutants
Title IV	Acid Deposition Control
Title V	Permits
Title VI	Stratospheric Ozone Protection
Title VII	Provisions Relating to Enforcement
Title VIII	Miscellaneous Provisions
Title IX	Clean Air Research
Title X	Disadvantaged Business Concerns
Title XI	Clean Air Employment Transition Assistance

IV, and V, which are summarized below along with other important concepts in the CAAA. Clean Air Act regulations are updated frequently in the *Federal Register* (FR), the *Code of Federal Regulations* (CFR), and the U.S. EPA Web site (*http:// www.epa.gov*). Additionally, state and local regions may have more strict air quality regulations than required by federal law.

3.3.1 Title I: Provisions for Attainment and Maintenance of NAAQS

Title I contains a summary of the provisions for nonattainment areas for the six *criteria pollutants* (Table 3.5) that have established National Ambient Air Quality Standards (NAAQS). Nonattainment areas are classified by the severity of their pollution problem and have attainment dates based on their nonattainment designation (Table 3.6). In 1990, Los Angeles was designated as *extreme* and had an attainment date of 2010. Baltimore and New York were classified with a rating of *severe* and given an attainment date of 2005. Chicago, Houston, Milwaukee, Muskegan, Philadelphia, and San Diego were also rated *severe,* but given attainment dates of 2007. Another 85 cities were also designated nonattainment with lower classification ratings and earlier attainment deadlines.

State and local agencies in nonattainment regions are required to develop comprehensive emissions inventory tracking systems to monitor air quality and progress toward reaching attainment. For ozone nonattainment areas, these inventories include VOCs, NO_x, and CO—all of which have been found to be precursors to ozone formation in the troposphere. All sources within an area must be inventoried, including stationary point sources, area sources, on-road and off-road motor vehicles, and biogenic (plant) sources. Specific control technologies are required for sources within the nonattainment area, with each class of severity requiring more stringent control measures. All ozone nonattainment areas must limit emissions of VOCs and NO_x to meet reasonably available control technology (RACT) guidelines and must update existing inspection and maintenance programs.

TABLE 3.5   National Ambient Air Quality Standards (NAAQSs)

Pollutant	Averaging time	Primary standard	Secondary standard	Measurement method
Carbon monoxide	8 h 1 h	10 mg/m^3 (9 ppm) 40 mg/m^3 (35 ppm)	Same	Nondispersive infrared photometry
Nitrogen dioxide	Annual average	100 µg/m^3 (0.05 ppm)	Same	Chemiluminescence
Sulfur dioxide	Annual average 24 h 3 h	80 µg/m^3 (0.03 ppm) 365 µg/m^3 (0.14 ppm) None	None None 1300 µg/m^3 (0.5 ppm)	Pararosaniline pulsed fluorescence
PM$_{10}$	AAM* 24 h	50 µg/m^3 150 µg/m^3	Same Same	PM$_{10}$ sampler
PM$_{2.5}$†	AAM* 24 h	15 µg/m^3 65 µg/m^3	Same Same	PM$_{2.5}$ sampler
Ozone†	1 h 8 h	240 µg/m^3 (0.12 ppm) 160 µg/m^3 (0.08 ppm)	Same Same	Chemiluminescence, UV absorption
Lead	3 months	1.5 µg/m^3	Same	Extraction/ AA spectroscopy

* AAM—annual arithmetic mean, determined by averaging PM concentrations for the past 3 calendar years. A violation occurs when the expected annual arithmetic mean is greater than the standard (rounded to the nearest 1 µg/m^3).
† Implementation pending action on court decision [American Trucking Association Inc. versus U.S. EPA, No. 97-1440 and 97-1441 (D.C. Cir. May 14, 1999)].

3.3.2   National Emission or Performance Standards

Emission standards place a limit on the mass or the concentration of a pollutant emitted from a source. Emission standards are used to maintain or improve air quality within a region by regulating individual sources or entire industries. Emission standards can be promulgated by federal or state government, but state regulations must be as strict or stricter than federal regulations. Performance standards are usually based on the maximum control technology presently available within an industry, and

TABLE 3.6   Classifications for Nonattainment Areas

	Classification	Design value, ppmv	Attainment deadline	Major source, tons VOC or CO/year
Ozone	Marginal Moderate Serious Severe I Severe II Extreme	0.121–0.138 0.138–0.160 0.160–0.180 0.180–0.190 0.190–0.280 >0.280	Nov. 15, 1993 Nov. 15, 1996 Nov. 15, 1999 Nov. 15, 2005 Nov. 15, 2007 Nov. 15, 2010	100 100 50 25 25 10
Carbon monoxide	Moderate Serious	9.1–16.4 >16.4	Dec. 31, 1995 Dec. 31, 2000	— 50

incorporate both technological and economic feasibility. Emission standards tend to become more strict over time as technology and economics improve. Emission standards can be grouped into the following types:

- *Visible emission standards*—Measurement of opacity of stack plumes or areas.
- *Concentration standards*—Maximum allowable emission rate of pollutant in units of mass/volume (e.g., grams per dry standard cubic meter, g/dscm) or volume/volume (parts per million by volume, ppmv). Concentrations may be corrected for combustion conditions and reported at a fixed O_2 or CO_2 content, so that dilution cannot be used as a means of lowering concentration.
- *Mass standards*—Maximum allowable emission rate of pollutant in terms of mass of material processed or produced, e.g., g particulate material per kg of product.
- *Zoning restrictions*—Limits emissions in a certain area by regulating the types of facilities that can be constructed.
- *Fuel standards*—Certain types of fuel may be specified to lower emissions, e.g., low-sulfur fuels or nonleaded fuel.
- *Dispersion-based standards*—Emissions may be limited on the basis of their contribution to the overall ambient air quality within a region.

The main sections of the *Code of Federal Regulations* that contain sections about emission standards are:

40 CFR 60: *Standards of Performance for New Stationary Sources (NSPS)*

40 CFR 86: *Control of Air Pollution from New Motor Vehicles and New Motor Vehicle Engines (Mobile Source Emissions)*

40 CFR 87: *Control of Air Pollution from Aircraft and Aircraft Engines*

40 CFR 61: *National Emission Standards for Hazardous Air Pollutants (NESHAP) (Title III, section 112)*

40 CFR 70: *State Operating Permits (Title V)*

40 CFR 52: *Prevention of Significant Deterioration (PSD)*

3.3.3 Best Available Control Technology

Best available control technology (BACT) is an emission limitation based on the maximum degree of reduction for each pollutant considering energy, environmental, and economic impacts, and is implemented through the application of production processes or available methods, systems, and techniques. BACT must be applied to any major source subject to prevention of significant deterioration (PSD) requirements in attainment areas. BACT is determined on a case-by-case basis, but in no case shall application of BACT result in emissions of any pollutant that would exceed emissions allowed under NSPS (40 CFR 60) or NESHAPs (40 CFR 61).

3.3.4 Reasonably Available Control Technology

Reasonably available control technology (RACT) refers to air pollution control devices or process modifications that are considered reasonably available when accounting for social, economic, and environmental impacts. RACT is applied when a

State Implementation Plan (SIP) calls for reduction in emissions of existing sources in nonattainment areas in order to progress toward attainment. RACT requirements are never more strict than BACT, and are usually less stringent, since they apply to retrofits of existing sources.

3.3.5 Lowest Achievable Emission Rate

The lowest achievable emission rate (LAER) refers to the level of control required of a major source subject to new source review (NSR) requirements for nonattainment areas. The LAER requirement applies only to the criteria pollutants for which the region is designated as nonattainment. LAER employs the most stringent emissions limitation contained within the implementation plan of any state for that stationary source category. The LAER must be met by the operator of the facility unless it can be demonstrated that such emissions limitations are not achievable.

3.3.6 Maximum Achievable Control Technology

Maximum achievable control technology (MACT) standards are covered in Sec. 112(d) of the Clean Air Act. They are developed by the U.S. EPA and are based on emissions levels already achieved by the best-performing similar facilities. MACT standards are performance-based and are deemed to be a reasonable and effective approach to reduce emissions of HAPs. When developing a MACT standard for a source category, the EPA examines the pollutant emissions achieved by the best-performing similar sources using any combination of control devices, clean processes, or other methods. These emission levels are then taken as the baseline for the new standard. The MACT standards must achieve at least the MACT baseline emission level throughout the industry. MACT standards are determined for a category or subcategory of sources in their entirety and not on an individual bases. The EPA may establish a more stringent standard depending on economic, environmental, and health considerations. For categories with 30 or more existing sources, the MACT baseline must equal the average emissions limits achieved by the best-performing 12 percent of sources within the source category. For categories with fewer than 30 existing sources, the MACT baseline must equal the average emission limits achieved by the best-performing five sources in the category. For new sources, the MACT baseline must equal the controlled emissions level currently achieved by the best-controlled similar source. Currently, MACT control levels are generally 90 percent or greater for organic species and 95 percent or greater for particulate material.

3.3.7 New Source Performance Standards

New source performance standards (NSPS) are the maximum allowable emissions for a new or modified source. The NSPS program was authorized in Sec. 111 of the 1970 Clean Air Act. The goal of NSPS is to prevent new air pollution problems, and it results in improvements in air quality as older existing plants are replaced by new facilities. NSPS were promulgated for categories of stationary sources that significantly cause or contribute to air pollution that could reasonably be anticipated to endanger public health or welfare. Currently there are about 75 categories of NSPS standards (Table 3.7), and modifications are published in 40 CFR 60 and on the EPA Web site.

TABLE 3.7   New Source Performance Standards Source Categories

Ammonium sulfate manufacture	Petroleum refinery wastewater systems
Asphalt processing and asphalt roofing manufacture	Phosphate fertilizer industry: diammonium phosphate plants
Automobile and light-duty truck surface coating operations	Phosphate fertilizer industry: granular triple superphosphate storage facilities
Beverage can surface coating industry	Phosphate fertilizer industry: superphosphoric acid plants
Bulk gasoline terminals	Phosphate fertilizer industry: wet-process phosphoric acid plants
Calciners and dryers in mineral industries	Phosphate rock plants
Coal preparation plants	Polymer manufacturing industry
Electric utility steam-generating units	Polymeric coating of supporting substrates facilities
Equipment leaks from on-shore natural gas processing plants	Portland cement plants
Equipment leaks in petroleum refineries	Pressure-sensitive tape and label surface coating operations
Equipment leaks in synthetic organic chemical manufacturing industry	Primary aluminum reduction plants
Ferroalloy production facilities	Primary copper smelters
Flexible vinyl and urethane coating and printing	Primary emissions from basic oxygen process furnaces
Fossil-fuel-fired steam-generating units	Primary lead smelters
Glass manufacturing plants	Primary zinc smelters
Grain elevators	Rubber tire manufacturing industry
Graphic arts industry: publication rotogravure printing	Secondary brass and bronze production plants
Hot asphalt facilities	Secondary emissions from basic oxygen process steel-making facilities
Incinerators	Secondary lead smelters
Industrial surface coating of plastic parts for business machines	Sewage treatment plants
Industrial surface coating of large appliances	Small industrial-commercial steam-generating units
Industrial-commercial-institutional steam generating units	Stationary gas turbines
Kraft pulp mills	Steel plants: electric arc furnaces
Lead-acid battery manufacturing plants	Storage vessels for petroleum liquids
Lime manufacturing plants	Sulfuric acid plants
Magnetic tape coating facilities	Sulfuric acid production units
Metal coal surface coating	Surface coating of metal furniture
Metallic mineral processing plants	Synthetic organic chemical manufacturing industry air oxidation and unit processes
Municipal solid waste landfills	Synthetic organic chemical manufacturing industry distillation operations
Municipal waste combustors	Sythetic organic chemical manufacturing reactor processes
New residential wood heaters	Volatile organic liquid storage vessels
Nitric acid plants	Wool fiberglass insulation manufacturing plants
Nonmetallic mineral processing plants	
On-shore natural gas processing; SO_2 emissions	
Petroleum dry cleaners	
Petroleum refineries	

3.3.8 Prevention of Significant Deterioration

Prevention of significant deterioration (PSD) air quality requirements are designed to ensure that air quality in clean areas (areas in attainment) will not degrade even as new pollution sources are constructed. The PSD program applies to new major sources and major modifications to existing sources. Under the PSD program, a major source must be one of the presently listed 28 categories of sources, and have the potential to emit 100 tons or more per year of any regulated air pollutant, or be any other type of source with the potential to emit 250 tons or more per year. Any new project that is subject to PSD rules must apply for a PSD permit. Applications for PSD permits include extensive information about the proposed project and must demonstrate that the air quality impacts from the project are within PSD guidelines. To comply with PSD provisions, the new source must employ best available control technology (BACT) for all regulated air pollutants, and show that introduction of the new source or modification of the existing source will not cause any violations of NAAQS.

PSD regulations also established the concepts of classes of air quality control regions (AQCR) and of incremental pollution. All air quality regions in the United States are designated one of three classes. These classes determine allowable PSD pollution increments within those regions. The three AQCR classes are:

- *Class I:* Pristine areas, including national parks and wilderness areas, where very little deterioration of air quality is allowed.
- *Class II:* Areas where moderate change in air quality is allowed, but where stringent air quality constraints are desirable.
- *Class III:* Areas where major growth and industrialization are allowed; typically, large metropolitan areas.

The concept of incremental pollution specifies the amount of additional ambient pollution that would be allowed in an area while still retaining its classification (Table 3.8). One source cannot use all of the total pollution increment provided to a given area, where the total increment is the sum of all new growth or pollutant sources.

TABLE 3.8 PSD Increments

	Maximum allowable increase, $\mu g/m^3$		
Pollutant	Class I	Class II	Class III
Particulate matter:			
TSP, annual arithmetic mean	5	19	37
TSP, 24-h maximum	10	37	75
Particulate matter:			
PM_{10}, annual arithmetic mean	4	17	34
PM_{10}, 24-h maximum	8	30	60
Sulfur dioxide:			
Annual arithmetic mean	2	20	40
24-h maximum	5	91	182
3-h maximum	25	512	700
Nitrogen dioxide:			
Annual arithmetic mean	2.5	25	50

3.3.9   Title III: Hazardous Air Pollutants

Hazardous or toxic air pollutants are covered in the Clean Air Act, Sec. 112(b), and in the *Code of Federal Regulations,* 40 CFR 61. As mentioned earlier, substances can be added or deleted from the HAPs list by the U.S. EPA. Additionally, state and local air pollution agencies can add substances to the HAPs list for their jurisdictions. Pollutants on the HAPs list must be primary pollutants, meaning that they are emitted directly into the atmosphere and not formed via chemical reaction in the atmosphere. They may exist either in particulate or gaseous form.

3.3.10   Title IV: Acid Deposition Control

Title IV of the 1990 CAAA requires the regulation of air pollutants that are precursors to acid rain. This includes regulating emissions of oxides of sulfur (SO_2) and oxides of nitrogen (NO_x). The acid deposition control program is designed to achieve a 10 million U.S. short ton reduction in SO_2 emissions from 1980 levels. This reduction is to be fully implemented in the year 2000, at which time it will place a cap of 8.9 million U.S. short tons on SO_2 emissions. Plants emitting SO_2 are allowed some flexibility as to how they will achieve reductions in SO_2 emissions. They may use a combination of control technologies, switching to lower-sulfur fuels, and trading of emission allowances.

The final rule for NO_x reduction was published in 1994 (40 CFR 76) and is designed to achieve a 1.8 million U.S. short ton per year reduction in NO_x emissions. NO_x emission limits depend on boiler type, but they are generally about 0.5 lb NO_x per million Btu. As with SO_2 control, NO_x emissions may be limited by changing combustion practices, e.g., using low-NO_x burners, or by end-of-pipe control measures, e.g., catalytic control.

3.3.11   Title V: Air Pollution Permits

The 1990 Clean Air Act Amendments established a national permitting program to control air pollution emissions from an estimated 40,000 stationary sources. The U.S. EPA has the authority for assuring compliance under Title V, but in most states, programs have been delegated by the EPA to state, regional, or local agencies. If the program is at least as stringent as what federal law requires, EPA can, and usually does, delegate the program to the state or regional agency. It is up to the facility obtaining the permit to determine whether or not the agency granting the permit has been delegated the authority by the EPA. If the agency has not been delegated authority, it is probably necessary to obtain an additional permit from EPA.

Title V requires the following sources to submit permits:

- Major sources as determined under Title I: (1) ≥ 100 U.S. short tons per year and a listed pollutant, excluding CO, or (2) ≥ 10 to 100 U.S. short tons per year of sources in nonattainment areas, depending on the classification of marginal to extreme.
- All NSPS, PSD, and NESHAPs sources.

- Major sources as determined under Title III: (1) ≥ 10 U.S. short tons per year of any air toxic, or (2) ≥ 25 U.S. short tons per year of multiple air toxics.
- Sources affected under Title V.
- Sources emitting ≥ 100 U.S. short tons per year of ozone-depleting substances regulated under Title VI.
- Other sources required by state, federal, or regional agencies to have operating permits.

Title V permits are issued for a period of up to five years. They must include all applicable Clean Air Act requirements, a compliance schedule, and monitoring and reporting requirements. In general, there are five major steps in the air permit application process (Shrock, 1994):

1. Develop a comprehensive emissions inventory for the affected source, including point, area, and fugitive emissions.
2. Develop a database management system to record, report, and periodically update the information collected in step 1.
3. Perform a compliance audit to determine which pre-1990 Clean Air Act Regulations and programs are applicable.
4. Prepare an enhanced monitoring and compliance certification protocol for applicable emissions
5. Prepare and submit an application for the Title V operating permit to a state or regional agency.

A flowsheet of the air permitting process is presented in Fig. 3.2.

3.3.12 Regulatory Direction for Hazardous Air Pollutants

Guided by the 1990 CAAA, the U.S. EPA has focused most of its initial HAPs control efforts on reducing emissions by setting technology-based MACT standards. After MACT standards have been developed for an industry, existing facilities have 3 years from the date a MACT standard is finalized to comply with its requirements. New sources must be in compliance at start-up. As MACT standards are developed and implemented for more industries, it is estimated that by 2010, emissions of toxic air pollutants will be reduced by about 75 percent from 1990 levels. EPA anticipates that a technology-based approach will continue to prove to be successful at reducing HAPs. The 1990 CAAA calls for the EPA to supplement its technology-based approach by assessing the effectiveness of MACT standards at reducing health and environmental risks posed by air toxics. On the basis of this assessment, the EPA may implement additional standards to address remaining residual risk posed by air toxics. After the EPA sets a MACT standard, it has 8 years to review the risk posed by continued emissions from MACT-regulated facilities. During that time, it issues requirements for additional air pollution control measures, if they are necessary to reduce an unacceptable residual risk.

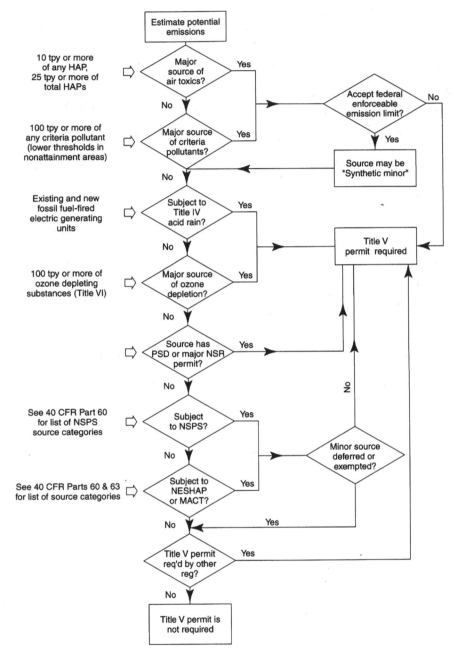

10 tpy or more
of any HAP,
25 tpy or more of
total HAPs

100 tpy or more of
any criteria pollutant
(lower thresholds in
nonattainment areas)

Existing and new
fossil fuel-fired
electric generating
units

100 tpy or more of
ozone depleting
substances (Title VI)

See 40 CFR Part 60
for list of NSPS
source categories

See 40 CFR Parts 60 & 63
for list of source categories

FIGURE 3.2   Determination of applicability of CAAA Title V permitting requirements. *(Source: Shrock, J., "Five Steps Toward Title V Permitting," Environmental Protection, pp. 41–44, July, 1994. Copyright 1994 by Stephens Publishing Corp.)*

3.4   RISK ASSESSMENT

In many jurisdictions, health risk assessments are used to determine whether impacts from hazardous or toxic air pollutant emissions are acceptable. A multipathway health risk assessment is conducted to evaluate the potential for adverse health effects, such as cancer and chronic noncancerous effects. In the case of air pollution emissions from a facility, the health risk assessment should determine the risk of cancer due to pollutants emitted from the facility, determine the potential for chronic noncancerous illnesses, and evaluate the potential for acute noncancerous effects. To account for uncertainties, conservative assumptions are made during the risk analysis process. These assumptions are made to provide large safety factors. When estimating risk, values used in calculations are likely to be overestimated, meaning that it is unlikely that the risk will be underestimated. Risk assessments are usually made for a total 70-year average lifetime exposure, even though exposures to risks are usually for much shorter time periods, and assume maximum exposure during that time period.

Risk assessment can be divided into four components: contaminant identification, exposure assessment, dose-response assessment, and risk characterization. The first step in the risk assessment process is to identify the toxic air pollutants likely to be emitted from the source of interest and their estimated emissions rates. Sometimes a surrogate pollutant is selected, because it has a higher potential for causing cancer than other similar compounds in the group. For example, one could select benzo-a-pyrene as a surrogate for polyaromatic hydrocarbons emitted from low-temperature combustion processes. Using a surrogate simplifies calculations and requires less data. It also overestimates health risk, because it provides a conservative estimate.

The exposure assessment portion of the risk assessment processes includes dispersion modeling, environmental fate, and exposure estimates. Dispersion modeling is used to estimate pollutant concentrations in the ambient air due to the addition of the source and maximum ground-level concentrations. Fate models are used to predict the behavior of pollutants after they have been deposited in the soil or water. They include how a pollutant may decompose in soil or water and how it may concentrate in plants and animals. The exposure assessment is an estimate of total daily intake of pollutants from inhalation, ingestion, and dermal contact. It is usually expressed in units of mg toxic substance/kg body weight/day of exposure.

Dose-response relationships provide toxicological factors that predict the likelihood of cancer or other health effects occurring because of contaminant exposure. Dose-response relationships are developed in participation with the U.S. EPA, the Occupational Safety and Health Administration (OSHA), the Food and Drug Administration (FDA), the International Association for Research on Cancer (IARC), et al. Dose-response assessment determines the correlation between the magnitude of exposure in terms of the dose administered to the probability of occurrence of a health effect. To obtain this information, experiments are usually performed on animals with concentrations much higher than found in the environment. Extrapolation is then used to predict dose-response at much lower concentrations. In addition to uncertainties present in the extrapolation, uncertainties are also introduced when trying to relate health effects found in laboratory animals to equivalent health effects in humans. The reason that dose-assessment tests must currently be conducted in this manner is that one or more lifetimes, or several lifetimes in the case of genetic mutations, may be required to observe health impacts. When using models to extrapolate health effects from high to low dose and from animals to humans, large margins of safety are used to assure that actual risks are not underestimated.

Risk characterization is a quantitative measurement used to estimate the health effects from an individual pollutant or a number of pollutants. Risk characterization usually includes lifetime cancer risk, cancer incidence, acute noncancerous effects, and chronic noncancerous effects. Cancer risk for exposure to toxic air pollutants can be calculated for inhalation exposure by multiplying the pollutant concentration by a unit risk factor (URF):

$$\text{Risk} = (\text{concentration, } \mu\text{g/m}^3)(\text{Unit risk factor, m}^3/\mu\text{g}) \qquad (3.2)$$

where the unit risk factor is the product of the cancer potency and body weight (assumed to be 70 kg) divided by the human inhalation rate of 20 m³/day. Unit risk factors determined by the EPA are shown in Table 3.9. It should be noted that URFs are subject to change, and one should always consult a recent reference before making calculations. Risk is calculated for each pollutant emitted from a facility, and the total risk is the sum of the individual risks for each pollutant and each pathway. Although this approach for determining total risk is recommended by the EPA, it does not account for synergistic or antagonistic effects that occur when several pollutants are present. To compare risk due to toxic air pollutants with everyday risks, Table 3.10 contains activities estimated to increase a person's chance of dying in any year by one in a million.

TABLE 3.9 EPA Unit Risk Factors

Chemical	CAS number	Unit risk factor, $\text{m}^3/\mu\text{g}$
Acetaldehyde	75-07-0	3.96×10^{-6}
Acrylonitrile	107-13-1	4.80×10^{-5}
Arsenic	7740-38-2	2.70×10^{-1}
Benzene	71-43-2	2.65×10^{-5}
Benzo-a-pyrene		1.75×10^{-2}
1,3-Butadiene	106-99-0	1.90×10^{-5}
Carbon tetrachloride	56-23-5	9.44×10^{-5}
Chloroform	67-66-3	1.12×10^{-4}
1,1-Dichloroethylene	540-59-0	1.98×10^{-4}
Ethylene oxide	75-21-8	1.80×10^{-4}
Formaldehyde	50-00-0	1.60×10^{-5}
Gasoline		1.69×10^{-6}
Methylene chloride	75-09-2	1.63×10^{-6}
Trichloroethylene	79-01-6	9.28×10^{-5}
Vinyl chloride	75-01-4	1.05×10^{-5}

3.5   DISPERSION MODELING

Air impact analyses are required as part of the PSD permitting processes. As part of that process, dispersion modeling is performed in order to conduct a health risk as-

TABLE 3.10 Activities Estimated to Increase a Person's Chance of Dying in Any Year by 1 in a Million (10^{-6}), or a Lifetime Exposure of 7×10^{-5}

Activity	Risk
Smoking 1.4 cigarettes	Cancer, heart disease
Drinking 0.5 L of wine	Cirrhosis of the liver
Spending 1 h in coal mine	Black lung disease
Spending 3 h in coal mine	Accident
Traveling 10 mil by bicycle	Accident
Traveling 150 mil by car	Accident
Flying 1000 mil by airplane	Accident
Flying 6000 mil by airplane	Cancer caused by cosmic radiation
One chest x-ray	Cancer caused by radiation
Living 2 months with a cigarette smoker	Cancer, heart disease
Eating 40 tablespoons of peanut butter	Liver cancer caused by aflatoxin B, a natural carcinogen
Living 150 years within 20 miles of a nuclear power plant	Cancer caused by radiation
Eating 100 charcoal-broiled steaks	Cancer from benzo-a-pyrene

Source: Wilson, Richard, "Analyzing the Daily Risks of Life," *Technology Review,* February 1979.

sessment. This section will examine the basics of dispersion modeling. It should be noted that dispersion modeling is an area that requires substantial training, and although powerful and easy-to-use computer software is now available, it is best to seek professional expertise in this area.

When air pollutants are released into the atmosphere, the pollutants mix with the surrounding air. Turbulence in the atmosphere determines how pollutant plumes will rise and disperse within the atmosphere. Turbulence can be due to temperature gradients that cause the pollutants to move because of differences in the buoyancy of the plume and surrounding air, or turbulence may be generated by aerodynamic forces resulting from wind. Under light wind conditions, buoyancy effects dominate the formation of turbulence, while under higher wind speeds, aerodynamic forces dominate. Atmospheric turbulence is generally not measured, because of measurement difficulties. Instead, atmospheric turbulence is estimated by using other meteorological measurements. The most common surrogate for atmospheric turbulence is a parameter called *atmospheric stability class.* One common method of estimating stability class was developed by Turner (1964). Turner's method uses measurements of wind speed, cloud cover, time of day, and incoming solar radiation to estimate stability class (Table 3.11). In Turner's method, Class A is the most unstable condition and results in the greatest turbulence, while Class F, which occurs at night under low wind speed, is the most stable.

The most commonly used model to simulate the transport and dispersion of a pollutant plume in the atmosphere is the *Gaussian* dispersion model. The Gaussian model is derived from an analytical solution to the equations of motion in a fluid, and is considered semiempirical and semitheoretical. It is based on the assumption that pollutant concentrations in a dispersing plume obey a Gaussian or normal dis-

TABLE 3.11 Turner Conditions for Determining the Stability Class of the Atmosphere

Wind speed* u, m/s	Day, solar radiation			Night, cloudiness[†]	
	Strong[‡]	Moderate[§]	Slight[¶]	>4/8 cloud	<3/8 cloud
<2	A	A–B	B	E	F
2–3	A–B	B	C	E	F
3–5	B	B–C	C	D	E
5–6	C	C–D	D	D	D
>6	C	D	D	D	D

* Surface wind speed is measured 10 m above the ground.
[†] Cloudiness is defined as the fraction of sky covered by clouds.
[‡] Corresponds to clear summer day with the sun higher than 60° above the horizon.
[§] Corresponds to a summer day with a few broken clouds, or a clear day with sun 35–60° above horizon.
[¶] Corresponds to a fall afternoon, a cloudy summer day, or a clear summer day with the sun 15–35° above the horizon.
   Note: A, very unstable; B, moderately unstable; C, slightly unstable; D, neutral; E, slightly stable; F, stable. Regardless of wind speed, Class D should be assumed for overcast conditions, day or night
   Source: Turner, D. B., "Workbook of Atmospheric Dispersion Estimates," AP-26, Office of Air Programs, U.S. Department of Health, Education, and Welfare, Research Triangle, N.C., 1970.

tribution. In the Gaussian model, dispersion is a function of downwind distance, with greater dispersion occurring at greater downwind distances (Figure 3.3). Using the Gaussian plume equation, pollutant concentrations at any downwind distance can be estimated knowing the release rate of the pollutant, the release height of the pollutant, the wind speed, and the dispersion parameters. The Gaussian dispersion model can be expressed mathematically as:

$$C(x, y, z{:}H) = \frac{Q}{2\pi\sigma_x\sigma_y u} \exp\left[-\frac{1}{2}\left(\frac{y}{\sigma_y}\right)^2\right]$$
$$\times\left\{\exp\left[-\frac{1}{2}\left(\frac{z-H}{\sigma_z}\right)^2\right] + \exp\left[-\frac{1}{2}\left(\frac{z+H}{\sigma_z}\right)^2\right]\right\} \qquad (3.3)$$

where C = pollutant concentration (g/m³) as a function of position
        Q = pollutant emission rate (g/s)
        u = wind speed (m/s)
        σ_y = dispersion parameter in the y direction (m)
        σ_z = dispersion parameter in the z direction (m)
        H = effective stack height (m)

Since wind direction is not a variable in the above Gaussian dispersion model equation, it is assumed that pollutant concentration is measured downwind of the source, such that the x direction is always the downwind direction.

The derivation of the Gaussian model includes several assumptions (Viegele and Head, 1978):

• Airflow is continuous, steady, and uniform.

• The pollutant is released from a single elevated point source, and the emission rate is constant and continuous.

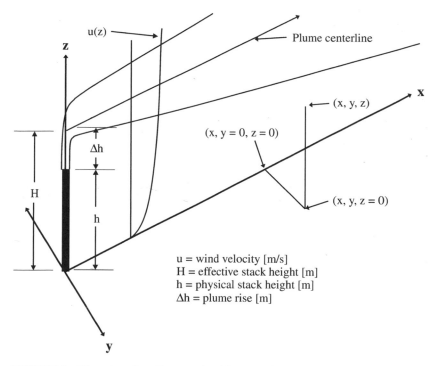

FIGURE 3.3 Dispersion of a pollutant emitted from a stationary source.

- Turbulence is uniform in the horizontal and vertical directions, but varies with downwind distance.
- Downwind transport on the mean wind dominates downwind dispersion and the wind speed cannot approach zero.
- Pollutants are not lost from the plume by chemical reactions, deposition, or other processes.

With these above limitations, the Gaussian model works best for areas of flat terrain, over short time periods, and up to distances of about 50 km.

As can be seen in the mathematical representation of the Gaussian model, stability class and downwind distance are not direct inputs. Instead, these variables are incorporated into the model through the dispersion parameters σ_y and σ_z. These parameters are functions of downwind distance and they are the assumed standard deviations of the horizontal and vertical Gaussian plume. Typically, Pasquill-Gifford dispersion coefficients are used for rural areas. The dispersion coefficients can be estimated by using coefficients that fit the experimental results. The form of the equations are presented below, and the values of the coefficients are presented in Table 3.12.

$$\sigma_y = ax^b \tag{3.4}$$

$$\sigma_z = cx^d + f \tag{3.5}$$

For urban sites, the values in Table 3.13 can be used to estimate the dispersion coefficients.

TABLE 3.12 Empirical Constants Used to Determine Values for Rural σ_y and σ_z*

Stability	a	$x \leq 1$ km			$x \geq 1$ km		
		c	d	f	c	d	f
A	213	440.8	1.941	9.27	459.7	2.094	−9.6
B	156	106.6	1.149	3.3	108.2	1.098	2.0
C	104	61.0	0.911	0	61.0	0.911	0
D	68	33.2	0.725	−1.7	44.5	0.516	−13.0
E	50.5	22.8	0.678	−1.3	55.4	0.305	−34.0
F	34	14.35	0.740	−0.35	62.6	0.180	−48.6

* The value of b is always 0.894, and x must be expressed in kilometers.

TABLE 3.13 Empirical Coefficients for Dispersion Cofficients for Urban Sites*

Stability class	σ_y, m	σ_z, m
A–B	$0.32x(1 + 0.0004x)^{-1/2}$	$0.24x(1 + 0.0001x)^{-1/2}$
C	$0.22x(1 + 0.0004x)^{-1/2}$	$0.20x$
D	$0.16x(1 + 0.0004x)^{-1/2}$	$0.14x(1 + 0.0003x)^{-1/2}$
E–F	$0.11x(1 + 0.0004x)^{-1/2}$	$0.08x(1 + 0.0015x)^{-1/2}$

* Downwind distance x measured in meters.
Source: Griffiths, R. F., "Errors in the Use of Briggs Parameterization for Atmospheric Dispersion Coefficients," *Atmospheric Environment,* vol. 28, no. 17, pp. 2861–2865, 1994.

Because a buoyant plume will initially rise, there will be some downwind distance before the pollutant can disperse to the ground. In general, the magnitude of the peak downwind pollutant concentration decreases with increased plume height and with decreased atmospheric turbulence (more stable conditions). For very unstable conditions, a pollutant plume can reach the ground close to the source, causing very high pollutant concentrations on the ground, since little dilution has occurred.

For most regulatory applications, the two desired results from dispersion modeling are the magnitude of the peak downwind concentration and its location. The peak concentration is compared to an ambient standard or allowable PSD increment. Additionally, dispersion modeling may be used to examine pollutant concentrations at areas in the community that may be susceptible to increased health risks, such as schools, hospitals, and nursing homes. It is important to understand that the Gaussian model only estimates downwind pollutant concentrations, and that it is an imperfect representation of reality. As a rule of thumb, and assuming correct implementation, the Gaussian model is accurate to within a factor of 2. In application of the Gaussian model, conservative estimates are usually made so that final downwind concentrations are overestimated. But care must be taken, because even with conservative estimates, the Gaussian model can still underpredict actual ambient pollutant concentrations.

More sophisticated applications of the Gaussian dispersion model have been implemented by using computer programs that overcome some of the limitations of the Gaussian model. The EPA Industrial Source Complex (ISC) model can accommodate multiple sources, complex terrain, nonuniform flow conditions, building wakes, and loss of pollutants through deposition. A variety of air pollution dispersion models are available on the EPA Technology Transfer Web site (*http://www.epa.gov/ttn*) and from commercial vendors. Commercial implementations of the Gaussian dispersion model are based on the EPA models, but include an MS Windows® interface, the ability to overlap maps and terrain, and the importation of large amounts of actual meteorological data. Commercial air pollution dispersion modeling software includes: Trinity Consultants' *Breeze Software* (*http://www.breeze-software.com/*), Lakes Environmental's *ISC-AEROMOD* (*http://www.lakes-environmental.com/*), and products offered by Bee-Line (*http://www.beeline-software.com/*).

3.6 AIR POLLUTION CONTROL TECHNOLOGIES: PARTICULATE CONTROLS

Particulate emissions are typically removed from industrial gas streams by one or more of the following control devices:

- Cyclones (centrifugal force)
- Wet collectors (diffusion, interception, and impaction)
- Fabric filters (diffusion, interception, and impaction)
- Electrostatic precipitators (electrostatic force)

The type of particulate control device that should be used for a specific application depends on many factors, including: gas flow rate, particle size distribution, particle loading, particle composition, gas temperature and pressure, desired collection efficiency, and acceptable capital and operating costs. For many applications, more than one type of control device may provide the desired collection efficiency, but they will most likely differ with regards to capital or operating costs. Therefore, economic factors must be considered along with technical issues. The following sections provide a general overview of the major types of particulate control devices. The reader is referred to the references section and to equipment manufacturers for more specific information.

3.6.1 Cyclones

Cyclones are inertial separators that use centrifugal force to remove particles from gas streams (Fig. 3.4). Standard cyclone designs can provide high removal efficiency (~100 percent) for particles greater than 20 μm, and particles with diameters greater than about 5 μm are usually removed with efficiencies greater than 50 percent. As particle size decreases from 5 μm, collection efficiency rapidly decreases, making cyclones ineffective for the removal of very small particles (< ~1 to 5 μm). Cyclones

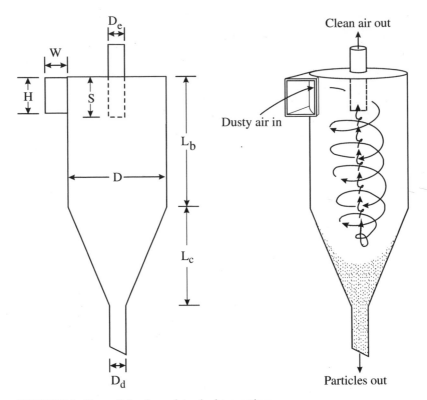

FIGURE 3.4 Tangential cyclone of standard proportions.

can be used as a stand-alone particulate control device or in conjunction with another device, such as an electrostatic precipitator or a fabric filter. When cyclones are used in series with another control device, they are used to pretreat the gas stream by removing larger particles. The second particulate control device is then used to collect the smaller particles at a much reduced particulate loading, and at a higher collection efficiency.

In a cyclone, the gas stream enters in a manner that causes it to spiral around the inside of the cyclone, causing particles with greater inertia to be forced to the outside walls of the cyclone. Upon collision with the outside walls of the cyclone, particles fall down the cone to the collection hopper below. There are two main methods to cause the gas to spin within the cyclone body: (1) introduce the gas to the cyclone in a tangential manner so that the gas curves around the inside body of the cyclone and (2) use axial vanes at the inlet of the cyclone, causing the gas to spin as it flows past the vanes. An advantage of the vane-axial cyclone is that it can be produced in small sizes, which causes the gas stream to make tighter turns, improving particle collection efficiency. A disadvantage of the vane-axial cyclone is that the pressure drop through the cyclone increases rapidly as the tangential velocity increases.

One way to maintain high collection efficiencies with a moderate pressure drop is to use a large number of small cyclones placed in parallel. When multiple cyclones are used in parallel, each cyclone can be of a smaller diameter than if only one cyclone were used. For a multicyclone assembly, the collection efficiency for each individual cyclone is identical, but it is greater than if only one larger cyclone were used.

Standard Cyclone Configuration. The cyclone removal efficiency for a given particle size is largely dependent on cyclone dimensions. Extensive research has been performed to determine how relative cyclone dimensions affect particle collection efficiency. Some general observations about cyclone design include:

• Pressure drop at a given volumetric flow rate is most affected by cyclone diameter.
• The overall length of the cyclone determines the number of turns of the gas stream, and the greater the number of turns, the greater the collection efficiency.
• As the size of the cyclone inlet decreases, the inlet velocity increases, thereby increasing particle collection efficiency, but also increasing pressure drop.

Several standard cyclone configurations have been proposed to make design calculations easier (Table 3.14). Cyclones are designed with geometric similarity such that the ratio of the dimensions of the cyclone remains constant and those dimensions are expressed in terms of the cyclone body diameter D. In addition to the conventional cyclone (also referred to as the Lapple standard conventional cyclone), standard designs have been developed for high-throughput, high-efficiency, and ultrahigh-efficiency cyclones. Performance data for a variety of cyclones can be obtained from Heumann (1997) and from equipment manufacturers. Typical collection efficiency curves for conventional, high-efficiency, and high-throughput cyclones are presented in Fig. 3.5.

TABLE 3.14 Standard Cyclone Dimensions*

Description	Conventional[†] (Lapple)	High throughput[‡]	High efficiency[§]
Body diameter D	$1.0D$	$1.0D$	$1.0D$
Height of inlet (H)	$0.5D$	$0.75D$	$0.5D$
Width of Inlet (W)	$0.25D$	$0.375D$	$0.2D$
Diameter of gas exit (D_e)	$0.5D$	$0.75D$	$0.5D$
Length of vortex finder (S)	$0.625D$	$0.875D$	$0.5D$
Body length L_b	$2.0D$	$1.5D$	$1.5D$
Cone length L_c	$2.0D$	$2.5D$	$2.5D$
Diameter of dust exit (D_d)	$0.25D$	$0.375D$	$0.375D$

* Adapted from Cooper and Alley (1994).
[†] Adapted from Lapple (1951).
[‡] Adapted from Stairmand (1951).
[§] Adapted from Swift (1969).

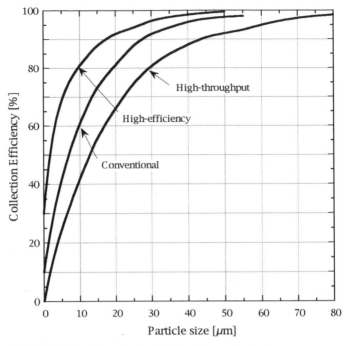

FIGURE 3.5 Particle collection efficiency for standard cyclones.

The collection efficiency for a cyclone of standard proportions can be obtained using a calibration curve or a curve-fitted equation. Cyclone calibration curves typically plot collection efficiency η as a function of $d_p/d_{p,50}$, where $d_{p,50}$ is the particle diameter that is collected at $\eta = 0.50$ (50 percent). As shown in Eq. (3.6), $d_{p,50}$ can be calculated given the geometry of the cyclone, density of the particles (ρ_p), gas viscosity μ_g, and gas volumetric flow rate Q_g.

$$d_{p,50} = \left(\frac{9\mu_g W^2 H}{2\pi \rho_p Q_g N_e} \right)^{1/2} \tag{3.6}$$

N_e represents the number of revolutions of the gas stream in the main outer vortex and can be calculated approximately from

$$N_e = \frac{1}{H}\left[L_b + \left(\frac{L_c}{2} \right) \right] \tag{3.7}$$

$d_p/d_{p,50}$ values can then be determined for the particle size distribution of interest. Values of $\eta(d_{pi})$ can then be read from a calculation curve for the calculated values of $d_p/d_{p,50}$, which are available in air quality handbooks and from equipment manufacturers, or by using a curve-fitted equation. Theodore and De Paola (1980) have fitted an algebraic equation to the Lapple standard conventional cyclone particle collection efficiency curve, making calculations much more convenient. The collection efficiency for a Lapple cyclone at any given particle size can be expressed as

$$\eta_i = \frac{1}{1 + (d_{p,50}/d_{pi})^2} \tag{3.8}$$

where η_i is the collection efficiency for a particle of size d_{pi} for a standard conventional cyclone, and d_{pi} is the average diameter of the ith particle size range.

Since emission limits for particulate matter are usually specified in terms of the total amount of particulate material released, it is useful to calculate the overall collection efficiency, η_T, for a distribution of particle sizes. The following formula for total collection efficiency can be used for any type of particulate control device:

$$\eta_T = \sum_i \frac{\dot{m}_{in}(d_{pi})}{\dot{m}_T} \eta(d_{pi}) \tag{3.9}$$

where $\quad \eta_T$ = total collection efficiency for the control device
$\dot{m}_{in}(d_{pi})$ = mass of particles at size d_{pi} entering the pollution control device
$\dot{m}_T$ = total mass of particles entering the pollution control device
$\eta(d_{pi})$ = collection efficiency for d_{pi}

Cyclone Pressure Drop. Besides collection efficiency, the other major design consideration for cyclones is pressure drop. A high pressure drop will increase the operating cost of a cyclone, because the fan will have to perform more work. In general, higher particle collection efficiencies are obtained by forcing the gas through the cyclone at higher velocities, resulting in increased pressure drop. Although cyclones are relatively easy to design, maintain, and install, and they are typically regarded as a low capital cost device, excessive pressure drop may make a cyclone prohibitively expensive when compared to other collection devices. This economic tradeoff must be considered in the design process.

Many equations have been developed to estimate the number of velocity heads or pressure drop in cyclones. An equation developed by Shepard and Lapple (1939; 1940) provides a reasonable estimate of the pressure drop across a cyclone. The pressure drop is presented as a function of K, which is an empirical constant and depends on cyclone configuration and operating conditions. K can vary considerably, but for standard tangential or involute cyclones, K is in the range of 12 to 18, and for vane-axial cyclones, K is usually taken to be 7.5. The pressure drop can be calculated from

$$\Delta P = \frac{u_g^2 \rho_g}{2} K \frac{HW}{D_e^2} \tag{3.10}$$

where $\quad \Delta P$ = pressure drop (N/m^2 or Pa)
u_g = superficial gas velocity at the cyclone inlet (m/s)
ρ_p = gas density (kg/m^3)
H, W, D_e = cyclone dimensions as described in Table 3.14

CYCLONE EXAMPLE *A Lapple standard conventional cyclone with a body diameter of 0.50 m is used to collect particles from a gas stream with a size distribution as follows:*

Size range, μm	Fraction in size range, % mass
0–4	5.0
4–10	8.0
10–20	32.0
20–40	35.0
40–80	18.0
>80	2.0

The particle density is 1200 kg/m³, the gas density is 1.183 kg/m³, the gas viscosity is 0.0666 kg/m-h, and the inlet gas velocity is 25 m/s. Determine the collection efficiency for each particle size range and the overall particle collection efficiency.

SOLUTION   For a Lapple standard cyclone,

$$H = 0.5D = 0.5(0.50 \text{ m}) = 0.25 \text{ m}$$

$$W = 0.25D = 0.25(0.50 \text{ m}) = 0.125 \text{ m}$$

The volumetric flow rate of the gas is

$$Q_g = U_g HW = (25 \text{ m/s})(0.25 \text{ m})(0.125 \text{ m}) = 0.78 \text{ m}^3/\text{s}$$

$d_{p,50}$ for the cyclone can be determined using Eq. (3.6)

$$d_{p,50} = \left(\frac{9\mu_g W^2 H}{2\pi\rho_p Q_g N_e} \right)^{1/2}$$

$$= \left[\frac{9\left(0.0666 \dfrac{\text{kg}}{\text{m-h}}\right)\left(\dfrac{1 \text{ h}}{3600 \text{ s}}\right)(0.25 \text{ m})^2(0.125 \text{ m})}{2\pi\left(1200\dfrac{\text{kg}}{\text{m}^3}\right)\left(0.78 \dfrac{\text{m}^3}{\text{s}}\right)(6)} \right]^{1/2} \left(\frac{10^6 \text{ μm}}{1 \text{ m}} \right) = 4.29 \text{ μm}$$

The collection efficiency for each particle size can be determined from (3.8) by using the average particle size for each range, and the overall collection efficiency can be determined from Eq. (3.9):

Size range, μm	d_{pi}, μm	m_i, mass fraction	η_i, %	$\eta_i m_i$, %
0–4	2	0.05	17.9	0.895
4–10	7	0.08	72.7	5.82
10–20	15	0.32	92.4	29.6
20–40	30	0.35	98.0	34.3
40–80	60	0.18	99.5	17.9
>80	80	0.02	99.7	1.99

$$\eta_T = \sum_i \eta_i m_i = 90.5\%$$

3.6.2 Particulate Scrubbers

Particulate or wet scrubbers contact a particle-laden gas stream with a liquid spray (Fig. 3.6). Particles are collected in wet scrubbers by liquid drops, or by being impacted on wetted surfaces. Soluble gases can also be removed by liquid droplets, and the gas temperature is lowered while the gas is humidified. Wet collection of particles has several advantages when compared to dry particle removal methods, including removal of sticky particles, removal of liquids, ability to handle hot gases, simultaneous gas and particle removal, and a reduced risk of dust explosion. Wet scrubbers have been used to control particles and gases from a variety of sources, including medical, hazardous and municipal waste incinerators, industrial boilers, acid plants, and limekilns.

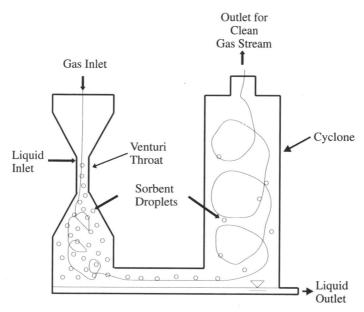

FIGURE 3.6 Venturi particle scrubber with cyclone.

Types of particulate scrubbers include spray tower, cyclonic, and venturi. Counter-current spray towers typically operate with scrubber droplets traveling downward and the gas stream containing the particulate matter traveling upward. Cyclonic scrubbers atomize droplets of water with a spray bar located along the centerline of the cyclone. These droplets then collect particles as they are transported to the outer edge of the cyclone. The liquid also allows cleansing of the walls of the cyclone. Venturi scrubbers work by accelerating the gas stream through a constricted duct to velocities of about 50 to 150 m/s. As the gas stream is flowing through the venturi, liquid is injected into the beginning of the venturi or at the throat entrance of the venturi. The liquid is then atomized by the high-velocity gas stream flowing past the inlets for the fluid. The high relative velocity between the scrubber droplets and particles allows for impaction of the particulate contaminants onto the scrubber droplets. Particulate material then becomes part of the scrubber droplets, allowing much easier removal of the contaminants from the gas stream because of the large particle size of the scrubber droplets. Scrubber droplets are typically removed from the gas stream by a centrifugal separator, such as a cyclone. The primary design parameters for wet scrubbers are pressure drop, particle size distribution, and liquid-to-gas flow rate. Typical operating characteristics for particulate wet scrubbers are presented in Table 3.15.

Although particles can be removed in wet scrubbers by using diffusion, interception, and impaction, wet scrubber models, and in particular venturi scrubber models, assume that the dominant particle collection mechanism is impaction. The optimal scrubber droplet diameter is about 500 to 1000 μm for impaction.

The particle penetration of a venturi scrubber (Fig. 3.6) can be calculated by an equation developed by Calvert et al. (1972):

$$Pt = \exp\left(\frac{6.3 \times 10^{-4}\rho_L\rho_p K_c d_p^2 u_g^2 \left(\dfrac{Q_L}{Q_G}\right)f^2}{\mu_g^2}\right) \tag{3.11}$$

where
Pt = particle penetration through the venturi scrubber
ρ_L = density of the liquid (g/cm^3)
ρ_p = density of the particles (g/cm^3)
K_c = Cunningham correction factor
d_p = particle diameter (cm)
u_g = gas velocity (cm/s)
Q_L/Q_G = liquid-to-gas flow rate ratio
μ_g = gas viscosity (g/cm-s)
f = experimental coefficient (varies from 0.1 to 0.5; is usually taken as 0.25 for hydrophobic particles and 0.5 for hydrophilic particles)

The collection efficiency is then equal to 1 minus the particle penetration.

The pressure drop across a venturi scrubber can be estimated by the equation

$$\Delta P = 1.03 \times 10^{-3}\, u_g^2 \left(\frac{Q_L}{Q_G}\right) \tag{3.12}$$

where ΔP is the pressure drop across the venturi (cmH$_2$O), u_g is the gas velocity (cm/s), and Q_L/Q_G is the liquid-to-gas flow rate ratio. Pressure drop in venturi scrub-

TABLE 3.15 Typical Operating Parameters for Particulate Wet Scrubbers

Type of scrubber	Pressure drop	Liquid-to-gas ratio	Liquid inlet pressure (gauge)	Particle cut diameter
Spray tower	1.2–20 mbar (0.5–8 in H$_2$O)	0.07–2.7 L/min-m^3 (0.5–20 gal/min per 1000 acfm)	0.7–28 bar (10–400 psig)	2–8 μm
Cyclonic	3.7–25 mbar (1.5–10 in H$_2$O)	0.3–1.3 L/min-m^3 (2–10 gal/min per 1000 acfm)	2.8–28 bar (40–400 psig)	2–3 μm
Venturi	12.4–124 mbar (5–50 in H$_2$O)	0.4–2.7 L/min-m^3 (3–20 gal/min per 1000 acfm)	0.07–1 bar (1–15 psig)	0.2 μm

bers can be very high, and normally ranges from 25 to 125 mbar. An increase in pressure drop generally correlates to an increase in particle removal efficiency at smaller particle sizes.

Experimental data shows that a venturi scrubber is essentially 100 percent efficient in removing particles larger than 5 μm, so in designing a venturi scrubber, it is only necessary to examine the penetration of particles less than 5 μm. Disadvantages of using wet scrubbers include corrosion problems, large amounts of liquid waste generated, potential for liquid freezing at low temperatures, and possibly expensive disposal of the waste sludge.

3.6.3 Fabric Filters (Baghouses)

Fabric filters are often used in indoor air ventilation systems and to control particles from industrial gas streams (Fig. 3.7). Fabric filters can achieve very high collection efficiency for a wide range of particle sizes. The collection efficiency of a properly operating fabric filter is often above 99.9 percent. A fabric filter operates by passing

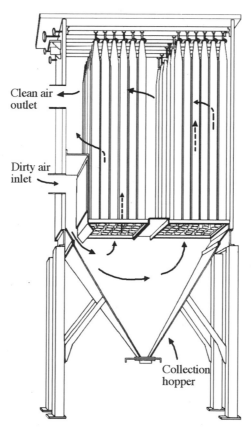

FIGURE 3.7 Mechanical shaker baghouse.

a particle-laden gas stream through a medium (filter) that allows for penetration of the gas stream and the capture of the particulate material. The medium can consist of a single sheet of woven material, such as fiberglass, cotton, or nylon. For high-temperature applications, filters can be manufactured from stainless steel or ceramics. Cartridge filters, such as high-efficiency particulate air (HEPA) filters, are often used in ventilation systems. The choice of fabric is dependent on the composition of the gas stream and the particulate material, gas temperature, the desired levels of particulate collection efficiency, and pressure drop.

Initially, the filter itself performs most of the filtration of the particles from the gas stream. As more particles are filtered from the gas stream, particle loading on the filter increases, forming a mat of particulate material referred to as a *filter cake*. The filter cake allows for more extensive particle filtration, mainly due to interception and impaction. This enhanced particle filtration is useful for achieving high particulate collection efficiencies, but as the thickness of the filter cake increases, so does the pressure drop. Eventually, the pressure drop will be prohibitively large, and the filter will need to be cleaned.

Mechanisms used to clean the filter cake from the filter include shaker, reverse air, and pulse-jet baghouses, and replacement of cartridge filters. Shaker and reverse air baghouses use bags with dimensions from about 6 to 18 in diameter and lengths up to 40 ft. Pulse-jet baghouses use bags with diameters of about 4 to 6 in and lengths of about 8 to 10 ft.

Fabric filter blinding occurs when the fabric pores are blocked and the fabric cannot be cleaned effectively. Blinding can result when sticky particles adhere to the fabric, when moisture blocks the pores and increases particle adhesion, or when a high gas velocity deeply embeds the particles into the fabric.

There are many models that attempt to describe the amount of particulate collection efficiency that can be achieved by filtration. These models have a difficult time realistically describing the contribution of the filter cake to particulate collection efficiency, in part because of the irregular manner in which the filter cake develops. Models that predict collection efficiency will not be discussed here, because collection efficiencies for filters are typically very high (>99 percent).

To design a baghouse, the filtration velocity, the cloth area, the pressure drop, and cleaning frequency need to be estimated. The filtration velocity is expressed as

$$v_f = \frac{Q}{A} \tag{3.13}$$

where v_f is the filtration velocity (m/min), Q is the gas volumetric flow rate (m³/min), and A is the area of the cloth filter (m²). The air-to-cloth (A/C) ratio is defined as the ratio of the gas volume filtered to the cloth filter area [m³/s-m² (ft³/min-ft²)]. Typically, shaker baghouses have A/C ratios of 2 to 6 ft³/min-ft², reverse air have A/C ratios of 1 to 3 ft³/min-ft², and pulse-jet have A/C ratios of 5 to 15 ft³/min-ft². Lists of recommended A/C ratios for various dusts, cleaning methods, and industries have been compiled by Noll (1999).

Equation (3.13) can be used to estimate the number of bags needed, given a process flow rate and a filtration velocity. The number of bags required in the baghouse can be determined from

$$A_b = \pi dh \tag{3.14}$$

where A_b is the bag area, d is the bag diameter, and h is the bag height. Then the total number of bags in the baghouse can be determined from

$$N = \frac{\text{total cloth area}}{\text{bag area}} \tag{3.15}$$

The pressure drop across a baghouse is a function of the individual pressure drops of the filter and the filter cake. Darcy's law can be used to estimate the pressure drop across a filter; it is given as

$$\Delta P_f = K_1 v_f \tag{3.16}$$

where ΔP_f is the pressure drop across a clean fabric [Pa (inH$_2$O)], K_1 is the fabric resistance [Pa/m-min (in H$_2$O/ft-min)], and v_f is the filtration velocity [ft/min]. K_1 is a function of gas viscosity and filter characteristics, such as thickness and porosity. The pressure drop across the filter cake can be estimated from

$$\Delta P_c = K_2 C_i v_f^2 t \tag{3.17}$$

where ΔP_c = pressure drop across the filter cake [Pa (inH$_2$O)]
   K_2 = resistance of the filter cake [Pa-min-m/kg (inH$_2$O-min-ft/lb)]
   C_i = particle loading [kg/m^3 (lb/ft^3)]
   t = the filtration time (min)

K_2 is determined experimentally from the particle loading, filtration velocity, and the pressure drop. Methods for estimating K_2 based on laboratory and pilot plant data are presented in Noll (1999). The total pressure drop across the filter and filter cake is then given as:

$$\Delta P_T = \Delta P_f + \Delta P_c = K_1 v_f + K_2 C_i v_f^2 t \tag{3.18}$$

The filter drag is the resistance across the fabric-particle layer. It is a function of the particle loading on the filter and is given as:

$$S = \frac{\Delta P}{v_f} \tag{3.19}$$

where S is the filter resistance (drag) [Pa-min/m (inH$_2$O/ft-min)] and ΔP is the pressure drop across the filter and filter cake [Pa (inH$_2$O)].

A typical filter performance curve for a single bag or compartment of a fabric filter baghouse is shown in Fig. 3.8. The filter cake resistance S is plotted as a function of the aerial dust loading, where S_R is the residual drag in a single compartment when it is first brought back on line after cleaning, S_T is the drag in a single compartment at the end of the filtration cycle, S_E is the effective residual filter drag, and W_{max} is the maximum particle loading on the fabric filter before it begins a cleaning cycle. The filter drag first increases exponentially and then linearly. The exponential portion of the curve is the period of cake repair and initial cake buildup. The slope of the straight line portion of the curve is equal to K_2 and represents the resistance to filtration through the filter cake. When the pressure drop across the compartment reaches the maximum allowable design pressure drop, the filter bags are cleaned. Once the bags have been cleaned, the pressure drop will decrease to its initial point, and the filter cycle will begin again.

The total pressure drop in a multicompartment baghouse is analogous to the voltage drop across a set of electrical resistances in parallel. The total filter drag across a multicompartment baghouse can be expressed as:

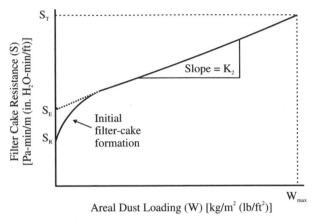

FIGURE 3.8 Dependence of filter cake resistance on aerial dust loading for a typical single-compartment fabric filter.

$$\frac{n}{S_e} = \frac{1}{S_1} + \frac{1}{S_2} + \cdots + \frac{1}{S_n} \tag{3.20}$$

where $S_e = \Delta P/v_{\text{ave}}$ = total multicompartment baghouse drag at any time
v_{ave} = average gas velocity, which is equal to the total volumetric flow rate divided by the total cloth area
n = total number of filter compartments
$S_1, S_2, \ldots$ = individual drag components

3.6.4 Electrostatic Precipitators

In an electrostatic precipitator (ESP) (Fig. 3.9), particles pass through a volume that has a large electrical potential (~50 kV) applied across a channel spacing of about 10 cm. This large electric field strength (electric potential/electrode displacement) causes the release of electrons from the discharge electrode. These high-energy electrons migrate away from the discharge electrode and impact with gas molecules, causing ionization of the gas molecules and the release of more electrons. This generation of a large amount of free electrons is termed *electron avalanche.* Eventually, the free electrons are transported beyond the localized high field strength near the discharge electrode, causing a decrease in their kinetic energy. As the electrons decelerate, they attach themselves to particles instead of generating more free electrons.

The mechanisms responsible for deposition of electrons onto aerosol particles are diffusion and field charging. Diffusion charging dominates for particles with diameters smaller than 0.3 μm, and field charging dominates for particles larger than 0.3 μm. The basic design equations for an ESP ignore diffusion charging, and therefore underpredict particle collection efficiencies for particles smaller than about 0.3 μm.

ESPs are typically designed in one of two basic configurations: flat plate or tubular. Tubular ESPs are much less common than flat-plate ESPs, and they are typically used to remove liquid droplets (e.g., sulfuric acid droplets) from gas streams. Flat-plate ESPs are more commonly used, e.g., in coal-fired power plants. They can be configured as a two-stage or a single-stage design. Two-stage ESPs charge the parti-

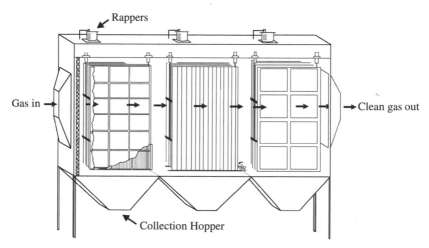

FIGURE 3.9   Flat-plate electrostatic precipitator (ESP).

cles first and then remove them in a different section of the ESP. Single-stage ESPs charge the particles and then remove them in the same section. In these ESPs, typically charging is from vertical wire electrodes and collection is on vertical flat plates placed in parallel rows. Gas flow is largely horizontal in the channels between the rows. Periodically the plates are rapped, and the dust falls down in the hoppers.

ESPs are typically operated under turbulent flow conditions, and the particulate collection efficiency for the ESP can be described by the Deutsch-Anderson equation, below. Assumptions in the Deutsch-Anderson equation include constant migration velocity for a particle of a given diameter, uniform mass concentration in the transverse (horizontal) and vertical directions, no re-entrainment of particles from the collection electrodes, no transport of particles above or below the collection electrodes, and a constant gas stream velocity in the horizontal direction (u_g = constant).

The Deutsch-Anderson equation is represented by

$$\eta = 1 - \exp\left(-\frac{w_p A}{Q_g}\right) \tag{3.21}$$

where η = collection efficiency for a flat-plate, turbulent-flow ESP
  w_p = particle migration velocity
  $A = 2nLH$ = total collection area for the particles
  n = number of channels
  L = length of the collection electrode
  H = height of the collection electrode
  Q_g = volumetric flow rate of the gas

The particle migration velocity w_p is sensitive to the electric field strength, particle composition, particle electrical resistivity, particle charging, and particle size. For particles of essentially homogeneous chemical composition, the migration velocity can be calculated by the following equation:

$$w_p = \frac{6.64 \times 10^{-18} E^2 d_p}{\mu} \tag{3.22}$$

where w_p = migration velocity (m/s)
        E = average electric field strength (V/m)
        d_p = particle diameter (μm)
        μ = gas viscosity (kg/m-s)

When collecting a range of particle sizes using an ESP, a migration velocity can be calculated for each particle size range of interest by using Eq. (3.22). A collection efficiency for each particle size can then be calculated by using Eq. (3.21), and an overall collection efficiency for the ESP can be calculated by using Eq. (3.9). When a gas stream contains particles of varying chemical composition, or a distribution of particle sizes, the effective migration velocity w_e is often used to determine collection efficiency or specific collection area. Values for w_e obtained from field measurements are presented in Table 3.16 for various ESP applications.

TABLE 3.16   Typical Field Measurement
Effective Migration Velocities w_e for ESPs
Operating at 90–95% Efficiency

Application	w_e, cm/s (ft/s)
Utility fly ash	4–20.4 (0.13–0.67)
Pulp and paper mills	6.4–9.5 (0.21–0.31)
Gypsum	15.8–19.5 (0.52–0.64)
Catalyst dust	7.6 (0.25)
Cement (wet process)	10.1–11.3 (0.33–0.37)
Cement (dry process)	6.4–7.0 (0.19–0.23)

Source: Noll (1999) and Oglesby and Nichols (1975).

An important design parameter for sizing an ESP is the specific collection area (SCA). The SCA is defined as

$$SCA = \frac{A}{Q} = -\frac{\ln(1-\eta)}{w_e}$$ (3.23)

For design purposes, the effective migration velocity w_e is often used for calculating SCA, because w_e is taken to represent the collection behavior of the total distribution of particles under a specific set of operating conditions. SCA is typically expressed in units of m^2 of collection area per m^3 of gas or ft^2 of collection area per 1000 acfm.

While the Deutsch-Anderson equation can theoretically predict the collection efficiency of an ESP under a certain set of conditions, uncertainties in the parameters can result in error by a factor of two or more. Some problems include the fact that the equation assumes a uniform gas flow rate, particles are not re-entrained and do not sneak past the collection plates, and particle size and composition are well known and invariable. With these failings of the Deutsch-Anderson equation, it is recommend that it be used only as an estimate, and it is recommended that empirical values from similar facilities be used whenever possible.

ELECTROSTATIC PRECIPITATOR EXAMPLE   *A particle-laden gas stream flows from a combustor at a total gas flow rate of 1500 m³/min. Calculate the specific and total col-*

lection areas needed to remove 99.9 percent of the particles in a single-stage ESP, assuming that the effective migration velocity $w_e = 0.12$ m/s.

SOLUTION   The specific collection area (SCA) can be calculated using Eq. (3.23):

$$SCA = -\frac{\ln(1-0.999)}{0.12 \text{ m/s}} = 57.6 \frac{s}{m} \cdot \left(\frac{1000 \dfrac{m^3}{min}}{1000 \dfrac{m^3}{min}}\right)\left(\frac{1 \text{ min}}{60 \text{ s}}\right) = \frac{960 \text{ m}^2}{1000 \dfrac{m^3}{min}} \quad (3.24)$$

The total collection area can then be calculated from:

$$A = SCA \times Q = \left(\frac{960 \text{ m}^2}{1000 \dfrac{m^3}{min}}\right)\left(\frac{1500 \text{ m}^3}{min}\right) = 1440 \text{ m}^2$$

3.6.5   Comparison of Particulate Control Equipment

The selection of a particulate collection device is based on required collection efficiency, capital and operating costs, pressure drop, operating temperature, gas flow rate, and particle loading. Although it is not possible to recommend a type of particulate control equipment for every type of application within this limited space, several general rules of thumb for choosing particulate control equipment are stated in Tables 3.17 and 3.18. Equipment manufacturers can provide information about current initial costs, as a function of variables such as total gas flow rate, particle loading, and materials selection. Cost spreadsheets developed by W. M. Vatavuk and the U.S. EPA Office of Air Quality Planning and Standards (OAQPS) are available from the EPA Web site at *http://www.epa.gov/ttn/catc/products.html#cccinfo.*

TABLE 3.17   When to Use Specific Particulate Control Equipment

Cyclones	Fabric filters
• Gas contains mostly large particles	• Very high collection efficiencies required
• High dust loadings	
• High efficiency for smaller particles not required	• Particles do not adhere to fabric (filter blinding)
• As a pretreater for another particle control device	• Gases will not condense
	• Relatively low gas temperature
• Particle classification is desired	• Gas volumes are reasonably low

Wet scrubbers	Electrostatic precipitators
• High particle removal efficiency for particles greater than 1 μm	• Very high collection efficiencies required for small particles
• Soluble gases, as well as particles, need to be removed	• Very large gas volumes need to be treated
• Humidification or cooling of gas stream is desired	• Valuable material needs to be recovered
• Gas stream contains a combustible gas mixture	• Low pressure drop required
	• Acceptable particle resistivity

TABLE 3.18 Operating Characteristics of Particulate Collectors

Type	Typical capacity	Pressure drop
Cyclones	2500–3500 ft³/min per ft² of inlet area	2.5–40 mbar (1.0–16 inH₂O)
Venturi Scrubber	6000–30,000 ft³/min per ft² of throat area	25–125 mbar (10–50 inH₂O)
Fabric Filter	1–6 ft³/min per ft² of fabric area	5–15 mbar (2.0–6.0 inH₂O)
Electrostatic Precipitator	2–8 ft³/min per ft² of collection area	0.5–1.3 mbar (0.2–0.5 inH₂O)

3.7 AIR POLLUTION CONTROL TECHNOLOGIES: GAS CONTROLS

Common gaseous air pollutants from stationary sources include CO, NO$_x$, SO$_2$, H$_2$S, and hydrocarbons. On a mass basis, about 90 percent of the air pollutants emitted in the United States are gases, with carbon monoxide contributing about 48 percent. In general, the concentrations of gaseous pollutants in effluent streams are relatively low, but they need to be reduced to some emission standard, which is regulated by the state or federal government. Gas streams may contain multiple pollutants, as well as particles, which can complicate any potential gas cleanup method. This section will examine some unit operations for removing gas contaminants: absorption, adsorption, thermal oxidation (incineration), and biofiltration.

3.7.1 Absorption

Absorption is a widely used unit operation for transferring one or more gas-phase inorganic species into a liquid. Absorption of a gaseous component by a liquid occurs because the liquid is not in equilibrium with the gaseous species. This difference between the actual concentration and the equilibrium concentration provides the driving force for absorption. Absorption can be physical or chemical. *Physical absorption* occurs when a soluble gaseous species, e.g., SO$_2$, dissolves in the liquid phase, e.g., water. In the case of SO$_2$ absorbing into water, the overall physical absorption mechanism is:

$$SO_2 \text{ in air} \leftrightarrow H_2O_{(l)} + H_2SO_3 + HSO_3^- + SO_3^{2-}$$

Chemical absorption occurs when a chemical reaction takes place in the liquid phase to form a new species. In the case of SO$_2$ absorbing into water containing calcium, the overall chemical absorption mechanism is:

$$SO_2 \text{ in air} \leftrightarrow Ca^{2+} + SO_3^{2-} + 2H_2O_{(l)} \leftrightarrow CaSO_3 \cdot 2H_2O_{(s)}$$

Several types of absorbers are used in practice, the packed tower absorber being one of the most common. *Spray towers* atomize droplets, typically consisting of water, at the top of the tower. The pollutant-laden gas in then passed up the tower, so that the droplets contact the gas stream in a countercurrent manner. Soluble species in the gas phase then absorb into the liquid droplets, thereby transferring them from the gas phase into the liquid phase. The liquid droplets are then collected at the bottom of the tower, and the clean gas passes out the top of the tower.

Plate/tray tower (Fig. 3.10) absorbers allow gas-liquid mixing as the gas flows upward through the liquid and as the liquid flows horizontally over each plate or tray. A plate/tray tower absorber is usually operated in a countercurrent (or cross-current) manner, with the liquid flowing down the tower and the gas flowing up.

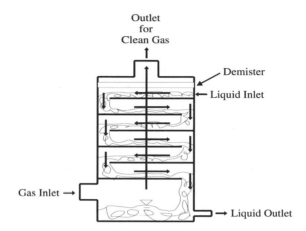

FIGURE 3.10   Plate/tray tower absorber.

Bubble absorbers (Fig. 3.11) intimately mix the gas and liquid by bubbling the gas stream through a large container of liquid, often containing a dissolved species such as Ca^{2+}. When a bubble absorber is used to remove SO_2 from a gas stream generated by a fossil-fuel combustor, dissolved SO_2 is converted into SO_4^{2-} in the liquid phase, which then reacts with dissolved Ca^{2+} to form a solid byproduct, $CaSO_4$ (gypsum). Calcium is added to the water in the form of $Ca(OH)_2$ (calcium hydroxide) or $CaCO_3$ (limestone).

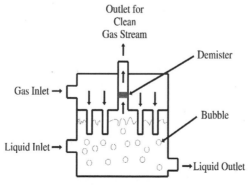

FIGURE 3.11   Bubbler absorber.

Spray-dryer absorbers, often referred to as wet/dry absorbers, atomize slurry droplets into the gas stream. As the water evaporates from the droplets, soluble gases absorb into the droplets and react with dissolved species, such as Ca^{2+} formed from $Ca(OH)_2$ in the slurry droplets. Eventually, all of the water evaporates, leaving dry particles consisting of $CaSO_3$ and unreacted $Ca(OH)_2$. These particles are then removed with a particulate control device.

Most of the remaining discussion about absorbers will be focused on packed tower absorbers, because they are commonly used in a wide variety of absorption applications and the design procedures are very well documented. *Packed tower absorbers* (Fig. 3.12) have an inert packing material with liquid flowing over the packing. The gas stream is then passed through the wet packing, allowing a large amount of interfacial surface area between the gas and liquid. Packed tower absorbers can be operated cocurrently or countercurrently, but countercurrent operation is the most common. Common types of packing material are shown in Fig. 3.13.

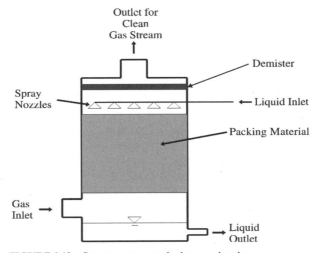

FIGURE 3.12 Countercurrent packed tower absorber.

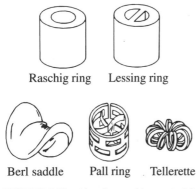

FIGURE 3.13 Absorber packing materials.

The basic model for mass transfer within an absorber is two-film theory. Two-film theory proposes that a mass transfer zone exists across a gas-liquid interface. The mass transfer zone is composed of two very thin films (~0.1 mm): a gas film and a liquid film. The theory assumes that there is complete mixing in both the gas and liquid bulk phases and that the gas-liquid interface is in equilibrium with respect to chemical species being transferred across the interface. Therefore, all resistance to molecular diffusion occurs when molecules are diffusing through the gas and liquid films.

The molar flux of species A diffusing across the gas-liquid interface can be expressed as:

For the liquid phase:

$$N_A = k_L(C_{Ai} - C_{AL}) \tag{3.25}$$

$$N_A = k_x(x_{Ai} - x_{AL}) \tag{3.26}$$

For the gas phase:

$$N_A = k_G(P_{AG} - C_{Ai}) \tag{3.27}$$

$$N_A = k_y(y_{AG} - y_{Ai}) \tag{3.28}$$

where N_A = molar flux of component A $(gmol/m^2\text{-s})$
k = interfacial mass transfer coefficient corresponding to the appropriate phase and driving force $(gmol/m^2\text{-s-Pa})$
C_A = liquid-phase concentration of component A (mol/L)
x_A = liquid-phase mole fraction of component A
y_A = gas-phase mole fraction of component A
P_A = partial pressure of component A in the gas phase

The mass transfer coefficients k for the gas and liquid phases represent the resistance the solute encounters while diffusing through the gas or liquid film. In practice, the above equations for molar flux of species A across the gas and liquid films are difficult to use because of the small scale of the film thickness ($\sim$0.1 mm) where the concentrations would have to be measured. Therefore, overall mass transfer coefficients are used instead of interfacial mass transfer coefficients. Overall mass transfer coefficients describe the mass transfer system at equilibrium conditions by combining the individual film resistances into an overall resistance.

For a system exhibiting a linear equilibrium line, the molar flux of component A can be expressed as:

For the liquid phase:

$$N_A = K_L(C_A^* - C_{AL}) \tag{3.29}$$

$$N_A = K_x(x_A^* - x_{AL}) \tag{3.30}$$

For the gas phase:

$$N_A = K_G(P_{AG} - P_A^*) \tag{3.31}$$

$$N_A = K_y(y_{AG} - y_A^*) \tag{3.32}$$

where K = overall mass transfer coefficient corresponding to the appropriate phase and driving force
C_A^* = equilibrium liquid phase concentration of component A
P_A^* = equilibrium gas phase partial pressure of component A
y_A^* = equilibrium gas phase mole fraction of component A
x_A^* = equilibrium liquid-phase mole fraction of component A

Packed-Tower Absorbers. When designing a packed-tower absorber, it is of interest to determine the liquid-to-gas flow rate ratio L/G, the pressure drop ΔP, the height of the packing material (Z), and the cross-sectional area of the tower (A_{CSA}). Absorbers typically operate in a countercurrent manner with the liquid entering from the top of the tower and the gas entering from the bottom. Applying a material balance around a packed-tower absorber yields the following relationship (Fig. 3.14):

$$G_{m,c}Y_1 + L_{m,s}X_2 = G_{m,c}Y_2 + L_{m,s}X_1 \tag{3.33}$$

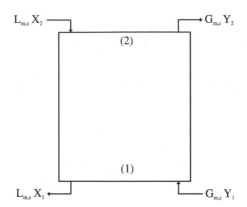

FIGURE 3.14   Material balance for a countercurrent absorber.

where $L_{m,s}$ = total liquid solvent molar flow rate (gmol/min)
$\quad\quad\ \ G_{m,c}$ = total carrier gas molar flow rate (gmol/min)
$\quad\quad\ \ \ X$ = mole ratio of contaminant in the liquid phase (moles contaminant/moles liquid solvent)
$\quad\quad\ \ \ Y$ = mole ratio of the contaminant in the gas phase (moles contaminant/moles carrier gas)

The above relationship for the material balance assumes that the total mass of the gas and liquid streams does not change appreciably during the absorption process. This is typically the case for most air pollution control systems, since the mass flow rates of the contaminant are usually very small compared to the gas and liquid flow rates. This approach is generally considered valid for dilute concentrations (<1 percent by volume) of soluble contaminants in the gas phase.

Equation (3.33) can be rearranged to solve for Y_2, the mole ratio of the contaminant in the gas phase exiting the packed tower absorber.

$$Y_2 = \frac{L_{m,s}}{G_{m,c}} (X_2 - X_1) + Y_1 \tag{3.34}$$

This expression for Y_2 is called the *operating line* for a countercurrent absorber. The operating and equilibrium lines for the system can be used to determine the liquid-to-gas flow rate ratio $L_{m,s}/G_{m,c}$. For a typical absorption application, Y_1, Y_2, and X_2 are specified, with X_1 depending on the liquid-to-gas ratio (Fig. 3.15). The minimum liquid-to-gas ratio is determined by the slope of the operating line that passes through (X_2, Y_2) and the point where Y_1 intersects the equilibrium line (Fig. 3.15). To satisfy the minimum liquid-to-gas ratio, the absorption tower would have to operate at equilibrium conditions at the inlet of the gas stream and at the outlet of the liquid stream. These conditions minimize the consumption of liquid, but they are impractical to achieve, because a very tall absorption tower would be required. In practice the minimum liquid-to-gas ratio is multiplied by a design factor ε_1, typically ranging from 1.3 to 1.7; i.e., as a general rule, an absorber is typically designed to operate at liquid flow rates that are 30 to 70 percent greater than the minimum rate:

$$\left(\frac{L_{m,s}}{G_{m,c}}\right)_{\text{design}} = \varepsilon_1 \left(\frac{L_{m,s}}{G_{m,c}}\right)_{\text{min}} \tag{3.35}$$

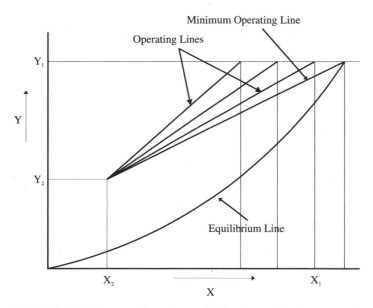

FIGURE 3.15 Minimum and actual operating lines for an absorber.

Pressure Drop in Packed Towers.   Extensive experimental research has resulted in empirical correlations describing pressure drop dependence on the liquid and gas flow rates in a countercurrent packed-tower absorbers (Fig. 3.16). The uppermost line in Fig. 3.16 represents the *flood point* of the tower. For a given packing size and type and a given liquid flow rate in an absorption tower, the pressure drop across the tower is a function of gas velocity. As the gas velocity is increased, liquid is retarded in its flow down the absorber column. As the gas flow rate is further increased, the quantity of the liquid holdup increases more rapidly, as does the pressure drop. Ultimately, the liquid will tend to fill the entire void space, and a layer of liquid will appear on top of the packing. At this point, the tower is said to be *flooded,* and it is marked by the term *flood point.* A packed tower should not operate at or near the flood point because of excessive pressure drop and poor utilization of the surface area of the packing material. As a rule of thumb, an absorption tower operates at gas velocities that are 40 to 70 percent of those that cause flooding.

The flooding condition also varies as a function of the packing factor F (Table 3.19). For packing material having an F value greater than 60, the flooding condition can be taken as 2.0 inH_2O/ft packing. When F is in the range of 10 to 60 (packing size of 2 to 3 in), the pressure drop associated with the flooding condition is expressed by

$$\Delta P_{\text{flood}} = 0.115 F^{0.7} \tag{3.36}$$

where ΔP_{flood} is the pressure drop at the flood point (inH_2O/ft of packing).

The following procedure can be used to determine the required mass flux of the gas stream G' (lb$_m$/ft^2-s) by Fig. 3.16:

1. The term

$$\left(\frac{L'}{G'} \sqrt{\frac{\rho_G}{\rho_L - \rho_G}} \right)$$

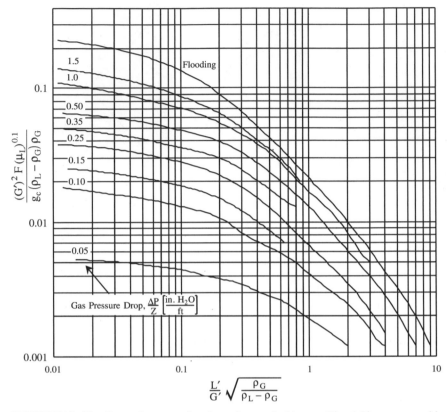

FIGURE 3.16 Flooding and pressure drop in random packed towers. G' and L' are expressed in lb/s-ft^2, g_c has a value of 32.2 lb$_m$-ft/lb$_f$-s^2, μ_L has units of centipose and ρ_L and ρ_G have units of lb$_m$/ft^3. *(Source: Wark, et al., 1998, and the Norton Company.)*

can be determined because the cross-sectional area of the tower drops out of (L'/G').

2. The value of the terms describing the ordinate can then be determined at the flood point. All values for the variables should be known except for G'.

3. Solve for G', the gas mass flux that would occur at the flood point (G'_{flood}).

4. The absorber should not operate at the flooding condition, but rather at some $G'_{\text{design}} < G'_{\text{flood}}$. So,

$$G'_{\text{design}} = \varepsilon_2 G'_{\text{flood}} \tag{3.37}$$

where ε_2 is a flooding safety factor ranging from 0.4 to 0.7.

Once G'_{design} has been determined, the cross-sectional area of the absorption tower (A_{CSA}) can be determined by the following equation:

$$A_{\text{CSA}} = \frac{G}{G'_{\text{design}}} \tag{3.38}$$

TABLE 3.19　Packing Factor F for Random Packing Materials

Packing type	Nominal size, in								
	¼	⅜	½	⅝	¾	1	1.25	1.5	2
Raschig rings:									
Ceramic	1600	1000	580	380	255	155	125	95	65
Metal (½-in wall)	700	390	300	170	155	115			
Pall rings									
Plastic				97		52		40	25
Metal				70		48		28	20
Ceramic Intalox saddles	725	330	200		145	98		52	40
Berl saddles	900		240		170	110		65	45

Source:　Wark et al. (1998) and the Norton Company.

This approach is valid for dilute concentrations (<1 percent by volume) of soluble contaminants in the gas phase. If the absorber is used to treat a more concentrated gas stream, A_{CSA} should be determined at the top and bottom of the absorber and then the largest value for A_{CSA} should be used as the design value.

The pressure drop per unit height of packing material can now be determined by using G'_{design} and L'_{design} to calculate the terms along the abscissa and the ordinate of Fig. 3.16, where G' and L' are the gas and liquid fluxes. The pressure drop per unit height for the absorber can be determined by reading its value from the line of constant pressure drop per unit height of packing that intersects the lines originating from the values describing the abscissa and ordinate.

The height of the absorption tower (Z) is equal to

$$Z = H_{tOG} N_{tOG} \tag{3.39}$$

where H_{tOG} is the overall height of a mass transfer unit based on the gas phase mass transfer resistance, and N_{tOG} is the overall number of mass transfer unit based on the gas phase mass transfer resistance. H_{tOG} is determined by using correlations with the dimensionless Schmidt number (Sc,) and for dilute systems it is equal to

$$H_{tOG} = H_{tG} + H_{tL}\left(\frac{H'G'_m}{L'_m}\right) \tag{3.40}$$

where H_{tG} is the height of a gas-phase transfer unit, H_{tL} is the height of a liquid-phase transfer unit, and H' is the Henry's law constant. H_{tG} and H_{tL} can be estimated from

$$H_{tG} = \frac{\alpha(G')^\beta}{(L')^\gamma}\sqrt{Sc_g} \tag{3.41}$$

$$H_{tL} = \phi\left(\frac{L'}{\mu_L}\right)^\eta \sqrt{Sc_l} \tag{3.42}$$

where α, β, γ, ϕ, and η are constants for a given type of packing material and liquid and gas flow rates (Tables 3.20 and 3.21), μ_L is the dynamic viscosity of the liquid [kg/m-s (lb$_m$/ft-h)], and Sc is the Schmidt number (dimensionless). Values for Schmidt numbers of gases and liquids are given in Tables 3.22 and 3.23.

TABLE 3.20 Constants for Use in Determining Gas Phase Height of a Transfer Unit (H_{tG})

Packing type	α	β	γ	Gas flow rate G', lb/h-ft^2	Liquid flow rate L', lb/h-ft^2
Raschig rings					
⅜ in	2.32	0.45	0.47	200–500	500–1500
1 in	7.00	0.39	0.58	200–800	400–500
1 in	6.41	0.32	0.51	200–600	500–4500
1.5 in	17.30	0.38	0.66	200–700	500–1500
1.5 in	2.58	0.38	0.40	200–700	1500–4500
2 in	3.82	0.41	0.45	200–800	500–4500
Berl saddles					
½ in	32.40	0.30	0.74	200–700	500–1500
½ in	0.81	0.30	0.24	200–700	1500–4500
1 in	1.97	0.36	0.40	200–800	400–4500
1.5 in	5.05	0.32	0.45	200–1000	400–4500

Source: Wark, et al. (1998).

TABLE 3.21 Constants for Use in Determining Liquid-Phase Height of a Transfer Unit (H_{tL}).

Packing type	ϕ	η	Liquid flow rate L', lb/h-ft^2
Raschig rings			
⅜ in	0.00182	0.46	400–15,000
½ in	0.00357	0.35	400–15,000
1 in	0.0100	0.22	400–15,000
1.5 in	0.0111	0.22	400–15,000
2 in	0.0125	0.22	400–15,000
Berl saddles			
½ in	0.00666	0.28	400–15,000
1 in	0.00588	0.28	400–15,000
1.5 in	0.00625	0.28	400–15,000

Source: Wark et al. (1998).

N_{tOG} can be calculated by one of several methods (Perry et al., 1997). For the case where the solute concentration is very low and the equilibrium line is straight, N_{tOG} can be determined from

$$N_{tOG} = \frac{\ln\left[\dfrac{Y_1 - mX_2}{Y_2 - mX_2}\left(1 - \dfrac{mG_m}{L_m}\right) + \dfrac{mG_m}{L_m}\right]}{1 - \dfrac{mG_m}{L_m}} \qquad (3.43)$$

where m = slope of the equilibrium line
G_m = molar flow rate of gas (kgmol/h)
L_m = molar flow rate of liquid (kgmol/h)

TABLE 3.22 Schmidt Numbers for Gases in Air at 25°C and 1 atm

Substance	Sc, $\mu/\rho D$
Ammonia	0.66
Carbon dioxide	0.94
Water	0.60
Methanol	0.97
Ethyl alcohol	1.30
Benzene	1.76
Chlorobenzene	2.12
Ethylbenzene	2.01

Source: Wark et al. (1998) and Perry (1997).

TABLE 3.23 Schmidt Numbers for Liquids in Water at 20°C

Substance	Sc, $\mu/\rho D$
Ammonia	570
Carbon dioxide	570
Chlorine	824
Hydrogen chloride	381
Hydrogen sulfide	712
Sulfuric acid	580
Nitric acid	390
Methanol	785
Ethyl alcohol	1005

Source: Wark et al. (1998) and Perry (1997).

X_2 = mole ratio of solute entering the column
Y_1 = mole ratio of contaminant in entering gas
Y_2 = mole ratio of contaminant in exit gas

For the case when a solute dissociates in the liquid phase, or when a chemical reaction occurs, the solute exhibits almost no partial pressure, and therefore the slope of the equilibrium line (m) approaches zero. For these cases Eq. (3.43) reduces to

$$N_{tOG} = \ln\left(\frac{Y_1}{Y_2}\right) \tag{3.44}$$

To determine N_{tOG} for more concentrated solutions, or for cases when the equilibrium line is not straight, a graphical method can be used (Fig. 3.17). N_{tOG} can be estimated by drawing an intermediate line vertically equidistant between the equilibrium and operating lines ($A = B$ and $C = D$). Units are then counted off from the lowest point of the operating line (G) with horizontal distances such that the horizontal distance between the operating line and intermediate line is equal to the length of the line surpassing the intermediate line ($E = F$). A vertical line is then drawn to the operating line and another step is drawn until point H is surpassed horizontally. Once N_{tOG} is determined, then the height of the column can be determined from

$$Z = H_{tOG}\, N_{tOG} \tag{3.45}$$

The total pressure drop of the tower can be determined from

$$\Delta P_{tot} = \left(\frac{\Delta P}{\Delta Z}\right) Z \tag{3.46}$$

3.7.2 Adsorption

An adsorber unit operation consists of a system containing one or more adsorbents and one or more adsorbates. The *adsorbent* is the solid adsorbing medium that provides active sites for the selective removal of gaseous contaminants, e.g., activated carbon, zeolites, and silica gel.

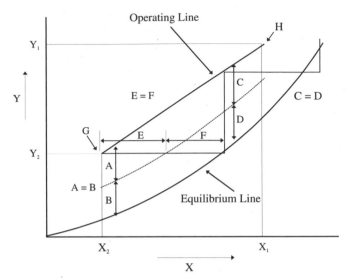

FIGURE 3.17 Schematic for graphical determination of the number of overall gas-phase transfer units (N_{tOG}).

The *adsorbate* is the gaseous material that can be selectively removed from the gas stream by the adsorbent. Adsorption is used for odor control, removal of hazardous air pollutants from a variety of industrial sources (chemical manufacturing, food processing, rendering plants, sewage treatment plants, pharmaceutical plants), purification of indoor air, dehumidification of gases, and the recovery of valuable solvent vapors. Adsorption is usually applied for pollution control of organic compounds, but some inorganic pollutants (SO_2, NO_x, H_2S) can also be adsorbed. In general, adsorption processes will work well for any organic compound with a molecular weight greater than 45. Adsorption is typically used when: (1) the pollutants are in dilute concentrations in the gas phase, making condensation uneconomical; (2) solvent vapors are to be recovered and reused; and (3) the pollutant is not combustible or difficult to burn. Once an adsorbent has been saturated with adsorbate, the adsorbent must be either regenerated or discarded.

There are two mechanisms for adsorption: physical and chemical. *Physical adsorption* occurs when gaseous molecules attach themselves to the surface of the adsorbent with an intermolecular attractive force (van der Waals force). Physical adsorption is reversible and exothermic with about 2 to 20 kJ per g-mol of adsorbed adsorbate, e.g., water on silica gel or benzene on activated carbon. Chemical adsorption occurs when gaseous molecules attach themselves to the surface of the adsorbent with valence forces. Chemical adsorption is typically irreversible and exothermic with about 20 to 40 kJ of energy released per g-mol adsorbed adsorbate, e.g., H_2S adsorption on activated carbon at high temperature (400 to 600°C).

Adsorbents are highly microporous materials with most of their pores less than 2 nm. This high porosity creates a very large surface area (100 to 2000 m^2/g adsorbent) for adsorption. Adsorbents can be manufactured from a variety of materials both organic and inorganic. Activated carbon, one of the most popular adsorbents, can be made from coal, wood, vegetable material, and polymers. For activated carbon, the pore size distribution can be tailored for a specific process by varying the carbonization and activation procedures.

Adsorption Isotherms.   The amount of gas adsorbed per gram of solid adsorbent is a function of concentration (partial pressure) of the adsorbate, temperature of the system, and the properties of the adsorbate and adsorbent. In general, for physical adsorption, the amount of adsorbate adsorbed decreases with increasing temperature, and increases with increasing molecular weight or boiling point. Measuring the amount of a compound adsorbed on an adsorbent versus concentration or partial pressure at constant temperature results in an *adsorption isotherm*. Adsorption isotherms represent equilibrium conditions and they are useful for characterizing adsorbents with respect to different adsorbates (Fig. 3.18). Adsorption isotherms have been produced for thousands of adsorbate-adsorbent combinations. Methods for predicting adsorption isotherms based on the properties of the adsorbate and adsorbent have also been developed (Cal, 1995; Noll, 1999; Vatavuk, 1990). While these predictive relationships are useful for developing estimates of adsorption capacity, laboratory or pilot plant data should be used whenever possible.

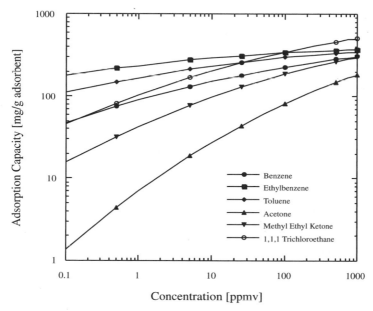

FIGURE 3.18   Adsorption isotherms for various organic species adsorbed onto activated carbon cloth. *(Source: Cal, 1995.)*

Many adsorption isotherm equations have been developed in the past 100 years. One of the most common isotherm equations in industrial use is the Freundlich equation. The Freundlich equation is an empirical expression used to describe adsorption isotherms where there is a linear response for adsorption capacity as a function of adsorbate concentration (or partial pressure) when this function is plotted on log-log scales. The valid concentration range for the Freundlich equation varies, depending on the adsorbate-adsorbent combination. The Freundlich equation is expressed as:

$$C_e = \alpha X^{\beta} \quad \text{or} \quad X = kC_e^m \tag{3.47}$$

where α, β, k, and m are constants determined from the adsorption isotherm plot and are dependent on the units used to describe the adsorbate concentration, C_e is the equilibrium gas phase contaminant concentration, and X is the amount of adsorbate adsorbed per unit mass adsorbent at C_e (g/g). Values for Freundlich constants have been compiled by Vatavuk (1990). When the Freundlich equation is used, the isotherms should not be extrapolated outside of the ranges provided, because the isotherm may not behave linearly beyond the range.

Analysis of an Adsorber.   It is useful to be able to describe how quickly an adsorption zone moves through the length of an adsorption bed and also to determine the length of the adsorption zone (Fig. 3.19). This allows for the prediction of how long an adsorption bed should be used before it must be regenerated or replaced. The velocity of the adsorption zone (V_{ad}) is determined by using a control volume approach and a material balance. The adsorbent is assumed to be flowing through a stationary adsorption zone with 100 percent of the adsorbate removed from the gas stream in the adsorption zone. The mass balance yields:

$$\frac{m_a C_0}{\rho_a} = \rho_{ad} A V_{ad} X_{sat} \tag{3.48}$$

where m_a = mass flow rate of the carrier gas (kg/s)
           ρ_a = density of the carrier gas (kg/m^3)
           C_0 = inlet pollutant concentration (kg/m^3)
           ρ_{ad} = apparent bulk density of the adsorption bed (kg/m^3), taking into account void spaces
           A = cross-sectional area of the adsorption bed (m^2)
           V_{ad} = velocity of the bed as it passes through the adsorption zone (m/s)
           X_{sat} = saturation adsorption capacity corresponding to a concentration C_0 (g adsorbate/g adsorbent)

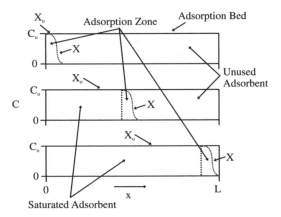

FIGURE 3.19 Schematic of adsorption zone as it moves through an adsorption bed.

Since the inlet concentration C_0 and the saturated adsorbent concentration X_{sat} are related by experimental data, an adsorption isotherm equation can be substituted into Eq. (3.48). If the Freundlich equation is substituted, the following equation is obtained for the mass balance on an adsorber:

$$\frac{m_a C_0}{\rho_a} = \rho_{ad} A V_{ad} \left(\frac{C_0}{\alpha} \right)^{1/\beta} \tag{3.49}$$

where β is unitless and α must have the same units as C_0. Solving for the velocity of the adsorption wave (V_{ad}) yields:

$$V_{ad} = \frac{m_a}{\rho_a \rho_{ad} A} \, \alpha^{1/\beta} C_0^{(\beta-1)/\beta} \tag{3.50}$$

In order to determine the thickness of the adsorption zone, a relationship is needed between the pollutant concentration anywhere in the adsorption zone (C) and any distance within the adsorption zone (x). This can be found by examining the mass transfer from the gas phase to the adsorbent within the adsorption zone. Defining the adsorption zone length such that C approaches 1 percent of its limiting values of 0 and C_0, then the length of the adsorption zone (δ) can be determined from (Wark et al., 1998):

$$\delta = \frac{m_a}{KA\rho_a} \left[4.595 + \frac{1}{\beta-1} \ln \left(\frac{1-0.01^{\beta-1}}{1-0.99^{\beta-1}} \right) \right] \tag{3.51}$$

where K is the mass transfer coefficient (s^{-1}), which can vary greatly, but is usually in the range of 5 to 50 s^{-1}. Assuming that the time required to establish the adsorption zone to its complete length at the adsorber inlet is zero, then the time to breakthrough can be calculated from:

$$t_B = \frac{L-\delta}{V_{ad}} \tag{3.52}$$

where L is the total length of the adsorption bed. Once breakthrough has been achieved, the adsorption bed should be regenerated or replaced. Since there is uncertainty in the calculation of the time to breakthrough, a safety factor of about 2 to 3 is recommended unless the effluent pollutant concentration is being monitored frequently to determine the actual point of breakthrough.

For most applications, it would be cost-prohibitive to discard the adsorbent after one use, so the adsorbent is usually regenerated. Desorbed vapors are much more concentrated and can therefore be recovered more easily and economically than before the adsorption step. Several regeneration methods are available: steam, reduction of gas pressure (pressure swing), heating the adsorption bed (thermal swing), and passing a clean gas over the adsorption bed. Since adsorption is essentially an equilibrium-based process, all regeneration methods rely on the same principle of shifting the adsorbate equilibrium so the adsorbate leaves the adsorbent and returns to the gas phase. After regeneration, the adsorption capacity of the adsorbent may decrease, because regeneration processes are usually not complete, and some adsorbate is retained within the adsorbent pores after the regeneration process.

If the system is designed properly, adsorbers can have high collection efficiencies for gaseous contaminants, even at very low gas-phase concentrations. The lifetime of the adsorbent is usually measured on the order of years for processes that use regeneration. Adsorbers may fail to perform well in humid environments, because water vapor can compete for adsorption sites. When water vapor competition is probable, the gas stream should first be dried before entering the adsorption bed, or a hydrophobic adsorbent should be used to minimize competitive adsorption effects. One possible disadvantage of using adsorption is that the adsorbent material is usually tightly packed in the adsorber, meaning that pressure drop can be significant, which leads to increased operating costs.

ADSORPTION EXAMPLE   *A gas stream containing acetone and air needs to be treated before it is emitted into the atmosphere. The partial pressure of acetone in the gas stream is 5 mmHg. The gas stream is at 100°C and a total pressure of 1 atm. The total mass feed rate of the gas stream is 1.5 kg/s. The acetone will be removed from the gas stream by adsorption with granular activated carbon. The values for ρ_{ad}, α, β, K, L, and A_{CSA} are 380 kg/m³, 17.3 kg/m³, 2.23, 40 s⁻¹, 3.5 m, and 7 m², respectively. The values for α, β, and K are experimentally determined. Determine the breakthrough time for acetone, in units of hours.*

SOLUTION   The first step in the solution is to convert the inlet partial pressure of acetone (C_0) into units of concentration (kg/m³). This is accomplished by using the ideal gas law:

$$C_0 = \frac{\text{ppmv}_a}{10^6} \frac{\text{PMW}_a}{RT}$$

$$= \frac{\left(\frac{5\ \text{mmHg}}{760\ \text{mmHg}}\right)(10^6)}{10^6} \frac{(1\ \text{atm})\left(\frac{58\text{g}}{\text{g-mol}}\right)}{\left(0.0821\ \frac{\text{atm-L}}{\text{g-mol-K}}\right)(373\ \text{K})}$$

$$= 0.0125\ \text{g/L} = 0.125\ \text{kg/m}^3$$

Now, the average molecular weight ($\overline{MW}$) and the density ρ_g of the gas stream must be calculated:

$$\overline{MW} = \sum_i \left(\frac{n_i}{n_T}\right)(\text{MW}_i) = \left(\frac{5\ \text{mmHg}}{760\ \text{mmHg}}\right)\left(58\ \frac{\text{g}}{\text{g-mol}}\right) + \left(1 - \frac{5\ \text{mmHg}}{760\ \text{mmHg}}\right)\left(29\ \frac{\text{g}}{\text{g-mol}}\right)$$

$$= 29.2\ \frac{\text{g}}{\text{g-mol}}$$

$$\rho_g = \frac{P(\overline{MW})}{RT} = \frac{(1\ \text{atm})\left(29.2\ \frac{\text{g}}{\text{g-mol}}\right)}{\left(0.0821\ \frac{\text{atm} \cdot \text{L}}{\text{g-mol} \cdot \text{K}}\right)(373\ \text{K})} = 0.954\ \frac{\text{g}}{\text{L}} = 0.954\ \frac{\text{kg}}{\text{m}^3}$$

The velocity of the adsorption zone (V_{ad}) can be calculated from Eq. (3.50):

$$V_{ad} = \frac{\left(1.5\ \frac{\text{kg}}{\text{s}}\right)\left(17.3\ \frac{\text{kg}}{\text{m}^3}\right)^{1/2.23}\left(0.125\ \frac{\text{kg}}{\text{m}^3}\right)^{(2.23-1)/2.23}}{\left(0.954\ \frac{\text{kg}}{\text{m}^3}\right)\left(380\ \frac{\text{kg}}{\text{m}^3}\right)(7\ \text{m}^2)} = 1.89 \times 10^{-4}\ \text{m/s}$$

The length of the breakthrough zone (δ) is calculated from Eq. (3.51):

$$\delta = \frac{1.5 \, \frac{\text{kg}}{\text{s}}}{\left(0.954 \, \frac{\text{kg}}{\text{m}^3}\right)(40 \text{ s}^{-1})(7 \text{ m}^2)} \left[\ln\left(\frac{0.99}{0.01}\right) + \left(\frac{1}{1.23}\right) \ln\left(\frac{1 - 0.01^{1.23}}{1 - 0.99^{1.23}}\right) \right] = 0.0459 \text{ m}$$

Finally, the time to breakthrough (t_B) can be calculated from Eq. (3.52):

$$t_B = \frac{(3.5 - 0.0459) \text{ m}}{1.89 \times 10^{-4} \text{ m/s}} = 1.84 \times 10^4 \text{ s} = 5.08 \text{ h}$$

3.7.3 Incineration

Incineration or afterburning is a combustion process used in thermal oxidizers to remove combustible air pollutants (gases, vapors, and odors) by oxidizing them. Complete oxidation of organic species results in CO_2 and H_2O, while reduced inorganic species are converted to an oxidized species, e.g., the conversion of H_2S to SO_2. The presence of inorganic species, such as Cl, N, and S, in the waste airstream can result in the production of acid gases after the incineration process. These acid gases, if present in high enough concentrations, will need to be scrubbed from the airstream before being emitted to the atmosphere.

There are two basic types of incinerators: direct thermal and catalytic thermal. A direct *thermal incinerator* usually consists of a refractory-lined chamber that is equipped with one or more sets of burners. The contaminant-laden airstream is passed through the burners, where it is heated above its ignition temperature. The hot gases then pass through a residence chamber, where they are held for a certain length of time to ensure complete combustion. Depending on the particular needs of the system, additional fuel and/or excess air can be added through the burners. To achieve high contaminant destruction efficiency in thermal incinerators, the contaminant must be held at a uniform temperature, generally between 1200 and 1500°F for 0.3 to 0.5 seconds. For some applications, higher temperatures direct thermal afterburners are used (e.g., pesticides destruction at 1800°F and PCB destruction at 2200+ °F). *Catalytic thermal incinerators* permit the use of lower temperatures (e.g., 600 to 750°F at the catalyst inlet) than the direct thermal incinerators for complete combustion, and therefore use less fuel and are made of lighter construction materials. The lower fuel cost may be offset by the added cost of catalysts and the higher maintenance requirements for catalytic units. As a general rule, direct thermal units operating above 1400°F achieve higher destruction efficiencies than catalytic thermal units.

In an incinerator, time and temperature are interrelated, so that a relatively short contact period and high temperature can produce an efficiency equivalent to a time/temperature unit with long contact time and low temperature. The design residence time for an incinerator generally applies only to the reaction zone. Additional reactor volume must be provided for initial combustion and mixing. Also, since the flue gases are discharged at elevated temperatures, a system to recover heat may be included in the incinerator design. The simplest method of heat recovery is to use the hot cleaned gases exiting the incinerator to preheat the cooler incoming gases. Design efficiency for the heat recovery system is usually 35 to 90 percent.

The destruction rates of volatile organic compounds (VOCs) are very sensitive to temperature. Therefore, sufficient time must be provided at the design temperature to allow the reactions to reach the desired degree of destruction. Turbulence ensures sufficient mixing of oxygen and VOCs during the process. This leads to the importance of the *three T's of combustion*—temperature, time, and turbulence. The three T's relate to three characteristic times: chemical time (t_c), residence time (t_r), and mixing time (t_m). The three characteristic times can be calculated from the following equations:

$$t_c = 1/k \tag{3.53}$$

$$t_r = V/Q = L/u_g \tag{3.54}$$

$$t_m = L^2/D_e \tag{3.55}$$

where V = volume of the reaction zone (m^3)
Q = volumetric flow rate at the temperature of the afterburner (m^3/s)
L = length of the reaction zone (m)
u_g = gas velocity in the afterburner (m/s)
D_e = effective turbulent diffusion coefficient (m^2/s)
k = rate constant (s^{-1})

The ratio of the mixing time to the residence time is called the *Peclet* number, Pe, and the ratio of the chemical time to the residence time is the inverse of the *Damkoler* number, Da. If Pe is large and Da is small, then mixing is the rate-controlling process. If Pe is small and Da is large, then the chemical kinetics are rate controlling. At the temperatures of most incinerators, provided a reasonable flow velocity is maintained, the mixing process will not be the limiting factor and chemical kinetics will be important.

For chemical reactions, the reaction rates are typically expressed by equations of the form:

$$\frac{dC_A}{dt} = r = -kC_A^n \tag{3.56}$$

where r = reaction rate
k = kinetic rate constant whose value is strongly dependent on the reactants and temperature
C_A = concentration of A
n = reaction order

The overall order of the reaction is determined experimentally. If the reaction is first order, then

$$-\frac{dC_A}{dt} = kC_A \tag{3.57}$$

and the solution to the differential equation is

$$C_A = C_{A0} \exp(-kt) \tag{3.58}$$

where C_{A0} is the initial concentration of A at $t = 0$. The assumption that allows incineration to be considered first order is that the VOC to be burned is much less than the concentration of oxygen in the contaminated air stream.

For most chemical reactions, the relation between the kinetic rate constant k and the temperature T is given by the Arrhenius equation:

$$k = A \exp(-E/RT) \qquad (3.59)$$

where A = experimental constant (pre-exponential factor) (s^{-1})
E = activation energy (J/mol, cal/g-mol)
R = universal gas constant (8.314 J/mol-K or 1.987 cal/gmol-K)
T = absolute temperature (K)

Cooper, Alley, and Overcamp (1982) combined collision theory with empirical data and proposed a method for predicting an effective first-order rate constant k for hydrocarbon incineration over the range of 940 to 1140 K. Once k is found, the design temperature can be obtained. By evaluating kinetic constants for hydrocarbon incineration reported in the literature, Cooper et al. (1982) determined a relationship for the pre-exponential factor A:

$$A = \frac{ZSy_{O_2} P}{R} \qquad (3.60)$$

where Z = collision rate factor
S = steric factor
y_{O_2} = mole fraction of oxygen in the incinerator
P = absolute pressure (atm)
R = ideal gas constant (0.08205 L-atm/mol-K)

The steric factor S accounts for some collisions that are not effective in producing reactions because of molecular geometry. The steric factor is calculated from

$$S = \frac{16}{MW} \qquad (3.61)$$

where MW is the molecular weight of the hydrocarbon. The activation energy and collision factor were correlated with molecular weight by Cooper et al. (1982), and the following relationships were proposed:

$$E = 193{,}020 - 40.45 \, (MW) \qquad (3.62)$$

$$Z = (0.5 + MW/32) \, 10^{11} \qquad \text{Alkanes} \qquad (3.63)$$

$$Z = (0.25 + 0.03 \, MW) \, 10^{11} \qquad \text{Alkenes} \qquad (3.64)$$

$$Z = (-0.60 + 0.0275 \, MW) \, 10^{11} \qquad \text{Aromatics} \qquad (3.65)$$

Once A and E have been estimated, k can be calculated at any temperature by using Eq. (3.59). In an isothermal plug-flow reactor, the hydrocarbon destruction efficiency, the rate constant, and the residence time are interdependent, and are related as

$$\eta = 1 - \frac{[HC]_{out}}{[HC]_{in}} = 1 - \exp(-kt_r) \qquad (3.66)$$

where η is the hydrocarbon destruction efficiency, [HC] is the hydrocarbon concentration, and t_r is the residence time within the incinerator.

INCINERATION EXAMPLE *Determine the value of k for benzene at 800 K and estimate the time required to destroy 99.9 percent of the benzene in a waste gas stream at 800 K and 1100 K.*

SOLUTION   *Using Eq. (3.59) and Table 3.24, the rate constant k for benzene can be calculated:*

$$k = 7.43 \times 10^{21} \exp\left(\frac{-95,900 \text{ cal/g-mol}}{(1.987 \text{ cal/mol-K})(800 \text{ K})}\right)$$

$$k = 4.67 \times 10^{-5} \text{ s}^{-1}$$

TABLE 3.24   Thermal Oxidation Parameters

Compound	A, s^{-1}	E, cal/g-mol
Benzene	7.42×10^{21}	95,900
Carbon tetrachloride	2.80×10^{5}	26,000
Chloroform	2.90×10^{12}	49,000
Dichlorobenzene	3.00×10^{8}	39,000
Nitrobenzene	1.40×10^{15}	64,000
Toluene	2.28×10^{13}	56,500
Trichloroethane	1.90×10^{8}	32,000
Vinyl chloride	3.57×10^{14}	63,300

Calculate the incineration time necessary to destroy 99.9 percent benzene at 800 K.

$$t_{800 \text{ K}} = \frac{1}{k} \ln \frac{C_{A0}}{C_A} = \frac{1}{4.67 \times 10^{-5} \text{ s}^{-1}} \ln \frac{1}{0.001} = 148,000 \text{ s} = 41.1 \text{ h}$$

At 1100 K, k = 654 s^{-1}. Calculate the incineration time necessary to destroy 99.9 percent benzene at 1100 K.

$$t_{1100 \text{ K}} = \frac{1}{k} \ln \frac{C_{A0}}{C_A} = \frac{1}{654 \text{ s}^{-1}} \ln \frac{1}{0.001} = 0.011 \text{ s}$$

As can be seen in the above example, destruction time is very sensitive to reaction temperature. Running an incinerator at a higher temperature requires less residence time to achieve the same destruction.

Catalytic Incinerators.   In a catalytic incinerator, gas is passed through a preheater chamber and then over a catalyst bed, which promotes oxidation at a lower temperature than does thermal incineration. Incinerator catalysts may be cylindrical or spherical porous pellets, ranging in size from $\frac{1}{16}$ to $\frac{1}{2}$ in in diameter. Other shapes include honeycombs, ribbon, and wire mesh. Fluidized-bed units are sometimes employed, using finer catalyst particulate sizes and achieving high destruction efficiencies at shorter residence times than the more common fixed-bed units. The catalysts are usually metals or metal salts. Platinum, palladium, cobalt, copper, chromium, and molybdenum are common catalyst materials. The catalysts can be placed on inert supports, such as alumina or porcelain, or they may be used directly in the unsupported state. Metals such as aluminum and iron readily react with oxygen, forming a strong oxide layer, which rules them out as catalyst materials. Platinum and similar metals work well as catalysts because they do not form strong oxides, but they do adsorb other atoms and molecules on their surfaces. Platinum catalyst is the one most commonly used for hydrocarbon destruction, and can be deactivated by halides or poisoned by certain heavy metals (e.g., lead).

Catalyst fouling is one of the most common problems with catalytic incinerators. If the catalyst temperature is not maintained at optimum levels, then a coating of organic material or carbon can be deposited on the catalyst surface, reducing its activity. The formation of metal oxides can also reduce catalyst activity. Organic contaminants can usually be removed by increasing the catalyst bed temperature and burning off the deposits. If the contaminants are inorganic, the catalyst must be cleaned with an acid or detergent wash. If a catalyst is too severely fouled, it will need to be replaced.

Catalytic incineration is essentially a flameless combustion process that occurs at much lower temperatures [260 to 540°C (500 to 1000°F)] than traditional incineration processes. The catalyst increases the rate of reaction and permits the reaction to occur at lower temperatures. A large amount of heat is released during the catalytic reaction, which causes a temperature rise in the gas as it passes through the catalyst bed. The amount of temperature rise increases with inlet gas VOC concentration. The basic problem in the design of a catalytic reactor is to determine the quantity of catalyst required for a given conversion and flow rate. To achieve destruction efficiencies between 90 and 95 percent, about 1.5 to 2.0 ft^3 of catalyst per 1000 standard ft^3/min of gas is required. The temperature of the gas stream can rise from 600°F at the catalyst bed inlet to 900°F at the bed outlet. At any given temperature, the catalytic reaction rate constant is considerably greater than for the thermal process. For example, to achieve 95 percent destruction of toluene in a thermal incinerator operating at 650°C (1200°F), a residence time of 0.104 s is required with a rate constant k of 0.93 s^{-1}. Alternatively, for the same destruction of toluene in a catalytic incinerator operating at 480°C (900°F), a residence time of only 0.024 s is required with a k of 128 s^{-1}.

3.7.4 Biofiltration

Biofiltration uses microbial reactions to treat contaminated airstreams. Biofiltration can be an economical treatment solution for large-volume airstreams with low concentrations of contaminants. In general, biofiltration is most economical for gas contaminant concentrations of less than 1000 ppmv and gas flow rates from 1000 to 50,000 m^3/h.

In the biofiltration process, contaminants are first sorbed from the gas stream to an aqueous phase. Microbes present in the aqueous phase then biologically degrade the contaminants through oxidative and occasionally reductive reactions. In the microbial process, contaminants are converted to carbon dioxide, water vapor, and organic biomass. Both organic and inorganic pollutants can be biodegraded, and the organisms used are typically naturally occurring.

Microorganisms in a biofilter grow on a biofilm present on the surface of the medium, or are suspended in the aqueous phase surrounding the support medium. The filter bed medium consists of a relatively inert substance, such as compost, peat, activated carbon, soil, wood chips, or a synthetic medium. The filter bed material provides a large contact surface area, which improves mass transfer (Fig. 3.20). Nat-

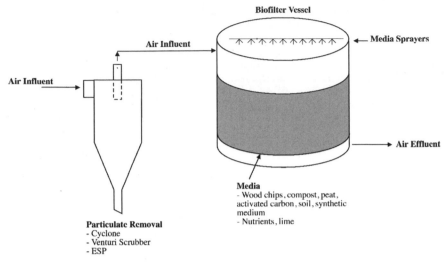

FIGURE 3.20   Typical biofilter configuration.

ural organic media (soil, peat, compost, etc.) may provide some or all of the essential nutrients required for microbial growth. Biofilters are classified as one of three types as listed in Table 3.25.

TABLE 3.25   Biofilter Classifications

Reactor type	Micro-organisms	Water phase
Biofilter	Fixed	Stationary
Biotrickling filter	Fixed	Flowing
Bioscrubber	Suspended	Flowing

Basic Biofilter Operation.   During operation of the biofilter, the contaminated gas stream is contacted with the filter bed, and contaminants are transferred from the gas phase to the biofilter media. Biodegradation of the contaminants can then occur when the contaminant is either adsorbed onto the organic media, adsorbed directly to the media biofilm, or dissolved into the aqueous phase. Once the contaminant is adsorbed onto the biofilter media, or dissolved into the water layer surrounding the biofilm, the contaminants can be used by micro-organisms as a food source to support microbial growth. After the contaminants have been removed from the airstream and deposited on the biofilter, cleaner air then exists from the exhaust gas of the biofilter.

   Biodegradability can vary widely from compound to compound, but a few general observations have been made about biodegradable organic species. Compounds with high degradability tend to:

• Have low molecular weights
• Be highly soluble

- Be nontoxic to the organisms at present concentrations
- Have simple chemical bond structures

Organic compounds such as alcohols, aldehydes, ketones, and some simple aromatics have exhibited good biodegradability (Table 3.26).

The moisture content of a biofilter medium is one of the most important parameters in biofilter operation. Obtaining an optimum moisture level is critical, and difficult to achieve in practice. If a biofilter is too wet, several operational problems can result, including:

- Nutrient washing from the biofilter medium
- Nutrients may already be present in medium, or may be added; typically sprayed in with the water
- Oxygen transfer problems due to reduced air/water interface per unit volume of biofilm
- Creation of anaerobic zones that promote odor formation and slow contaminant degradation rates
- Large pressure drop and low gas retention time, because pore spaces become filled with water
- Production of high strength, low-pH leachate that requires disposal

Alternatively, a dry biofilter can cause the following problems:

- Deactivation of contaminant-degrading organisms
- Contraction and cracking of biofilter medium
- Difficulty in rewetting the dry hydrophobic medium materials

High-velocity gas flow rates with low relative humidities can dry the biofilter by transferring water from the filter medium to the gas phase. Exothermic reactions that occur within the filter bed can increase the bed temperature, which speeds up

TABLE 3.26 Biodegradability of Various Contaminants in a Biofilter

Good biodegradability	Moderate biodegradability	Some biodegradability
Phenol	Hexane	Methane
Toluene	Benzene	Pentane
Dichloromethane	Xylene	Cyclohexane
Amines	Styrene	Carbon tetrachloride
Alcohols	Carbon disulfide	Chloroform
Aldehydes	Dimethyl sulfide	Tetrachloroethene
Ethyl, butyl, and isobutyl acetate	Vinyl acetate	Vinyl chloride
Ketones		Nitriles
Ammonia		Methyl mercaptan
Hydrogen sulfide		Nitrogen oxide
		Ethers

Source: Joseph S. Devinny, Marc A. Deshusses, and Todd S. Webster, *Biofiltration for Air Pollution Control,* Lewis Publishers, 1999, ISBN 1-56670-289-5.

the biodegradation processes, which can then further increase bed temperature. This situation, if left unchecked, can lead to moisture stripping from the bed due to the increase in water vapor pressure. This problem occurs most often at the inlet of the biofilter, where contaminant concentration is highest.

The optimal moisture content for most biofilters is in the range of 40 to 60 percent of total capacity. Moisture is maintained in biofilters by the following methods:

- Humidifying the inlet gas stream by a packed tower, atomizers, spray nozzles, or venturi
- Adding water directly to the biofilter by spray nozzles
- A combination of both humidification and direct water addition

The micro-organisms used in a biofilter to treat VOCs are typically heterotrophs, such as bacteria or fungi, while the micro-organisms used to treat gases containing inorganic contaminants are chemoautotrophs, which use CO_2 as a carbon source. The number density of organisms will vary with filter bed depth. Typically, a higher density of organisms will be present at the bed influent, where contaminant concentration is highest. Deeper in the bed, where lower contaminant concentrations are present, smaller populations of different organisms exist, which have adapted themselves to varying conditions found downstream of the bed entrance. Beds can be seeded with micro-organisms that have already adapted to the contaminants of interest, or the more general micro-organisms can be used, which may require additional time for acclimation to the bed contaminants and operating conditions.

Biofilters require a period of acclimation before they operate at greatest efficiency. The acclimation time is the start-up period of the biofilter, where removal efficiencies steadily increase until they reach some steady-state maximum value. Acclimation is required because it takes time for micro-organisms to develop adaptive enzymes and degradation pathways to metabolize the contaminates. Acclimation time is usually taken to be the time needed to reach 95 percent of the maximum removal capacity. Acclimation time is usually longer for gas streams containing multiple pollutants and for compounds that are difficult to degrade. The acclimation time varies, depending on the substrate, inlet contaminant concentration, filter medium type, and operational parameters, such as temperature and moisture content. If the medium does not contain large colonies of micro-organisms, acclimation times may be longer, because the organisms need time to multiply and distribute within the filter medium. After a bed has been operating for some time and the biofiltration process has been established, the restart acclimation time after shutdowns and other disruptions is usually much shorter than initial start-up, provided that the bed is kept moist.

Another important parameter in biofilter operation is the pH of the biofilter medium. As with most aerobic biological processes, the optimal pH for biofilter operation is about 7 to 8. Some biological processes can produce acids, which may not be washed out during operation, so special provisions must be made to treat acid discharges. For example, organic species containing sulfur can form H_2SO_4, and chlorinated organics can form HCl. If the buildup of acidic compounds is of concern, buffering solutions are added to prevent acid accumulation. Common buffers include: limestone, crushed oyster shells, and marl. Acid accumulation can also occur during periods of organic overload, which leads to the formation of organic acids, such as acetic acid. Acetic acid can be neutralized by adding sodium bicarbonate to the irrigation water.

The recommended temperature for biofilters containing mesophiles is 25 to 35°C. In general, rates of reaction and diffusion increase with increasing temperature, but the solubility of VOCs will decrease with increasing temperature, as will the sorption capacity of the filter. If the temperature of the biofilter is too cold, microbial growth may be inhibited, which effectively shuts down the biofilter. Restart of a biofilter after a cold shutdown is usually easily accomplished. To avoid cold shutdowns, it is generally a good idea to preheat the inlet gas to the biofilter if its temperature drops below 10 to 15°C.

Design of Biofilters. At high pollutant concentrations, the change in removal efficiency is linear with respect to the distance into the biofilter media, or with empty bed residence time (EBRT). At lower concentrations, the removal rate decreases and follows a power law function. EBRT is considered one of the primary design factors for a biofilter reactor, and determining it is one of the main objects of a pilot-scale biofilter test. For a given pollutant concentration and biofilter operating conditions, the pollutant removal efficiency or maximum outlet concentration allowed by regulations dictates a minimum EBRT. In typical biofilters, EBRT ranges from 15 to 60 seconds. This value corresponds to a required filter volume of 4.2 to 16.7 m^3 filter media per 1000 m^3/h. To avoid media compaction and uneven moisture distribution, individual biofilter beds are typically in the range of 1 to 3 m depth. To determine the biofilter reactor footprint area based on the EBRT and the total gas flow rate, the following equation is used:

$$A = \frac{Q}{v} = Q \frac{\text{EBRT}}{60h} \qquad (3.67)$$

where
$\quad A$ = cross-sectional area or footprint (m^2)
$\quad Q$ = volumetric gas flow rate (m^3/h)
$\quad v$ = superficial gas velocity (m/h)
$\quad h$ = filter bed height (m)
$\quad$ EBRT = empty bed residence time (min)

Another design parameter for biofilters is the bulk elimination capacity (EC), which is a measurement of the removal of target compounds per unit volume of media. It is measured in grams of pollutant removed per cubic meter of media per hour (g/m^3-h) and is represented by the following equation:

$$EC = (C_{in} - C_{out}) \frac{Q}{V} = C_{in} RE \frac{Q}{v} = \Delta C \frac{60}{\text{EBRT}} \qquad (3.68)$$

where
$\quad C_{in}$ = inlet concentration (g/m^3)
$\quad C_{out}$ = outlet concentration (g/m^3)
$\quad EC$ = elimination capacity (g/m^3/h)
$\quad RE$ = fractional removal efficiency
$\quad \Delta C$ = concentration difference (inlet − outlet)
$\quad V$ = volume of filter material (m^3)

3.7.5 Comparison of Gaseous Control Technologies

Absorption is typically used to remove soluble inorganic pollutants, while adsorption, incineration, and biofiltration are typically used to remove low concentrations of organic vapors. The presence of multiple contaminants in the gas stream will complicate the design of any pollution control device. Design of gaseous air pollution control devices usually requires reducing pollutant levels below some emission standard, so the inlet and outlet pollutant concentrations are known. Design requirements then include sizing the unit and determining reasonable gas and/or liquid flow rates. This section has covered the basics needed to design common units, but data from equipment manufacturers and/or laboratory or pilot-scale data should be used whenever possible. Information on providing costs estimates for gaseous control technologies can be obtained from equipment manufacturers or from the U.S. EPA Web site *http://www.epa.gov/ttn/catc/products.html#cccinfo* and Vatavuk (1990).

3.8 INDUSTRY PROFILE: PHARMACEUTICAL INDUSTRY

Pharmaceuticals are produced in about 1500 manufacturing plants in the United States and Puerto Rico, with the majority of facilities located in California, New Jersey, and New York (Fig. 3.21). The pharmaceutical industry uses hundreds of raw materials for chemical synthesis processes. Organic solvents, acids, and bases are used for extraction and purification processes, and other compounds, such as carbohydrates, carbonates, steep liquors, phosphorus compounds, and antifoam agents, are used in fermentation processes. Chemical releases from pharmaceutical facilities can occur as both point and fugitive air emissions. As noted in the 1995 Pharmaceutical Industry Toxic Release Inventory (TRI), many of the commonly reported chemicals are regulated hazardous air pollutants under the Clean Air Act (Table 3.27). Total

FIGURE 3.21 Distribution of pharmaceutical manufacturing plants in the United States (1992). *(Source: 1992 U.S. Census of Manufacturers.)*

TABLE 3.27 Total 1995 Air Releases for Pharmaceutical Facilities (SIC 2833 and SIC 2834) TRI (lb/year)

Chemical name	Number of facilities reporting	Fugitive air emissions	Point air emissions	HAPs under Clean Air Act
Acetonitrile	25	206,608	106,670	Y
Ammonia	42	772,824	380,822	N
Aniline	3	3,896	1,173	Y
Benzene	3	2,970	582	Y
n-Butyl alcohol	14	145,024	476,734	N
Chlorine	19	4,315	9,036	Y
Chloroform	14	55,536	88,826	Y
Chloromethane	6	28,840	97,844	N
Cyclohexane	9	47,574	147,052	N
Dichlorodifluoromethane	8	22,610	195,178	N
1,2-Dichloroethane	5	928	1,313	N
Dichloromethane	63	2,386,889	4,611,794	N
Dichlorotetrafluoroethane (CFC-114)	3	4,978	2,260	N
n,n-Dimethylformamide	20	63,972	10,598	N
Ethyl benzene	5	789	977	Y
Ethylene glycol	30	21,721	2,638	Y
Ethylene oxide	3	12,143	9,550	Y
Formaldehyde	9	2,662	3,772	Y
Formic acid	13	21,550	3,173	N
Freon 113	2	3,500	38,119	N
Glycol ethers	7	1,310	27,944	Y
n-Hexane	18	201,267	258,124	Y
Hydrochloric acid	62	68,269	532,143	Y
Isopropyl alcohol	2	61,250	140,250	N
Methanol	104	1,396,868	2,100,445	Y
2-Methoxyethanol	5	9,130	9,455	N
Methyl ethyl ketone	7	20,624	51,120	Y
Methyl isobutyl ketone	14	273,952	109,175	Y
Methyl tert-butyl ether	11	4,061	18,449	Y
Napthalene	7	515	1,014	Y
Nitric acid	13	8,029	12,928	N
Phosphoric acid	31	5,194	5,160	N
Pyridine	7	2,820	3,093	N
Sulfuric acid	11	22,283	3,091	N
Tert-butyl alcohol	7	26,713	19,473	N
Toluene	54	498,932	593,839	Y
Trichlorofluoromethane	6	59,306	61,801	N
Triethylamine	17	22,262	15,957	N
Xylenes	14	10,712	107,105	Y
Zinc compounds	16	765	11,169	N

1997 air pollutant releases compared to other major industry sectors are presented in Table 3.28.

Each pharmaceutical facility tends to be unique in the type and amounts of air pollutant emissions generated. Some bulk substances and intermediates are made frequently, while others may be made for only a few weeks over a period of several years. This makes it difficult to calculate emission rates for different types of pharmaceutical processes. However, in general, a majority of emissions from the bulk manufacture of pharmaceuticals occur from dryers, reactors, distillation units, and storage and transfer of materials. Condensers are widely used in the pharmaceutical industry to recover solvents used in process operations. Typically, the coolant in the condensers does not directly contact the solvent, so that the solvent is not contaminated and may be directly reused. This limits air pollutant emissions and reduces the cost of solvents. Wet scrubbers, adsorption, and incineration are also used to control air pollutant emissions from point sources.

TABLE 3.28 Total 1997 Air Pollutant Releases by Industry Sector (tons/year)

Industry sector	VOC	CO	NO$_2$	SO$_2$	PM$_{10}$
Dry cleaning	7,441	102	184	155	3
Electronics and computers	4,866	356	1,501	741	224
Fabricated metals	86,472	4,925	11,104	3,169	1,019
Furniture and fixtures	67,604	2,754	1,872	1,538	2,502
Ground transportation	104,824	128,625	550,551	2,569	8,417
Inorganic chemicals	65,427	153,294	106,522	194,153	6,703
Iron and steel	83,882	1,386,461	153,607	232,347	83,938
Lumber and wood production	55,983	122,061	38,042	9,401	20,456
Metal casting	19,031	116,538	11,911	6,513	10,995
Metal mining	915	4,670	39,849	17,690	63,541
Motor vehicles, bodies, parts, and accessories	96,338	15,109	27,355	20,378	1,048
Nonferrous metals	11,058	214,243	31,136	253,538	10,403
Nonmetal mining	4,002	25,922	22,881	18,000	40,199
Organic chemicals	180,350	112,410	187,400	176,115	14,596
Petroleum refining	313,982	734,630	355,852	619,775	27,497
Pharmaceuticals	37,214	6,586	19,088	21,311	1,576
Plastic resins and man-made fibers	74,138	16,388	41,771	67,546	2,218
Power generation	57,384	366,208	5,986,757	13,827,511	140,760
Printing	103,018	8,755	3,542	1,684	405
Pulp and paper	127,809	566,883	358,675	493,313	35,030
Rubber and miscellaneous plastics	132,945	2,200	9,955	21,720	2,618
Shipbuilding and repair	3,967	105	862	3,051	638
Stone, clay, and concrete	34,337	105,059	40,639	308,534	192,962
Textiles	27,768	8,177	34,523	43,050	2,028

Source: U.S. EPA Office of Air and Radiation, AIRS Database, 1997.

REFERENCES

Cal, M. P., 1995. "Characterization of Gas Phase Adsorption Capacity of Untreated and Chemically Treated Activated Carbon Cloths," Doctoral Dissertation, University of Illinois at Urbana-Champaign.

Calvert, S., Goldschmid, J., Leith, D., and Mehat, D., 1972. "Wet Scrubber System Study," in *Scrubber Handbook,* Vol. I, U.S. Department of Commerce, NTIS PB 213016, August.

Cooper, C. D., and Alley, F. C., 1994. *Air Pollution Control: A Design Approach,* 2d ed., Waveland Press, Prospect Heights, Ill., p. 130.

Cooper, C. D., Alley, F. C., and Overcamp, T. J., 1982. "Hydrocarbon Vapor Incineration Kinetics," *Environmental Progress,* vol. 1, no. 2, pp. 129–133.

Heumann, W. L., 1997. *Industrial Air Pollution Control Systems,* McGraw-Hill.

Noll, K., 1999. *Fundamentals of Air Quality Control Systems,* American Academy of Environmental Engineers, Annapolis, Md., pp. 170–171.

Oglesby, S., and Nichols, G., 1975. "Electrostatic Precipitators," in *Gas Cleaning for Air Quality Control,* Marchello, J., and Kelly, J., eds., Marcel Dekker, New York.

Perry, R. H., Green, D. W., and Maloney, J. O. eds., 1997. "Gas Absorption and Gas-Liquid System Design," Chap. 14 in *Perry's Chemical Engineers' Handbook,* 7th ed., McGraw-Hill, New York.

Shepard, C. B., and Lapple, C. E., 1939; 1940. "Flow Patterns and Pressure Drop in Cyclone Dust Collectors," *Ind. Eng. Chem.,* vol. 31, no. 8; vol. 32, no. 9.

Shrock, J., 1994. "Five Steps Toward Title V Permitting," *Environmental Protection,* June, pp. 41–44.

Stairmand, C., 1951. "The Design and Performance of Cyclone Separators," *Trans. Inst. Chem. Engrs.,* 29, p. 356.

Swift, P. "Dust Control in Industry," *Steam Heating Eng.,* vol. 38, p. 453, 1969.

Theodore, L., and De Paola, V., 1980. "Predicting Cyclone Efficiency," *J. Air Pollution Control Assoc.,* vol. 30, no. 10.

Turner, D. B., 1964. "A Diffusion Model for an Urban Area," *J. Appl. Meteor.,* vol. 3, no. 1, pp. 83–91.

U.S. EPA, 1995. *Compilation of Air Pollutant Emission Factors (AP-42),* 5th ed., PB 86-142906. U.S. EPA, Research Triangle Park, N.C., October.

U.S. EPA, 1997. Profile of the Pharmaceutical Industry, Office of Compliance Sector Notebook Project, EPA/310-R-97-005, September, *http://www.epa.gov/oeca/sector.*

Vatavuk, W. M., 1990. *OAQPS Control Cost Manual,* EPA 450/3-90-006, 4th ed. U.S. EPA OAQPS. Also available at *http://www.epa.gov/ttn/catc/products.html#cccinfo.*

Viegele, W. J., and J. H. Head, 1978. "Derivation of the Gaussian Plume Model," *J. Air Pollution Control Assoc.,* vol. 28, no. 11, 1139–1141.

Wark, K., Warner, C. F., Davis, W. T., 1998. *Air Pollution: Its Origin and Control,* 3d. ed., Addison-Wesley. Reading, Mass.

BIBLIOGRAPHY

Buonicore, A. J., and Davis, W. T., 2000. *Air Pollution Engineering Manual,* Wiley Interscience.

Cooper, C. D., and Alley, F. C., 1994. *Air Pollution: A Design Approach,* 2d Ed., Waveland Press.

Devinny, S., Deshusses, M. A., and Webster, T. S., 1999. *Biofiltration for Air Pollution Control,* CRC Lewis Press.

Heinsohn, R. J., and Kabel, R. L., 1999. *Sources and Control of Air Pollution,* Prentice Hall.

Leonard, R. L., 1997. *Air Quality Permitting,* CRC Lewis Press.

Mycock, J. C., McKenna, J. D., and Theodore, L., 1995. *Handbook of Air Pollution Control Engineering and Technology.*

Noll, K. E., 1999. *Fundamentals of Air Quality Systems,* American Academy of Environmental Engineers.

Perry, R. H., Green, D. W., and Maloney, J. O. eds., 1997. "Gas Absorption and Gas-Liquid System Design," Chap. 14 in *Perry's Chemical Engineers' Handbook,* 7th ed., McGraw-Hill.

Treybal, R. E., 1980. *Mass Transfer Operations,* McGraw-Hill.

Wark, K., Warner, C. F., Davis, W. T., 1998. *Air Pollution: Its Origin and Control,* 3d Ed., Addison-Wesley.

U.S. EPA Air Pollution Control Costs, *http://www.epa.gov/ttn/catc/products.html#cccinfo.*

U.S. EPA Office of Air Quality Planning and Standards, *http://www.epa.gov/airs/airs.html.*

U.S. EPA Emission Factors Database, *http://www.epa.gov/ttn/chief.*

U.S. EPA Technology Transfer Website, *http://www.epa.gov/ttn/f.*

CHAPTER 4

AN EVALUATION OF ENVIRONMENTAL DREDGING FOR REMEDIATION OF CONTAMINATED SEDIMENT

J. Paul Doody

Blasland, Bouck & Lee, Inc.
Syracuse, New York

Bradford S. Cushing

Applied Environmental Management, Inc.
Malvern, Pennsylvania

4.1 INTRODUCTION

This chapter examines the role of environmental dredging in the efforts to reduce risks and protect human health and the environment from chemicals in sediments. Bioaccumulative chemicals are a particular focus because reduction to levels acceptable to some regulatory agencies requires achieving low residual concentrations in water and sediments in contact with water. Achieving this goal now and in the future is problematic. It warrants careful analysis to determine which portion of the contaminants in sediments is bioavailable and an accurate assessment of the capabilities and limitations of the various remedial technologies, including dredging, to achieve these low levels. Despite increasing reliance upon dredging, to date there has been no systematic evaluation of how effective environmental dredging projects have been in controlling risks from contaminants in sediments. However, a number of projects have been undertaken that allow such an evaluation to be made, which provides an opportunity to learn what works and what does not.

To that end, this chapter reviews major sediment remediation projects undertaken in the United States and summarizes key aspects of these projects, such as the objectives of the sediment remediation projects, the technologies being employed, and the capabilities and limitations of those technologies. Finally, recommendations are provided on needed programmatic change. Project details are provided in the associated tables and appendices.

The key findings of this analysis are:

- Dredging has become the "default" remedy for contaminated sediments.
- The current approach for evaluating the ability of dredging remedies to control risk lacks rigor and is not based on a sound scientific understanding of contaminant dynamics in aquatic systems.
- There has not been a systematic experience-based review of the capabilities and limitations of dredging technology in reducing risks posed by contaminated sediments. Thus, an opportunity exists to apply lessons learned from the current base of experience that can help guide future decision making.
- From an evaluation of completed dredging projects, we now have useful (albeit limited) information on the capabilities and limitations of dredging technology. The data on postdredging residual contaminant levels in surface sediments, contaminant resuspension, production rates, and costs need to be more rigorously used in the evaluation of dredging technology in sediment remedy decisions.
- While much effort is dedicated to evaluating risk posed by contaminated sediments, there has been no equivalent effort to evaluate risks from implementing remedies. No guidance is available on how to perform such evaluations nor on how to compare the potential benefits of a project to the impacts. Given the potential impacts on local communities and the aquatic ecosystem, there should be confidence that the risk reduction benefits are real and outweigh the adverse impacts. In general, risks from site contaminants are often overstated because they are based on conservative assumptions under the guise of the precautionary principle and typically assume unrealistic exposure scenarios for these risks.

The national sediment remediation program needs to incorporate these findings and recognize the technical limitations and inherent disadvantages of dredging. This will require a decisional framework that incorporates the considerations identified and discussed in this chapter. It will also require coherent and thorough data collection and analysis. If conditions before and after a remedy are not measured, one cannot tell whether dredging has made conditions better or worse. In fact, in its report, "A Risk-Management Strategy for PCB-Contaminated Sediments" the National Research Council (NRC, 2001) has recognized the limitations of dredging and proposes the use of a risk-based approach.

4.2 BACKGROUND

Risk to human health and the environment from contaminants in sediments is a concern to both state and federal governments. Approximately 100 of the sites currently targeted for cleanup under the Comprehensive Environmental Response, Compensation, and Liability Act (CERCLA) involve aquatic-related contamination (NRC, 1997). The U.S. Environmental Protection Agency (EPA) estimates that about 10 percent of the sediment underlying our waterways, some 1.2 billion cubic yards, is contaminated and may need some form of cleanup or recovery effort (EPA, 1997a).

Dredging, a term used here to include both wet and dry excavation, for environmental restoration ("environmental dredging") is increasingly used in an attempt to manage the risks posed by contaminated sediments. In contrast, the goal of navigational dredging, which has long been used to create or maintain waterways for commercial shipping and other maritime purposes, is to remove large volumes of sediments, not to reduce risk.

This chapter evaluates current efforts by the government to manage risks from contaminated sediments in waterways, with particular focus on the effectiveness of dredging to control risks to human health and the environment—the method most commonly employed to control those risks. Although government policy states that the goal of sediment remediation is "risk reduction" to protect human health and the environment, this evaluation shows that cleanup decisions rarely contain a clear line of reasoning showing how the selected project will achieve these goals. Most of the decisions appear to be based on the simple, yet largely incorrect, assumption that removing a percentage of the contaminant mass from the sediment will result in a roughly equivalent reduction in risks. This approach is referred to as *mass removal*. Our review shows, however, that this approach is flawed.

The information underlying this review is taken primarily from the Major Contaminated Sediment Sites (MCSS) Database (Release 3.0), which was commissioned by General Electric Company (available at www.hudsonvoice.com). The MCSS database assembles, for the first time, available information concerning remedies at the major contaminated sediment sites in the United States and elsewhere. This chapter offers a review of experiences at several contaminated sediment sites and points to how this experience can be applied to develop a coherent framework for future decision making based on the goal of effectively reducing risks to human health and the environment.

4.3   UNDERSTANDING THE PROBLEM

An accurate understanding of contaminant fate in waterways is essential to devising an effective strategy to reduce risks posed by chemicals in sediment. We begin with a brief overview of how contaminated sediments create potential risks to human health and the environment. This involves two key concepts. First, it is only the contaminants within the biologically active, uppermost layer of the sediment bed that are available for uptake by sediment-dwelling organisms and fish or susceptible to migration downstream. Second, and a direct corollary to the first point, contaminants buried below the bioavailable zone present a risk only if the overlying sediment is subject to significant erosion or other mechanical disturbance, or if groundwater moves the contaminants upward through the sediments, thus creating the possibility that the buried contaminants might make their way to the surface and become bioavailable. Appendix 4A provides a more detailed review of sediment contaminant dynamics.

Consequently, if a buried chemical mass is stable and is not and will not become available to the water column or biota, the human health and ecological risks at that site will not be reduced by removing that mass. As obvious as this conclusion is, it is frequently overlooked because the greatest mass of contaminants is often found in buried sediments. It is important to remember that most of the contaminants in sediments are the result of waste disposal practices that began 50 to 60 years ago and largely ceased 20 to 25 years ago. The fact that the chemical mass remains buried 25 to 50 years after it entered the sediment is strong evidence that it is associated with stable sediments and is unlikely to migrate to the surficial bioavailable layer in any significant way. Dredging is effective in removing sediment mass to, for example, clear a clogged navigational channel. However, removing chemicals that are not available to the food chain or the water column does not reduce risks. In fact, removing the surface layers may expose otherwise stable buried sediments with contaminants at higher concentrations, making them bioavailable and thereby increasing risks.

Thus, although targeting sediment deposits with the highest chemical concentration through dredging (mass removal) may intuitively make sense, thorough analysis to test this intuition is critical. When evaluating remedial options, it is necessary to evaluate both the sources of contaminants to the bioavailable surface layers and the capabilities of different technologies to reduce risks posed by contaminated sediments. The analysis begins with the identification of contaminant sources to the bioavailable surface. If the sources are unstable deposits subject to erosion, then the focus should be on finding and remediating these deposits. If the bioavailable surface layer is not receiving contaminants from elsewhere, then methods for accelerating the remediation of the surface layer should occur. If the chemicals in the surficial sediments come from on-shore sources, those sources must be controlled. A particularly important consideration, largely overlooked in previous decisions, is the inability of dredging equipment to achieve low levels of contaminants in the bioavailable surface sediments. Last but not least, one needs to compare the potential benefits from dredging (or any other remedy) against the potential harm to the ecosystem and risks to workers and communities. A large-scale dredging project can have devastating impacts on sensitive ecological habitats, and, like any large construction project, carries with it both significant risks to workers and disruption to local communities.

Only after all of these factors are considered can one make a reasoned, well-informed remedy selection. Unfortunately, our review indicates that regulators are not adequately taking these fundamental considerations into account. The bottom line is that a rigorous analysis of the contaminant source and fate in the aquatic system is required before an effective remedy can be evaluated and selected.

4.4 CURRENT REGULATORY APPROACH

Most contaminated sediment sites are subject to one of the federal or state cleanup programs, such as the federal CERCLA, commonly known as Superfund, the federal Resource Conservation and Recovery Act (RCRA), or comparable state laws. Although differences exist among these laws, they all have the primary goal of ensuring that cleanups manage risks from contaminants so as to protect human health and the environment.

Although risk management is the stated goal of many sediment remedial projects, experience shows that dredging has become the default remedy for managing contaminated sediments, with little apparent consideration given to whether dredging actually reduces risks. The presumption appears to be that the dredging will effectively control risks even though objective analysis is usually not provided to support such a presumption. For example, of the 64 completed projects in the MCSS database (listed in Table 4.1), 58 have used dredging or excavation, as summarized below.

Types of Remedies Implemented for
64 Completed Projects

Remedy implemented	Times selected*
Dredging[†]*	32
Wet/dry excavation	27
Natural recovery/burial[‡]	4
Engineered capping[§]	2

* One project used dredging and wet/dry excavation.
[†] Includes diver-assisted/hand-held dredging.
[‡] Four others have natural recovery as a component of the overall remedy.
[§] Portions of four other sites were capped following removal because of elevated surface sediment concentrations.

For the purposes of this analysis, dredging is defined as the underwater removal of sediments using mechanical (e.g., clamshell mounted on a barge) or hydraulic (e.g., cutterhead dredge) means. Diver-assisted dredging, which involves a diver removing sediments using a flexible suction hose connected to a land- or barge-based pump, is included under the dredging category. Wet excavation involves removal of underwater sediments by using conventional excavation equipment (e.g., backhoe positioned on a barge or on shore). Dry excavation involves diverting water flow and dewatering the area targeted for removal. Once dewatered, the sediments are removed with conventional excavating equipment (e.g., bulldozers, backhoes).

It is not clear why dredging has become the default remedy at sediment sites because the basis for selecting dredging as the remedy is generally inconsistent and vague. Table 4.2 provides a detailed summary of the stated goals, apparent or known basis for decisions, and reported outcomes relative to remedial goals and specific objectives for 29 sites having 10,000 yd^3 or more removed. A review of the MCSS database shows that decisions at sediment sites rarely are based on site-specific, quantitative analysis of risk. Instead, regulators often use default sediment cleanup values or seek to remove a large mass of contaminants regardless of whether such approaches will actually reduce risk. The variability and absence of stated goals is symptomatic of the confusion surrounding sediment remediation and the absence of a clear and consistently applied decision-making framework.

Our analysis also shows that the agencies responsible for these decisions and for implementing or overseeing sediment cleanups have not implemented reasonably thorough programs to assess whether cleanup efforts have successfully reduced risks. Several years of high-quality and comparable data before and after remediation are essential to assess the effectiveness of sediment removal in reducing contaminant levels in fish and the associated reductions in contaminant bioavailability, exposure, and risk. An adequate sampling program, database, and evaluation methodology should include the ability to: (1) distinguish the effect of removal from the effects of other recovery processes such as the natural burial, transport, decomposition, or containment of chemicals; (2) reduce the uncertainties inherent in field sampling of biota; and (3) account for the long biological half-lives of strongly hydrophobic chemicals, such as PCBs, that can delay the response of fish tissue levels to changes in exposure. These important data are simply not available for virtually all of the sediment

TABLE 4.1 Summary of Remediated Contaminated Sediment Sites

Project	U.S. EPA region	Setting	Contaminant of concern	Methods of remediation and disposal	Volume removed, yd³	Total cost, millions of dollars	Total unit cost, $/yd³
Baird & McGuire, MA	I	3-mile sector of the Cochato River and several tributaries	As, DDT, chlordane, PAHs	Dry/wet excavation; on-site incineration[a]; natural recovery	4,712	0.9[a]	186
Bay Road Pond, NY	II	Slightly >1-acre pond (26,000 ft²)	PCBs	Dry excavation; commercial landfill	3,210	1.3	405
Bayou Bonfouca, LA	VI	Turning basin and 4000 ft of bayou	PAHs	Mechanical dredging, on-site incineration	169,000	115	680
Black River, OH	V	Two hot spots totaling 8 acres	PAHs	Hydraulic dredging and mechanical dredging; on-site landfill	60,000	5	83
Bryant Mill Pond, MI	V	22-acre 2500-ft-long Bryant Mill Pond area of Portage Creek	PCBs	Dry/wet excavation; on-site former dewatering lagoons	165,000	7.5[e]	45
Cannelton, Industries, MI	V	0.8-mile near-shore area of the St. Mary's River	Metals (Cd, Pb, As, Cr, Hg)	N/A (containment, monitoring, and natural recovery)	N/A	N/A	N/A
Cherry Farm, NY	II	Approximately 1600 ft of shore-line (full length of site) extending about 150 ft into river	PAHs	Hydraulic dredging; on-site existing disposal pond	42,445	2.2	52
Convair Lagoon, CA	IX	10-acre embayment	PCBs	Engineered three-layer cap over 5.7 acres	N/A	2.75[c]	N/A
Cumberland Bay, NY	II	34-acre contaminated sludge bed in the 75-acre Cumberland Bay	PCBs	Hydraulic dredging and diver-assisted removal; commercial landfill	195,000	34.5[c]	177
DuPont Newport Plant, DE	III	1.5-mile sector of the Christiana River	Metals (Pb, Cd, Zn); solvents	Mechanical dredging; on-site landfill disposal	11,870	2.3	194
Duwamish Waterway, WA	X	Slip	PCBs	Divers (hand-held dredging techniques); pneumatic dredging; off-site disposal ponds	10,000	N/A	—

TABLE 4.1 Summary of Remediated Contaminated Sediment Sites (*Continued*)

Project	U.S. EPA region	Setting	Contaminant of concern	Methods of remediation and disposal	Volume removed, yd^3	Total cost, millions of dollars	Total unit cost, $/yd^3$
Eagle (West) Harbor, WA	X	Puget Sound Embayment comprising about 200 acres of West Harbor	Mercury, PAHs	Mechanical dredging, wet excavation, thin-layer capping, and enhanced natural recovery; nearshore CDF, commercial landfill, and in situ capping	3,000	3	1000
Ford Outfall, MI	V	2.6-acre nearshore area (about 750 ft long by 150 ft wide)	PCBs	Mechanical dredging; on-site landfill	28,500	5.65	198
Former Messer Street MGP, NH	I	Multiple areas in the Winnipesaukee River; combined area is approximately 3 acres	PAHs	Mechanical dredging and dry/wet excavation; commercial thermal description	12,500[b]	N/A	—
Formosa Plastics, TX	VI	1.1 acres (about 150 by 350 ft) in corner of an active turning basin	EDC	Mechanical dredging; commercial landfill	7,500	1.4	187
Fox River, WI (SMU 56/57)	V	9-acre depositional area in river	PCBs	Hydraulic dredging; commercial landfill	81,000	20	247
Fox River, WI (Deposit N)	V	Approximately 3-acre depositional area	PCBs	Hydraulic dredging; commercial landfill	8,175	4.3	526
Gill Creek, NY (DuPont)	II	250-ft sector of Gill Creek near its confluence with Niagara River	PCBs, PAHs	Dry/wet excavation; commercial landfill	8,020	12[c]	1,496
Gill Creek, NY (Olin Industrial Welding Site)	II	About 1800-ft length of Gill Creek bed	BHCs, PAHs, mercury	Dry/wet excavation; use as on-site fill material	6,850	N/A	—
GM (Massena), NY	II	11-acre, 2500-ft-long near-shore area in the St. Lawrence River	PCBs	Hydraulic dredging, wet excavation, and capping; commercial landfill[a]	13,800	12.3	891

TABLE 4.1 Summary of Remediated Contaminated Sediment Sites (*Continued*)

Project	U.S. EPA region	Setting	Contaminant of concern	Methods of remediation and disposal	Volume removed, yd³	Total cost, millions of dollars	Total unit cost, $/yd³
Gould (Portland), OR	X	3.1-acre East Doane Lake remnant, a shallow impoundment	PAHs	Hydraulic dredging; on-site landfill	11,000	3	273
Grasse River, NY	II	1-acre near-shore hot spot in river	PCBs	Hydraulic dredging, wet excavation, and diver assisted; on-site landfill	3,000	4.9	1633
Hooker (102nd St.), NY	II	25 acres in an embayment in the Niagara River	VOCs, metals	Dry/wet excavation; on-site landfill	28,500	N/A	—
Housatonic River, MA	I	550-ft sector of the river	PCBs	Dry/wet excavation; commercial landfill	6,000 (sediment and banks)	4.5	750
James River, VA	III	81-mile-long estuary; 0.6 to to 7 miles in width	Kepone	In situ; natural recovery	N/A	N/A	N/A
Ketchikan (Ward Cove), AK	X	80 acres within the approximately 250-acre Ward Cove	Ammonia, sulfide, and 4-methylphenol	Mechanical dredging; industrial landfill	8,700 (paid); 11,865 (total)	1.4	159 (paid)
Lake Jarnsjon, Sweden	N/A	62-acre lake (bank-to-bank removal)	PCBs	Hydraulic dredging; on-site dedicated landfill	196,000	6.5	33
Lavaca Bay, TX	VI	One deep and one shallow bay area comprising about 7 acres	Mercury	Hydraulic dredging; on-site existing disposal ponds	79,500	2.1	26
LCP Chemical, GA	IV	13-acre tidally influenced marsh area; one-half mile of an outfall channel; a separate natural drainage channel	PCBs; mercury	Wet excavation; bucket-ladder dredge; commercial landfill	25,000	10	400
Lipari Landfill, NJ	II	18 acres of Alcyon Lake; 5 acres of Chestnut Branch Marsh; Chestnut Branch Stream	Multiple organics, inorganics	Dry/wet excavation; some thermal desorption and beneficial reuse; some stabilization and placement	163,500	50	306

TABLE 4.1 Summary of Remediated Contaminated Sediment Sites (*Continued*)

Project	U.S. EPA region	Setting	Contaminant of concern	Methods of remediation and disposal	Volume removed, yd³	Total cost, millions of dollars	Total unit cost, $/yd³
Loring AFB, ME	I	>2500-ft-long Flightline Drainage Ditch; 15-acre Flightline Drainage Ditch Wetland (about 2000 by 400 ft); >2500-ft-long East Branch Greenlaw Brook	PCBs, PAHs	Dry/wet excavation; on-site landfill	162,000	13.85	85
Love Canal, NY	II	About 10,000 linear ft of Black and Bergholtz Creeks	TCDD	Dry/wet excavation; commercial incineration[a]	31,000	14[a]	452
LTV Steel, IN	V	3500 ft of intake flume (width ranges from 96–467ft)	PAHs, oils	Hydraulic dredging and diver-assisted removal; commercial landfill	109,000	12	115
Mallinckrodt Baker, NJ (formerly J.T. Baker)	II	Near-shore hot spot (about one-half acre) in the Delaware River	DDT	Dry/wet excavation; on-site landfill	3750[b]	1.2	320
Manistique River, MI	V	One 2-acre hot spot in dead-end and backwater area; two other hot spots: one of 2 acres in the river and one of 15 acres in the 97-acre harbor	PCBs	Hydraulic dredging; commercial landfill	136,000	45.2	332
Marathon Battery, NY	II	200 acres of open cove and a small cove in the Lower Hudson River	Cadmium	Hydraulic dredging and mechanical dredging; natural recovery; commercial landfill	77,200	10[c]	130
Marathon Battery, NY[f]	II	340 acres of backwater marshes and sheltered cove	Cadmium	Dry/wet excavation; commercial landfill	23,000	N/A	—
Menominee River (Eighth Street Slip), WI	V	One to 2-acre slip adjacent to Menominee River	Arsenic	Hydraulic dredging; commercial landfill	10,000	N/A	N/A
National Zinc, OK	VI	5300 ft of the north tributary (unnamed) of Eliza Creek	PCBs	Dry/wet excavation; commercial landfill	6,000	N/A	—

TABLE 4.1   Summary of Remediated Contaminated Sediment Sites (*Continued*)

Project	U.S. EPA region	Setting	Contaminant of concern	Methods of remediation and disposal	Volume removed, yd^3	Total cost, millions of dollars	Total unit cost, $/yd^3
Natural Gas Compressor Station, MS	IV	2-mile length of Little Conehoma Creek	PCBs	Dry excavation; commercial landfill	71,700 (includes floodplain soils)	11	153
New Bedford Harbor, MA	I	5 acres of hot spots in the estuary	PCBs	Hydraulic dredging; commercial landfill[a]	14,000	28.1	2007
Newburgh Lake, MI	V	105-acre man-made lake	PCBs	Dry/wet excavation; commercial landfill	588,000	11.8	20
N. Hollywood Dump, TN	IV	40-acre man-made lake adjacent to the Wolf River	Pesticides	Hydraulic dredging; on-site burial in an isolated oxbow	40,000	2.4	60
Ottawa River (Capping with AquaBlok), OH	V	Three 1-acre areas in the Ottawa River	PCBs, PAHs, heavy metals	Capping using articulating arm conveyor, helicopter, and land-based dragline (clamshell bucket)	N/A	N/A	N/A
Ottawa River (unnamed trib.), OH	V	Unnamed tributary about 975 ft long and 90 ft wide at its mouth, and tapering to 10 ft wide at its origin	PCBs	Dry/wet excavation; commercial landfill	9,692	5	516
Pettit Creek Flume, NY	II	1-acre cove in the Durez Inlet of the Little Niagara River	DNAPLs (VOCs and semivolatiles)	Diver-assisted dredging; portion to commercial hazardous waste landfill	2,000	N/A	—
Pioneer Lake, OH	V	200 × 240 ft (depth: 0.5 to 3 ft) area of southern lake	PAHs	Hydraulic dredging; commercial landfill	11,100	2.5	225
Queensbury NMPC, NY	II	An area of the Hudson River extending 180 ft off shore and 800 ft downstream from site	PCBs	Dry/wet excavation; commercial landfill	4,750[b]	3.5	737
Ruck Pond, WI	V	800–1000 ft long by 75–100 ft wide impoundment in Cedar Creek	PCBs	Dry/wet excavation; commercial landfill	7,730	7.5	970

TABLE 4.1 Summary of Remediated Contaminated Sediment Sites (*Continued*)

Project	U.S. EPA region	Setting	Contaminant of concern	Methods of remediation and disposal	Volume removed, yd^3	Total cost, millions of dollars	Total unit cost, $/yd^3
Saginaw River, MI	V	6 near-shore areas totaling about 58 acres in the Saginaw River	PCBs	Mechanical dredging; confined disposal facility	345,000	9.7	28
Sangamo-Weston, SC	IV	7-mile sector of Twelvemile Creek and 730 acres of Lake Hartwell	PCBs	In situ; enhanced sedimentation and natural recovery	N/A	N/A	N/A
Selby Slag, CA	IX	Near-shore area of about 17 acres (fronting on 61.5 acres of shoreline and extending into the water about 280 ft)	Lead	Mechanical dredging; on-site disposal as fill	101,000^b	2.1	21
Sheboygan River, WI	V	17 small hot spot areas in the upper 3.2 miles of river immediately downstream of the PRP site	PCBs	Mechanical dredging, wet excavation, and capping; on-site storage (temporary)	3,800	7^a	1842
Shiawassee River, MI	V	1.5-mile stretch of the South Branch of the Shiawassee River	PCBs	Dry/wet excavation; commercial landfill	1,805	1.3	720
Starkweather Creek, WI	V	About 1 mile upstream of the confluence of the east and west branches of Starkweather Creek	Mercury (primary); also lead, zinc, cadmium, and oil and grease	Dry excavation; on-site disposal in former dewatering lagoons	15,000	1.0	67
Sullivan's Ledge, MA	I	12-acre area that includes Unnamed Stream sediment and floodplain soils and golf course water hazards; 7 acres of wetland in the Middle Marsh and adjacent wetland areas	PCBs, PAHs	Dry excavation; on-site burial	N/A	N/A	N/A
Tennessee Products, TN	IV	2.5-mile sector of the Chattanooga Creek	Coal tar	Dry/wet excavation; off-site fuel source and commercial landfill	24,100	12	498

TABLE 4.1 Summary of Remediated Contaminated Sediment Sites (*Continued*)

Project	U.S. EPA region	Setting	Contaminant of concern	Methods of remediation and disposal	Volume removed, yd^3	Total cost, millions of dollars	Total unit cost, $/yd^3
Terry Creek, GA	IV	3-acre Outfall Ditch and Mouth and 6 areas totalling 5 acres in Terry Creek	Toxaphene	Mechanical dredging; commercial landfill	30,000	3	100
Town Branch Creek, KY	IV	3.5-mile sector of the Town Branch Creek	PCBs	Dry/wet excavation; commercial landfill	76,000 (sediment and banks); 163,000 (flood-plains)	N/A	N/A
Triana/ Tennessee River, AL	IV	11-mile stretch of two tributaries of the Tennessee River	DDT	Rechannel-ization and in-situ burial	N/A	30	N/A
United Heckathorn, CA	IX	Lauritzen Channel ~1600 ft long by 200 ft wide; Parr Canal about 1000 ft long by 70 ft wide	DDT	Mechanical dredging; commercial landfill	108,000	12^e	111
Velsicol Chemical (Pine River), MI	V	3-acre hot spot in St. Louis Impoundment	DDT, HBB, PBB	Dry excava-tion following stabilization; commercial landfill	35,000	7.8	246
Waukegan Harbor (Outboard Marine), IL	V	10 acres of 37-acre harbor; abandoned boat harbor; aban-doned boat slip no. 3; and a north ditch that flowed directly into Lake Michigan	PCBs	Hydraulic dredging; near-shore CDF	38,300	15	392
Willow Run Creek, MI	V	Edison and Tyler Ponds— 21 acres com-bined; Willow Run Sludge Lagoon	PCBs	Dry/wet exca-vation; nearby new on-site landfill	450,000	70	156
Rounded totals					3,950,859^g	618.7^g	420^g (mean)

^{a} Does not include disposal cost. Several years delay to determine disposal method.
^{b} Final volume is a range; midpoint is listed.
^{c} Cost is a range; midpoint is listed.
^{d} Cost listed is a midpoint; actual not determined.
^{e} Cost is a minimum, actual not determined.
^{f} Listed twice since both dredging and dry excavation were used.
^{g} Does not include sites without either volume or cost data.
N/A = not applicable.

TABLE 4.2 U.S. Sediment Remediation Projects Implemented (>10,000 yd³)—Primary Goal versus Outcome

Project	Primary goal	Basis for primary goal	Sediment remedial target	Relationship of target to goal	Remediation method	Achievement of remedial target	Achievement of primary goal
Saginaw River, MI[a] (345,000 yd³)	Reduce PCB mass within the river by 80%.	Consent judgment based on NRD claims	Depth horizon	Direct	Mechanical dredging	Depth horizon achieved; no analytical verification	Likely accomplished, but not verified
Cumberland Bay, NY[a] (195,000 yd³)	Reduce PCB levels in fish and minimize potential direct human exposures.	Human health and ecological risk assessments	Depth horizon (remove sludge bed down to original bottom)	Direct	Hydraulic dredging and diver-assisted removal	Depth horizon achieved; 51 samples from 42 cores collected following dredging and analyzed for PCBs ranged from 0.04 to 18.0 ppm and averaged 6.82 ppm	Unknown; post-monitoring program yet to be implemented
Bayou Bonfouca, LA[a] (169,000 yd³)	Reduce PAH human contact risk to <10⁻⁴ and minimize threat to aquatic biota.	Human health risk assessment	Depth horizon to achieve <1300 ppm PAHs	Direct	Mechanical dredging followed by fill	Depth horizon achieved; no analytical verification	Likely accomplished, particularly since fill was added to the dredged areas. However, postmonitoring consists of the state annual monitoring program for water, sediment, and fish and seems hit or miss. Also, it is unclear if targeted surface PAH levels were achieved since a sediment contact and swimming advisory is still in effect because of PAHs in sediment samples exceeding EPA guideline values, but not verified.

TABLE 4.2 U.S. Sediment Remediation Projects Implemented (>10,000 yd³)—Primary Goal versus Outcome (*Continued*)

Project	Primary goal	Basis for primary goal	Sediment remedial target	Relationship of target to goal	Remediation method	Achievement of remedial target	Achievement of primary goal
Manistique River, MI[a] (136,000 yd³)	Reduce PCB in fish levels, reduce carcinogenic and noncarcinogenic risks to <10⁻⁴ and <1, respectively, except for high-end subsistence and some high-end recreational exposure from fish consumption.	Human health risk assessment	10 ppm PCBs	Default level after using biota to sediment accumulation factor (BSAF) to estimate a target sediment level, then increasing the estimate to 10 ppm PCBs, which EPA justified on the basis of cleanup levels at other EPA projects, the likelihood of achieving <10 ppm, and future natural burial	Hydraulic dredging	In progress; consistent achievement of 10 ppm or less proving difficult	Too soon to tell. No postmonitoring program defined yet.
LTV Steel, IN[a] (109,000 yd³)	Remove all oil-contaminated sediments from a 3500-ft man-made intake channel.	Clean Water Act Consent Decree	Depth horizon (removal down to original bottom)	Direct	Hydraulic dredging and diver-assisted removal	Depth horizon achieved; no analytical verification	Likely accomplished, but not verified
United Heckathorn, CA[a] (108,000 yd³)	Achieve EPA marine chronic water quality criteria of 1 part per trillion (ppt) DDT; achieve human health surface water criteria of 0.6 ppt DDT; achieve the National Academy of Sciences action levels for DDT in fish to protect	Ecological risk assessment	Remove all "young bay mud" to achieve <0.59 ppm DDT	Indirect (calculated in the ecological risk assessment)	Mechanical dredging	Depth horizon (penetration into "old bay mud") achieved; 20 samples for chemical analysis collected from top 6 in of final dredged surface for informational purposes (several exceeded 0.59 ppb DDT)	Too soon to tell; postmonitoring in progress

Site	Objective	Basis	Target/Cleanup Level	Prediction	Method	Achievement	Status
Battery, NY[a] (102,000 yd³)	ecological impacts by achieving 100 ppm cadmium in sediment in East Foundry Cove (EFC) Marsh and 10 ppm cadmium in other areas; allow natural recovery in over 300 acres of adjacent cove/marsh.	ment based on "weight of evidence," bioassay tests, and comparison with ambient water-quality standards	1 ft of sediment in areas targeting 10 ppm cadmium; remove to <100 ppm cadmium in EFC marsh; allow natural recovery in over 300 acres.	95% cadmium mass removal predicted	mechanical dredging; dry excavation	1 ft; decided to take verification samples for analysis in some areas; achieved an average of 25 ppm cadmium in EFC Marsh; achieved an average of <10 ppm cadmium in EFC and near pier	progress. Two years of reported results are inconclusive.
Black River, OH[a] (60,000 yd³)	Remove all PAH- and metal-contaminated sediments.	Clean Air Act Consent Decree	Depth horizon (removal down to "hard bottom" or "bedrock")	Direct	Hydraulic and mechanical dredging	Depth horizon achieved; no analytical verification	Likely accomplished, but not verified
Cherry Farm, NY[a] (Niagara River) (42,445 yd³)	Reduce PAH-related risks to benthic aquatic life and fish.	Ecological and biotoxicity testing; literature review for ecotoxicity of PAHs	Depth horizons based on characterization data to achieve 20 ppm PAHs in the top 1 ft; 50 ppm PAHs below 1 ft	Vague; target levels set by negotiation and by comparing PAH levels to upstream background levels	Hydraulic dredging	Achieved depth horizons based on bathymetry; no analytical verification	Unknown; postmonitoring program being negotiated
N. Hollywood Dump, TN[a] (40-acre lake) (40,000 yd³)	Restore the pesticide-contaminated fishery in the lake so that it is suitable for human consumption.	Human health risk assessment	Remove or isolate pesticide-contaminated surface sediments.	Direct	Fish harvesting first, then part hydraulic dredging/part direct burial	Achieved	Too soon to tell; long-term biannual fish and sediment sampling in progress
Outboard Marine, IL (Waukegan Harbor)[a] (38,300 yd³)	Eliminate PCB flux from the harbor into Lake Michigan.	Hydrodynamic modeling	50 ppm PCBs in the harbor; 500 ppm PCBs in slip no. 3	Direct for the harbor; unknown for the 500 ppm target in slip no. 3	Hydraulic dredging	Unknown. No analytical verification. Dredged to a predefined depth in the harbor to the reportedly uncontaminated sand layer.	Unknown. Some limited analysis of surface samples at undefined locations in the harbor over 4 years after dredging exhibited 3 to 9 ppm PCBs. PCB levels in harbor fish are trending downward.

TABLE 4.2 U.S. Sediment Remediation Projects Implemented (>10,000 yd³)—Primary Goal versus Outcome (*Continued*)

Project	Primary goal	Basis for primary goal	Sediment remedial target	Relationship of target to goal	Remediation method	Achievement of remedial target	Achievement of primary goal
Terry Creek, GA[a] (30,000 yd³)	Reduce toxaphene mass within the site boundaries by 90%.	Not identified	Depth horizon	Direct	Mechanical dredging	Depth horizon achieved; analytical verification performed but results not available	Likely accomplished, but not verified
Ford Outfall, MI (River Raisin)[a] (28,500 yd³)	Reduce PCB levels in fish.	Risk analysis by EPA	10 ppm PCBs after removal down to the native clay layer	Direct	Mechanical dredging	Partially achieved. Removal to refusal was accomplished. Verification by field test kits, then 14 samples (one per quadrant) for laboratory analysis; seven quadrants had insufficient sediment to collect; four quadrants exhibited 0.5 to 7 ppm PCBs; three quadrants exhibited 12 to 20 ppm PCBs.	Unknown. No formal postmonitoring program identified. Results of fish samples and caged fish studies from a monitoring program performed by MI Department of Environmental Quality (MDEQ) are not yet available. Two postremoval sediment core samples taken by MDEQ from the dredged area exhibited 60 and 110 ppm PCBs.
New Bedford Harbor, MA[a] (14,000 yd³)	Remove PCB mass at an optimum "residual concentration to volume removed" ratio and reduce PCB flux to the water column (interim measure).	Mass removal calculations; flux modeling studies conducted by PRPs; water column data	4000 ppm PCBs in 5 acres of hot spots	Direct	Hydraulic dredging	Achieved according to a limited number of verification samples (15 composite samples ranging from 67 to 2068 ppm PCBs)	Achieved mass removal. Water column data postdredging (if collected) not obtained. PCBs in surface sediment samples in the Upper Harbor increased 32% on average, following hot spot dredging.
GM (Massena), NY[a] (13,800 yd³)	Reduce PCB levels in fish.	Human health risk assessment	Achieve 1 ppm PCBs and remove as much sediment as technically feasible	Vague; 0.1 ppm PCBs desired, but 1 ppm selected on basis of technical feasibility	Hydraulic dredging	Not achieved. Average residual PCB levels at completion in six dredged quadrants across 11 acres ranged from 3 to 27 ppm with a maximum of 90 ppm.	Two annual postdredging fish monitoring programs completed. No discernible trends other than a slight increase in fish PCB concentration in year

Site	Remediation objective	relevant and appropriate requirement (ARAR)	Cleanup level		Remediation method		
(11,000 yd^3)	protect from direct contact risk and remove lead-contaminated surface (0 to 2 ft) sediments that exceed the extraction procedure (EP) toxicity concentration.				followed by filling in the 3.1-acre lake		verification sampling
Newburgh Lake, MI[b] (588,000 yd^3)	Restore 105-acre lake depth and restore fishery.	Not identified	Depth horizon that will both restore depth and remove the detectable PCBs	Direct	Dry excavation supplemented by hydraulic dredging in undrained bypass channel through the lake	Depth horizon achieved; no analytical verification	Achieved, but no anlytical verification. Fish harvested and restocked. Postmonitoring not identified.
Willow Run Creek, MI[b] (450,000 yd^3)	Eliminate adverse ecological impacts.	Ecological assessment based on ecological ingestion modeling, then feasibility and compliance with MI Environmental Response Act 307	Removal to 21 ppm or 1 ppm PCBs below waterline depending on locale; removal to 21 or 2.3 ppm PCBs above waterline	Direct	Dry excavation	Achieved according to verification sampling	Unknown. No formal postmonitoring is planned.
Lipari Landfill, NJ[b] (163,500 yd^3)	Reduce human health risk from direct contact with or air exposure to targeted VOCs to below 10^{-6}.	Human health risk assessment	Depth horizon 6 in into the underlying Kirkwood Clay layer to achieve nondetection for bis(2-chloroethyl)ether	Direct	Dry excavation	Depth horizon achieved except in areas where no Kirkwood Clay was encountered, in which instances excavation continued to 18 in below a level extrapolated from adjacent contiguous clay layers; no analytical verification.	Apparently achieved, particularly since clean fill was also placed. No postmonitoring identified.

TABLE 4.2 U.S. Sediment Remediation Projects Implemented (>10,000 yd³)—Primary Goal versus Outcome (*Continued*)

Project	Primary goal	Basis for primary goal	Sediment remedial target	Relationship of target to goal	Remediation method	Achievement of remedial target	Achievement of primary goal
Bryant Mill Pond, MI (Kalamazoo River)[b] (165,000 yd³)	Mitigate the public health threat posed by direct human and wildlife contact and mitigate threats posed to aquatic life and wildlife by ongoing releases (i.e., source control) to the Kalamazoo River.	Ecological risk assessment along with direct observation of continuing releases by erosion and sloughing from banks	10 ppm PCBs	Unknown	Dry excavation	Reportedly achieved according to verification sampling; sample results not obtained or reviewed	Unknown and probably too early to tell since removal was completed in June 1999. However, as stated in the Action Memorandum, "the nature of the removal is, however, expected to minimize the need for post-removal site control, at least in the Bryant Mill Pond area."
Loring AFB, ME[b] (162,000 yd³)	Reduce human health risk to below 10^{-6} and below a hazard index of 1 and eliminate adverse ecological impacts.	Human health and ecological risk assessments	Various for specific contaminants (e.g., 1 ppm Aroclor 1260, 35 ppm total PAHs)	Direct	Dry excavation	Apparently achieved based on verification sampling for PCBs and less rigorous testing for five other indicator compounds	Too soon to tell. A long-term environmental and wetlands monitoring plan was finalized in late 1998.
Love Canal, NY[b] (31,000 yd³)	Reduce human health risk from direct contact and from fish consumption.	Evaluation of various health advisories for dioxin from multiple sources such as NY Department of Health, Canadian agencies, and FDA	1 ppb 2,3,7,8-TCDD (CDC action level)	Direct	Dry excavation	No details obtained	Probably achieved, but no details obtained
Hooker (102nd Street), NY[b] (28,500 yd³)	Vague; apparently reduce risk from fish ingestion to below 10^{-4} to 10^{-6} and a hazard index of 1 ~~and a~~	Human health risk assessment and environmental endangerment assess~~ment by chemi~~	Remove out to a "clean" boundary line and to a depth horizon ~~dictated~~	Vague	Dry excavation	Areal and depth horizon achieved; no analytical verification	Too soon to tell. 1 ft of fill added to remediated areas. No post-monitoring identified.

Site	Cleanup goal	Basis	Cleanup level	Approach	Method	Status	Outcome
Tennessee Products, TN[b] (24,100 yd³)	Remove visual coal tar material from several thousand feet of the creek (interim measure).	Non-time-critical removal action	Remove all visual coal tar material	Direct	Dry excavation	Achieved. Visual confirmation only	Achieved. Visual confirmation only
Town Branch Creek, KY[b] (76,000 yd³)	Reduce PCB in fish levels to <2 ppm FDA limit.	State environmental agency evaluation and circuit court judgment	0.1 ppm PCBs	Direct	Dry excavation	Achieved sediment removal to extent practical but not always 0.1 ppm in 30% of 3.5 miles of creek so far. Work on remaining 2.5 miles on hold pending resolution of access issues.	Too soon to tell. Post-monitoring plan yet to be implemented
Former Messer Street MGP, NH[c] (12,500 yd³)	Reduce free product mass within the site by 80%.	Human health and ecological risk assessments	Depth horizon	Direct	Mechanical dredging and dry/wet excavation	Depth horizon achieved; backfill	Likely accomplished, but not verified
Triana/Tennessee River, AL[d] (no removal)	Reduce DDT in fish levels to <5 ppm FDA limit.	Negotiated agreement and consent decree to restore the fishery	Rechannelization and direct burial of the two isolated tributaries (2.5 miles) containing an estimated 93% of the DDT mass	Vague, basically a "try it and see what happens" approach	Stream diversion, direct burial, and some natural recovery	Achieved	Substantial progress. One target species reached the 5 ppm standard in the 10-year attainment period, two species did not but they exhibit 80 to 90% DDT reductions in the 10 years. Annual monitoring continuing.
James River, VA[d] (no removal)	Allow natural recovery of fish and biota to below FDA limit for Kepone (0.3 ppm in fish and 0.4 ppm in blue crabs).	Technical impracticability of achieving FDA limits in fish by remediation	None	N/A	Natural recovery	N/A	Natural burial by clean sediments is continuing to decrease the bioavailability of Kepone. Crab/oyster Kepone levels dropped from 0.8 to 0.1–0.2 ppm from 1976 through 1985. The commercial fishing ban was lifted in 1988; only a subsistence fish eating advisory remains.

TABLE 4.2 U.S. Sediment Remediation Projects Implemented (>10,000 yd³)—Primary Goal versus Outcome (*Continued*)

Project	Primary goal	Basis for primary goal	Sediment remedial target	Relationship of target to goal	Remediation method	Achievement of remedial target	Achievement of primary goal
Sangamo-Weston, SC[d] (no removal)	Reduce PCB in fish levels to <2 ppm FDA limit by natural recovery.	Technical impracticability of achieving risk-based concentrations in fish by remediation; existence of an ARAR (the FDA limit); and the voluntary nature of fish consumption	1 ppm PCBs	Default level, per the Record of Decision: "The time for two to eight year old largemouth bass to achieve 2 ppm for the range of sediment cleanup goals was compared to a baseline. It was determined that fish PCB levels decline at about the same rate regardless of sediment cleanup goal. Therefore, 1 ppm was selected based on technical feasibility. . . ."	Natural recovery; modeling predicts 2 ppm levels in fish will be reached by 2004	Too soon to tell	Too soon to tell. Annual monitoring in progress. No reports yet available for review.

Notes:
[a] True dredging projects.
[b] Dry excavation projects.
[c] Combined dredging and dry excavation projects.
[d] Natural recovery projects.

4.20

remediation projects compiled in the MCSS database. Even the relatively limited amount of data that does exist for a subset of projects does not indicate that the projects conducted to date have resulted in an acceptable level of risk control. What is particularly disturbing in light of this are claims by EPA regarding the success of dredging projects. In the March 7, 2000 update to an article originally appearing in *Engineering News Record* (Hahnenberg, 1999), it is stated: "Results from recent environmental remediation dredging projects demonstrate significant risk reduction is consistently achieved on environmental projects." Quite to the contrary, careful review of the existing data shows that: (1) dredging projects are not being carefully monitored and evaluated with respect to achieving risk reduction goals and (2) there are limitations to achieving risk-reduction goals.

4.5 A PROPOSED RISK-BASED DECISION FRAMEWORK

It is evident that a risk-based decision-making framework is needed. Such a framework would build from real-world experience at other sites and from an understanding of how contaminants in sediments have the potential to create risks to humans and the environment. This framework needs to answer the appropriate questions for remedial decision making and must be able to document through measurement whether stated remedial goals are achieved. With these concepts in mind, one can develop a simple and straightforward risk-based framework to guide decision making at sediment sites:

1. *Do chemicals present in bioavailable surface sediments pose an unacceptable risk to human health and the environment?*

2. *Are there active sources that are currently contributing contaminants to the surface sediments in quantities that cause unacceptable risks?* If these sources are not controlled or eliminated, they will greatly reduce the likelihood that any remedy directed at contaminants already in the sediments will be successful.

3. *Do the chemicals of concern that are buried below the bioavailable surface sediments have reasonable potential to materially increase contaminant concentrations in the bioavailable surface sediments?* Contaminated sediments that are stable and isolated below the surface sediment and not likely to become exposed during future events, such as flooding, do not warrant active remediation.

4. *If the system and bed are stable, would any active remedial effort (e.g., dredging, capping) materially accelerate natural recovery?* Natural recovery is the benchmark against which remedial options must be measured.

5. *If the answer to 4 is yes, is the accelerated risk reduction outweighed by the potential adverse impacts to human health, the community, and the environment from implementation of the remedy?* Decisions should maximize risk reduction and minimize the negative impacts of remedial technologies on the ecosystem and affected communities.

In answering these questions, evaluations of remedial options should be based on a comprehensive scientifically sound analysis:

- Decisions should be based on a thorough site assessment that is derived from well-conceived, statistically valid monitoring programs that allow a thorough understanding of chemical sources and fate. Where appropriate, these data should be utilized to construct a quantitative site model that will allow for evaluation of remedial alternatives.

• Decisions should be based on a thorough evaluation of *all* sediment management options. Such evaluations should incorporate experience gained from other sites as to the engineering capabilities and limitations of remedial technologies along with the benefits of natural processes and administrative controls to manage risks.

4.6 OBSERVATIONS FROM ENVIRONMENTAL DREDGING EXPERIENCE

A review of available information from contaminated sediment sites shows that the environmental dredging projects implemented to date have been relatively small (compared with traditional navigational dredging), costly, and difficult to implement. Moreover, the projects typically have vaguely or inconsistently defined cleanup targets and goals, and their ability to achieve risk control has not been documented or demonstrated.

Appendix 4B provides a summary of results from completed environmental dredging projects that have some postdredging data available (e.g., contaminant levels in surface sediment, fish, and water). The MCSS database provides additional site information. The primary conclusions drawn from a review of these data are presented below.

1. *Environmental dredging has not reduced surface sediment concentrations to acceptable levels.* Cleanup goals and their derivation vary considerably from site to site (e.g., 0.1 to 4000 ppm for PCBs). However, sediment cleanup goals selected by regulators for bioaccumulative chemicals, such as PCBs, typically are on the order of 1 ppm or less. However, experience has shown that PCB levels of 1 ppm or less in surface sediments have not been consistently achieved through dredging because of the limitations of dredging technologies. Average surface sediment PCB concentrations before and after dredging at several projects are plotted in Fig 4.1.

As can be seen, average PCB levels of 1 ppm or less have not been attained at dredging projects in the United States. At the St. Lawrence River in New York, the 1 ppm cleanup goal was not achieved in all six areas sampled; even though some locations were redredged up to 30 times, the average surface sediment PCB level after dredging was still 9.2 ppm. Similarly, after dredging at the Sheboygan River in Wisconsin and the Grasse River in New York (where the objective was to remove all sediment) average surface sediment PCB levels were 39 and 75 ppm, respectively. At Ruck Pond on Cedar Creek in Wisconsin, the pond was dewatered and excavated "in the dry" in an effort to remove all sediment to the extent practicable. Extensive efforts were employed (e.g., squeegees used on a bulldozer blade, vacuum trucks), yet surface sediment averaged 84 ppm PCBs after removal efforts were finished. Based on the experience to date, it has not been demonstrated that dredging will consistently achieve less than 5 ppm PCBs in surface sediments. The central reasons for these poor results are discussed in Sec. 4.7, "Technical Limitations of Environmental Dredging."

2. *In some cases, dredging has resulted in increased surface sediment contaminant levels.* As shown in Figs. 4.2 and 4.3, dredging at Manistique Harbor in Michigan and

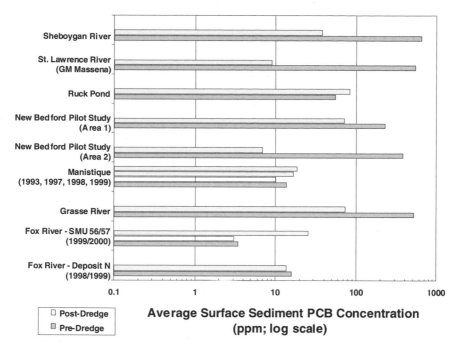

FIGURE 4.1 This summary figure shows how dredging has failed to reliably and consistently reduce average surface sediment contaminant levels (PCBs in this case) to typical acceptable levels (i.e., 1 ppm or less). Left at the surface, these contaminants may be available for exposure to biota or movement into the water column. *(Source: MCSS Database Release 3.0.)*

the Fox River (SMU 56/57) in Wisconsin resulted in increases in surface sediment contaminant levels. At Manistique Harbor, the increase occurred despite 3 years of dredging. While the project was apparently completed in 2000, data were not available for review prior to publication. However, it is doubtful that any substantive reduction in surficial PCB levels will be achieved by dredging alone. At both sites, conditions before dredging showed lower PCB concentrations at the sediment surface and the highest concentrations were observed in deeper sediment. In essence, dredging has exposed the buried sediments either directly or through sloughing in of the excavation wall, leading to increased surface sediment concentrations.

In Manistique Harbor, the average surface sediment PCB levels since 1993 have decreased in areas that have not been dredged, yet increased in areas that were dredged (see Fig. 4.2 and Fox River Group, 2000b). This suggests that a natural recovery remedy would have resulted in greater risk reduction than dredging, and that dredging actually has increased potential risks. EPA returned to the site in year 2000 to complete dredging in the harbor. Sediment data collected after dredging was completed have been requested but were not available for review in time for publication.

At the Fox River SMU 56/57 project, executed by the Wisconsin Department of Natural Resources in 1999, average surface sediment PCB concentrations were 3.6 ppm before dredging and 75 ppm after dredging. Because of schedule and budget constraints, only four small subareas were actually dredged "as designed" (i.e., with

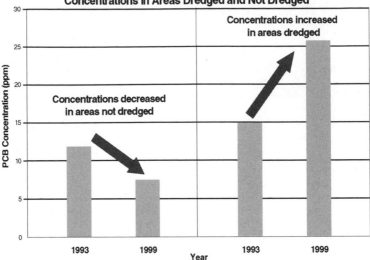

FIGURE 4.2 At Manistique Harbor, average surface sediment concentrations have declined since 1993 in areas where EPA has not dredged (i.e., data points outside dredging areas), but average concentrations have increased in areas where EPA has dredged since 1997 (i.e., data points within and bordering dredged areas). EPA has returned to the Harbor in 2000 for a fourth and final season of dredging. Data were not available prior to publication. *(Source: Fox River Group, 2000b.)*

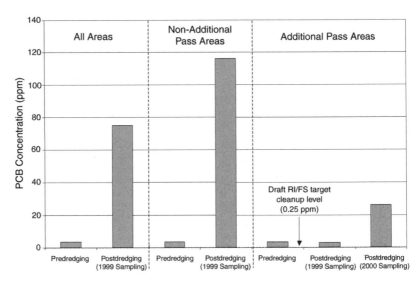

FIGURE 4.3 Fox River, WI—SMU 56/57: average pre- and postdredging surface sediment (0 to 4 in) PCB concentrations. In late 1999, approximately 30,000 yd³ of PCB-containing sediment was dredged from SMU 56/57 on the lower Fox River. Monitoring data for all areas dredged show that average surface sediment PCB concentrations rose sharply after dredging. For a short period after dredging in areas where additional passes were used, certain subareas remained at predredging average levels. Dredging was reinitiated and completed in 2000, with reported average PCB levels on the order of 2 ppm remaining in the entire dredged area. *(Source: Fox River Group, 2000a.)*

additional cleanup passes of the dredgehead). Samples obtained shortly after completion of dredging at these subareas showed average surface sediment PCB levels essentially unchanged (i.e., 3.5 ppm before and 3.2 ppm after dredging). However, as shown in Fig. 4.3 subsequent sampling conducted two months after completion of dredging (in early 2000) showed 26 ppm as the average surface sediment PCB levels in these areas. Fort James Corporation resumed dredging in the fall of 2000 to remove an additional 50,315 yd^3. As reported in Foth & VanDyke (2001), an average PCB concentration of 2.1 ppm remained in surface sediment in these four subareas.

3. *Dredging has not been shown to lead to quantifiable reductions in fish contaminant levels.* As noted previously, collection of several years of high-quality and comparable data before and after remediation is critically important to assess the effectiveness of sediment removal in reducing contaminant levels in fish, and the associated reductions in contaminant bioavailability, exposure, and risk. These data are generally not available.

What data do exist are usually inadequate to assess whether dredging has reduced risks from contaminants in sediments. At the Waukegan Harbor site in Illinois, for instance, the preremediation fish tissue data consist of one measurement. At the Ruck Pond site, the preremediation study included fish cages that were disturbed and one that was lost completely. Pre- and postdredging data for Ruck Pond are limited to data collected in only one event each. At Waukegan Harbor, multiple years of post-dredging carp data indicate an increasing trend in carp PCB levels. The uncertainties associated with these minimal monitoring data limit their utility for quantifying, and therefore demonstrating, whether reductions in fish contaminant levels were in fact achieved through dredging.

In addition, monitoring data collected at several sites before dredging indicate that natural processes were already reducing chemical concentrations in fish (e.g., Ruck Pond and Michigan's Shiawassee River), and at some sites other actions such as containment were taken in addition to dredging (e.g., Waukegan Harbor, Sheboygan River, St. Lawrence River, Ruck Pond). Distinguishing the effects of these elements on fish levels from the effects of dredging is not possible. At the Sheboygan River and Grasse River sites, where several years of fish data are available after dredging, trends in fish levels are not evident in the vicinity of the removal actions. The data do not support the conclusion that dredging reduced fish contaminant concentrations. Fox River Group (1999) presents additional discussion of this issue.

4. *Dredging releases contaminants.* Dredging unavoidably resuspends sediment and releases associated contaminants into the water column. Silt containment systems have been employed at many dredging sites in an effort to contain the suspended solids. Although one might think that, if suspended solids can be contained, associated contaminants could be as well, this is not always true. Again, there is a paucity of data to evaluate the importance of resuspension and the effectiveness of control. However, there are recent data from projects at Grasse River and Fox River showing that although silt containment systems generally were effective in containing resuspended solids, increased PCB levels were observed downstream of the dredging (see Fig. 4.4 for Deposit N on the Fox River). The U.S. Geologic Service (USGS) published a report (USGS, 2000) summarizing the results of extensive water sampling during dredging on the Fox River (SMU 56/57). USGS concluded that reliance on total suspended solids (TSS) and turbidity data for contaminant transport is inadequate and that over 2 percent of the PCB mass removed from the river was released and transported downstream.

In Manistique Harbor, PCB levels in water in the vicinity of the dredging operation were orders of magnitude higher than predredging levels, indicating PCBs were released during dredging (App. 4B).

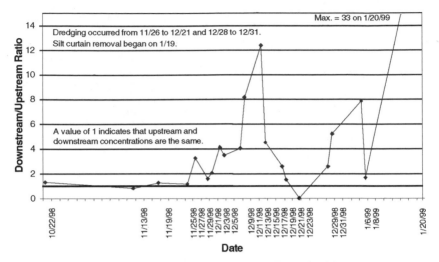

FIGURE 4.4 Fox River, WI—deposit N 1998 water column data: ratio of downstream to upstream total PCB concentration. This plot of the ratio between upstream and downstream surface water PCB concentrations in the lower Fox River during the Deposit N project shows that despite the use of silt curtains around the dredging area, PCBs were released to downstream waters during dredging. *(Source: Fox River Group, 2000a.)*

When released to the water column, the bioavailability of contaminants increases. For example, minnows placed in stationary cages in the Grasse River showed significantly higher PCB uptake during dredging (20 to 50 times higher) and up to 6 weeks following dredging (2 to 6 times higher) compared with PCB uptake before dredging. These results, combined with the water data, demonstrate increased exposure and potential risks. Given the scarcity of postdredging data, it is impossible to know how important these releases are in the long term. At a minimum, the release of contaminants will likely delay recovery of the system and therefore must be carefully considered. Further, as project size increases, so does project duration, resulting in prolonged impacts.

Contaminants can also be released to the atmosphere during dredging. At the New Bedford Harbor, Massachusetts, site, air monitoring documented elevated levels of PCBs downwind of the dredging operation, in some cases exceeding EPA's action level, requiring modifications to the dredge operation.

5. *Environmental dredging projects are costly and take a long time to complete.* A common theme observed in evaluating completed projects is that environmental dredging projects generally take longer to complete and cost more than originally anticipated. This is extremely important since cleanup decisions rely heavily on these estimates in weighing and justifying various remedial alternatives. Consequently, *actual* schedule and cost information available from completed projects (see Table 4.1 and MCSS database) needs to be thoroughly considered in making cleanup decisions. A graphic example of this issue is the Manistique dredging program. In 1995, it was anticipated that the project would take 2 years to complete at a cost of $15 million. After 5 years of dredging the harbor and lower river, expenditures grew to about $45 million, before the project was complete.

The costs for removal projects cover a wide range, as shown in Table 4.1. Costs are highly variable because of: (1) differences in goals from project to project; (2) differences in production (i.e., removal) rates, which are influenced by a wide variety of site-specific variables such as ease of access; and (3) wide differences in disposal costs, which are influenced by disposal method and location and type of contamination. Average unit costs are summarized below, and a more complete list of factors influencing sediment removal costs is provided in Table 4.3.

TABLE 4.3   Cost Factors Associated with Sediment Removal

A. Extent of sediment subject to removal
- Larger extent = larger costs
- Economies of scale advantages significantly diminish with larger projects

B. Dredge production rate, which is primarily dependent upon:
- Unique site conditions such as access, water depth, debris/vegetation, and free oil
- The targeted sediment depth or cleanup level
- Limitations in land-based water management facilities
- Operational controls imposed to limit resuspension
- Whether or not verification sampling is performed during dredging

C. Disposal cost, which is dependent on type of contaminant, and type and location of disposal facility. Commercial disposal facilities tend to be more costly, but may be appropriate for smaller projects or may be required under regulation (e.g., RCRA, TSCA)
- The disposal methods for 58 completed removal projects listed in Table 4.1 were: offsite commercial landfill (30); on-site landfill, CDF, or burial (18); off-site thermal treatment (3); on-site thermal treatment (3); other, such as stabilization and beneficial reuse (4); disposal method not selected or unknown (2). (Note: Two of the projects used a combination of two disposal methods)

D. Access: Availability of upland areas for staging, sediment processing, and disposal (if on-site) can significantly affect cost, and the absence of such areas in fact makes a project infeasible. Limited access can result in higher costs due to:
- More extensive river-based transport of sediment
- Costs to obtain access from property owners
- More extensive land-based transport of sediment

E. Presence of rocks, vegetation, and debris: The presence of obstructions not only impacts dredge selection, but may require multiple equipment types to be used, which will increase costs.

- The average cost for the 24 dredging-only projects with available volume and cost information is $240 per cubic yard of material removed. The overall cost is highly dependent on two primary factors: dredge production rates and disposal costs. Additional factors that affect the performance of sediment removal are summarized in Table 4.4. There are a number of uncertainties that also can affect the success of a sediment removal project. Several of the more common uncertainties are also summarized in Table 4.4, all of which can impact effectiveness, cost, and schedule.

- The average cost for the 20 wet or dry excavation-only projects with available volume and cost information is $412 per cubic yard of material removed. The high overall cost reflects the low production rates compared with traditional earth-moving projects (using similar equipment) because of difficulties with accessibility and wet terrain, additional water management requirements for maintaining dry conditions, and high costs for disposal.

TABLE 4.4 Performance Factors Associated with Sediment Removal

1. Performance metrics—primary risk-based measurements of the effectiveness of removal
   - Bioavailable surface sediment characteristics before and after removal
     - Chemical contamination levels
     - Organic carbon levels
     - Physical characteristics (affecting mobility)
     - Density
     - Geotechnical (cohesion, etc.)
     - Bathymetry (verify amount removed and geometry)
   - Biota concentrations before and after removal
     - Resident fish
     - Other site-specific species
     - Caged fish (controlled study bioavailability indicator); can also be used during removal
   - Water column data before, during, and after removal
     - Chemical contamination levels
     - Total suspended solids (TSS)
     - Turbidity (sometimes an indicator of TSS)
   - Ambient air concentrations before, during, and after removal; need for measurement is chemical and site specific

2. Factors affecting performance of sediment removal
   - Aquatic environment characteristics
     - Water body type (lake, river, harbor, estuary, bay)
     - Water level fluctuations (tides, seiche, etc.)—can affect accessibility to sediment
     - Water velocities—will affect selection and performance of dredge equipment and resuspension controls
     - Water depth—will affect accessibility and equipment selection
   - Sediment characteristics
     - Presence of debris (rock, timber, man-made objects)—will require removal or will limit effectiveness of removal; removal can create cavities which may limit removal of remaining sediment
     - Sediment depth—deeper sediment removal drives multiple dredge passes, more likely to leave furrows/windrows and higher removal volumes to account for side sloughing
     - Subbottom characteristics (below contamination)—bedrock, hard pan, and irregularity all act to reduce effectiveness of removal by inherently leaving material behind
     - Sediment type (sand, gravel, silt, clay)—fines will tend to be resuspended and either migrate, desorb contamination, and/or settle (in the removal area or elsewhere in system); also clays tend to clog hydraulic dredges
     - Type of contamination—highly sorptive chemicals will tend to stay with solids; less sorptive compounds more likely to be released to water column
     - Chemical concentration profile—higher contamination at depth will have a tendency to result in higher concentrations remaining after removal
   - Removal equipment selected—dredging (or removal through water column) inherently limits capability to accurately remove sediment since operator can't see sediment to be removed
     - Hydraulic dredges—(numerous types available)
       - Resuspension inevitable, although generally less than mechanical removal
       - Material left behind because of "furrowing," irregular subbottom, settling, or resuspended material
       - Releases with transport pipeline malfunctions/breaks
     - Mechanical dredges (primarily clamshells)
       - Resuspension inevitable; recent innovations (Cable Arm, Bonacavor) claim to reduce, but can't eliminate

TABLE 4.4 Performance Factors Associated with Sediment Removal (*Continued*)

- Material left behind due to "cratering," sloughing, irregular subbottom, settling of resuspended material
- Excavation in "dry" conditions
  - Air emissions (dust, chemical) may need to be controlled
  - Material left behind due to irregular subbottom, "smearing," equipment tracking, wet slurry conditions from infiltration
- Resuspension control system—suspended silt curtains, sheetpiling typically used to minimize migration of inevitable sediment resuspension. None are watertight, so releases are inevitable. The higher degree of containment will act to allow reuspended sediment to settle within removal area, less containment will allow material to settle outside removal area.
- Disposal method
  - On site (landfill, confined disposal facility) versus off-site commercial facilities
  - The method of disposal will affect the dredge technology selection, and limit sediment removal rates (because of dewatering and water treatment requirements)
- Predisposal processing—this factor is primarily defined by the disposal method and may include
  - Primary settling
  - Dewatering
  - Stabilization/solidification
  - Water treatment
  - The extent of preprocessing required will drive the need for space, affect dredge selection, affect production rates (may increase project duration), and increase risk of contaminant release (more unit processes)
3. Uncertainties associated with sediment removal
   - Unpredictability of sediment concentration after removal
   - Bioavailable surface sediment concentration affects biota levels and water column concentrations
   - Highly variable results achieved elsewhere (see Tables 4.1 and 4.2)
   - Numerous variables involved (see Table 4.3), which essentially prohibits prediction of results at a given site
   - This uncertainty must be recognized before embarking on a sediment removal project
   - Site conditions never entirely predictable
   - Underwater environment compounds this common uncertainty at all contaminated sites
   - Surprises are inevitable
     - Volumes tend to increase
     - Debris tends to be more extensive
   - Project schedule and cost (refer to cost factors in Table 4.3)
   - Weather unpredictability can affect schedule and cost
   - Extent of winter weather affects overall schedule
     - Freeze-up significantly reduces or prohibits removal productivity and interferes with land-based water handling and treatment
   - Items 1 and 2 above also impact schedule and cost

- Project duration and cost are heavily influenced by the effective production rates of environmental dredging (i.e., how quickly sediment can be removed). While the production rate is influenced by numerous site-specific factors, a review of completed projects shows that typical production rates of only 3000 to 7500 yd^3 per month have actually been achieved. These production rates are extremely low in comparison to navigational dredging, and extrapolation to large-scale projects involving hundreds of thousands of cubic yards of sediment indicates that such projects are likely to be decadal in duration.

6. *There is limited environmental dredging experience in large rivers.* Almost all of the projects completed to date have covered limited areas and had relatively straightforward access. Of the 32 completed dredging projects in the MCSS database (i.e., not including wet/dry excavation projects), the largest project was at Saginaw River and involved 345,000 yd^3, small by navigational or maintenance dredging standards. Two-thirds of the 32 projects involved removal of 40,000 yd^3 or less. In many of these smaller projects, access and space were available at a responsible party's property in close proximity to the areas to be dredged. This simplifies the implementation by eliminating the need to obtain access to unrelated properties, minimizing transport of sediment, and reducing the schedule and quantities that need to be removed, processed, and disposed of. In fact, projects where access to third-party properties has been required have experienced significant delays in implementation (i.e., Town Branch Creek in Kentucky and the Sheboygan River). For example, barges transporting removed sediment on the Sheboygan River had to travel relatively long distances between the removal areas and the limited number of available land-based access points. Also, shallow water limited the movement of equipment, making the operation inherently slow. In contrast, there is no experience with large-scale environmental dredging projects on extended rivers. With these larger projects, the access, waste management, and disposal issues are likely to be much more problematic. This means that experience on smaller projects (in terms of ease of implementation) may not apply to larger projects.

7. *Advances in dredging technology have been limited.* Specialty dredges, designed to overcome some of the shortcomings of conventional navigational dredges when applied to environmental dredging, have their own limitations with respect to remediating large contaminated sediment sites. Japan and the Netherlands have been leaders in developing specialty dredging systems suitable for removing fine-grained contaminated material from harbor and lake bottoms with minimum resuspension. The availability of foreign-made specialty dredges is limited both by law (e.g., the Jones Act) and demand in the United States. Furthermore, their production rates are low compared with production rates of conventional hydraulic dredges. Also, specialty dredges typically have narrow or shrouded dredgehead openings that are particularly susceptible to plugging by debris or vegetation.

Actual production rate data for specialty dredges are sparse, and available data are poorly documented with respect to site conditions and dredge operating parameters. Further, specialty dredges are subject to the same inefficiencies and logistical difficulties as are conventional dredges for environmental dredging.

Of the specialty dredges listed in the table below, the Cable Arm environmental bucket has been used on eight major environmental dredging projects in the United States, but it is relatively lightweight, and the absence of "digging" teeth limits its use to unconsolidated (soft) sediments. In addition, as noted in the table, although minimizing resuspension is an intended feature, actual experience has shown that sediment resuspension with the Cable Arm bucket is still a concern. For the major environmental dredging projects implemented in the United States to date, conventional hydraulic cutterhead and horizontal auger dredges or mechanical clamshells have traditionally been used, but with inconsistent results.

Features of Several Specialty Dredges

Dredge type	Feature
Matchbox Cleanup Refresher	Shielded auger or cutterhead to reduce resuspension
Soli-Flo Versi AgEm	High solids, underwater pump located at dredgehead to shorten suction line and allow passage of large solids/objects
Cable Arm Watertight Dry Dredge	Environmental bucket to maximize percent solids and minimize resuspension upon impact and minimize losses to water column removal
Pneuma Oozer	Compressed air piston/cylinder pump to minimize resuspension and maximize percent solids

4.7 TECHNICAL LIMITATIONS OF ENVIRONMENTAL DREDGING

Several technical limitations are inherent in environmental dredging. These limitations restrict the effectiveness of sediment removal in reducing contaminant levels in surface sediments. Although dredging can remove significant volumes of sediment and associated contaminant mass, dredging inevitably leaves behind residual materials at the sediment surface. These residuals are attributed to "missing," "mixing," and "messing," which are described below. In addition, dredging introduces new risks to the ecosystem and community.

4.7.1 Missing: Dredging Cannot Remove All Targeted Sediment and Contaminants

Even with careful operations, experience has shown that sediments are unavoidably left behind after dredging. According to the Army Corps of Engineers, "No existing dredge type is capable of dredging a thin surficial layer of contaminated material without leaving behind a portion of that layer and/or mixing a portion of the surficial layer with underlying clean sediment" (Palermo, 1991). Because surface sediments play a central role in transferring contaminants to fish and the wider food web, any action that leaves contaminants at the biologically active sediment surface is unlikely to achieve risk-based goals requiring low part-per-million concentrations of chemicals.

Dredging's inability to reliably remove all sediments and contaminants and create a clean sediment surface results from various factors, including: (1) incomplete spatial coverage in dredged areas as evidenced by cratering of the sediment bed from the action of a mechanical clamshell or creation of windrows and furrows between swaths of a hydraulic dredge; (2) inaccessibility of sediments located in shallow waters where barges and hydraulic dredging equipment cannot operate effectively, located adjacent to or under boulders and debris that cannot be removed, or resting on an irregular hardpan or bedrock bottom; and (3) performing work underwater and out of sight of the operator.

4.7.2 Mixing: Dredging Unavoidably Mixes Sediment Targeted for Removal with Underlying Materials

To remove sediments, a dredge must cut into the sediment bed, which mixes sediments targeted for removal with other sediments either above or below the targeted material. Whether higher-concentration sediments are present at depth and cleaner sediments are present at the surface, or vice versa, the mixing caused by dredging inevitably leaves behind contaminated sediment on the new sediment surface created by the dredge. Many sediment sites have lower concentrations of the target chemical in surface sediments than at depth. This is often due to previous implementation of source controls and ongoing natural recovery through sedimentation and burial. Thus, dredging mixes the lower concentration surficial sediments with deeper, higher-concentration sediments, which can result in elevated residual concentrations at the new sediment surface. This is particularly problematic at sites with stable sediments because dredging does what nature cannot, bringing contaminants once sequestered in deep sediments to the surface and exposing them to biota and the water column. It also is problematic at sites where deeper, more contaminated sediment rests on bedrock because one cannot overcut into cleaner sediments beneath the contaminated layers. For example, this underlying bedrock condition exists at the Manistique Harbor site.

4.7.3 Messing: Dredging Resuspends and Releases Contaminants into the Water Column

The physical mixing action of the dredge inevitably stirs up sediments, releasing both suspended and dissolved contaminants to the water column. Although there are devices to reduce resuspension and the dredge operator can modify certain operating parameters such as production rate, no dredging method has totally eliminated local sediment resuspension. Sediment resuspended during dredging will eventually settle on the surficial layer of the area dredged or be transported and redeposited outside or downstream of the removal area. Thus, for contaminants with an affinity for binding to sediments, surface sediments both within and outside the removal area may become more contaminated than before dredging.

The transport of suspended sediments outside the removal area along with increased turbidity can cause a variety of adverse effects in fish, including interference with gill function, enhanced fungal infections of fish embryos, and reduced resistance to disease. In addition, certain chemicals that may be acutely toxic to local biota (e.g., metals, ammonia) may be released during dredging or result in anoxic conditions. Other chemicals released when the sediment bed is disturbed (e.g., nitrogen compounds, phosphorus) may degrade water quality by stimulating algal blooms.

To reduce the negative impacts of downstream sediment transport, environmental dredging areas are typically isolated from the rest of the waterway by a silt curtain or other containment barrier. Silt curtains do not effectively control the transport of dissolved contaminants, and experience shows contaminants (especially in dissolved phase) typically migrate outside the silt curtains and downstream (see examples in App. 4B). Once contaminants are dissolved in the water, they also are more apt to volatilize into the atmosphere.* Further, the more effec-

---

* This situation was encountered at the New Bedford Harbor site where, according to EPA (1997b), "control of airborne PCB emissions did contribute to a slower rate of dredging and thus a longer project duration."

tive the barrier system is in containing resuspended sediment, the more contaminated sediment will resettle within the removal area. If sediments migrate outside the removal area, they can resettle over a larger surface area.* Chemicals in this resettled/residual sediment will be bioavailable, and the sediments will generally be more susceptible to scour than the pre-existing surface sediment since any natural armoring that may have occurred over time is removed during the dredging operation.

The impacts of resuspension are generally considered a short-term effect of dredging since most environmental dredging projects performed to date have been of limited duration. However, for large-scale, long-term dredging projects, the cumulative effect of these "short-term" impacts could be substantial and should be considered in remedial decision making.

4.7.4 Dredging Introduces New Risks to the Ecosystem and Community

In 1995, EPA posed the question, "How can dredging affect the environment?" The agency's response was that "impacts can include benthic disturbance, water quality degradation, impacts on aquatic organisms, and water and soil contamination from disposal of dredged materials" (EPA, 1995). EPA was right. Environmental dredging operations bring with them a myriad of risks and impacts not directly related to what is happening at the sediment surface. For example, dredging can destroy important ecological features of a site, such as vegetation, the benthic environment, and various fish spawning and nursery habitats, not to mention the communities of biota that inhabit the removal areas. Although some reconstruction of habitat can be attempted, impacts are typically observed until recolonization occurs, which may take years. As observed by Suter (1997), "the ecological risks related to remedial activity must be balanced against risks associated with the contaminant to the ecosystem components and against often hypothetical health risks." Unfortunately, these impacts are seldom evaluated with any rigor on environmental dredging projects despite the fact that they are carefully analyzed on proposals for navigational dredging projects.

In addition, environmental dredging operations, on-shore sediment handling and processing equipment (e.g., dewatering, treatment), and transportation of materials (via pipeline, barging, conveyance, trucking) to treatment or disposal facilities are inherently dangerous processes. Environmental dredging operations invariably cause normal commercial shipping and recreational boating near a site to become more hazardous and difficult or restricted. Indeed, large-scale environmental dredging projects could take decades and severely impair portions or all of a waterway during active operations. Such disruptions can have devastating economic impacts on a local community's use of the waterway for tourism or other commercial purposes. Again, the impacts from these types of projects in terms of injuries to workers and community members are real, not hypothetical.

As part of the planning process for all types of dredging projects, the Army Corps of Engineers evaluates the potentially detrimental effects of dredging on habitat to ascertain whether dredging must be confined to specific time periods to minimize its adverse environmental impacts. The most persistent concerns are (Dickerson et al., 1998):

---

* Studies of the Yazoo and Yalobusha Rivers in Mississippi indicated that turbidity plumes extended up to one-half mile downstream of dredging activities, even when containment measures were utilized (Wallace, 1992). Similar evidence was noted at the New Bedford Harbor site as discussed in App. 4B.

1. Disruption of avian nesting activities and destruction of bird habitat
2. Sedimentation and turbidity issues involving fish and shellfish spawning
3. Disruption of anadromous fish migrations
4. Entrainment of juvenile and larval fishes
5. Burial and physical destruction of protected plants
6. Disruption of recreational activities

It is sensible and prudent to consider and weigh the potential damage to habitat and disruption to ecosystem structure and functioning against whatever environmental benefits might accrue from removal of contaminated sediments.

Clear guidance is needed on the evaluation of actual risks to ecological resources and communities resulting from implementation of environmental dredging projects and how to balance these risks and impacts relative to any benefits achieved in risk reduction. Currently, detailed guidance does not exist on how to evaluate objectively and quantitatively the negative consequences of sediment remediation projects.

4.8 FINAL OBSERVATIONS AND RECOMMENDATIONS

Dredging has historically been used to remove bulk sediments from shipping channels and harbors. It is effective for that purpose. Dredging to reduce risks posed by contaminated sediments is relatively new, and its effectiveness has not been demonstrated. When viewed in the context of risk reduction, there is no sound justification for dredging stable, isolated sediments that contain contaminants that are not and will not migrate to the bioavailable surface sediment layer in any meaningful way. Decision makers often have not recognized the technical limitations of dredging and its potential for adverse ecological and community impacts. If this does not change, the contaminated sediment program will fall short of its goal of effectively reducing risks to human health and the environment. A number of conclusions can be drawn on the basis of our review of sediment remediation projects undertaken in the United States.

- There is no consistent framework for making cleanup decisions at contaminated sediment sites. The goal of any program should be to effectively control risks. There is a need for a clear, simple-to-apply risk reduction decision framework. This chapter proposes such a framework, which is based on an understanding of sediment dynamics using sound scientific principles.

- Appropriate data-collection programs to acquire the data necessary to measure the effectiveness of remedial techniques in adequately reducing risks at sediment sites have not been developed. As a result, substantial experience cannot be properly incorporated into remedial decisions. This chapter and the MCSS database should help fill this gap.

- The limited available data clearly show the limitations of environmental dredging technology:

  - Dredging has not reliably and consistently removed all contaminated sediment, restored a "clean enough" sediment surface, or decreased the bioavailability of contaminants. Dredging is unable to reliably and consistently achieve low residual concentrations typically sought in surface sediments, even after repeated passes with the dredging equipment. The residuals left behind after dredging may be at a higher concentration and more bioavailable than before dredging, resulting in increased risk.

- While environmental dredging typically employs controls to prevent resuspension and release of contaminants during operations, such releases to water, biota, and air occur. These releases could create unacceptable long-term risks due to redeposition of resuspended sediment and are particularly problematic at large projects, where such releases may occur over a multiyear implementation period.

- Dredging removes material that must then be handled and processed, typically on shore. This can increase the complexity of remediation. Dredging is inherently dangerous, a fact verified by insurance statistics, and poses serious short-term risks to workers and the community, and long-term risks to the extent the material must be permanently managed in a disposal facility. Dredging will disrupt or destroy the habitat and biota in the areas in which it is applied. These very real impacts and risks imposed by the remedy need to be balanced against the hypothetical risks posed by the sediment itself.

- Environmental dredging projects are costly and take a long time to complete.

Decision makers should select remedial alternatives that are protective, technically feasible, and cost-effective. Other options can be more effective than dredging with fewer negative impacts. Based on the evidence presented in this chapter and supporting documents, we offer the following recommendations regarding how environmental dredging should be viewed in managing risk:

- Regulators need to reaffirm that risk reduction is the proper goal of any remedial action.

- How contaminants move in the aquatic system must be evaluated during risk analysis and remedy selection. Risk reduction in aquatic systems is directly linked to a remedy's ability to decrease the probability that fish and other biota are actually or potentially exposed to sediment-bound contaminants. The first step is to control or eliminate active sources of contaminants to the surficial bioavailable sediments. The second step is to evaluate sediment deposit stability to assess whether normal erosion or some extreme events (e.g., high flows, flooding) could mobilize otherwise isolated contaminants being currently buried, thus moving nonbioavailable chemicals into the surface sediment layer. The final step is to evaluate methods to reduce surface concentrations of the contaminants now and in the future so as to minimize their bioavailability. Fair consideration must be given to less disruptive risk controls like natural recovery and administrative controls (e.g., fish consumption advisories).

- Regulators must recognize the technical limitations of dredging that result in the inability of dredging to reliably and consistently achieve low residual contaminant concentrations in surface sediments. They must consider the new and potentially higher risks that might occur from increases in contaminant concentrations in surface sediment, the water column, and ultimately fish tissue concentrations.

- Regulators must consider the real environmental and human impacts of environmental dredging projects. These impacts must be weighed against any hypothetical reduction in risk that might be achieved. Comprehensive policy and guidance in this area are needed.

- The experience at completed projects needs to be considered in making future decisions. Adequate monitoring data and formal plans for pre- and postremediation evaluation of risk reduction are essential elements in sediment remediation projects. These types of essential data can reduce uncertainty and allow one to draw sound conclusions regarding the relative effectiveness of remedial activities.

• Regulators must thoroughly consider actual schedule and cost information available from completed projects and incorporate this into their decisions. Experience shows that projects completed to date generally have taken longer to complete and cost more than originally anticipated.

REFERENCES

Dickerson, D. D., K. J. Reine, and D. G. Clarke. 1998. "Economic Impacts of Environmental Windows Associated with Dredging Operations." *DOER Technical Notes.* TN DOER-E3. Vicksburg, Miss.: U.S. ACE Research and Development Center.

EPA. 1995. *Pollution Prevention—Environmental Impact Reduction Checklist for NEPA/309 Reviewers.* Prepared for U.S. EPA Office of Federal Activities, contract no. 68-W2-0026.

EPA. 1997a. *The Incidence and Severity of Sediment Contamination in Surface Waters of the United States: Volume 1 of the National Sediment Quality Survey.* EPA 823-R-97-006, September.

EPA. 1997b. *Report on the Effects of the Hot Spot Dredging Operations—New Bedford Harbor Superfund Site,* October.

Foth & VanDyke, Hart Crowser, Inc. 2001. *Final Report, 2000 Sediment Management Unit 56/57 Project, Lower Fox River Green Bay, Wisconsin,* January.

Fox River Group. 1999. *Effectiveness of Sediment Removal: An Evaluation of EPA Region V Claims Regarding Twelve Contaminated Sediment Removal Projects,* September.

Fox River Group. 2000a. *Effectiveness of Proposed Options for Additional Work at SMU 56/57,* March.

Fox River Group. 2000b. *Dredging-Related Sampling of Manistique Harbor: 1999 Field Study,* June.

Hahnenberg, J. 1999. "Long-term Benefits of Environmental Dredging Outweigh Short-Term Impacts." *Engineering News Record.* March 22–29, 1999.

Major Contaminated Sediment Sites Database, Release 3.0. 2001. General Electric Company, Applied Environmental Management, Inc, and Blasland Bouck & Lee, Inc., April.

NRC Committee on Contaminated Marine Sediments. 1997. *Contaminated Sediments in Ports and Waterways: Cleanup Strategies and Technologies.* Washington, D.C. National Academy Press.

NRC Committee on Remediation of PCB-Contaminated Sediments. 2001. *A Risk Management Strategy for PCB-Contaminated Sediments,* March.

Palermo, M. 1991. "Equipment Choices for Dredging Contaminated Sediments." *Remediation,* Autumn, pp. 473–492.

Suter, G. W. 1997. "Integration of Human Health and Ecological Risk Assessment." *Environmental Health Perspective,* vol. 105, pp. 1282–1283.

USGS. 2000. *A Mass-Balance Approach for Assessing PCB Movement During Remediation of a PCB-Contaminated Deposit on the Fox River, Wisconsin,* USGS Water-Resources Investigation Report 00-4245, December.

Wallace, D. L. 1992. "Short- and Long-term Water Quality Impacts from Riverine Dredging." *Water Quality '92 Proceedings of the 9th Seminar.* USACE Paper W-92.

APPENDIX 4A: SURFACE SEDIMENTS PLAY KEY ROLE IN DRIVING RISK

Contaminants accumulate in sediments if they possess chemical properties that cause them to associate preferentially with the particulate matter that forms the sediment. These same properties tend to cause such contaminants to accumulate in biotic tissue and to become more concentrated as they are transferred through the food web. As a result, ingestion of fish is typically the prevailing human and ecological exposure pathway at contaminated sediment sites.*

The transfer of a contaminant from sediment to fish is initiated by direct transfer from sediments to benthic animals or by the flux of contaminant from the sediment to the water column and the transfer from water to animals living in the water column. Either way, the sediments involved in the transfer are those close to the sediment-water interface (Fig. 4.5). Sediments buried below the surface "mixed" layer that are subject to disturbance by hydrodynamic forces or inhabited by benthic animals typically provide almost no contribution to the transfer process. This is so because the contaminant's propensity to associate with the sediment particulate matter greatly inhibits its ability to migrate from below the mixed layer into the mixed layer.

The size of the surface mixed layer depends on the nature of the sediment particles, the magnitude of the forces placed on the sediments by currents and waves and the depth to which infaunal benthic animals mix sediments in a process termed *bioturbation.* In most cases, bioturbation is the controlling factor. Studies have shown that depths can range up to about 20 cm, but are typically on the order of 10 cm or

---

* Major transport mechanisms include downstream migration of contaminated fine-grained materials that are suspended within the overlying water column (carried as a portion of bed load), partitioning to dissolved organic matter, or available as dissolved phase in the water column (Paris et al., 1978; Valsaraj et al., 1997).

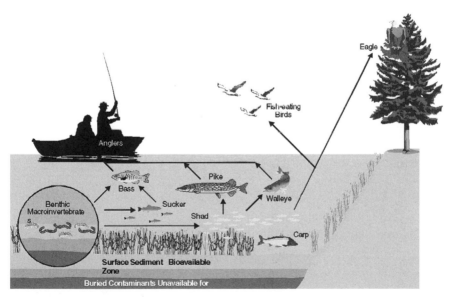

FIGURE 4.5 At most sites, the primary route of exposure for people or wildlife is consumption of fish that have accumulated contaminants from the surface of the sediment bed. Contaminants located at the sediment surface, as shown here, are "bioavailable" and thus prone to transfer up the food chain from benthic organisms.

less in sandy substrate (Palermo et al., 1998). Below this hydrologically and biologically active surface layer, contaminants may be locked in the consolidated deeper sediments and, according to the IJC (1997), "once buried in deep sediment, particles are often considered lost to the system" and thus unavailable for transport or exposure. In these cases, newer sediments with continually lower concentrations deposit on the surface and gradually bury those older sediments having higher concentrations representative of past discharges. These long-buried contaminated sediments remain unavailable for biological exposure and therefore pose no appreciable associated risks. In the words of a guidance document from EPA's Assessment and Remediation of Contaminated Sediments (ARCS) Program (U.S. EPA, 1998):

> Humans, aquatic organisms and wildlife will generally only be exposed to sediment contaminants in the uppermost active layer of the sediment deposits. Hence, contaminated sediments separated from the overlying water by a surface layer of relatively clean sediments may not represent an ongoing risk to humans, aquatic organisms or wildlife. [I]n fact, as ARCS and other coring studies have shown, the most contaminated sediments may be located well below the surface sediment (i.e., in older sediments).

These factors combine to suggest that, in order for dredging (or any other remedy) to be effective in reducing exposure and associated risks, it must "break the link" between the surface sediment source of contaminants and the fish and other receptors within the system's food webs. If remediation can effectively reduce surface sediment concentrations, bioavailability will be reduced and subsequent exposure to all receptors along the food chain from benthic organisms to fish and on to humans and wildlife also will be reduced. Remedial actions that do not address these linkages will not be effective in reducing bioavailability, exposure, and potential risks (IJC, 1997). Thus, any action that fails to create a sufficiently clean sediment surface will not be effective in achieving the desired risk reduction.

References

ATSDR. 1998. Toxicological Profile for PCBs.

Crivelli, A. J. 1983. "The Destruction of Aquatic Vegetation by Carp," *Hydrobiology,* vol. 106, pp. 37–41.

Edgington, D. N. 1994. "The Effects of Sediment Mixing on the Long-Term Behavior of Pollutants in Lakes," in *Transport and Transformation of Contaminants Near the Sediment-Water Interface,* J. V. DePinto, W. Lick, and J. F. Paul, eds., Lewis Publishing, pp. 307–328.

IJC. 1997. Sediment Priority Action Committee, "Overcoming Obstacles to Sediment Remediation in the Great Lakes Basin."

Jepsen, R., J. Roberts, and W. Lick, "Effects of Bulk Density on Sediment Erosion Rates," *Water, Air and Soil Pollution,* vol. 99, pp. 21–31.

Matisoff, G. 1982. "Mathematical Models of Bioturbation," in *Animal/Sediment Relations,* McCall and Tevasz, eds., Plenum Press, New York, pp. 289–330.

McNeil, J., C. Taylor, and W. Lick. 1996. "Measurement of Erosion of Undisturbed Bottom Sediments with Depth," *Journal of Hydraulic Engineering,* vol. 122, pp. 316–324.

Meijer, M. L., M. W. de Haan, A. W. Breukelaar, and H. Buiteveld. 1990. "Is Reduction of the Benthivorous Fish an Important Cause of High Transparency Following Biomanipulation in Shallow Lakes?," *Hydrobiology,* vol. 200/201, pp. 303–315.

Palermo, M., S. Maynord, J. Miller, and D. Reible. 1998. "Guidance for In-Situ Subaqueous Capping of Contaminated Sediments," EPA-905-B96-004, Great Lakes National Program Office, Chicago.

Paris, D. F., W. C. Steen, and G. L. Baughman. 1978 "Role of Physico-Chemical Properties of Aroclors 1016 and 1242 in Determining Their Fate and Transport in Aquatic Environments," *Chemosphere,* vol. 4, pp. 319–325.

Reible, D. D., V. Popov, K. T. Valsaraj, L. J. Thibodeaux, F. Lin, M. Dikshit, M. A. Todaro, and J. W. Fleeger. 1996. "Contaminant Fluxes from Sediment Due to Tubificid Oligochaete Bioturbation," *Water Research,* vol. 30, pp. 704–714.

Robbins, J. A. 1982. "Stratigraphic and Dynamic Effects of Sediment Reworking by Great Lakes Zoobenthos," *Hydrobiologica,* vol. 92, pp. 611–622.

U.S. EPA. 1998, "EPA's Contaminated Sediment Management Strategy," EPA 823-F-98-001, Office of Water, Washington, D.C., April.

Valsaraj, K. T., L. J. Thibodeaux, D. D. Reible. 1997. "A Quasi-Steady-State Pollutant Flux Methodology for Determining Sediment Quality Criteria," *Environmental Toxicology and Chemistry,* vol. 116, pp. 391–396.

Vanoni, V. A. 1975. *Sedimentation Engineering,* ASCE Manuals and Reports on Engineering Practice—no. 54, American Society of Civil Engineers.

APPENDIX 4B: ENVIRONMENTAL DREDGING— SITE PROFILES

This appendix summarizes several case-study examples of dredging. Among the many sites referenced or mentioned in this chapter, the following sites are reviewed in greater detail within this appendix:

- Grasse River, New York
- St. Lawrence River, New York
- Sheboygan River, Wisconsin
- Lake Järnsjön, Sweden
- Fox River, Wisconsin (two projects)
- Duwamish Waterway, Washington
- River Raisin, Michigan
- Manistique River/Harbor, Michigan
- Shiawassee River, Michigan
- Ruck Pond, Wisconsin
- Waukegan Harbor, Illinois
- New Bedford Harbor, Massachusetts

Compared to navigational dredging, environmental dredging is in its infancy. Through mid-2001, only about 58 sediment removal projects had been completed, compared with the many hundreds of navigational dredging projects completed over many decades. These 58 projects largely exclude small projects [i.e., less than 3000 cubic yards (yd^3)], since these smaller projects typically represent spill cleanups, interim measures, or "hot spot" removal actions that are much less representative of larger-scale dredging. Monitoring data at these 58 sites is typically lacking and sporadic. Indeed, the International Joint Commission (IJC, 1999) notes that for 38 remediation projects in the Great Lakes region, "only two currently have adequate data and information on ecological effectiveness." Further, the IJC suggests that "much greater emphasis be placed on postproject monitoring of effectiveness of sediment remediation," that "a high priority be placed on monitoring ecological benefits and beneficial use restoration," and that "additional research is essential to . . . forecast ecological benefits and monitor ecological recovery and beneficial use restoration in a scientifically defensible and cost effective fashion" (IJC, 1999). Of the 58 completed projects, 30 are polychlorinated biphenyl (PCB) sites (see Table 4.1), and of these 30, 13 have some data that are usable for assessing how effective dredging has been. Each of these sites is discussed below.

As described in App. 4A, the level of PCBs accumulated by fish depends on the concentration of PCBs found in surface sediment and the water column. Although PCB concentrations in fish may be the most important source of potential risks to humans and wildlife, it can take years for PCB concentrations in fish to respond to a dredging project. In addition, there are limited fish data available for completed environmental dredging projects. Thus, PCB concentration in residual surface sediment provides a more immediate and the most important measurement of the effectiveness of dredging in reducing human and ecological risks. This appendix discusses the available data for residual PCB concentrations in surface sediment, the water column, and fish tissue for several environmental dredging projects. A more thorough evaluation of fish data at many of these sites is provided in the paper titled "Effectiveness of Sediment Removal: An Evaluation of EPA Region 5 Claims Regarding Twelve Contaminated Sediment Removal Projects" (Fox River Group, 1999). Additional information on these sites and other sediment removal projects can be found in the Major Contaminated Sediment Sites (MCSS) database.

4B.1   Grasse River—Massena, New York

Between July and September 1995, Alcoa, Inc. removed approximately 3000 yd^3 of sediment and boulders/debris from two areas of the Grasse River because of elevated levels of PCBs (up to 11,000 mg/kg). The removal areas covered approximately 1 acre of the Grasse River (i.e., a river area and adjacent outfall structure). The goal of the removal action was to remove all sediment within these areas to the extent practicable. Nearly 400 yd^3 of boulders were removed from a boulder zone with a mechanical long-stick excavator (with a specialized perforated bucket) mounted on a barge. The sediments were removed with a horizontal auger hydraulic dredge. Sediments were dewatered and disposed with the boulders and debris in an on-site landfill (BBL, 1995b). Sediments within the outfall structure were removed with small manually directed plain-suction hydraulic hoses.

Sediment Data.   As shown in Fig. 4.6, preremoval PCB surficial sediment concentrations (i.e., in this case the top 12 in) ranged from 12 to 1780 parts per million (ppm) (average of 518 ppm). After hydraulic dredging was completed in an effort to remove all sediment, an average sediment depth of 4 in (up to a maximum of 14 in) remained even after multiple dredge passes. On the basis of these results, U.S. Environmental Protection Agency (EPA) and its representatives, Alcoa, and the contractors determined that sediment had been removed to the extent practicable (BBL, 1995b). Conditions such as the rocky nature of the river bottom and the presence of hardpan reduced the dredge's effectiveness in removing sediment. It was estimated that approximately 84 percent of the sediments were removed (along with 27 percent of the PCB mass in the lower Grasse River). Following removal, residual (surficial) PCB concentrations ranged from 1.1 to 260 ppm (average of 75 ppm). Moreover, at 30 percent of postremoval sample locations, residual surface sediment PCB concentrations

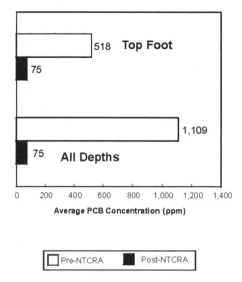

FIGURE 4.6 Average sediment PCB concentrations.

increased relative to preremoval concentrations (BBL, 1995b). Even in the outfall structure, where operators were able to manually direct vacuum hoses to remove sediment, surface sediment remained with PCB concentrations of 108 ppm (388 ppm PCBs in surface sediment before removal).

Water Data.   During removal activities, a triple-tiered silt curtain system was used in an attempt to contain suspended PCB-containing sediments. The curtains were quite effective in containing suspended sediments, with only one action level exceeded for total suspended solids (TSS) and turbidity. However, elevated PCB water column concentrations were observed; that is, PCBs were present in 88 percent of the samples collected at a location 2300 ft downstream of the removal area, while PCBs were detected only once at the upstream location. Also, two of the downstream fixed-station filtered samples had quantifiable PCB levels, whereas quantifiable levels were never observed at this location in the preremoval monitoring.

Fish Data.   In addition to water column PCB level increases during removal, increases in fish levels also were noted during removal. Figure 4.7 shows both caged fish and spottail shiner data before, during, and after removal. Although limited data are available before removal, it is obvious that sediment removal increased PCB levels in fish during removal, and levels remained elevated for several years following removal.

Other resident fish (i.e., brown bullhead and smallmouth bass) also were collected and analyzed for PCBs as part of pre- and postremoval monitoring (through 1998) of the Grasse River project. Resident fish collected in 1995 immediately following removal exhibited an increase in PCB concentrations. PCB concentrations in resident smallmouth bass and brown bullhead samples collected prior to the re-

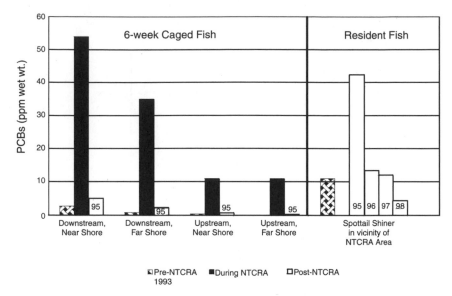

FIGURE 4.7   PCB levels in caged and resident fish before, during, and after PCB sediment removal.

moval activities are similar to those collected in 1997, and concentrations increased slightly in 1998. Overall, the apparent negative effect of the removal was greater for smallmouth bass than for brown bullhead and was most significant for spottail shiners, with the most significant differences observed in the vicinity of the removal area.

4B.2   St. Lawrence River—Massena, New York

Between May 8 and December 22, 1995, General Motors (GM) removed approximately 13,250 yd^3 of PCB sediment and associated boulders/cobbles from an approximately 11-acre area of the St. Lawrence River. These materials were dewatered and stockpiled at the GM Powertrain facility for subsequent off-site disposal.

EPA selected a 1 ppm sediment cleanup goal in the St. Lawrence River because it believed it was achievable and provided an acceptable measure of human health protection. In doing so, EPA believed it had balanced its desire for a very low cleanup level to minimize residual risk with the constraints posed by the limitations of dredging as a means of removing sediment [in Turtle Creek, an applicable or relevant and appropriate (ARAR) cleanup level of 0.1 ppm was set]. However, EPA recognized that technical limitations may preclude removal of sediments to this level (EPA, 1990b).

After efforts to utilize a silt curtain containment system failed (because of excessive water velocities), a sheetpile wall was installed around the removal area as a suspension containment measure. Prior to sediment removal, the initial footprint of the sheetpile wall was modified to exclude a cobble and boulder zone. It was agreed by the EPA and GM that the removal of sediment from this area was technically impractical because of large boulders and the potential for slope failures. Within the removal area, boulders and debris were removed mechanically prior to hydraulic dredging.

Sediment Data. Preremoval surficial sediment PCB concentrations ranged from 0.08 to 8800 ppm (average of 548 ppm) (ERM, 1993).

Even after significant passes with a hydraulic dredge were performed (up to 15 to 30 passes in some areas), residual surface sediment in all six removal quadrants remained above the cleanup goal of 1 ppm, with an overall average PCB concentration of 9.2 ppm (average PCB concentration was up to 27 ppm in one quadrant). EPA determined that sediments were removed to the maximum extent possible. Consequently, EPA "determined that installation of a cap over Quadrant 3, effectively isolating this area from the rest of the river, was the only remaining technically practicable remedial alternative." This area was subsequently capped with a multi-layer granular cover (BBLES, 1996).

Water Data. Early on in the sediment removal process, turbidity action levels were exceeded because of turbid water escaping over the top of low sheetpiling sheets. The low sheets were installed according to the design and assured stability of the containment system during storms and high waves from passing ships. To compensate for the low sheets, the contractor installed filter fabric over the low sheets and installed short steel sheets over some of the low sheetpiles. At one point during sediment removal activities, elevated water column turbidity and PCB levels were reported outside the sheetpile wall. Because of the high concentrations, a silt curtain was installed along the inside of the sheetpile wall. PCBs were also released via air, as PCBs were detected at levels exceeding the project action level at the closest downwind sample location.

Fish Data. Figure 4.8 shows total PCB concentrations in spottail shiner (the only species monitored) whole-body composite samples collected from the GM site. PCB levels may have decreased since the late 1980s, but comparison of the pre- and postremediation data are complicated by factors such as fish sizes, lipid contents, species, mobility, and uncertainties about sampling locations (especially the 1988–1989 and 1992 data relative to all other years). Previous sampling locations are important for data comparability over time. Note that remediation occurred in 1995.

The annual monitoring reports describe an anomaly to the apparent general downward trend since the late 1980s: two spottail shiner samples collected by New York State Department of Environmental Protection (NYSDEC) in 1992. The wide difference in concentrations for these two samples (total PCB concentrations of 5.7 mg/kg and 65 mg/kg) is difficult to explain. Similar variability, although not as great, is also evident in the data collected by the Ontario Ministry of the Environment (OME) in 1989. The variability of the data may be due to several factors, including differences in sampling locations, fish lengths and sizes, fish lipid content, or species mobility. In fact, discussions with both NYSDEC and OME regarding sampling locations indicate that the specific sampling locations cannot be determined. This is extremely important given the relative size of the St. Lawrence River (about 2000 ft wide, flowing at 250,000 ft³/s) compared to the area dredged (about 200 ft wide in an embayment). Postdredging sampling locations are well documented, but without predredging location details, one cannot consider the data truly comparable. Regardless, the variability of the data precludes a more detailed evaluation and interpretation of the overall spottail shiner data. Therefore, the monitoring reports

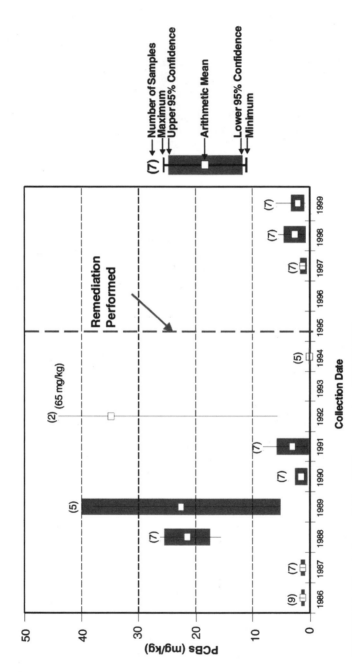

FIGURE 4.8 Historical spottail shiner PCB concentrations *(BBLES, 2000)*.

conclude that the significance of the 1997, 1998, and 1999 PCB data, and any apparent trends, will need to be more thoroughly evaluated following the collection of additional data over the next several years.

4B.3 Sheboygan River—Sheboygan Falls, Wisconsin

Approximately 3800 in situ yd³ of PCB-containing sediments were removed from the Sheboygan River by Tecumseh Products Company (Tecumseh), the only participating potentially responsible party (PRP), from 17 discrete sediment deposits in the Upper River from 1989 through 1991, using a modified "sealed" clamshell mechanical dredge. Dredging was performed within the confines of a silt containment system composed of an internal geotextile silt screen and external geomembrane silt curtain. In general, a minimum of two dredge passes (and up to four passes in some areas) were performed in each area, followed by sampling and analysis. The first dredge pass was performed in an effort to remove as much sediment as possible (i.e., to hard subgrade material). Following the first pass, the resuspended sediment within the silt containment system was allowed to settle, and a second dredge pass followed. Additional dredge passes were utilized if postdredging sampling results exhibited elevated PCB levels (BBLES, 1992; BBL, 1995a, 1998).

Sediment Data.   Preremoval surficial sediment concentrations ranged from 0.2 to 4500 ppm (average 640 ppm) in 1987. Postremoval surficial sediment concentrations ranged from 0.45 to 295 ppm (average 39 ppm). After four dredge passes, one sediment deposit exhibited residual PCB concentrations up to 295 ppm. The EPA and Wisconsin Department of Natural Resources (WDNR) agreed that the sediment had been removed to the extent practicable and directed Tecumseh to cap and armor the deposit to contain the sediment and residual PCBs (BBL, 1995a). At another Upper River deposit, preremoval surficial sediment PCB concentrations ranged from 2.6 to 8.2 ppm (average of 5 ppm) with 1.6 to 1400 ppm (average of 376 ppm) present in subsurface sediment. After several removal passes, up to 136 ppm remained in a portion of this deposit. Again, the EPA and WDNR directed that that portion of the deposit be capped/armored. Two other deposits also required capping and armoring to contain elevated residual PCB concentrations following dredging. Removed sediments remain in on-site facilities pending final disposal.

Water Data.   Water-column monitoring activities were conducted before, during, and after sediment removal activities by measuring total suspended solids (TSS) and/or turbidity and PCBs. Monitoring data indicated an increase in PCB concentrations in the water column during dredging. As a result, dredging was halted several times during the project because of increased turbidity, PCB water-column concentrations, or visual observations of sediment migration. Specifically, PCBs were detected in one or more fixed downstream sampling stations during 19 of 29 sampling events, with the highest measured concentration of 0.47 ppb detected at a location approximately 500 ft downstream of removal activities. No PCBs were detected at the upstream location during that sampling round. Typical causes of elevated PCB or turbidity levels were water disturbances from boats, breaking ice, barges in motion upstream of the sample locations, damaged silt curtains due to high flows, etc. In addition, PCB concentrations within the silt control system were as high as 8.3 ppb (measured 11 days after dredging activities were completed) (BBL, 1995a).

Fish Data. Figure 4.9 shows the smallmouth bass data collected during and after removal activities. Note that no preremoval data are available because of a laboratory problem. There is no apparent downward trend, and therefore no apparent risk reduction, in the Rochester Park vicinity (area where removal activities were concentrated), despite removal of over 95 percent of the PCB mass from the targeted deposits and 70 percent overall mass removal from the Upper River. In addition, although a slight downward trend is evident between the Kohler Dams and in the vicinity of Kiwanis Park after sediment removal, both locations show an increase in 1991, possibly a result of removal activities.

4B.4 Lake Järnsjön—Sweden

Lake Järnsjön is a 62-acre lake located 72 miles upstream of the mouth of the Emån River in Sweden. In 1993–1994, approximately 196,000 yd^3 of PCB sediments were removed from the lake.

Sediment Data. Preremoval PCB concentrations in sediment in 1990 and 1992 ranged from 0.4 to 30.7 ppm (average 8.1 ppm) in the top 1.3 ft and 0.18 to 2.9 ppm (average 1.5 ppm) in the top 0.1 ft (Bremle, Okla, and Larsson, 1998). Sediment remained following dredging with postremoval concentrations ranging from 0.01 to 0.85 ppm (average 0.13 ppm) from the top 0.66 ft (Bremle, Okla, and Larsson, 1998).

Water and Fish Data. Although this project appears to have been successful in reducing surficial sediment PCB concentrations, review of the fish data indicates that PCBs in the lake continue to influence fish concentrations.
Figure 4.10 depicts total lipid-normalized PCB concentrations in fish (1-year-old perch) and water from the Emån River, comparing 1991 preremediation levels with 1996 postremediation levels. Spatial trends are also apparent and indicate that while PCB concentrations decreased by approximately 50 percent in Lake Järnsjön, upstream and downstream concentrations were also on the decline, likely due to ongoing systemwide natural recovery processes. Finally, it is apparent that even after dredging an estimated 97 percent of PCB mass from the entire bottom of Lake Järnsjön, lake sediments remain a dominant source of PCBs to fish and the water column (Fox River Group, 1999).

4B.5 Fox River Deposit N—Kimberly, Wisconsin

Sediment Data. Approximately 8200 yd^3 of sediment was removed from a 3-acre area at Deposit N in the Fox River located near Little Chute and Kimberly, Wisconsin beginning in November 1998 as part of a demonstration project. [*Note:* This volume includes 1000 yd^3 of sediment from a nearby sediment area (Deposit O).] The project specification for the demonstration project was to remove the majority of the contaminated sediments from the 3-acre area deposit efficiently and in a cost-effective manner, realizing that a thin layer of sediment would be left behind because of the presence of bedrock and the limitations of dredging (Foth & VanDyke, 2000). The sediment volume targeted for removal was approximately 65 percent of the 11,000 yd^3 present in Deposit N (Foth & VanDyke, 2000). Two rounds of dredging were conducted at Deposit N, the first during November and December 1998 and the second between August and October 1999, since dredging could not be completed in 1998. After the removal of approximately 7200 yd^3 of sediment from Deposit N, funds

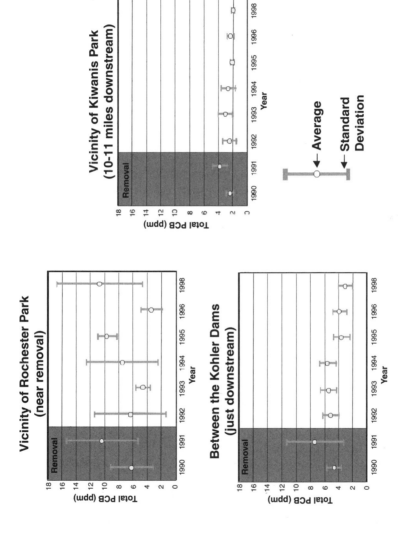

FIGURE 4.9 Sheboygan River—smallmouth bass mean total PCB concentrations (1990–1996, 1998).

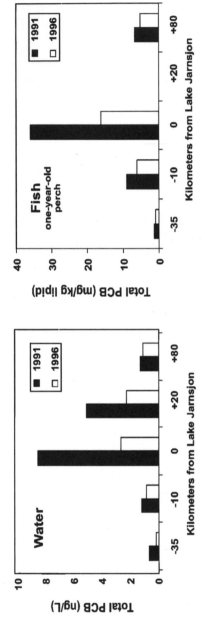

FIGURE 4.10 Total lipid-normalized PCB concentrations in fish and water in Emån River.

and good weather allowed the removal of approximately 1000 yd³ from Deposit O in October and November 1999. The overall cost of the demonstration project was $4.3 million, which equates to unit cost of $525/yd³ (Foth & VanDyke, 2000).

As shown in Fig. 4.11 the predredge average surface sediment PCB concentration for Deposit N in 1998 was 16 ppm (BBL, 2000). The 1998 postdredge average surface PCB concentration was calculated by BBL to be approximately 9 ppm. The 1999 postdredge average surface PCB concentration is 14 ppm as reported by Foth & VanDyke (2000). Independent calculations by BBL resulted in a 1999 postdredge average surface PCB level of 21 ppm.

The predredging average sediment thickness was 2 to 3 ft over fractured bedrock in water depths of approximately 8 ft (Foth & VanDyke, 2000). Shallow bedrock at the site prevented overcutting beneath the sediment and resulted in residual sediment left behind. Postdredge 1999 probing data collected from the west lobe of Deposit N showed that an average of 5 in of PCB-containing sediment remained, with as much as 15 in remaining in one portion of the deposit.

Resuspension Data.   Two rounds of dredging were conducted at Deposit N, the first during November and December 1998 and the second between August and October 1999. In 1998, the dredging area was surrounded by a silt containment system including an 80-mil high-density polyethylene (HDPE) flexible plastic barrier and a silt curtain. In addition, two deflection barriers were used to direct water around the local paper mill water intake. No turbidity barrier was used during the 1999 dredging. However, a silt curtain was placed approximately 150 ft or less downstream of the dredge (Foth & VanDyke, 2000). Generally speaking, data from both Deposit N dredging events indicate higher PCB concentrations downstream of the dredging site during dredging, while predredging upstream and downstream PCB concentrations are similar.

In 1998, the predredging PCB concentrations in upstream and downstream samples were similar, averaging 15 nanograms per liter (ng/L) upstream and 15 ng/L downstream. As indicated in Fig. 4.12, evaluating the changes in the downstream to upstream PCB concentration (*D/U*) ratio indicates that downstream PCB concen-

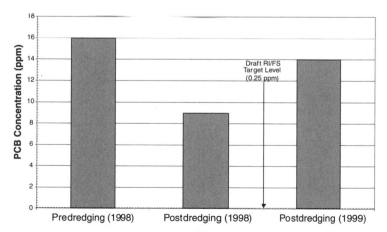

FIGURE 4.11   Fox River Deposit N—West Lobe. Average pre- and postdredging surface (0 to <6 in) sediment PCB concentrations.

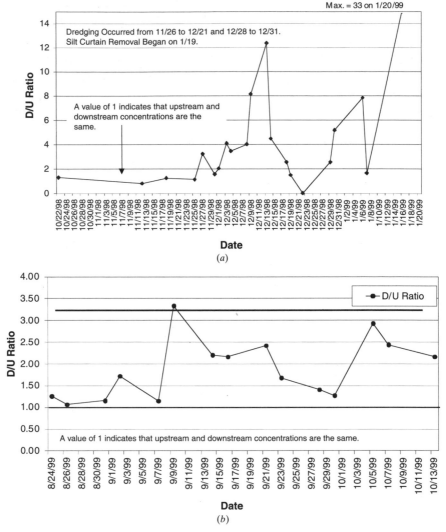

FIGURE 4.12   Ratio of downstream to upstream total PCB concentration. (*a*) 1998 water column data. (*b*) 1999 water column data, during dredging.

trations during dredging exceeded upstream concentrations in both 1998 (by a factor of 1.5 to 12.4) and 1999 (by a factor of 1.1 to 3.3) (BBL, 2000). This trend was not evident in the predredging samples. On average, downstream PCB concentrations were 4.3 times higher than upstream PCB concentrations during 1998 dredging and 1.9 times higher during 1999 dredging (BBL, 2000).

4B.6   Fox River Sediment Management Unit 56/57—Green Bay, Wisconsin

Sediment Data.   Sediment Management Unit (SMU) 56/57 is a 9-acre area located along the west bank of the Fox River in Green Bay, Wisconsin. Of the 117,000 yd^3 of sediment with PCB concentrations greater than 1 ppm, 80,000 yd^3 were targeted for removal. In August 1999, dredging began and removed approximately 31,500 yd^3 of sediment (mainly from eleven 100- by 100-ft subunits), using a hydraulic horizontal auger dredge. The goal of this demonstration project was to understand the implementability, effectiveness, and cost of a large-scale sediment removal project. Dredging continued through mid-October 1999, when review of survey information indicated that the dredging process was leaving a very uneven surface on the river bottom. WDNR directed the contractors to stop disturbing new areas and instead redredge areas that had already been disturbed. In December 1999, additional dredging passes were performed on small (30- by 30-ft) sections of four subunits designed to remove ridges in the sediment bed left from previous dredging. On average, the additional dredge passes targeted the removal of an additional 6 in of sediment.

Pre- and postdredge PCB data were collected by BBL and Montgomery Watson (Fig. 4.13). Predredge surface PCB concentrations collected in the 11 dredged subunits averaged 3.6 ppm and ranged from 1.7 to 5.9 ppm (BBL, 2000). Two rounds of postdredging sampling were conducted, the initial round in December 1999/January 2000 immediately following dredging and the second round in February 2000. The average surface PCB concentration in the 11 subunits increased to 75 ppm (range: 0.03 to 280 ppm) in the December 1999/January 2000 sampling event. A subset of seven of the eleven subunits were sampled during the February 2000 events and the resulting average surface PCB concentration was 43 ppm (range: 16 to 110 ppm).

In those four subunits where an additional "cleanup" pass was performed, predredge surface PCB concentrations were 3.5 ppm (range: 2.7 to 4.7 ppm). In December 1999/January 2000, surface PCB levels decreased slightly to an average of 3.2 ppm (range: 0.03 to 10.8 ppm), while the February 2000 sample results indicated an increase in PCB surface concentration to 26 ppm (range: 16 to 34 ppm) in these four subunits (BBL, 2000).

The predredge surface PCB concentration in those seven subunits that did not receive a cleanup pass was 3.7 ppm (range: 1.7 to 5.9 ppm). Results of the December 1999/January 2000 sampling indicate that average surface PCB concentration in these seven subunits was 116 ppm (range: 32 to 280 ppm). Only three of these seven subunits were sampled in February 2000, and the resulting average surface PCB concentration was 65 ppm (range: 40 to 110 ppm) (BBL, 2000). Surface sediment concentrations before, during, and after dredging are shown in Fig. 4.14. Dredged sediments were dewatered and disposed (as an in-kind service) at a landfill operated by the Fort James Corporation. Fort James returned to the site in 2000 to complete the dredging, with residual PCB levels of about 2 ppm remaining in surface sediment.

Resuspension Data.   The SMU 56/57 dredge area was enclosed by a silt curtain. PCB levels in the water column were monitored pre-, during, and postdredging. Generally speaking, PCB concentrations were higher downstream of the removal area than upstream during dredging.

As shown in Fig. 4.15, water column PCB data were analyzed through an evaluation of the downstream to upstream PCB concentration (D/U) ratio. Samples collected during coal boat delivery times were removed to eliminate downstream bias,

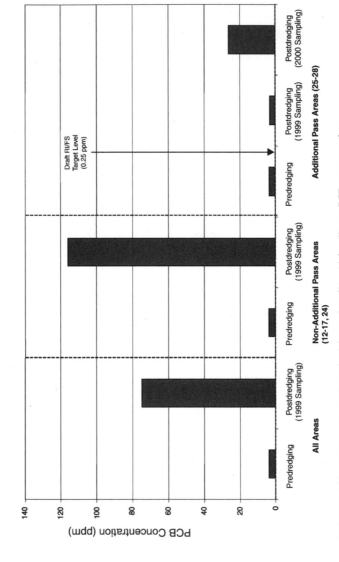

FIGURE 4.13 Average pre- and postdredging surface (0 to 4 in) sediment PCB concentrations.

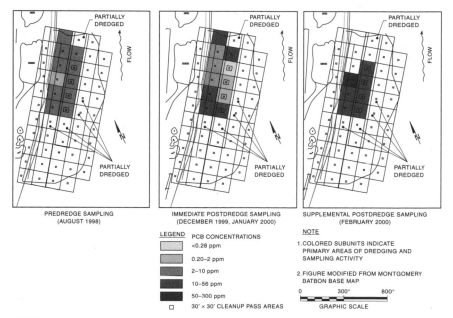

FIGURE 4.14 Surface sediment concentrations before, during, and after dredging in 1999.

which may be caused by resuspension due to coal boat travel. The predredging upstream and downstream average PCB concentrations were 53 and 52 ng/L, respectively (resulting in a *D/U* ratio of approximately 1.0). The overall *D/U* ratio during dredging indicates that, on average, PCB concentrations were higher in downstream samples by 2.6 times after removal of sampling dates that coincided with coal boat arrivals and departures.

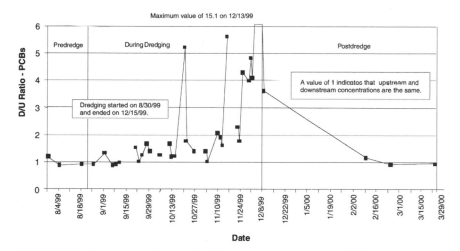

FIGURE 4.15 Water column data—ratio of downstream to upstream total PCB concentration.

4B.7   Duwamish Waterway—Seattle, Washington

Sediment Data.   A dredging effort was implemented at Slip 1 of the Duwamish Waterway to clean up sediment from a 255-gallon PCB spill that occurred on September 12, 1974. Preremoval PCB concentrations at the spill site were detected in excess of 30,000 ppm (Blazevich, 1977). The first phase of remediation was conducted in October 1974 using divers with hand-held dredges to remove approximately 50 yd^3 of sediment (Willmann, 1976). Post–Phase I removal concentrations ranged from 1200 to 1900 ppm (Blazevich, 1977). Prior to implementation of Phase II dredging activities in 1976, surficial (top 1 ft) PCB concentrations ranged from nondetect to 42 ppm (average of 4 ppm). Extensive dredging was performed with a Pneuma pump dredge in an effort to achieve maximum PCB removal near the spill source. After the first dredging pass, sediment PCB concentrations increased to as much as 2400 ppm. Thus, several passes were employed to achieve maximum removal. According to Willmann (1976), it was originally thought that 4 ft of dredging would be required to sufficiently reduce the concentrations. However, it was found that surface sediment still contained about 200 ppm after 6 ft of material had been removed, so additional dredging to hardpan (a depth of about 10 to 12 ft) was performed and resulted in residual PCB concentrations of about 10 ppm (Willmann, 1976). Overall, the postdredge surficial sediment PCB concentrations ranged from 0.2 to 140 ppm (average of 7 ppm), which were higher than the Phase II preremoval concentrations of nondetect to 42 ppm (average of 4 ppm).

4B.8   River Raisin—Monroe, Michigan

Sediments were removed from an embayment area of the River Raisin adjacent to a former outfall of the Ford Monroe facility. Approximately 27,000 yd^3 of soft sediment were removed from the embayment between April and October 1997 in a mechanical clamshell operation. A silt containment system was also used at the work area perimeter (Metcalf & Eddy, 1998).

Sediment Data.   Preremoval surface concentrations ranged from 11 to 28,000 ppm (average of 4130 ppm) and subsurface concentrations ranged from 0.78 to 29,000 ppm (average of 6510 ppm) (Metcalf & Eddy, 1993). The cleanup goal for this site was removal of PCBs >10 ppm. Despite removal efforts, potential exposure and risk may not have been reduced because, according to Metcalf & Eddy (1998), "confirmatory sample collection activities in many dredge-cells were revealing that sediment remained, even though prior dredging to refusal had occurred." Postremoval PCB levels ranged from 0.54 to 20 ppm (arithmetic average of 9.7 ppm), where only 4 of the 14 data points were usable for the postdredging calculation. The other 7 had immunoassay results >50 ppm and were redredged; however no sediment reportedly remained from which to obtain a final confirmatory sample. Two of the suspected sources of sediment were "a 0–0.5 foot layer of sediment deposited following resuspension during dredging" and "sloughing of sediment outside of the SRA (sediment removal area) into the SRA along the base of the silt curtain" (Metcalf & Eddy, 1998). Cells not meeting the 10 ppm cleanup goal in surficial sediments were redredged until PCB concentrations were less than 10 ppm in the cells.

Fish Data.   As shown in Fig. 4.16, the Michigan Department of Environmental Quality (MDEQ) performed preremoval caged fish studies at the mouth of the River Raisin in 1988 and 1991 (remediation occurred in 1997). The total PCB concentration

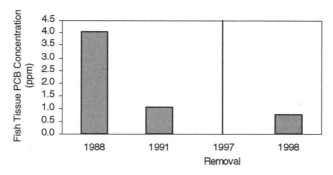

FIGURE 4.16   Caged fish net PCB uptake.

was 4.06 ppm in 1988 and 1.07 ppm in 1991 (Michigan Department of Environmental Quality, 1998). In comparison, the PCB concentration after removal in 1998 was approximately 0.77 ppm. The 1991 concentration was about 25 percent of the 1988 concentration (a decrease of about 1 ppm/year), and the 1998 concentration was about 72 percent of the 1991 concentration (a decrease of about 0.04 ppm/year), thus indicating that natural recovery was taking place prior to removal activities and that removal activities did not have a marked effect in reducing the postremoval caged fish concentrations.

4B.9   Manistique River and Harbor—Manistique, Michigan

At the Manistique River and Harbor site in Michigan, dredging has been performed in three areas (the North Bay, an area in the river, and the harbor) to remove PCB sediments. Dredging at the site has been performed with a combination of diver-assisted and hydraulic cutterhead dredging. EPA's goal is to achieve a PCB concentration of 10 ppm at all depths in sediments.

Through the end of 1999, according to the U.S. EPA, a total of less than 100,000 yd^3 of sediment has been dredged and 41,800 tons of dewatered sediments have been shipped to off-site landfills for disposal. The table below summarizes the volumes removed by year.

Year	Volume removed,* yd^3	Tons disposed
1995	10,000[†]	1,200[†]
1996	12,500[†]	2,100[†]
1997	62,000[‡]	12,000[‡]
1998	31,200[§]	12,600[§]
1999	25,000[¶]	13,900[¶]
TOTAL	97,000	41,800

* The volumes are based upon U.S. EPA Pollution Reports; volume to date modified by EPA in 1999 to 72,000 yd^3 through 1998.
[†] Quantities removed from Area B, POLREP 15 and 20.
[‡] Quantities removed from Areas C and D, POLREP 40.
[§] Quantities removed from Area D, POLREP 56.
[¶] Quantities removed from Areas B and D, POLREP 70.

As of November 2000, the cost for the project is over $45 million. The original budget in 1995 was $15 million. Initially, EPA expected the dredging to be completed by the end of 1997. Currently, EPA estimates that dredging will be completed by the end of 2000.

Sediment Data. *North Bay (Area B).* Preremoval surficial sediment PCB concentrations in the North Bay ranged from nondetect to 62 ppm (average of 8.8 ppm), according to data collected in 1995.

The EPA originally dredged the North Bay in 1995 and 1996. These activities were initially performed by using diver-assisted dredging to remove sediment along with a layer of wood chips. Subsequent removal was then accomplished by using a horizontal auger cutterhead dredge. In September 1996, the EPA declared that dredging operations were completed in the North Bay (Nied, 1996a). Postdredging sampling of the North Bay by EPA in the fall of 1996 revealed that sediment with PCB concentrations greater than 10 ppm remained. In response, the EPA placed washed gravel in the North Bay in October 1996 to "improve the river bottom in this area as habitat for aquatic species as well as enhance containment of the contaminated residuals which could not be cost effectively recovered from beneath the debris layer during dredging" (Nied, 1996b).

In October 1998, BBL collected five sediment cores in the North Bay to confirm whether EPA had reached the 10 ppm PCB cleanup level. PCB concentrations in surficial (0 to 3 in) sediment samples ranged from 1.3 to 1300 ppm, with two of the five detections being greater than 10 ppm, and an overall arithmetic average of 270 ppm. Some of the subsurface intervals sampled also had PCB concentrations greater than 10 ppm. In April 1999, prior to dredging, EPA collected five cores in the North Bay. PCB concentrations in the surficial samples (0 to 1 ft) ranged from 16 to 116 ppm, and averaged 48 ppm. On the basis of the results of these sampling efforts, EPA decided additional dredging was needed in the North Bay, which was conducted in May and June 1999.

After the additional dredging had ceased for the season in 1999, BBL collected nine sediment core samples from the North Bay. In the surficial interval (0 to 3 in), PCB concentrations ranged from 0.25 to 15 ppm. One sample had a PCB concentration greater than 10 ppm. Six out of 13 subsurface (deeper than 3 in) samples had PCB concentrations greater than 10 ppm, with a maximum PCB concentration of 620 ppm.

River Area (Area C). In 1993, an interim geomembrane cap was installed as a temporary measure near an outfall. In 1997, the temporary cap was removed and the sediment was dredged. Sediment PCB concentrations were determined by using immunoassay tests to assess whether the cleanup goal of 10 ppm was reached. The data document that sediment PCB concentrations remained above 10 ppm. In fact over 20 percent of the samples showed that sediment above 50 ppm was left behind.

Harbor (Area D). Preremoval surficial sediment PCB concentrations in the Harbor ranged from nondetect to 340 ppm (average of 14 ppm) according to data collected during the Engineering Evaluation/Cost Analysis (EE/CA).

After EPA completed its dredging activities in 1997, 1998, and 1999, BBL collected between 24 and 46 core samples within the harbor. In all years, the samples were distributed throughout the harbor area without bias toward dredged or undredged areas. The average surface sediment PCB data is summarized in Fig. 4.17.

In addition, data from 1993 were compared to data from 1999 to determine whether there was any difference between areas which were dredged and those which

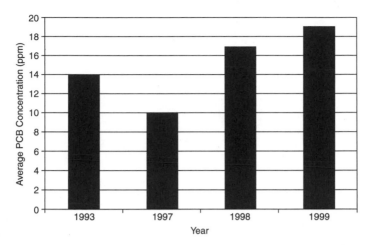

FIGURE 4.17   Area D—average PCB concentration in surface sediments (0 to 3 in).

were not dredged. The delineation of areas dredged (as provided by EPA) was overlaid with the sampling locations in 1993 and 1999 to categorize locations as either within or outside dredged areas.

Given potential mapping inaccuracies, it is possible that some sample locations may be interpretable either way (hereinafter called *border samples*). According to best judgment, the border samples would be considered within the dredged areas. However, for completeness, both scenarios have the average surface sediment concentrations plotted in Fig. 4.18.

The figure shows that while the average PCB concentrations in undredged areas in 1999 was roughly twofold lower than in 1993, this was not the case in dredged areas. The apparent decline in undredged areas may be evidence of natural recovery.

In addition to sampling by BBL, EPA conducted predredging surveys of the harbor in 1998 and 1999. In 1998, EPA collected 112 samples in the harbor, and PCB concentrations ranged from nondetect to 1250 ppm and averaged 16 ppm. In 1999, EPA collected 124 cores in the harbor. PCB concentrations in the surficial (0 to 1 ft) sediments ranged from nondetect to 1096 ppm and averaged 30 ppm. The average concentration both years was greater than 10 ppm and increased from 1998 to 1999, generally consistent with BBL data.

EPA continues to have difficulties achieving the 10 ppm cleanup goal in the harbor. At the end of the 1999 dredging season, EPA collected sediment samples in the harbor that showed an average PCB concentration greater than 10 ppm. In the 151 grab samples collected by EPA, PCB concentrations ranged from nondetect to 340 ppm and averaged 20 ppm (compared to 19 ppm average for BBL data). EPA returned in 2000 to complete the dredging, however, data are not yet available for review.

Water Data.   PCB data are available for surface water samples from the Manistique River and Harbor Site from the early 1980s to 1998. In the early 1980s, Marti and Armstrong (1990) collected five surface water samples from the mouth of the river, and in April to May 1994, EPA collected three surface water samples at the site

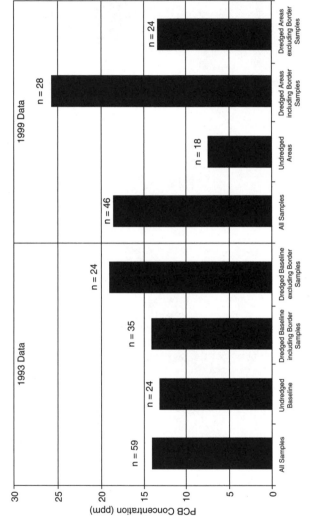

FIGURE 4.18 Manistique Harbor (Area D) Surface Sediment (0 to 3 in) average PCB concentrations.

as part of the Lake Michigan Mass Balance Study. These sample results are presented below.

Water Column Total PCB Concentrations, ppb

Sampling period	Range	Mean	No. of samples	Reference
Early 1980s	0.007–0.043	0.024 ± 0.015	5	Marti and Armstrong, 1990
April/May 1994	0.0002–0.0021	0.0009	3	EPA; LMMB Study
1995	ND–0.49	0.10	102	EPA
1996	ND–3.5	0.62	23	EPA
1997	ND–0.81	0.26	10	EPA
1998	ND–0.14	0.081	17	EPA

ND = not detected.

The average total water column PCB concentrations in 1994 were an order of magnitude lower than the early 1980s data. In EPA's surface water PCB data for 1995 through 1998 (during dredging), the mean PCB concentration was 0.19 ppb (range of 0.042 to 3.5 ppb), an order-of-magnitude or more higher than the preremediation concentrations. The annual means are as reported in the table above. Of all the years with water column data, the periods during dredging show the highest mean PCB detections.

Silt containment has been used during dredging of all three areas. In the North Bay, silt containment included plastic sheeting with wooden shoring at the mouth of the Upper Bay and silt barrier (filter fabric). In the river area, silt containment included silt barrier constructed from surplus wet felt from a nearby paper mill. In the harbor, a silt barrier was used for containment.

In 1998, BBL performed sediment trap sampling in Manistique Harbor. The results were generally low; however, three of the higher detections observed (9.5, 42, and 84 ppm) suggest resuspension of bottom sediments that may have been due to dredging-related activity, including dredged sediment transport by barges to and from the work area. Since no predredging data are available, comparisons with preremoval conditions are not possible.

4B.10   South Branch of the Shiawassee River—Howell, Michigan

In 1982, a backhoe was used to remove PCB-containing sediment from around a factory discharge, and a dragline was used to remove PCB-containing sediments near Bowen Road, 1.2 miles downstream from the plant site. Small pockets of oily sediments also were vacuumed from this stretch. As discussed by Malcolm Pirnie Engineers, "although intended to clean up a total of eight miles of the river, the remediation project stopped at the end of 1982 with only 1.5 miles of river remediated. Cost overruns and the presence of contamination extending farther than initially anticipated were identified as reasons for the incomplete removal action" (Malcolm Pirnie, 1995). No postremoval verification sampling was performed to determine if the 10 ppm cleanup goal was achieved. Only visual and olfactory observations were used to determine the extent of dredging (Environmental Research Group, 1982).

Water Data.   Rice et al. (1984) investigated changes in PCB concentrations in surface water before, during, and after dredging. The results are summarized in Fig. 4.19.

The two downstream locations show increases in PCB concentrations during dredging; however, the samples collected 6 months later do not show a significant decrease in PCB concentration when compared to the predredge concentrations. In fact, it was recognized that "dredging of sediments is likely to cause temporary resuspension of contaminants into the water column which can cause a temporary increase in tissue contaminant concentrations of aquatic biota. Dredging also removed indigenous benthic fauna, which can take years to reestablish" (Malcolm Pirnie, 1995).

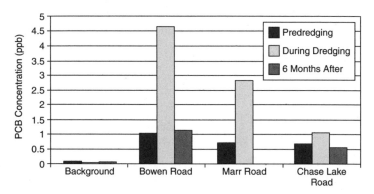

FIGURE 4.19 Arithmetic average PCB concentration in surface water.

Sediment and Fish Data.   Figure 4.20 shows total PCB concentrations in sediment and white sucker fillet samples from the Shiawassee River. Twenty years of data indicate that PCB levels in fish and sediment were undergoing a decline prior to and after the 1982 remediation, which limits the ability to differentiate the effects of remediation versus other processes such as natural attenuation or source control. Note that data are plotted on a log scale.

To assess the effectiveness of the cleanup, the University of Michigan (UM) performed caged fish and clam studies in the Shiawassee River on behalf of MDEQ (formerly Michigan Department of Natural Resources) before, during, and after the 1982 dredging effort (Rice and White, 1987). At all locations downstream from the plant site and in the area of removal, the UM study indicated an increase in the bioavailability of PCBs following dredging (Rice et al., 1984). For example, at the Bowen Road location (1.2 miles downstream of the source), the PCB levels in caged fathead minnows increased from 64.5 ppm (before removal) to 87.95 ppm dry weight after dredging. PCB concentrations in caged clams collected approximately ¼ mile downstream from the plant site ranged from 13.82 ppm before dredging to 18.30 ppm after dredging, and averaged 59.1 ppm during dredging (Malcolm Pirnie, 1995; Rice et al., 1984), indicating that dredging actually increased exposure rather than decrease it as intended.

4B.11   Ruck Pond—Cedarburg, Wisconsin

Ruck Pond is one of a series of mill ponds created on Cedar Creek, just upstream of the low-head Ruck Pond Dam. In 1994, an impounded 1000-ft section of the creek

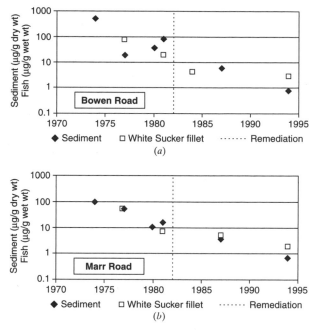

FIGURE 4.20 Total PCB concentrations in sediment and white sucker fillet samples from the Shiawassee River.

(Ruck Pond) was drained after a temporary dam was installed on the upstream end and flow was bypassed through siphon piping. The project goal was to remove all soft sediment (contaminated with PCBs) down to bedrock, to the extent practicable.

Sediment Data.   A total of 7730 yd³ of sediment was removed by dry excavation and disposed of at commercial landfills. After removal efforts were completed, clean materials used for access to the pond were spread along portions of the pond bottom. Although not intended for capping, these materials inevitably provided some containment of the residual sediment, and likely would have reduced (via burial) the relatively high PCB concentrations remaining at the sediment surface that the dredge equipment could not effectively remove (Praeger, Messur, and DiFiore, 1996).

The maximum PCB concentration measured within the sediments was approximately 150,000 ppm, with an average concentration of 474 ppm (EPA, 1999b). However, 60 soft-sediment surface samples collected from the top 0.5 to 2 ft just before remediation exhibited PCB concentrations ranging from nondetectable to 2500 ppm (arithmetic average 76 ppm). Despite 5 months of intensive removal efforts (e.g., use of squeegees attached to a bulldozer blade and vacuum truck), some residual sediment was left on the bedrock surface of the creek bed (Baird and Associates, 1997). Even though 96 percent of the PCB mass was removed, 7 postremediation surficial sediment samples exhibited PCB concentrations ranging from 8.3 to 280 ppm (arithmetic average 84 ppm) (Baird and Associates, 1997).

Fish Data. The Wisconsin Department of Natural Resources (WDNR) measured whole-body PCB congener concentrations in caged fathead minnows at three locations before and after the sediment removal operation (Amrhein, 1997). Three cages were placed at each of three stations: a site in Cedar Creek upstream of Ruck Pond called Cedarburg Pond, a site within the downstream end of Ruck Pond, and a site downstream of the Ruck Pond Dam, located just upstream of Columbia Dam.

In July 1994, just before the start of removal, PCBs were measured in caged fathead minnows at the three stations. The average PCB concentrations were 0.12 ppm upstream, 24 ppm at the Ruck Pond station, and 12 ppm at the downstream station (7.1, 1700, and 630 mg/kg lipid-normalized PCB, respectively). The average PCB concentrations measured in caged fish in August and September 1995, about 1 year after remediation, were 0.09 ppm upstream, 4.2 ppm within the pond, and 11 ppm downstream (2.2, 170, and 360 mg/kg lipid-normalized PCB, respectively). These PCB levels in the caged fish collected in Ruck Pond would, at face value, appear to have declined 75 to 85 percent* on a wet-weight basis and approximately 90 percent on a lipid basis after remediation. However, caged fish PCB concentrations at the upstream "background" location also declined 25 percent wet weight and 70 percent on a lipid basis 1 year after remediation, and caged fish concentrations downstream of Ruck Pond declined 10 percent wet weight and 40 percent on a lipid basis. The declines upstream of Ruck Pond would indicate that other factors, such as natural recovery processes or metabolism/feeding differences were occurring.

The other more important issue is that construction activities were taking place in the pond (e.g., siphon installation, work boat traffic, etc.) during the premediation sampling. In fact, all three cages in the pond were displaced from their original locations, with one cage unrecovered. This all indicates that the premediation cages in Ruck Pond should not be considered representative of premedial conditions.

4B.12   Waukegan Harbor—Waukegan, Illinois

Waukegan Harbor is approximately 37 acres in size and is located on Lake Michigan approximately 25 miles north of Chicago. Remediation areas in the harbor included boat slip no. 3 and the 10-acre Upper Harbor. For the Upper Harbor, EPA concluded that, on the basis of modeling, residual sediment PCB concentrations of between 100 ppm and 10 ppm would result in a negligible PCB influx to Lake Michigan. EPA therefore set a 50 ppm PCB cleanup level for the Upper Harbor and calculated that 96 percent of the PCB mass would be removed from the Upper Harbor if the 50 ppm goal was met (EPA, 1984, 1989).

The original goal of the Record of Decision (ROD) was elimination of PCB flux to Lake Michigan (restoration of the harbor fishery was not a specific objective). Regarding the effectiveness of sediment removal, EPA stated in the ROD's Responsiveness Summary that "Remedial alternatives based on a sediment cleanup level below 50 ppm raise technical and cost-effectiveness concerns. EPA had to consider the technical limitations inherent in the available dredging technology. Any dredging technique would involve some resuspension of sediment into the water column,

---

* Two exposure periods occurred in Ruck Pond, 29 and 37 days. Average PCB levels were greater in the longer exposure, indicating that the fish were not at steady state with respect to their exposure sources. Therefore, pre-and postremediation comparisons were carried out independently for each exposure period. The range of values given reflects the two comparisons.

and resettling back into the sediment. It may be difficult to assure that lower sediment levels could be achieved given the technological limitations. . . . As further explained, implementation of the proposed remedy essentially eliminates PCB influx to the Lake from the site."

In late 1991 and early 1992, a total of 6300 yd^3 of sediment with PCB concentrations greater than 500 ppm were hydraulically dredged from Slip no. 3, and 32,000 yd^3 were hydraulically dredged from the Upper Harbor. Slip no. 3 was abandoned and prepared as a permanent containment cell. The 6300 yd^3 were treated by thermal desorption to remove PCBs and then placed in the cell. The 32,000 yd^3 from the Upper Harbor were pumped from the dredge directly to the cell, and then the cell was capped. The dredging of sediments (primarily organic silts) in 10 acres of the Upper Harbor was completed to a designated depth and to a designated sediment layer such as clay till or sand. Characterization data had shown the underlying clay till and sand layers were only slightly contaminated with PCBs. Sampling was performed during dredging to determine sediment consistency (i.e., to determine if the clay or sand layer had been reached), but not to measure residual PCB concentrations (Canonie Environmental, 1996).

Sediment Data. No formal postremoval monitoring program was implemented following completion of the dredging, but in April 1996 (over 4 years after dredging was completed) Illinois EPA reported the results of "Harbor sediment samples collected to document the effectiveness of dredging." Thirty surface sediment samples (3-in depth) were collected from 29 locations. Eleven of the samples were archived in a freezer and not analyzed, and two sample bottles were broken in transit. Results for the other 17 samples (one duplicate) showed PCB concentrations ranging from 3 to 9 ppm. Six of the 17 samples were from within the 10 acres of harbor that were dredged and had PCB concentrations of 5 to 8 ppm.

Fish Data. Preremediation fish data from Waukegan Harbor are extremely limited. For example, only one carp composite sample consisting of two fish and one alewife composite sample consisting of five fish were collected and analyzed in 1991 by the EPA. EPA also concluded that the 1991 alewife data (as well as additional carp data from 1983) should not be used to assess temporal trends because of technical problems associated with the data. Postremediation data include several fish species collected in the Upper Harbor and in Lake Michigan in the vicinity of the Waukegan Harbor between 1992 and 1998.

Figure 4.21 provides average total PCB concentrations in carp collected from the Upper Harbor (with range representing 2 standard errors). While these graphs seem to indicate that PCB levels were lower in 1993 (compared to 1991), they also indicate a general increasing trend since dredging. The lack of adequate preremediation data and the fact that fish tissue concentrations have generally been rising since 1994 indicate the presence of other factors that limit the ability to differentiate the effects of various remedial activities (removal and/or containment) in the harbor. In addition, such a significant drop in PCBs from 1991 is inconsistent with expected trends in tissue PCB levels due to the rate of natural depuration of PCBs by fish.

4B.13 New Bedford Harbor—New Bedford, Massachusetts

Starting in 1976, the EPA detected high concentrations of PCBs in marine sediments over a widespread area of New Bedford Harbor (e.g., PCB concentrations up to

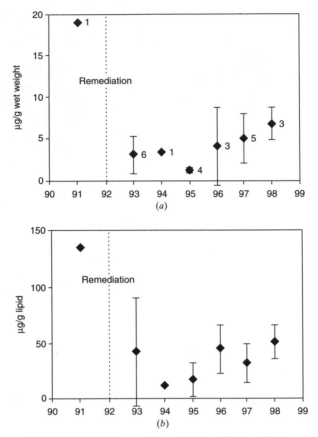

FIGURE 4.21 PCBs in carp from the Upper Harbor, Waukegan, Illinois. (*a*) Wet weight; (*b*) lipid basis.

250,000 ppm were reported in 1982). From May 1988 to February 1989, the United States Army Corps of Engineers (U.S. ACE) performed a full-scale dredging pilot study at the site to assess the performance of dredge equipment, the suitability for the removal of contaminated sediments, and the recommended procedure for operation (U.S. ACE, 1990). Three hydraulic dredges were evaluated: hydraulic cutterhead, horizontal auger (mudcat), and matchbox. The study used two small shallow (water depth less than 5 ft) dredging areas, and approximately 10,000 yd³ of sediments were removed (U.S. ACE, 1990).

Sediment Data. Prior to removal, both test areas contained higher concentrations in the surface (top 6 in) sediments (i.e., average of 226 ppm in Area 1 and 385 ppm in Area 2) compared to subsurface concentrations, which were 1 to 3 orders of magnitude lower. Postremoval average residual sediment (top 3 in) concentrations for each of the dredges tested were as follows:

- Cutterhead (Area 1): 80 ppm
- Horizontal auger (Area 1): 66.4 ppm
- Cutterhead (Area 2): 8.6 ppm
- Matchbox (Area 2): 5.4 ppm

Note that a theoretical versus actual residual PCB concentration evaluation also was performed, which showed that actual postremoval concentrations were much higher than those theoretically predicted.

Following performance of the pilot study, the remediation for the New Bedford site was split into two operable units. EPA issued an ROD for the first operable unit (hot-spot areas, those areas with greater than 4000 ppm PCBs) in April 1990. The 1990 ROD called for dredging of approximately 10,000 yd^3 of sediment with PCB concentrations greater than 4000 ppm, dewatering (with effluent treatment), incineration of dewatered sediment, and stabilization of the incineration remains (EPA, 1990a). The dredging portion of this phase was initiated in April 1994 and was completed in September 1995. Over the 1994–1995 construction period, a total of about 14,000 yd^3 was dredged and placed in a confined disposal facility (CDF) nearby, pending determination of final treatment and/or disposal. Predredging surficial sediment samples (upper 2 ft) had PCB concentrations ranging from 4000 to 200,000 ppm, with an arithmetic average of 25,000 ppm (EPA, 1999a). Initial postdredging sampling showed up to 3600 ppm PCBs remained after dredging (personal communication with P. L'Hreaux of U.S. ACE, 1996). After the completion of the project, it was estimated by Ebasco Services and the EPA that only about 45 percent of the PCBs in the harbor had been removed by dredging (EPA, 1997).

Water Data.   Water-column monitoring was performed during the hot-spot removal initiated in 1994 to assess and limit the amount of cumulative transport of PCBs to the lower harbor. For the entire removal operation, EPA calculated that a mass of approximately 57 kg (24 percent of the maximum allowable cumulative transport) was transported into the lower harbor (EPA, 1997).

Air Data.   During dredging operations, ambient air PCB concentrations were monitored at 16 monitoring locations to characterize impacts from dredging operations. If the airborne PCB concentrations exceeded predetermined action levels (i.e., 0.05, 0.5, or 1 μg/m^3), then modifications or additions of engineering controls were implemented to dredging operations, with respect to severity. Of 4041 total samples collected over the course of remedial actions, 1063 (26 percent) exceeded the 0.05 μg/m^3 action level, 49 (1 percent) exceeded the 0.5 μg/m^3 action level, and 10 (0.25 percent) exceeded the 1 μg/m^3 action level. Because of the exceedences, operational changes were implemented to minimize airborne PCB levels, leading EPA to conclude that "control of airborne PCB emissions did contribute to a slower rate of dredging and thus a longer project duration" during the hot spot removal operation (EPA, 1997).

References

Amrhein, J. 1997. Memorandum, Cedar Creek Cage Fish Study, September 22.

Baird and Associates. 1997. *Final Report, Milwaukee River PCB Mass Balance Project,* prepared for Wisconsin Department of Natural Resources, September 4.

BBL. 1994. *Engineering Evaluation/Cost Analysis, Manistique River and Harbor Site,* April.

BBL. 1995a. *Alternative Specific Remedial Investigation Report—Sheboygan River and Harbor,* October.

BBL. 1995b. *Non-Time-Critical Removal Action Documentation Report—Grasse River Study Area, Massena, New York,* December.

BBL. 1998. *Feasibility Study for the Sheboygan River and Harbor Site,* April.

BBL. 2000. *Effectiveness of Proposed Options for Additional Work at SMU 56/57: Lower Fox River, Green Bay, Wisconsin,* March.

BBLES. 1992. *Removal Action Construction Documentation Report for the Sheboygan River and Harbor Site,* March.

BBLES. 1996. *St. Lawrence River Sediment Removal Action Completion Report,* June 1996.

BBLES. 2000. *St. Lawrence River Monitoring and Maintenance Annual Inspection Report,* January 2000.

Blazevich, J. N., A. R. Gahler, G. J. Vasconcelos, R. H. Rieck, and S. V. W. Pope. 1977. *Monitoring of Trace Constituents During PCB Recovery Dredging Operations, Duwamish Waterway,* EPA/9109/9-77-039, August.

Bremle, G., L. Okla, and P. Larson. 1998. "PCB in Water and Sediment of a Lake after Remediation of Contaminated Sediment," *Ambio,* vol. 27, no. 5, pp. 398–403.

Canonie Environmental. 1996. *Construction Completion Report: Waukegan Harbor Remedial Action: Waukegan, Illinois,* July 3.

EPA. 1984. *Superfund Record of Decision: Outboard Marine Corporation Site.*

EPA, 1989, *Record of Decision Amendment—Outboard Marine, IL,* March 30.

EPA. 1990a. *Record of Decision—New Bedford Harbor Superfund Site Hot Spot,* April 6.

EPA. 1990b. *Record of Decision—General Motors Powertrain, Massena, New York Superfund Site,* December.

EPA. 1997. *Record on the Effects of the Hot Spot Dredging Operations—New Bedford Harbor Superfund Site,* October.

EPA. 1999a. *Amended Record of Decision—New Bedford Harbor Superfund Site Hot Spot,* April.

EPA, 1999b, www.epa.gov/glnpo/sediment/realizing/realpast.html#ruckpond.

Environmental Research Group. 1982. *Polychlorinated Biphenyl–Contaminated Sediment Removal from the South Branch Shiawassee River.* Ann Arbor, Mich.

ERM. 1993. *Sediment Cores in the St. Lawrence River,* November 17.

Foth & Van Dyke. 2000. *Summary Report: Fox River Deposit N,* April.

Fox River Group. 1999. "Effectiveness of Sediment Removal: An Evaluation of EPA Region 5 Claims Regarding Twelve Contaminated Sediment Removal Projects," September 27.

IJC. 1999. Sediment Priority Action Committee, "Identifying and Assessing the Economic Benefits of Contaminated Aquatic Sediment Cleanup," IJC Biennial Forum, Milwaukee, September.

Malcolm Pirnie. 1995. *Development of Sediment Quality Objectiveness for PCBs for South Branch Shiawassee River,* June.

Marti, E. A., and D. E. Armstrong. 1990. "Polychorinated Biphenyls in Lake Michigan Tributaries," *Journal of Great Lakes Research,* vol. 16, no. 3, pp. 396–405.

Metcalf & Eddy. 1993. *Summary Report of Field Activities, Analytical Results and Remedial Alternatives: Ford Outfall Site, River Raisin Sediments,* August 6.

Metcalf & Eddy. 1998. *Completion of Removal Action/Completion of Work Report for River Raisin Sediment and Soil Removal: Fort Outfall Site, Monroe, Michigan,* September 23.

Michigan Department of Environmental Quality. 1998. *Michigan Fish Contaminant Monitoring Program: 1998 Annual Report,* MI/DEQ/SWQ-98/091, December.

Nied, W. 1996a. *Pollution Report No. 24—Manistique Harbor Site, September 9 through September 27,* U.S. EPA, September 27.

Nied, W. 1996b. *Pollution Report No. 25—Manistique Harbor Site, September 28 through October 16,* U.S. EPA, October 11.

Praeger, T. H., S. D. Messur, and R. P. DiFiore. 1996. "Remediation of PCB-containing Sediments Using Surface Water Diversion—Dry Excavation: A Case Study," *Water Science & Technology,* vol. 33, no. 6, pp. 239–245.

Rice, C. P., D. S. White, M. S. Simmons, and R. Rossman. 1984. *Assessment of Effectiveness of the Cleanup of PCBs from the South Branch of the Shiawassee River—Field Results,* University of Michigan, prepared for Michigan Department of Natural Resources, October.

Rice, C. P., and D. S. White. 1987. "PCB availability assessment of river dredging using caged clams and fish." *Environmental Toxicology and Chemistry,* vol. 6, no. 4.

U.S. ACE. 1990. *New Bedford Harbor Superfund Pilot Study—Evaluation of Dredging and Dredged Material Disposal,* May.

Willmann, J. C. 1976. "PCB Transformer Spill, Seattle, Washington," *Journal of Hazardous Materials,* vol. 1, pp. 361–372.

CHAPTER 5

HAZARDOUS CONTAMINANTS IN MARINE SEDIMENTS

Jack Q. Word
Lucinda S. Word

MEC Analytical Systems
Sequim, Washington

5.1 INTRODUCTION: SOURCES AND CAUSES OF HARBOR SEDIMENT CONTAMINATION

Permitted and nonpermitted point and nonpoint sources of contamination enter the air, land, or aquatic environments and are then transported to rivers and harbors, with many of the contaminants flocculating with suspended particles and ultimately settling into the sediment of marine harbors and estuaries. Some of these contaminants are persistent and remain buried in harbor sediment while others are less persistent and may rapidly biodegrade to "inert" or nonbiologically available materials. The objectives of national and international legislation and agreements are to protect the marine environment so that it can be used by future generations for all current and projected uses (commercial as well as aesthetic). As a result, agreements have been made to protect the environment from persistent chemicals of ecological concern. These are the chemicals that, at concentrations in excess of "trace" levels, are predicted or known to cause adverse ecological impacts either directly, because of acute or chronic toxicity, or to cause indirect impacts on food webs through their uptake into tissues of exposed organisms.

Over the past 45 years, the study of marine pollution has extended beyond the simple description of changes that occur in association with sources of pollution to better understanding and the ability to predict the impacts of these persistent chemicals. However, precise estimates of the extent of exposure to toxic chemicals in the environment, and the projection of adverse health or ecological effects, have been difficult to achieve. Exposure is difficult to assess because of the wide diversity of potential routes of exposure (air, soil, water, and food web), the large differences in the biological availability of contaminants associated with the different environmental media, and individual and species-specific differences in the pharmacody-

FIGURE 5.1 Watershed and river influence on Commencement Bay, Port of Tacoma, Washington. *(Photo: Kemer Nelson, 1997.)*

namic pathways and uptake rates of different contaminants (McCarthy and Shugart, 1990). Recent sediment evaluations have isolated contributors to biological effects on the basis of persistent sediment features (e.g., sediment grain size) and nonpersistent contributors to biological impacts (e.g., salinity, ammonia, sulfides). These latter features of biological impacts are termed *confounding factors* (CFs). The key to understanding the implications of biological effects in marine systems is to be able to separate the effects of persistent chemical contaminants from the CFs. The separation of CFs from persistent chemicals of ecological concern is an area of study that is at the leading edge of many assessments of port and harbor contamination and will serve as a focus point for this chapter.

5.2 HISTORICAL PERSPECTIVES

At least 37 percent of the U.S. population is located in counties adjacent to the oceans or major estuaries (National Research Council, 1993). Many of these people live in large urban areas such as New York/New Jersey harbors and Los Angeles/ Long Beach and San Diego harbors, which represent well-studied eastern and western urban ports. As early as 1910, it was recognized by New York and New Jersey that waste drainage into local rivers was an unacceptable system that required transport of waste materials to the sea (DeFalco, 1967). The solution to these problems in the rivers was to construct transport mechanisms away from the rivers and into the estuaries, resulting in the contamination of estuaries, which began to be recognized in the early 1950s. On the West Coast, studies were performed in the early 1950s that

FIGURE 5.2   Receiving waters of Commencement Bay, Port of Tacoma, Washington. *(Photo: Kemer Nelson, 1997.)*

documented highly anaerobic sediment with little to no visible macrobenthic life in subtidal sediment (Reish, 1955, 1959, and San Diego Regional Water Pollution Control Board, 1952). DeFalco (1967) described the estuary as the septic tank of the megalopolis as the country established the legislation to control waste and storm water discharges to aquatic environments through the 1972 amendments to the Federal Water Pollution Control Act, reauthorized in 1977 and 1987.

Over the next 20 years the contamination of estuaries and coastal environments decreased rapidly (NRC, 1993). The National Research Council (2000) in a review of clean coastal waters evaluated our understanding of and plans to reduce the effects of nutrient pollution to aquatic systems. They indicated that, while many improvements have occurred over the past 20 years since the passage of the Clean Water Act, there are still issues that need to be understood and controlled, and encouraged our legislature to address the Clean Water Act that has been waiting for reauthorization since 1990.

FIGURE 5.3 Clam shell dredge in Long Beach Harbor, California, 1972.

5.3 INTERNATIONAL AND NATIONAL LEGISLATION AND AGREEMENTS

Dredging activities remedy several common problems encountered in ports and harbors, especially the siltation of channels and provision of deeper navigation channels required by increasing ship sizes. However, negative impacts on marine flora and fauna can occur as a result of these activities, including disturbance of benthic community habitats at the dredging and disposal sites, physical smothering of these communities, and potential chemical contamination of the sediment and/or biota. Inappropriate selection of disposal sites can impact fisheries, recreation, and navigation. In recognition of potential negative effects of dredging activities, international congresses have established several international conventions to ensure proper management of dredging activities and of dredged material disposal practices (Burt and Fletcher, 1997). In the 1970s, protocols for the control of dredged material disposal practices, addressed by the London Dumping Convention and the Oslo Convention, were predicated on regulation of disposal of noxious substances into the oceans and regulation of disposal of dredged sediment. Two fundamental principles were established:

The precautionary principle: Preventive measures are to be taken when there are reasonable grounds for concern that substances or energy introduced into the marine environment may bring about hazard, harm, damage, or interference, even when there is no conclusive evidence of a causal relationship between inputs and effects. A secondary tenet is a "reverse list" process; i.e., only substances that have been proved not to cause harm are permitted for ocean disposal.

The polluter pays: The costs of pollution prevention, control, and reduction measures are to be borne by the polluter.

5.3.1   The London Convention 1972 and 1996 Protocol to the LC 72

The original London Dumping Convention 1972 (LC 72) has 10 main articles that address the obligations of the members to ensure that properties of dredged material disposed at sea are in accordance with the convention requirements, to encourage cooperation between members, and to ensure that measures are taken to prevent and punish any conduct in contravention of these articles. A new protocol to the LC 72 adopted in 1996 introduced the "reverse list" approach and specifies which substances are permitted for ocean disposal. Further amendments include: (1) promotion of sustainable use; (2) inclusion of the sea bed in the definition of the marine environment, which effectively brings most dredging activities (not just disposal) under control of the Convention; and (3) consideration that uncontaminated dredged materials are a valuable resource.

5.3.2   Dredged Material Assessment Framework (DMAF)

The Convention adopted a new method of assessment of the suitability of material for disposal: the Waste Assessment Framework (WAF). Implementation of this framework is left to the individual countries. A dredged material guideline offers generic guidelines for decision makers. Additionally, disposal of uncontaminated material is also subject to audit and is permitted only subject to consideration of beneficial use options and assessment of disposal site impacts. Specific components of the DMAF are listed below:

> *Annex 1.* A list of contaminants known to cause harm to aquatic organisms even in low concentrations, such as organohalogens, mercury, cadmium, oil and oil products, radioactive substances, materials for biological warfare, and persistent plastics. Under the revisions by the new protocol, Annex 1 now outlines the reverse list principle, which specifically identifies materials and concentrations of those materials that are believed to not result in adverse environmental impacts.

> *Annex 2.* A list of the contaminants that should not be present in concentrations higher than 1000 ppm, with the exception of lead, which should not be present in concentrations higher than 500 ppm. Listed contaminants are: arsenic, lead, copper, zinc, organosilicons, cyanides, fluorides, and pesticides. Annex 2 under revision of protocols (1996) outlines WAF and takes into consideration waste minimization at the source, assessment of other disposal options, characterization of dredge materials, action lists, dump site selection, assessment of potential effects, monitoring, and permit procedures.

Under the London Convention, if any of the Annex 1 or 2 substances were found in significant concentrations, a special permit would be required to dispose of the dredged sediment. A determination of potential for undesirable effects, i.e., chronic, acute, or toxic to marine or human life, should also be made.

5.3.3   OSPAR

The Oslo Convention for the Prevention of Marine Pollution by Dumping from Ships and Aircraft, 1972, and the Paris Convention of the Prevention of Marine Pollution from Land-based Sources, 1974, were revised and combined in 1992 and are now know as OSPAR. As with the LC 72, dredged material guidelines were produced and presented in two parts; the first part deals with assessment and management of dredged material disposal, while the second part deals with the design and monitoring of marine and estuary disposal sites. These guidelines suggest that specific information on density, percent solids, grain size fractions, and total organic carbon should be obtained in addition to mandatory analysis of substances listed in Annex 1 and 2 of the amended LC 72 protocols.

The OSPAR guidelines and the LC guidelines are very similar in both structure and content. However, OSPAR offers more flexibility in instances when concentrations of contaminants exceed trace levels; disposal activity may be accepted if it is demonstrated that it is the option of least detriment to the environment, whereas under LC 72 guidelines, marine disposal of listed contaminants exceeding trace levels is prohibited. Another distinction is that oil and its products are not included in OSPAR guidelines. As the reverse list is implemented fully, these distinctions will be resolved.

5.3.4 Implications

1. *Quantitative assessment of dredged material properties and contaminant potential.* Sediment-associated contaminants are dispersed through numerous pathways: dredging, as well as storm, current, bioturbation-related resuspension, desorption, ingestion by benthic biota and epibenthic feeders, and adsorption to or uptake through membranes during sediment contact (e.g., fish embryos; Burton, 1992). Adequate characterization of the dredged material is a prerequisite to proper assessment of the environmental impacts of disposal. Three criteria are fundamental to any execution of these guidelines:

- Application of standards that define the quality of dredged material in terms of contaminants present (either as concentrations or total loads).
- Definition of ecotoxicological effects (determines the impact of contaminated sediments on marine ecosystems, commonly measured with solid-phase and elutriate bioassay testing).
- Assessment in terms of quantity and quality of sediment materials and specific characteristics of the receiving site. Implementation of these guidelines varies, depending on historical evolution of legislation, philosophy, and independent attitudes to best management practices defined by each country's legislation and regulatory systems. Quantitative assessment of dredged material properties varies; most are based solely on the chemical composition of the dredged sediment (known toxic heavy metals, organic contaminants such as hydrocarbons, polychlorinated biphenyls, and pesticides). The United States was the first country to incorporate both chemical composition and ecotoxicological assessment.

2. *Requirement to explore beneficial uses.* Physicochemical properties of sediment materials often define contaminant loading; soft sediments are known to have an increased affinity for fine particles (<63 μm) that promote contaminant loading. Generally there has been more success in finding beneficial uses for granular material, sand, and gravels than fine silts that compose the dredged materials most often encountered during maintenance dredging operations. Use of fine-grained materials when compacted for building materials is a recent application.

3. *Assessment of ocean disposal sites and impact hypotheses.* Existing conventions require that if ocean disposal is proposed, appropriate assessment of the dredged material and the proposed disposal site needs to be undertaken so that any likely impacts of the disposal operation can be identified. This assessment process generally includes: chemical and physical analysis, biological testing, formulation of an impact statement, development and use of quality standards, and monitoring. Disposal site evaluations should include studies of the seabed, water column, and delimitation of any sensitive areas in the vicinity. Modeling studies are to assist examination of the disposal site and estimation of longer-term consequences of contaminant dispersion.

The London Convention DMAF states that impact assessments should lead to a concise statement of the expected consequences of a disposal operation, summarizing potential effects on human health, living resources, amenities, and other legitimate uses of the sea. It should define the nature, scale, and duration of expected impacts on the basis of conservative assumptions. Alternative disposal options, such as open water disposal followed by capping, upland confined disposal, and controlled beneficial use need to be evaluated case by case. A framework for such an assessment has been developed by the Permanent International Association of Navigation Congresses (PIANC, 1996). An impact hypothesis forms the basis of the required monitoring program (Fig. 5.4); the monitoring program includes baseline monitoring and postdisposal monitoring.

5.4 U.S. MANAGEMENT FRAMEWORK FOR CONTAMINATED SEDIMENT

Decades of rapidly developing industry after World War II and serious neglect of water resources resulted in severe impairment of numerous commercially important harbors and estuaries. The need for restraint became obvious: pollution was commonly evident in major waterways, harbors, and most urban areas across the nation. Important stakeholders represented the broad spectrum of public, governmental, and private sectors. For instance, the U.S. Army Corps of Engineers (USACE) maintains more than 400 ports and more than 25,000 miles of coastal and inland waterways for the purpose of safe navigation; numerous military bases operate within the protected waters of the nation's harbors and bays. Moreover, population densities cluster around coastal lands and intensify fishing, boating, and recreational uses of these waters. Therefore, task forces developed rapidly to address pollution issues and developed technologies to solve the very complex interactions of chemicals and biological systems. Milestones of criteria development and a summary of U.S. regulatory legislation are summarized below and presented in more detail in Apps. 5A and 5B.

5.4.1 Clean Water Act, 1972

The Clean Water Act (CWA) is the single most important law dealing with the environmental quality of all U.S. surface waters, both marine and fresh; with this act a national goal was established to restore and maintain the physical, chemical, and biological integrity of the nation's waters. At the time, sediments were not a focal point. The Environmental Protection Agency (EPA) worked with the states to monitor the quality of surface water; data on sediment quality generally were derived from intensive surveys or special studies and were not routinely monitored. The EPA's Storage and Retrieval Data system (STORET) manages the data collected from EPA surveys and also bulk sediment chemistry data from the U.S. Geological Survey (USGS) and National Oceanic and Atmospheric Administration (NOAA). At this time, no national directive required the states to monitor sediments for contamination.

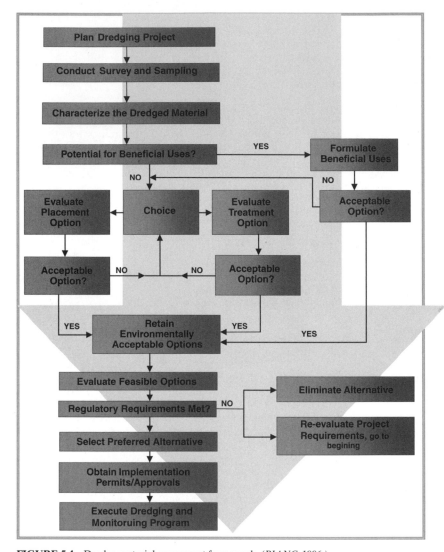

FIGURE 5.4 Dredge material assessment framework. *(PIANC, 1996.)*

5.4.2 Setting Standards for Sediment Criteria

Many harbors and navigational channels silt up and require maintenance dredging to remain open. Estimates of the amount of sediment dredged by the U.S. Army Corps of Engineers range from 300 to 450 million cubic yards per year. Disposal of dredged material is, in many parts of the country, a constant and increasingly difficult problem. Under Section 304 of the CWA, the EPA broadened its authority to develop chemical-specific criteria to identify hazardous contaminant levels in sediments and management standards to control broader dissemination of these sediments, which are burdened with contaminants. Various disposal options were considered, depending on the outcome of chemical assessment of contaminant concentrations or loading within target sediment:

- Open ocean disposal
- Confined ocean disposal
- Unconfined land disposal
- Confined land disposal in leveed containment areas (dredged material containment areas, DMCAs)
- Beneficial use—using a combination of methods that results in creation or enhancement of habitat (e.g., building islands or creating marshes).

Initially, sediment standards were based solely on chemical evaluations. However, as of 1991, the EPA had not published national sediment criteria and the states had not adopted chemical-specific standards for sediment, except for the state of Washington. Puget Sound was one of the first areas in the country where dredged materials were studied extensively for sediment contamination. In March 1991, the state of Washington adopted sediment management standards for Puget Sound (PSDDA). More recently, the EPA has become active in the development of chemical-specific criteria for sediment as a relatively inexpensive and quick way to identify which contaminants may cause chronic effects in aquatic life. Chemicals of known or potential concern (COPC) were monitored in proposed dredged material programs, and various techniques and information (e.g., dilution models, toxicity data bases) were used to predict exposure and subsequent hazard to aquatic organisms. However, this approach proved to be of limited value, as it may inaccurately estimate toxicity and be over- or underprotective of the aquatic ecosystem. All toxic chemicals of consequence may not be identified and measured, and this approach does not allow for evaluation of bioavailability or toxic potential of the complex mixtures of contaminants often found in sediment. Moreover, it became apparent that toxicity was influenced by the extent to which chemical contaminants bind to other constituents in sediment, and that similar concentrations of a chemical can produce widely different biological effects in different sediments.

This transition of approach was reiterated by the International Joint Commission policy recommendation that assessment of dredged materials incorporate evaluation of biological effects because of the shortcomings of the chemical inventory approach (International Joint Commission, 1988). Several regulatory programs have been initiated more recently to implement the CWA and inject specific sediment criteria. These more recent programs include: Dredged Material Disposal Programs included under CWA Section 404, MPRSA, the Resource Conservation and Recovery Act (RCRA), and the Toxic Substances Control Act (TSCA), as well as remediation programs under RCRA and the Comprehensive Environmental Response, Compensation, and Liability Act (CERCLA, 1980), and the Superfund Amendments and Reauthorization Act (SARA, 1986). They are summarized in App. 5B. Biological assessment techniques were further refined for use in conjunction with chemical investigations, and thus formed an approach based on:

- Field assessments (chemical contamination, benthic community structure analysis, fish tissue analyses, caged animal studies to measure bioaccumulation and toxic effects)
- Biological testing (to measure toxicity and bioaccumulation)
- Chemical-specific measurements, a less expensive surrogate for bioassay and field work (Burton, 1992).

5.5 STATE OF THE ART: THE SCIENCE OF SEDIMENT

5.5.1 Characterization of the Sediment Substrate

Sediment stratigraphy is an important feature; sediments are generally sampled to a predetermined depth dictated by program requirements (a characteristic sediment profile collected with a coring device is shown in Fig. 5.5). Important features include:

FIGURE 5.5 Sediment stratigraphy in core sample. *(Photo: Jack Word.)*

- Grain size
- Small-scale surface boundary roughness
- Depth of apparent redox potential discontinuity (RPD)
- Erosional or depositional features such as ripples, mud clasts, and laminates of bedded intervals
- Subsurface methane gas pockets
- Epifauna
- Tube density of benthic infauna
- Thickness of pelletal layers
- Surface aggregations of bacteria
- Infaunal successional stage (Germano, personal communication)

5.5.2 Contaminant Profiles

The focus of scientific and regulatory concern centers on five major types of contaminants associated with sediments:

1. *Nutrients* (phosphorus and nitrogen compounds such as ammonia)
2. *Organic hydrocarbons* (e.g., oil and grease)
3. *Halogenated hydrocarbons* [compounds very resistant to decay such as dichloro-diphenyltrichloroethane (DDT), polychlorinated biphenyls (PCBs), and dioxins]
4. *Polycyclic aromatic hydrocarbons* (petroleum and petroleum by-products)
5. *Metals* (e.g., iron, manganese, lead, cadmium, zinc, and mercury, and metalloids such as arsenic and selenium)

5.5.3 Sediment Grain Size and Bioavailability

Total organic carbon (TOC) ranges from <0.1 percent in sandy sediments to 1 to 4 percent in silty harbor sediment, and <20 percent in navigation channel sediments (Clarke and McFarland, 1991). The toxicity of chemical contaminants is governed by the interaction with other sediment constituents such as organic ligands and inorganic oxides and sulfides that control the bioavailability of accumulated contaminants. Toxicant binding, or sorption, to sediment particles defines the toxic mode of action with respect to biological systems. Because the binding capacity of sediment varies, the degree of toxicity exhibited also varies for the same total quantity of toxicant. The bioavailability of divalent metals in sediment can be represented by the comparison of the molar concentration of sulfide anions—i.e., acid-volatile sulfide [AVS]—to the molar concentration of metals—i.e., simultaneously extracted metals [SEM]. The [SEM]-[AVS] difference is most applicable as an indicator of when metals are not bioavailable. If [AVS] exceeds [SEM], there is a sufficient binding capacity in the sediment to preclude metal bioavailability. However, if [SEM] exceeds [AVS], metals might be bioavailable or other nonmeasured phases might bind up the excess metals (U.S. EPA, 1997).

Biological effects concordance approaches such as effects range medians (ERMs) or probable effects levels are based on the evaluation of paired field and laboratory data to relate incidence of adverse biological effects to the dry-weight sediment concentration of a specific chemical at a particular sampling station. Researchers use these data sets to identify level-of-concern chemical concentrations by the probability of observing adverse effects. Exceedance of the identified level-of-concern concentration is associated with a likelihood of adverse organism response, but it does not demonstrate that a particular chemical is solely responsible, only the action of the complex mixtures that were present during testing. These correlative approaches tend to result in screening values that are lower than the theoretical sediment quality criteria (SQCs) and sediment quality advisory levels (SQALs), which address the effects of a single contaminant. These approaches will be better at predicting toxicity in complex mixtures of contaminants in sediment as we attain better understanding of the effects of these mixtures. The effects range approaches to assessing sediment quality also do not generally account for such factors as organic matter content and AVS, which can mitigate the bioavailability and, therefore, the toxicity of contaminants in sediment (U.S. EPA, 1997).

5.5.4 Contaminant Analyses

Sediment chemical analyses are useful because they provide an indication of the concentrations of contaminants of concern that may cause adverse effects in the receiving environment. Sediment chemistry is the first screening level in the new framework and is useful in sediment categorization because it provides a relatively low-cost and rapid (certainly the case for metals) means of assessing the potential for toxicity. Many contaminants are strongly bound to sediment particles and unlikely to be released into ambient waters. Contaminants tightly bound to sediments may not only be prevented from release to the ambient water, but also, when ingested by marine organisms, pass through the alimentary canal relatively unchanged without inducing a biochemical lesion. In cases where partition coefficients are high (i.e., contaminants are tightly bound), adverse effects in the receiving environment are unlikely and classification of such sediments may be overprotective (Nicholson et al., 2000). Chemical analyses also permits evaluation of the source of contamination through examination of spatial or temporal trends and, recently, the use of forensic chemistry and fingerprinting sources of contamination.

5.5.5 Exposure Pathways

A chemical becomes potentially available to sediment-dwelling organisms through a variety of processes and pathways (Fig. 5.6). Organisms can come into contact with sediment when the sediments are in place, while they are undergoing natural disturbance created by waves or burrowing infauna (bioturbation), or during the process of dredging and disposal. They may come in contact with the chemical as it is transported through the interstices of sediment grains (pore water exposure), after the chemical initially escapes the sediment and is in the water directly overlying the sediment's surface (sediment-water interface exposure), when the chemical is contained within the water column, or when the chemical is attached to food particles that the organism ingests.

Two factors need to occur prior to an organism being exposed to a contaminant. First, the contaminant needs to be in a bioavailable form and, second, the organism needs to come in contact with the contaminant. The behavior of the organism limits the potential contact. The appropriate organisms need to be evaluated in order to assure contact with the bioavailable fraction of the contaminant. Organisms that oc-

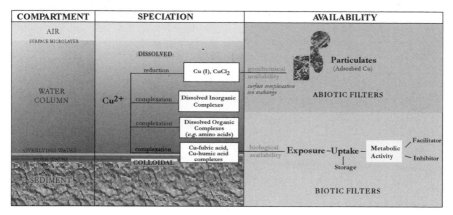

FIGURE 5.6 Factors governing bioavailability illustrated by speciation of copper. [*Based on Tessier and Turner (1995).*]

cur in the water column (as larvae or adults), those that live and are exposed at the sediment water interface, and those that live within and are exposed to pore water and buried sediment are the three predominant forms to be tested.

The water column species evaluate the influence of contaminants that are released at the dredging site and/or the disposal site. The standard testing protocol uses sediment and water collected at the dredging site to create a 1:4 sediment-to-water solution. This mixture represents the maximum concentration of contaminants believed to occur during dredging. The elutriate is then diluted into disposal site water to represent the conditions that occur during disposal. The test organisms are then exposed to these elutriates and dilutions (typically 1, 10, and 50 percent) to evaluate potential risk associated with dredging and disposal. The test organisms that are used for these toxicological tests are generally juvenile fish or arthropods and planktonic larvae of mollusks (oysters, clams, mussels) or echinoderms (sea urchins). These organisms and the evaluation end points are among the most sensitive of our present toxicological tests.

The species that live at the sediment-water interface are used to evaluate the influence of contaminants that are being released from sediment by the organisms' own activities, the activities of other species or weather conditions, or disturbance created by people or physical/chemical transport processes. The organisms generally live at the surface of the sediment, often in mud- or detrital-walled tubes and capture food particles at the sediment-water interface or just above the sediment-water interface. These species evaluate the risk of contaminants that are fluxing out of the sediment and into the overlying water. Typical species used in toxicological tests with this type of behavior include tube-dwelling amphipods that live near the sediment-water interface, mysid crustaceans, tube-dwelling polychaete worms living at the sediment-water interface, and many bivalve mollusks that live at depths but have siphons that feed on the sediment-water interface and in the water column. These organisms are effective at evaluating the potential risk of contaminants fluxing out of undisturbed sediment (e.g., in situ assessment of ongoing risk at a Superfund site) or those species that live near dredging or disposal sites where newly settled sediment would come in direct contact with these organisms.

The species that live in the sediment are used to evaluate the influence of contaminants that are present within the pore water or attached to buried sediment particles. These organisms have more direct contact with the bedded sediment contaminants than either the water column or the sediment-water interface species. They include species of burrowing amphipods or polychaetes and deeper-burrowing species of crustaceans or echinoderms that feed at depth within the sediment.

5.6 SELECTION OF TEST SPECIES AND TESTING CONDITIONS FOR SPECIFIC ASSESSMENT QUESTIONS

There are three basic assessment questions that need to be addressed in ecological assessments for dredged material evaluations:

1. *In situ.* Are there any "unacceptable adverse ecological effects" associated with leaving the sediment and any associated contaminants in place? Generally this is termed the *no-action alternative* in ecological risk assessments.

2. *Removal.* Are there any "unacceptable adverse ecological effects" associated with the removal of sediments and associated contaminants from a specific site? If it has been determined that the no-action alternative is unacceptable, sediment will be removed from a site. This question addresses the effects that would occur adjacent to the contaminated site during the removal process only.

3. *Placement.* Are there any "unacceptable adverse ecological effects" associated with the placement of sediment removed from one location and placed at a new location (e.g., unconfined or confined disposal sites or beneficial use applications)?

Figure 5.7 addresses the selection of test species and testing conditions that are appropriate for the three separate assessment questions. Effects-based testing can provide important data that address each of these questions but the appropriate species and test conditions need to be selected to make the assessments more useful. Figure 5.7 provides a diagrammatic representation of the three assessment questions

	IN SITU	REMOVAL	PLACEMENT
Organism Exposure	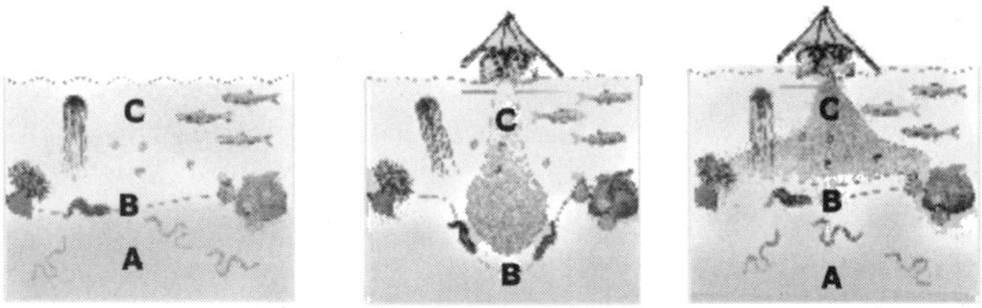		
Sediment Disturbance	Minimize	Maximize	Maximize
Storage	Minimize	Minimize	Minimize

FIGURE 5.7   Types of organisms to be tested for three types of assessments (C = water column; B = sediment-water interface; A = burrowing).

and the types of species that should be include in an assessment of the sediment. The three categories of organisms represent the water column, sediment water interface, and burrowing infaunal organisms (C, B, and A respectively).

5.6.1 In Situ Assessments

Sediment proposed to remain in place can influence organisms that live within the sediment, at the sediment-water interface, and in the overlying water. All three of the exposure conditions are then necessary to attain an estimate of the potential for "unacceptable adverse ecological effects" at or adjacent to the site. The potential for effects can be direct and based on the maximum exposure that can occur within pore waters for species that live within the sediment, somewhat reduced for those species that live at the sediment-water interface from fluxes of contaminants into and out of the sediment, and much reduced from the flux of contaminants out of the sediment and into the overlying water. The effects can be less direct through the uptake of contaminants into the tissues of organisms during their feeding on bedded sediment, sediment at the sediment-water interface, suspended particulate materials above the sediment-water interface, and dissolved and particulate contaminants that are fluxed into the water column, where they may be ingested by organisms that live only in the water column. Therefore, the effects-based assessment for the no-action alternative will consist of effects on: organisms that live within the sediment and are exposed directly to pore water and contaminated sediment particles, organisms that live at the sediment-water interface and obtain exposure of contaminants that flux through the sediments surface and into the water just overlying the surface of the sediment, and those species that live in the water column and are exposed only to contaminants that flux into the water column.

In addition to selecting those species that provide the appropriate exposure assessment profiles, in situ assessment of toxicity from sediment contaminants needs to maintain equilibration between chemical contaminants and sediment pore waters that is as similar as possible to the conditions at the site. Therefore, the sediments being tested need to be disturbed as little as possible and the conditions of the site (in terms of salinity, water hardness, alkalinity, dissolved oxygen, or hydrogen sulfide content, etc.) need to be maintained under conditions that mirror the in situ conditions. This requirement limits the number of species within each of the exposure profiles to those species that are capable of tolerating the in situ conditions and sediment characteristics. Appropriate selection of test species means that each of the exposure profiles be included and that the physical conditions of the sediment and water at the site further control the selection of potential species.

5.6.2 Removal Assessments

Sediment that has been determined to be removed either to reduce toxicity at a site or to increase bottom depth to handle larger vessels also needs to be evaluated in terms of the potential "unacceptable adverse ecological effect(s)" that may occur during the process of removal. Under this scenario the process of removing sediment from one location assumes that those organisms living in the removed sediment will be lost. As a result, the organisms that live within the sediment being removed are of less interest than those organisms that live adjacent to the removal operations. The species that live at the sediment-water interface and in the overlying water are of more importance in this assessment than the burrowing infauna. These two exposure profiles are then necessary to attain an estimate of the potential for "unacceptable adverse ecological effects" at or adjacent to the site. The potential for effects can be direct and based on the maximum exposure that can occur when pore waters and particles are released into the water column surrounding the removal action. Settling particles can then influence the adjacent sediment-water interface species while the suspended and dissolved materials in the water column can influence those species in the water column. The effects can also be less direct through the uptake of contaminants into the tissues of organisms during their feeding on settled particles at the sediment-water interface, suspended particulate materials above the sediment-water interface, and dissolved, and particulate contaminants that are placed into the water. Therefore, the effects-based assessment for the removal alternative should be based on organisms that live at the sediment-water interface and those species that live in the water.

Those species that provide the appropriate exposure assessment profiles then need to be reduced to those species that can also accommodate the conditions at locations adjacent to the removal site. Since the removal action is likely to disturb the equilibria between contaminants and sediment, it is not necessary to maintain those equilibria; in fact it is best to disrupt them. Since the disruption of equilibria will occur under conditions at the removal site, it is also best to select species that can accommodate the physical conditions of the site in terms of salinity, water hardness, alkalinity, dissolved oxygen or hydrogen sulfide content, etc. This requirement limits the number of species within each of the exposure profiles to those species that are capable of tolerating the in situ conditions and sediment characteristics. Appropriate selection of test species means that each of the two exposure profiles should be included and that the physical conditions of the sediment and water at the site further control the selection of potential species.

5.6.3 Placement Assessments

Sediment proposed to be remediated through disposal or beneficial use can influence organisms that live within the sediment, at the sediment-water interface, and in the overlying water. All three of the exposure conditions are then necessary to attain an estimate of the potential for "unacceptable adverse ecological effects" at or adjacent to the placement site. The potential for effects can be direct and based on the maximum exposure that can occur within pore waters for species that live within the sediment, somewhat reduced for those species that live at the sediment-water interface from fluxes of contaminants into and out of the sediment, and much reduced from the flux of contaminants out of the sediment and into the overlying water. The effects can be less direct through the uptake of contaminants into the tissues of organisms during their feeding on bedded sediment, sediment at the sediment-water interface, suspended particulate materials above the sediment-water interface, and dissolved and particulate contaminants that are fluxed into the water column where they may be ingested by organisms that live only in the water column. Therefore, the effects-based assessment for the placement options are based on organisms that live within the sediment and are exposed directly to pore water and contaminated sediment particles, organisms that live at the sediment water interface and obtain exposure of contaminants that flux through the sediments surface and into the water just overlying the surface of the sediment, and those species that live in the water column and are exposed only to contaminants that flux into the water column.

In addition to selecting those species that provide the appropriate exposure assessment profiles, assessment of toxicity from sediment contaminants that are placed at a new site will disturb chemical equilibria with the sediment, and therefore the need for lack of disturbance as required in the in situ assessment is not required. In this situation, however, the sediments being tested need to be modified to accommodate placement site conditions in terms of salinity, water hardness, alkalinity, dissolved oxygen or hydrogen sulfide content, etc. This requirement again limits the number of species within each of the exposure profiles, but in this case to those species that are capable of tolerating the placement site conditions and sediment characteristics. Appropriate selection of test species means that each of the exposure profiles should be included and that the physical conditions of the sediment and water at the placement site further control the selection of potential species.

5.7 WATER SAMPLING STRATEGIES

There are three types of water that can be sampled for assessment of sediment contamination. The first is the pore water that is contained in the interstices of the sediment grains. The second type is the water overlying the sediment surface that may be a source of additional contamination to the sediment or a receiver of sediment contaminants as they are released into the overlying water during natural or man-made disturbances. The third is the disposal site water that is located at the placement site, generally with very low contamination, and that receives the mixed sediments and water that are released during placement.

Collection of Pore Water Samples.   A number of procedures have been used to isolate pore water from sediment samples. These include compression techniques such as displacement of water from sediment by the use of inert gases, centrifugation of bulk sediment, direct sampling of pore water through the use of dialysis membranes, and microsyringe sampling. There are advantages of each of these techniques as well as weaknesses that need to be evaluated prior to selecting one technique over another. One example of a problem with extraction through filters is the removal not only of particles but also nonpolar contaminants that will adsorb to the surfaces of the filter. The majority of pore water assessments used for dredged material evaluations rely on centrifugation techniques to squeeze pore water out of the interstices.

Collection of Overlying Water.   Water that overlies the sediment surface that is being evaluated can be collected in a variety of ways. The water can be sampled with water sampling bottles, peristaltic pumps from specific depths, or buckets from the surface. These waters are used to produce elutriates of sediment that are used to evaluate the potential risks associated with suspensions of sediment into the local environment. For dredged material evaluations, the elutriates are prepared by suspending bulk sediment into this water at a ratio of 1:4 followed by aeration for 30 min and centrifugation for 10 min at 6000 to $7000g$ (Baudo et al., 1990). The resulting elutriate is termed the *100 percent suspended particulate phase* (SPP) or *modified elutriate* for both biological and physical/chemical evaluations.

Collection of Disposal Site Water.   As with the collection of overlying water at the dredging site, the collection of water from the disposal site can be obtained by a variety of methods that are the same as those indicated above. This water is used to dilute the 100 percent SPP (= modified elutriate) into 50, 10, 1, and 0% solutions of the SPP. The 0% solution represents the control conditions at the disposal site, while the dilution series represents different potential effects levels as the disturbed dredged material falls through the water column. Assessments of the concentrations of contaminants and the biological effects are determined on these dilutions, and the data are then plugged into equations, which represent the mixing zones of the disposal sites. If the contaminant levels or toxicity exceeds proscribed levels at the edge of the mixing zone then the potential for "unacceptable adverse effects" is indicated.

5.8 SEDIMENT SAMPLING STRATEGIES

Three basic types of sediment are found in harbors and estuaries. Unconsolidated, fine-grained, potentially contaminated sediments of relatively recent origin are the principal type of materials removed during the maintenance dredging process. These sediments have generally settled in the harbor or port over the last year or so and may contain contaminants from upstream or harbor sources. Underlying this material are sediments that are generally compacted, of various grain sizes from sands to clays, with little likelihood of anthropogenic contamination, and of much greater age. These compacted materials are the principal sediment that is removed during deepening projects. These sediments were generally deposited during earlier geologic periods and have been compressed by overburden sediment so that they are generally very dry and hard packed. The final sediment type is hard, rocky substrate.

One objective in evaluating sediment for dredging and disposal projects is to represent the entire depth of mud that resides over the depth at which the harbor or port is maintained. Each sediment type is sampled with different tools. For example, the typical lengths of sediment cores sampled in maintenance dredging programs are from 4 to 10 ft. Any shallower and the need for dredging is not as high; any deeper and the dredging event has been delayed and the harbor or port is not functioning as it would desire. As stated above, these sediments are composed of softer, finer particles and these dredge materials are relatively easy to sample with a variety of sampling instruments. Gravity cores are used for the shallower samples, <5 to 6 ft, and longer corers as well as vibracore or piston core devices are used to sample greater depths. The coring tools are often lined, in order to protect the sediment from being contaminated with the inner surfaces of the core tool. Sediment is collected from mud line to project depth, and composites are made of the sediment to characterize the chemical, physical, and toxicological properties of the sediment proposed for dredging.

Maintenance versus Deepening Projects. Sampling for deepening projects is generally conducted by coring through the loosely consolidated maintenance dredging material and into the harder packed sediment that has not been disturbed for extensive periods of time. In some cases these sediments are Pleistocene, or older, in origin. The coring tools required to sample these sediments are generally larger and require more mass and momentum than can be attained with gravity cores. Coring tools that can accommodate these requirements and take intact cores to project depth are vibracores and kasten cores that produce a core length <20 ft long, and vibratory hammer cores that produce cores up to 40 ft long (Fig. 5.8). These latter cores are 4 to 12 in in diameter, depending on required volume of sediment for testing, and are driven by sheet pile drivers. Other sampling devices are available that can take sections of sediment after drilling to a specific depth, but use of this group of samplers also introduces potential contamination from the upper layers of the sediment. Of greatest concern in deepening assessment projects is the need to minimize or eliminate potential contamination of deeper layers by the sediment that

FIGURE 5.8 Vibracore and vibratory hammer core operations. *(Photos: Jack Word.)*

overlies the older materials. Contamination can be introduced by carrying contaminants down the inside of the coring tube or by disturbance. Both sources of potential contamination need to be addressed prior to sampling deeper sediments.

Characterizing hard bottom substrate (rock outcropping and bedrock formations) is a specialized task. This substrate type is not conducive to established testing methodologies for the effects of contaminants; however, it is also unlikely that contaminants contained in the mineral matrices of the rocks are readily available to the indigenous or laboratory test organisms. Sampling of this substrate type for biological testing of toxicity or chemical contamination evaluation is not usually necessary.

5.9 BIOLOGICAL TESTING IN THE LABORATORY

Bioassays are controlled experimental procedures used to determine the toxicity of test sediments. As sediment bioassays use standard controlled protocols, the test results are reasonably reproducible and are an effective means of assessing likely impacts in receiving waters. Assessing sediment toxicity through bioassays is, however, complicated by differences in contaminant binding characteristics between different sediment types, grain-size effects on test organisms, and the presence of nonpersistent toxicants such as ammonia in the test chambers. These complex interacting properties can contribute to the confounding of toxicological assessment of dredged material, i.e., an apparent toxic response that is actually due to conditions created in the test vessel rather than due to levels of contaminants present in the test sediments. Other examples of the factors that can potentially influence the outcome of bioassays are changes in the sensitivity or condition (health) of the test population and differences in the time allowed for the test organism to acclimate to laboratory conditions (Nicholson et al., 2000).

5.9.1 Solid-Phase (SP) and Suspended Particulate (SPP) Bioassays and Bioaccumulation

These tests are the backbone of laboratory biological testing. Various marine test species are exposed to field-collected sediment in the laboratory, to establish acute or chronic end points of toxicity or growth and reproductive viability. An attempt to mirror naturally occurring organisms is made with the selection of species to be used. Commonly used organisms include amphipod crustaceans (*Rhepoxynius abronius, Eohaustorius estuarius, Ampelisca abdita, Grandidierella japonica*), mysid crustaceans (*Mysidposis* spp, *Holmesimysis costata, Neomysis mercedis*), the polychaetous annelids (*Neanthes arenaceodentata, Nephtys caecoides, Nereis virens*), larvae of echinoderms or mollusca (*Strongylocentrotus purpuratus, Dendraster excentricus, Crassostrea gigas, Mytilus galloprovincialis*), mollusca and polychaetes for bioaccumulation testing (*Macoma nasuta, Nereis virens, Nephtys caecoides*), and various species of fish.

Results of various solid-phase and aqueous-fraction toxicity tests are generally part of the decision-making process for selecting disposal options for dredged sediment. Specific compounds or sediment characteristics that are responsible for sediment toxicity may be identified by toxicity identification evaluation (TIE) procedures. The U.S. EPA toxicity identification and evaluation scheme is divided into three tiered phases:

- *Phase I.* Specific compounds or classes of compounds are removed or rendered biologically unavailable prior to toxicity testing; toxicants potentially present in the sample are characterized via a standard series of chemical/physical manipulations and toxicity tests.

- *Phase II.* Subsamples from Phase I are chemically analyzed, and lists are compiled of all compounds identified in each subsample. Concentrations of the identified chemicals and their LC_{50} values (the toxicant concentration causing lethality in 50 percent of the test treatments) are then compared and a list of potential suspect toxicants is generated.

- *Phase III.* Confirmation of suspect toxicant by a variety of confirmation techniques, such as correlating observed and expected toxicity, observations of poisoning symptoms characteristic of a given compound, standard additions (spiking) of suspect toxicant, and mass balance techniques.

TIE procedures developed by the U.S. EPA for complex effluents can be adapted to identify the compounds responsible for toxicity observed in pore water or elutriates from contaminated sediments. Previous studies have demonstrated that toxicity and/or bioaccumulations of sediment-associated contaminants, such as cadmium, zinc, mercury, kepone, fluoranthrene, chlorobenzenes, and various organochlorines, by benthic macroinvertebrates is highly correlated with the concentrations of these chemicals in pore water (Baudo et al., 1990).

5.9.2 Confounding Factors

There are many influences on the interpretation of toxicity and adverse biological effects with dredged materials not related to COPECs. The identification of confounding factors (CFs) has greatly improved the validity of sediment testing results, and has saved unneeded cleanup expenditures. Some of the factors that influence the outcome of the toxicity tests are summarized below and a matrix of questions is included in Fig. 5.9 that will help address the potential for CF influences to occur in test samples.

Ammonia.   Ammonia toxicity is one of the most common confounding factors in sediment bioassays (Sims and Moore, 1995). Release of sediment ammonia into test vessels during bioassays may occur as a result of organism excretion and/or the decay of organic components within the sediment. Under natural conditions, ammonia does not normally accumulate because it is kept in balance by microbial populations. Perturbation of sediment microflora can, however, lead to the accumulation of ammonia to concentrations that are toxic in small test vessels. In marine waters, ammonia dissipates rapidly with water circulation and thus toxicity would not likely be observed. For this reason, it is important to establish whether observed toxicity is due to ammonia levels in the test chamber and to use this information in evaluating disposal options. In the United States, where bioassays are extensively used for dredged material management and also under the London Dumping Convention, ammonia is not considered a persistent contaminant of concern and elevated levels do not prevent the open sea disposal of otherwise environmentally benign sediments.

Persistent Physical Features of the Sediment.   Sediment physical characteristics (e.g., grain size or total organic carbon) may induce mortality. For example, owing to native sediment preferences, the test material may not provide an appropriate medium to maintain test organism viability. Organisms may, therefore, show high mortality in sediments that have an unsuitable grain size rather than death that is attributable to con-

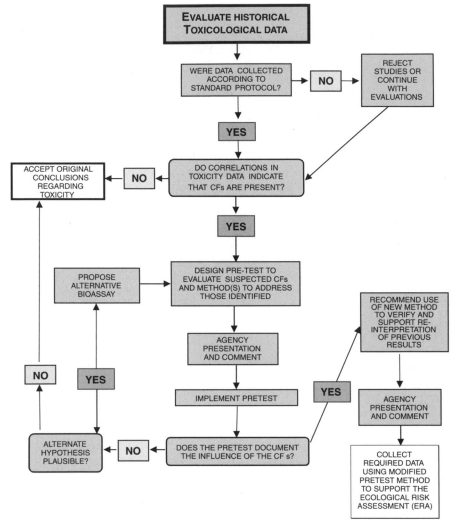

FIGURE 5.9 Protocol for evaluation of existing toxicity/bioassay data for identifying the possible interference of confounding factors.

taminant-induced effects. In the simplest example, the lack of high quality organic carbon can result in starving test organisms, with resultant death or reduced growth.

Acclimation and Holding of Test Organisms. Acclimation is the amount of time and rate of change required for a given test organism to adjust to the water quality parameters required for laboratory testing that differ from those in their native environment. Holding time is the period over which the test organisms are held following acclimation prior to test initiation. Test organisms collected from field populations require a period of time to physiologically adjust to laboratory conditions. Failure to allow test organisms sufficient time to acclimate can increase stress and hence sensitivity to both the laboratory and test conditions. Inadequate acclimation or excessive holding time may result in higher than expected observed mortality.

pH and Water Hardness. Both pH and water hardness can influence the survival of test organisms. In seawater, pH is buffered more effectively than in freshwater, but it can be a cause of toxicity if the pH of the sediment or water strays too far from normal conditions. It can also influence the toxicity of other compounds (e.g., ammonia) by changing the ionic state of the chemical and its bioavailability. Water hardness is a characteristic of freshwaters that influences toxicity of chemical components. As an example, it takes more ammonia to create a specific toxicity at higher water hardness concentrations. Test organisms have a preferred hardness level, and when the harness exceeds these preferred values, the health of the test organisms can be influenced.

Sediment Preparation. Sediment preparation may, depending on the degree of perturbation in the test vessels, lead to the mobilization of water-soluble contaminants from interstitial water. Failure to standardize test sediment handling and preparation protocols could result in differences in water-soluble contaminant release to the test vessels. Variation in laboratory sediment handling procedures may, therefore, result in differential mortality responses (Nicholson et al., 2000).

5.10 STATE OF THE ART: TECHNOLOGICAL ADVANCES IN DREDGING AND DEFINITION OF NAVIGABLE WATERS/CHANNEL

5.10.1 Advances in Positioning Systems

Modern positioning systems for vessels [differential global positioning system (GPS), long-range kinematic techniques] dramatically increase the accuracy of positioning of sediment-removal tools, e.g., dragheads, cutters, backhoes. Rapid monitoring of equipment positioning within well-defined sediment beds has greatly increased the perfomance of dredging operations, approaching a level of "surgical precision."

5.10.2 Silting and Measurement of in Situ Density

Estuarine zones are generally faced with major silting problems; assessment of navigable depth in a channel that is subjected to silting carries a constant degree of uncertainty when traditional depth-sounding devices (echo sounders) are used. An echo sounder identifies liquid silt formations as solid; this leads to misinformation and possibly unnecessary dredging. The presence of "fluid mud," or loosely packed silt layers, in maritime access channels and harbors causes problems, and traditional acoustic hydrographic survey methods are insufficient. Fluid mud results in unpredictable changes in the registered depth, caused mainly by hydrometerological conditions and seasonal variations. The Navitracker system and Trackersoft software can graphically plot navigable depths. Towed bodies penetrate only in mud with a density of up to 1.3 t/m³. An H-shaped inclinometer equipped with a vertical density profiler may be added to allow collection of data from areas with more consolidated substrate sediment.

Innovations include: high-resolution side-scan sonar, nuclear transmission ultrasonic density meters, and chirp profilers [tunable high-resolution acoustic subbottom profilers (1.5- to 10-kHz computer-generated acoustic pulses with a theoretical resolution of 0.1 m and a penetration of over 15 m in soft mud)]. For the investigation of high concentrations of suspended solids and the occurrence and behavior of fluid mud layers, a rapid-drop profiling siltmeter (pulsed infrared) may be used. Seismic profiling is a geophysical tool that portrays spatial disturbance of stratification layers, and incorporates digital, shallow water high-resolution seismic profiling and data processing. Not only the seafloor return is observed from generation of acoustic sound echo, but also reflections from subbottom strata, such as silts, sands, and gravel deposits as well as rock strata. However, seismic profiling does not identify constituents of strata, and this methodology should be used in conjunction with small-scale coring validation.

5.10.3 Field Release of Dredged Materials

Offshore dredge disposal activity causes environmental effects both on the seabed and in the water column surrounding the dredge. These effects are the following:

- Disturbance to benthic community
- Increased turbidity and release of nutrients
- Morphological and hydrodynamic changes due to the alteration of the seabed
- Other less serious effects such as noise and visual disturbance

Dredge overflow occurs as excess water combined with fine solid particles is discharged overboard at the upper part of the hopper used to receive dredge materials. A turbid plume is formed which is then moved by local currents. A heavy increase of turbidity can damage fish respiratory systems and interfere in photosynthesis required by phytoplankton. Furthermore, it affects larvae by reducing light penetration and also interferes with nourishment of filter feeders by clogging their filters. Overflow is considered to have minor local influence with respect to other operations during the dredging process; however, it can affect a much wider area because of the spreading of suspended material by the action of waves and currents. Mathematical models for evaluating the areal extent and relative concentration of solid particles can be applied (e.g., a Lagrangian model). Overflow of fine sediment should be avoided, as it causes a turbid plume that expands in response to weather and sea conditions and vanishes only after a long time. In contrast, the release of medium- or large-grained sediment does not present particular problems, since the turbidity created affects a smaller area and the particulates have a faster settling rate.

Mitigating measures are now commonplace, such as: (1) instituting a temporal limit (imposed on an overflow operation to limit the quantity of released particles; a duration of 100 minutes has been suggested as a good compromise between environmental and economic needs), (2) observance of weather conditions (even wind intensity influences grain dispersion), (3) use of hopper dredges provided with anti-turbidity overflow systems (strongly recommended), and (4) deployment of silt curtains to retain suspended materials behind barriers during the dredging process. These systems allow a fast settlement of sediments by reducing the turbulence of the overflowed mixture.

5.10.4 Environmental Monitoring

Dredging companies now survey for turbidity, water quality, and sediment transport analyses in addition to current and wave monitoring. It has become common practice, largely due to recent dredge studies in Hong Kong and the Øresund Fixed Link Project connecting Kastrup, Denmark, with Lernacken, Sweden. Dredge platforms now are equipped with data logging systems and turbidity, current, conductivity, salinity, and temperature instruments. Vessel-mounted devices include turbidity, conductivity-temperature-depth meters (CTDs), and water sampling apparatus. The measurements are integrated with acoustic doppler current profilers to estimate mass transport of particles in suspension generated by dredging operations. Upgrades of monitoring apparatus include integration of in situ particle size profilers and triaxial magnetometers in towed bodies.

5.10.5 Seafloor Reconnaissance and Assessment

Traditional methods of characterizing bottom communities and sediments were very labor intensive. Newer technologies such as sediment profile cameras (REMOTS) and laser-line scan (LLS) systems are very useful for reconnaissance and baseline monitoring studies. REMOTS (Remote Ecological Monitoring of the Seafloor) integrates physical, physicochemical, and biological parameters scanned from individual images. Information on biological structure-successional status of bottom fauna, sediment transport and fluid mud layers, erosion structures, and organic enrichment are obtained. The organism-sediment index (OSI) characterizes overall benthic quality.

Studies from Hong Kong Harbor produced 785 images. The majority of the area surveyed was determined to be of high benthic habitat quality, with mature, deep-burrowing bottom communities being ubiquitous except in areas of fluid mud and in urban areas where chronic pollution was well documented and not related to dredging and dumping activities. Six percent of the images indicated substantial disturbance to the benthic fauna. Recolonization was demonstrated in a disused dumping site. The aftermath of storm events was also documented, and results showed that active scouring occurred that changed benthic depositional patterns (Rhoads and Germano, 1982). This study concluded that the seabed is in a state of dynamic equilibrium. REMOTS data are augmented with traditional benthic grab sampling and taxonomy that provide quantitative data on biomass, diversity, and community structure.

A sediment profile imaging system (SPI) is a rapid, cost-effective method for mapping changes in the surface of the seafloor. This technique can image, measure, and analyze physical, chemical, and biological parameters over large areas of the bottom of lakes, rivers, estuaries, and oceans (Fig. 5.10).

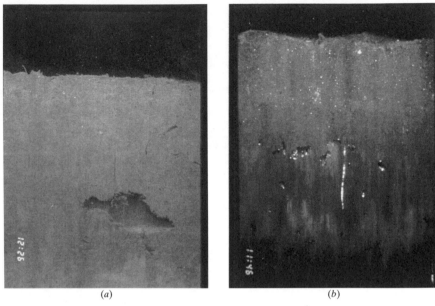

(a) (b)

FIGURE 5.10   Sediment profile image (SPI): (a) taken from a healthy mud bottom showing a sub-surface feeding devoid of deposit-feeding marine worms; (b) taken near a sewage outfall showing anoxic sediment that lacks an oxidized surface layer and has numerous methane gas bubbles being produced at depth. *(Photo: Joe Germano.)*

5.11   STATE OF THE ART: DEVELOPMENT OF MANAGEMENT STRATEGIES

A field validation program (FVP) funded by the U.S. Army Corps of Engineers (ACE) in the 1980s was designed and conducted to test predictive capabilities of various methods compared with results from field trials using identical dredge test materials. The program demonstrated that effluent and surface water quality prediction methods have good utility for predisposal evaluation of dredged material proposed for upland disposal. Methods for testing toxicity and bioaccumulation in wetland plants showed good predictive ability. However, predictive evaluations of the upland and wetland animal bioassays were deemed not entirely reproducible (Peddicord, 1988). It is possible the bioassay studies were biased by unidentified confounding factors. An equilibrium partitioning methodology (AVS/EqP) and apparent-effects threshold (AET) approach have been used to model toxicity responses and have been used to develop sediment management standards.

5.11.1   Superfund Sites

Today, serious sediment contamination is being confirmed in waterways through-out the United States, and each year more sites are being placed on the Superfund national priorities list (NPL). There have been important shifts in both regulatory focus and methodologies used to abate these hot spots of sediment contamination. In evaluating management alternatives for a Superfund site, the following screening factors are employed:

1. Environmental acceptability [i.e., overall protection of human health and ecological processes, compliance with applicable or relevant and appropriate regulatory requirements (ARARs), state agency acceptance, and community acceptance]
2. Technological feasibility (i.e., long-term effectiveness and permanence; reduction of toxicity, mobility, or volume through treatment, short-term effectiveness, implementability)
3. Economic viability (i.e., cost)

In general, regulatory and enforcement initiatives have broadened their focus to encompass watersheds and comprehensive ecological risk assessments rather than the more narrow emphasis on human health issues. Another critical issue is how to equitably distribute the financial burden of costly cleanup programs: downstream constituents have begun to use the citizen suit authority under the Superfund authority, RCRA, and other statutes to recover the increased costs imposed on them by upstream discharge practices (Kamlet and Shelley, 1997).

5.12 STATE OF THE ART: REMEDIATION, TREATMENT, DECONTAMINATION/REUSE OPTIONS

Remediation of contaminant loading in marine sediment requires attention to the issue of "how clean is clean?" Cost-effective technologies are under development; often problem solving is confounded by a mosaic of regulatory agencies with somewhat differing perspectives. Port authorities and regional water quality control boards are developing threshold levels of contaminant loading and identifying situations of high levels of contamination requiring expedited cleanup actions and rule-making criteria for whether sediments should be capped, excavated, or treated, or no action is required. The three principal disposal options considered include the open ocean, wetland augmentation, and upland confinement. Assessment of the potential for hazardous wastes to leach from sites is a prime consideration when dredged material is determined to be highly contaminated. Any option selected for the disposal of highly contaminated dredge materials requires long-term maintenance to maximize degree of sediment isolation (U.S. Congress, 1987). Protocols and standards for classification and management of contaminated sediments are inconsistent and still being refined. It is hoped that further funding and research will expand the knowledge and use of alternative disposal strategies and site remediation successes.

Changes in the perception and classification of dredged materials as a resource and not as a waste have occurred relatively recently. Several reuse options are currently viable: wetland creation and restoration, upland levee maintenance and construction fill, landfill daily cover, aquaculture, and beach restoration. Several new technologies have been developed recently that provide remediation of sediments to higher-grade uses.

5.12.1 Capping

Contaminated dredged material can be isolated from aquatic organisms by covering it with a layer of uncontaminated material; approximately 3 ft of cover material is usually required to minimize the possibility of organisms burrowing into the contaminated material. In selected locations there are organisms that can burrow to depths greater than 3 ft. In these cases, capping needs to become deeper or the process is less efficient. In relatively quiescent marine environments, caps appear to be stable and subject to little erosion. As long as the cap is not disturbed, the contaminated material remains in a relatively unoxidized state, thereby minimizing the upward migration of contaminants. In some cases, natural or artificial pits have been used that restrict the lateral movement of the contaminated dredge materials. Some disadvantages of capping include:

1. Excavation and movement of contaminated sediments to a disposal site can result in water column exposure to contaminants.
2. If the disposal site is greater than 100 ft in depth, it is difficult to prevent lateral spreading of the dredged materials.
3. Capping requires a large volume of clean cover material, leading to increased costs.
4. Contaminated water may be released as the sediment consolidates prior to capping or after capping as the sediment materials merge.
5. Storm events or currents can erode the cap; requiring periodic maintenance.
6. Capping must occur at a rate that is sufficient to minimize interim bioturbation between capping episodes.

Capping has been used at depths of 100 ft or less in Long Island Sound; in the New York Bight; off the coast of Maine; in the Duwamish/Waterway in Commencement Bay, Tacoma; in Alaska and is being tested to treat a large area of contaminated sediments offshore of Palos Verdes, California (U.S. Congress, 1987).

5.12.2 Confined Disposal Facility (CDF)

Dredged materials treated in CDFs are screened to determine if standard dewatering techniques will be applicable, how leachable the contaminants are, at what temperature the contaminants volatilize, and whether biological treatment is a possibility. These treated sediments are commonly reused in landfill covers, land reclamation projects, manufactured soils, gravel and brick production. Water quality resulting from the dewatering processes is a major concern because it may contain both dissolved and particle-associated contaminants. The U.S. ACE and U.S. EPA have produced guideline documents that address these issues (U.S. ACE, 1987; U.S. EPA, 1982). Impact of leaching water on the groundwater depends on the density difference along the freshwater/saltwater interface; freshwater in upland areas moves upward as it approaches the more dense saltwater. Use of liners to prevent leaching into groundwater may be recommended. In order to protect colonizing plants and animals, the site may be covered with clean soil or a surface liner, or a cover crop having shallow roots to eliminate penetration of surface soils. Gas relief wells may be included in the design of the facility to facilitate dispersion of gases (e.g., methane).

5.12.3 Decontamination Technologies

Both organic and inorganic contaminants tend to adhere to fine grains in sediment and become immobilized in a reduced environment. Therefore, maintaining contaminated dredge materials in a reduced, or oxygen deficient, condition is advantageous. Treatment technologies currently under investigation use several different strategies to isolate, immobilize, or neutralize contaminants as outlined in Fig. 5.11.

Pretreatment: Separation of clean versus contaminated or water versus solid fractions; produces need for further treatment, but maximizes usability of the majority of materials

- Dewatering
- Size separation (using hyrocyclone screens, cleaning with flotation, or fluidized bed classifiers)
- Washing
- Froth flotation (frothing chemicals plus forced air result in froth that floats contaminants, both metals and organics, away from solid particles
- Density separation (hydrocyclone, dense media settling basins, screw classifier)
- Magnetic separation

Biological: Based on degradation of organic substances by micro-organisms. This approach offers good prospects for treatment of petroleum hydrocarbons and PAHs.

- Land farming
- Bioslurry
- Decontamination using plant cultivation

Chemical: Based on chemical physical interactions of contaminants such as: adsorption/ desorption, oxidation/reduction reactions, pH adjustment, ion exchange

- Destruction of organics. Chlorinated organic contaminants (PCBs) can be subjected to chemical treatment that removes chlorine from the molecular structure. There are a variety of patented techniques; most use an earth metal such as sodium or potassium to react with the chlorine atoms and render them harmless.
- Organic contaminants can also be destroyed by addition of strong oxidizing agents such as hydrogen peroxide, ozone, or "wet air." However, oxidizing agents are not specific to contaminants and a large quantity of treatment chemicals are consumed by oxidation of naturally occurring organic materials.
- Organic contaminants can be extracted by washing with organic solvents. This method is time intensive, and is used for special projects.
- Extraction of metals using acids or complexing agents added or produced by micro-organisms may considerably reduce heavy metal content. This method is costly, and only applicable if contamination is from one or two metals, and is applicable to highly contaminated sites.

Thermal: Dredge materials heavily contaminated with organic compounds may be treated by:

- Thermal desorption (application of heat to volatize and remove the organic contaminants and mercury present in a solid matrix).
- Incineration (destroys all of the organic material by oxidation at a very high temperature).
- Thermal reduction (uses high temperatures coupled with hydrogen gas which reduces the organic molecules into lighter, less toxic products),

(Continued)

- Vitrification (acts to thermally desorb organic contaminants and mercury, and also immobilizes metals). Common end products are building materials such as gravel or bricks. A disadvantage of this process is that it has a high energy requirement and produces flue-gas emissions.

  Immobilization: Accomplished by either chemically binding the contaminants to the solid particles (fixation) or physically preventing the contaminants from moving (solidification). The principal disadvantage is that contaminants reside in the contaminated dredged material and may be subject to weathering or leaching over a number of years.

- Fixation: large quantities of hydroxyl-forming agents are added that increase the pH of the dredged material and immobilize most of the metal species. A silica solution may also be used to bind contaminant/particle agglomerations.

- Solidification: may be accomplished by adding a cementing substance or by high-temperature melting (as above).

FIGURE 5.11 Various treatment strategies used to detoxify contaminated sediments. *(continued)* *(PIANC 1996.)*

5.12.4 Decontamination/Reuse Technologies

Several small- and large-scale demonstration projects are currently investigating various decontamination processes that also combine with development of value-added products such as construction-grade cement, lightweight aggregate, and clean topsoil to help defray the high costs of decontamination treatments (Mouché, 2000).

1. *Sediment Washing.* Uses high pressure water jets and proprietary chemical additives to extract organic and inorganic contaminants from the sediment. Cleaned sediment will be used as manufactured soil. A demonstration project is ongoing; it is forecasted that the facility will have a capacity of processing 250,000 yd³ per year. (Vendors are BioGenesis Enterprises, Roy E. Weston, Inc., and NUI Environmental.)

2. *Cement-Lock.* Contaminated sediments are mixed with inorganic modifiers, then melted in the presence of oxygen to destroy organic contaminants and immobilize nonvolatile heavy metals within the solidified matrix. The resultant material is converted to construction-grade cement. It is expected that the plant facility will be able to process 30,000 yd³ per year. (Vendor is Endesco Clean Harbors LLC.)

3. *High-Temperature Drying.* Uses existing rotary kilns for high-temperature processing to dewater dredged material. The dried materials are then turned into pellets, heated and fused in kilns, and then used as lightweight construction aggregates. (Vendor is JCI/Upcycle.)

4. *Georemediation.* Testing of enhanced mineralization technology wherein an additive speeds up the natural attenuation of metals and acts as a catalyst for destruction of organic compounds. (Vendor is BEM Systems.)

PCBs: A Special Case. To date, options available for treatment of heavily contaminated sediments characterized by high concentrations of PCBs are not governed by federally mandated screening regulations, and each site must be reviewed on a site-specific basis. The toxic life span of this class of chemicals can be over 250 years; management efforts are increasingly directed toward finding solutions to successfully reduce environmental and human health hazards generated by these compounds.

DMCAs and Aquaculture.   The U.S. Army Corps of Engineers has sought to identify ways by which the landowner can use disposal site acreage for activities that produce income, but not interfere with periodic disposal of dredge material. Aquaculture has been identified as a potential beneficial use of containment areas. The Containment Area Aquaculture Program (CAAP) was developed with the objective of demonstrating technical and economic feasibility of this application of aquaculture techniques. Over 400 species have been identified as potentially capable of producing commercially viable industries, especially adaptive and valuable species such as various shrimp species.

Beach Restoration.   The potential of this use of CDM has been hampered by its consequent loss of sediment, principally fine materials that are left in a rather steep profile. Most materials of <0.15 mm are rapidly lost. However, options are available that may result in improved beach fill stability. Profile geometry is an important feature of the fill process; proper profiling (i.e., simultaneous amendments to the beach as well as nearshore bottom) with regard to grain size can improve its longevity. Profile nourishment should exclude materials finer than those naturally occurring, except possibly as a sublayer. Profile nourishment may be more economical than beach nourishment because cheaper material may be used as part of the offshore profile, although this process requires a greater diversity of equipment such as hopper dredgers and shallow draft split hull dredge (Bruun, 1990).

Advancement in remediation strategy is exemplified by an instance that occurred in San Diego Harbor between the years 1979–1985; during this time, 30,000 tons of copper were dumped onto sediments surrounding the PACO Terminals. Several remediation steps were undertaken. Sediments were dredged from the bay onto barges and off-loaded ashore; the hazardous sediments were screened and passed through a commercial scale sediment treatment facility that separates materials by particle size with a series of screens and cyclone filters. Those sediments with elevated copper levels were recycled; those having low concentrations were placed in an adjacent on-land repository. Wherever practicable, the sediments were treated with a process that binds heavy metals in an insoluble silicate structure. These silicate materials are largely inert and can then be used in construction projects.

Military installations have become environmentally proactive. Major issues revolve around management of particular pollution sources such as stormwater discharges, oil/water separators, bilge water collection systems, alternatives for the present antifouling coatings, oil spill response readiness, and data acquisition of biological constituents, particularly distribution and population dynamics. Operations of routine base facilities as well as new construction, maintenance, or capital improvement dredging activities all hinge on proper management and remediation of contaminants.

5.13 CASE STUDIES: PORT AND HARBOR ASSESSMENT PROGRAMS

There are many examples of port and harbor assessment programs that have occurred over the past 15 to 20 years. Some of these and a summary of special issues associated with each of these examples are highlighted in the following list, with details contained in the following paragraphs.

- *Oakland Harbor, California.* Appropriate combination of identifying confounding factors and contaminants of concern resulted in acceptable agreements for disposal of dredged materials in beneficial as well as ecologically protective manners.

- *Richmond Harbor (Superfund Program and Dredged Material), California.* Combination of programs that have Superfund as well as maintenance and deepening dredged material programs.

- *New York Harbor.* Combination of high levels of chemical contamination and also many confounding factors that have yet to be addressed completely.

- *Hong Kong.* Use of borrow pits for confined aquatic disposal of contaminated sediment.

- *U.S. Navy sites in San Francisco Bay.* Ongoing program that has a history of complicated interaction of federal and state agencies; slow down in Base Realignment and Closure (BRAC) handovers as a result of conflicting biological and chemical information.

5.13.1 Oakland Harbor

In the early 1970s, the Port of Oakland recognized that it needed to deepen its harbors substantially to accommodate the deeper-draft container vessels that were proposed for use in the future. At the same time the U.S. ACE and U.S. EPA developed guidelines to assess contamination and biological effects of sediment proposed for dredging (U.S. EPA/U.S. ACE 1977), and the U.S. ACE San Francisco District in 1979 reported on studies of sediment throughout San Francisco Bay for the Pollution Distribution Study for dredged material assessments. The report on these studies indicated that Inner Oakland Harbor sediments of fine-grained nature had variable levels of contamination but that there were areas within the port that were highly contaminated with numerous industrially derived contaminants. Further studies by the National Oceanic and Atmospheric Administration (NOAA) (Long et al. 1988) indicated that significant contamination occurred in Oakland Harbor areas. The problem resulting from these analyses was that the ability of the Port of Oakland to deepen its harbors to accommodate the larger vessels was hindered by the contamination at selected locations within the harbor that were believed by many to be more widespread within the harbor. This problem in perception was a critical stumbling block that restricted the development of the port. As time moved on, the ACE/EPA introduced new testing requirements that contained more sensitive end points, which added to the perception of "potential unacceptable adverse risk" in disposing of the sediment at ocean disposal sites.

The port, in association with the resource agencies in the San Francisco Bay area, developed strategies for long-term management of dredged material disposal alternatives and implemented a series of detailed chemical, physical, and biological evaluations and critical studies that permitted better understanding of the causes of biological effects and refinement in the distribution of chemical contaminants within the harbor area. The ultimate outcome of these studies in combination with the partnership of the resource agencies resulted in characterizing the extent of true "bioavailable" contamination, the causes of toxicity that were the result of factors other than chemicals of concern, alternative dredging and disposal plans (application of beneficial uses) and the successful deepening of the harbors to depths of −35, −38, and −42 ft mean lower low water and recently the approval of the −50 ft project.

The success of these deepening projects relied on separating areas of the harbors into areas of potential concern and areas with less concern on the basis of the studies that were performed. Specific conclusions were:

• Confined areas of contamination were identified within upper layers of sediment at specific industrial locations within the harbor. These locations are being handled separately from the remaining locations within the harbor. Some will be taken to confined upland disposal sites, some were remediated, and others are still being considered for various options. The ready acceptance of handling these sediments separately from the rest of the harbor helped improve the overall trust in the port and the agencies that were reviewing the data.

• Hard-packed sediment in deeper layers of the harbor was found to have elevated levels of metals in biologically unavailable states. These sediments had elevated levels of Cr and Ni that were much higher than concentrations that were expected to have effects from prior studies. The results of the biological assessments that were performed on these sediments showed the lack of toxicity in solid-phase tests, suspended particulate phase tests, and bioaccumulation testing. Where toxicity occurred in these older bay mud sediments, the factor that controlled the toxicity was the available quantity and quality of the food in the sediment.

• Biological effects that were observed at selected locations in the outer portion of inner Oakland Harbor had elevated concentrations of the nonpersistent chemical ammonia, which was increasing in concentration because of the presence of lower saline content of interstitial water in the sediment.

• Other biological effects that had been observed in the past were also due to the use of organisms that could not accommodate the fine-grained nature of the sediment that was present in Oakland Harbor. Appropriate selection of test species that could accommodate those conditions resulted in the removal of apparent toxicity.

• A majority of hard-packed sands and muds were planned for use in the construction of a subtidal wetland environment within Oakland Harbor combined with improved public access and use of the site. This option provided a beneficial-use alternative for sediment that had been established as clean and not only reduced the ultimate cost of the project but provided a localized improvement to the environment that will help improve conditions for many years to come.

5.13.2 Richmond Harbor

Richmond Harbor is a smaller port within San Francisco Bay that concentrates its activities on offloading bulk cargo and cars. As for Oakland Harbor, the development plans for Richmond Harbor included continuing to operate at its current authorized depths and ultimately to deepen the harbor to accommodate deeper-draft vessels. In contrast to Oakland Harbor, a Superfund site was located in the middle of the harbor. This site repackaged and distributed the chlorinated pesticides DDT, DDD, and Dieldrin for approximately 50 years. As a result of the handling activities on site, there were locations that had very high concentrations of these pesticides. At some locations there were even layers of essentially pure product. The problem in the maintenance and development of this port was therefore complicated by the contamination of sediments within the Superfund site. The contaminant levels in two blind channels that serviced these activities were sufficiently high to continually contaminate sediments as they were cleaned up. As a result, the Superfund cleanup activities had to be completed prior to moving on the continued maintenance and development of the harbor. Special characterizations were needed for this project and they consisted of the following:

• Vertical and horizontal characterization of the concentrations of pesticides within and adjacent to the Superfund site.

• Development of an acceptable cleanup standard that would be protective of the environment and human health at the completion of the cleanup activity within the harbor.

• This level of acceptable cleanup would also have to be such that it could permit disposal at an offshore site. The definition of *acceptable for disposal* would be that contaminant levels would be less than trace in order to exclude "unacceptable adverse effects."

• These last two points are key in that acceptable cleanup values and acceptable disposal levels need to essentially be the same and result in no "unacceptable adverse effects." If they are not the same, the port could not be developed while using the option of ocean disposal but would have to treat or isolate sediment that had already been determined to be acceptable through the Superfund process.

• Sediment within Richmond Harbor at deeper depths had compacted sediment with low concentrations of food (total organic carbon), elevated metals levels, and the presence of toxicity with selected species.

• As with Oakland Harbor, the elevated metals levels were unavailable for uptake into the tissues of organisms that fed on the sediment and were not correlated to biological effects.

• Also, as with Oakland Harbor sediments, these harder-packed, low-food-value sediments were unable to support certain types of species. These burrowing forms that fed on buried organic materials were unable to burrow into the sediment and unable to find sufficient nutrients to survive the exposure periods. Specialized tests that softened sediment and provided food resulted in survival that was correlated to the increased food value. Thus the confounding factors of sediment compactness and food value were determined to be a principal cause of the toxicity of the sediment that was to be removed for the deepening project.

5.13.3 New York/New Jersey Harbors

The New York/New Jersey Harbor complex is an essential component of the shipping industry of the northeastern portion of the United States. Hundreds of years of development within the region surrounding these harbors and the presence of streams and rivers that carry contaminants to the harbors have resulted in localized areas of chemical contamination. Maintaining and improving the harbor environment through dredging and disposal is necessary if the nation wants to retain this harbor as a major shipping center of the world. The combination of high levels of chemical contamination and many confounding factors have yet to be addressed completely. Disposal sites have also been reduced dramatically and as a result, the combination of all of these factors has resulted in slowing harbor development. Issues that have been identified and that require evaluation include:

• Special localized contamination of sediment with chlorinated dioxins produced since the 1960s. The sources are generally known and have been halted, but the residual materials in the sediments are present throughout the system at various concentrations. The cleanup levels and methods to handle the remediation or destruction of dioxins contained in sediment are not well developed.

* Generalized contamination of various reaches and waterbodies with metals, petroleum compounds, pesticides, etc. from local businesses and business upriver is recognized; sources are sometimes known but not always, thus permitting the potential for contamination of remediated sites.

* Red clays that are buried at depth are generally hard packed, have elevated metal levels, low concentrations of food (total organic carbon), and are generally considered to be free from man-made contamination may not be able to accommodate test organisms. That lack of accommodation of test organisms is not generally considered to be a result of chemical toxicity but the confounding factors of hard-packed sediment and lack of food.

* The influence of freshwaters on the production of elution of ammonia from particles and the harm that occurs to bacterial populations on those particles when they reach seawater can and often does result in toxic levels of ammonia. Ammonia is important to the observations of toxicity within these environments, and the decision of whether ammonia will control dredging and disposal needs to be made. Without that decision the influence of ammonia on toxicity tests within the New York/New Jersey Harbors will control decisions on the acceptability of dredged material disposal.

* New disposal sites for dredged materials need to be identified and designated.

5.13.4   Hong Kong

The harbor of Hong Kong is one of the first large-scale uses of sediment borrow pits for confined aquatic disposal of contaminated sediment. This large-scale project is disposing of millions of cubic yards of sediment, many of which have elevated levels of contaminants in subtidal holes that were produced during the mining of sands for construction projects. The contaminated dredged materials are placed at the bottom of these pits and filled to a specified level before clean sediments are placed on top to cap these sediments. The Environmental Protection Division of the Chinese government in Hong Kong has contracted numerous studies of the use of these filled-in borrow pits by grab sampling and trawling to evaluate benthic community recovery, use of the area by fish and invertebrates, their contaminant levels, and ecological risk assessments for the endangered Chinese white dolphin. These studies will be critical in determining the potential for use of borrow pit sites as locations for the isolation of contaminated dredged materials.

5.13.5   U.S. Navy Sites In San Francisco Bay

The Navy in San Francisco Bay has many sites that are now undergoing Base Realignment and Closure (BRAC) proceedings. During the past 20 years, the Navy has had an ongoing program evaluating the sediment contamination levels and biological effects at these sites. The Navy has had a history of complicated interaction with federal and state agencies, which is slowing down the BRAC process as a result of conflicting biological and chemical information. Issues related to confounding factors in toxicity studies that have arisen in the Navy programs include:

- The influence of ammonia on toxicity tests
- The influence of acclimation and holding times on toxicity tests
- The influence of alternative end points with different sensitivities
- The variability in repeated assessments of the same sediments both chemically and biologically
- The use of inappropriate assessment organisms for sediment grain size
- The use of inappropriate end points for specific assessment questions
- The variability in sensitivity of analytical chemistry methods

5.14 FUTURE OF PORT AND HARBOR EXPANSION PROGRAMS

The importance of the marine transportation system (MTS) to the economy of the United States is demonstrated by the $742 billion contributed by the waterborne cargo that is shipped and the more than 13 million jobs that are created annually. This contribution to the United States gross domestic product is created from the movement of more than 2 billion tons of domestic and international freight, 3.3 billion barrels of petroleum imports, servicing 78 million recreational users, hosting 5 million cruise ship passengers, and supporting 110,000 commercial fishing vessels and recreation fishing that contribute an additional $111 billion to state economies annually (MTS Task Force, 1999). The future shipping industry in the United States and throughout the world is progressing toward the handling of longer, wider, and deeper draft vessels than at any time in the past. It has been predicted that within the next 20 years the amount of U.S. trade is expected to double or even triple. While there is still a role for the smaller ports and harbors for localized transshipment, the major carriers are all progressing to larger vessels coupled with use of ports that are closely tied to intermodal shipment processes that can handle the increased cargo potential inherent in these larger vessels. Providing the harbors and infrastructures to handle the existing commerce as well as the increased commerce potential in an environmentally sound and economically feasible way is the challenge for future port and harbor development. Examples of issues that will need to be addressed in the future include:

1. Improve the infrastructure of the aging as well as the modernized ports and harbors. This includes increasing water depths, widths of channels, and sizes of turning basins and maneuvering areas as well as on-land improvements for handling the existing and increasing amounts of materials to be shipped into and out of the ports.

2. Improve the communication between the shipping community and the environmental regulator community to provide for more efficient decisions on development and environmental protection for the maintenance and improvement of dredged channels.

3. Improve methods of minimizing the settlement of sediment, which results in the need for dredging and disposal, or beneficial reuse of uncontaminated sediment.

4. Proactively work to control sources of contaminant release from port and harbor activities.

5. Help modify current regulations and licensing of permitted and unpermitted discharges into navigable rivers that ultimately deposit in ports and harbors, creating contaminated sediments that impinge on the ability of the harbor development projects. Modification of these regulations would help to decrease the amount of contaminants that are deposited into port and harbor sediment (license application requirements are summarized in App. 5C).

6. Develop procedures that will effectively eliminate the transport of nonindigenous species into harbor waters through ballast water.

7. Investigate the issues associated with current and potential listings of endangered or threatened species and how they may be influenced positively or negatively by current or future port activities. These Endangered Species Act (ESA) issues will become more visible in the future and may substantially change the environmental controls that are placed on development.

8. Assure that toxicity of sediment proposed for dredging is related to chemicals of potential environmental concern and not confounding factors. This will become of even greater importance as the evaluation methods become more sensitive to adverse conditions.

9. Identify alternative methods of contaminated or uncontaminated dredged material disposal, reuse, remediation, and beneficial applications of clean or remediated sediment.

Ports and harbors will be needed in the future to support not only the U.S. but also the global economy. Whether we address the issues of port development and environmental protection as competing or cooperative issues will determine the success of those efforts.

APPENDIX 5A: MILESTONES OF CRITERIA DEVELOPMENT IN MANAGEMENT OF DREDGED MATERIAL

A summary of some of the more noteworthy advancements in regulatory standards and protocols developed for the management of contaminants is presented below:

- Early ecological studies
  - San Diego Bay (San Diego Regional Water Pollution Control Board, 1952)
  - Los Angeles–Long Beach Harbors, 1959 (Reish, 1959). Dr. D. J. Reish pioneered the use of benthic animals in monitoring the marine environment.
  - New York Bight (Mayer, 1982)
- Dredged Materials Assessment Program. Joint EPA/ACE biological assessment program guidelines developed for use with dredged material assessments (U.S. EPA/U.S. ACE, 1977).
- Commencement Bay Superfund Program and the development of a triad approach that led to apparent-effects threshold (AET) chemical-based criteria for Puget Sound, Washington, 1980s.
- NOAA ERL/ERM studies in Oakland Harbor, California, 1980s. Assessment of biological impacts using a triad approach for evaluating impacts of chemical contaminants in the marine environment (Long et al., 1995; Long and Morgan, 1990).

- A field validation program (FVP) led by the ACE during the 1980s investigated the capabilities of laboratory biological testing to document and predict both short- and long-term effects that would occur under similar field conditions (Gentile et al., 1988).

- Revision of dredged material assessment program produced the Implementation Manual for Assessment of the Offshore Disposal of Dredged Materials—1990 (U.S. EPA/U.S. ACE, 1991). "The Ocean Testing Manual" addressed test procedures and evaluation guidance for bioaccumulation of contaminants as mandated by regulatory criteria. Each EPA region or ACE district involved in ocean dumping is to use the national guidance in developing local testing and evaluation procedures based on the COPECs and species existing in a given area, and the levels of contaminants in sediment already existing in the area of the disposal site (the reference site).

- National inventories of point and nonpoint sources of sediment contaminants, National Sediment Inventory (NSI), as mandated by the Water Resources Development Act (WRDA) of 1992.

- Revision of dredged material assessment program produced the second Inland Testing Manual (U.S. EPA/U.S. ACE, 1994, updated 1998).

- Evaluations of the effectiveness of remediation/restoration efforts.

- Equilibrium partitioning studies with nonpolar organic contaminants and SEM/AVS studies of metal bioavailability (U.S. EPA, 1997).

APPENDIX 5B: A SUMMARY OF U.S. REGULATORY LEGISLATION TO DATE

Statute	Year	Provisions	Navigation enhancement	Remediation/ restoration	Water quality improvement	Waste disposal	Beneficial uses
Rivers and Harbors Act (RHA)	1899; 1970	Regulates dredging and other construction activities in navigable waters; Section 10 is administered by the U.S. ACE	X				
National Environmental Policy Act (NEPA)	1969	Requires the analysis and documentation of potential primary and secondary impacts, including those associated with dredging and dredged material discharges.	X	X	X	X	
Federal Water Pollution Control Act (FWPCA) Clean Water Act (CWA)	1972 1977	1. Section 301(h): Waivers for publicly owned treatment works (POTWs) discharging to marine waters.					

(Continued)

Statute	Year	Provisions	Navigation enhancement	Remediation/ restoration	Water quality improvement	Waste disposal	Beneficial uses
		2. Section 402: Established the NPDES (National Pollutant Discharge Elimination System); permitting, especially under best available technology (BAT) in water-quality-limited water.					
		3. Section 403(c): Criteria for ocean discharges; mandatory additional requirements to protect marine environment.					
		4. Section 404: Permits for dredge and fill activities (administered by U.S. ACE): Guidelines for specifying dredged or fill material disposal sites based on contaminant status of the material as determined with "Gold Book" procedures; determines suitability for unrestricted open ocean disposal. Management actions, such as capping or treatment, may be used to bring the sediment disposal activity into compliance with guidelines; no focus on wetlands.*	X	X	X		
Toxic Substances Control Act (TSCA)	1976	Section 5: Premanufacture notification reviews for new industrial chemicals. Sections 4, 6, and 8: Review for existing industrial chemicals.				X	
Resource Conservation and Recovery Act (RCRA)	1976, 1984 amendments	Assessment of suitability and permitting of on-land disposal or beneficial use of contaminated sediments considered hazardous; amendments, "hammer provisions," prohibit land-based disposal of highly toxic materials.				X	

(Continued)

Statute	Year	Provisions	Navigation enhancement	Remediation/ restoration	Water quality improvement	Waste disposal	Beneficial uses
Marine Protection, Resources and Sanctuary Act (MPRSA)	1980	"The Ocean Dumping Act," permits for ocean dumping; Section 103 regulates the transport of dredged material to the ocean for the purpose of disposal. Criteria based on Ocean Testing Manual procedures (administered by the U.S. ACE).*	X				
Comprehensive Environmental Response, Compensation, and Liability Act (CERCLA)	1980	Assessment of need for remedial action with contaminated sediments; assessment of degree of contaminated sediments; assessment of degree of cleanup required, disposition of sediments.		X		X	
Superfund Amendments and Reauthorization Act (SARA)	1986	Evaluates remedial actions for highly contaminated area (~1300 listed or proposed for inclusion on Superfund NPL).		X	X	X	
Water Resources Development Acts (WRDAs)	Biennial, 1986, 1992, 1996	Established a comprehensive cost-sharing scheme for distributing construction costs for water resource development projects between the U.S. government and nonfederal interests; deals with dredged material management options; cost sharing whether new dredge or maintenance dredge projects.		X			X

* Uses tiered testing.

APPENDIX 5C: LICENSE APPLICATION REQUIREMENTS

1. Dredged material
  • Source, total amount, and average composition (e.g., per year)
  • Form (solid, sludge, liquid, gaseous)

- Properties: physical (solubility and density), chemical and biochemical (e.g., oxygen demand, nutrients), and biological (e.g., presence of viruses, bacteria, yeasts, and parasites)
- Toxicity
- Persistence: physical, chemical, and biological
- Accumulation and biotransformation in biological materials or sediments
- Susceptibility to physical, chemical, and biochemical changes and interaction in the aquatic environment with other dissolved organic and inorganic materials
- Probability of production of taints or other changes reducing marketability of resources (fish, shellfish, etc.)

2. Method of disposal

- Details of dredgers or barges to be used
- Details of the proposed method of placement
- Rate of disposal
- Proposed monitoring system

3. Characteristics of disposal site location

- Dilution and dispersion characteristics
- Water characteristics
- Sea bed characteristics
- Existence of other dump sites in the area

4. General considerations

- Possible effects on amenities
- Possible effects on marine life
- Effects on other uses of the sea
- Availability of alternative land disposal or treatment of dredged materials
- Consideration of possible beneficial uses of dredged materials

REFERENCES

Ankley, Gerald T., M. K. Schubauer-Berigan, and R. A. Hoke. 1992. Use of Toxicity Identification Evaluation Techniques to Identify Dredged Material Disposal Options: A Proposed Approach. *Environmental Management,* 16(1):1–6.

Baudo, R., J. Giesy, and H. Muntau (eds.). 1990. *Sediments: Chemistry and Toxicity of In-Place Pollutants.* Lewis Publishers, Chelsea, Mich., 405 pp.

Bruun, Per. 1990. "Beach Nourishment—Improved Economy through Better Profiling and Backpassing from Offshore Sources." *Journal of Coastal Research,* 6(2): 265–277.

Burt, T. N., and C. A. Fletcher. 1997. "Dredged Material Disposal in the Sea." In: Terra et Aqua, *International Journal on Public Works, Ports & Waterways Developments.* Intl. Assoc. of Dredging Companies (IADC). The Hague, The Netherlands. pp. 3–13.

Burton, G. A. 1992. *Sediment Toxicity Assessment.* Lewis Publishers, London. 457 pp.

Clarke, J. U., and V. A. McFarland. 1991. *Assessing Bioaccumulation in Aquatic Organisms Exposed to Contaminated Sediments.* Long-Term Effects of Dredging Operations Programs, misc. pap. D-91-2. U.S. Army Corps of Engineers, Waterways Experiment Station, Vicksburg, Miss.

De Falco, P. 1967. "The Estuary—Septic Tank of the Meagalopolis." In *Estuaries.* American Association for the Advancement of Science. Publication No. 83, Washington, D.C. 757 pp.

Gentile, J. H., G. G. Pesch, J. Lake, P. P. Yevich, G. Zaroogian, P. Rogerson, J. Paul, W. Galloway, K. Scott, W. Nelson, D. Johns, and W. Munns. 1988. "Synthesis of Research Results: Applicability and Field Verification of Predictive Methodologies for Aquatic Dredged Material Disposal." Technical report D-88-5, U.S. ACE Waterways Experiment Station, Vicksburg, Miss.

Germano, J. D. Personal communication. Germano & Associates, Seattle, Wash.

International Joint Commission. 1988. *Procedures for the assessment of contaminated sediment problems in the Great Lakes.* Windsor, Ontario, 140 pp.

Kamlet, K. S., and Peter Shelley. 1997. "Regulatory Framework for the Management and Remediation of Contaminated Marine Sediments" *ELR News and Analysis.* Adapted from *Contaminated Sediments in Ports and Waterways: Cleanup Strategies and Technologies,* National Academy Press, Washington, D.C., 1997, 21 pp.

Long, Edward, D. MacDonald, M. B. Matta, K. VanNess, M. Buchman, and H. Harris. 1988. "Status and Trends in Concentrations of Contaminants and Measures of Biological Stress in San Francisco Bay." NOAA Tech. Mem. NOS OMA 41. National Oceanic and Atmospheric Administration, Seattle, Wash., 268 pp.

Long, E. R., D. D. MacDonald, S. L. Smith, F. D. Calder. 1995. "Incidents of Adverse Biological Effects Within Ranges of Chemical Concentration in Marine and Estuarine Sediments." *Environmental Management* **19**, 1:81–97, January/February.

Long, E. R., and L. G. Morgan. 1990. "The Potential for Biological Effects of Sediment-Sorbed Contaminants Tested in the National Status and Trends Program," NOAA Technical Memo NOS DMA 52, U.S. National Oceanic and Atmospheric Administration, Seattle, Wash., 175 pp.

McCarthy, J. F. and L. R. Shugart (eds.). 1990. *Biomarkers of Environmental Contamination.* Lewis Publishers, Boca Raton, Fla., 457 pp.

Mayer, Garry F. (ed.). 1982. *Ecological Stress and the New York Bight: Science and Management.* Estuarine Research Federation, Columbia, S.C., 715 pp.

Mouché, Carol. 2000. "Waterways Benefit from Sediment Decontamination," *Pollution Engineering,* "Federal Focus," October 2000.

MTS Task Force. 1999. An Assessment of the U.S. Marine Transportation System. A Report to Congress, September 1999. U.S. Dept. Transportation. Washington, D.C., 113 pp.

National Research Council. 1993. *Managing Wastewater in Coastal Urban Areas.* Committee on Wastewater Management for Coastal Urban Areas; Water Science and Technology Board; Commission on Engineering and Technical Systems. National Academy Press, Washington, D.C., 477 pp.

National Research Council. 2000. *Clean Coastal Waters: Understanding and reducing the effects of nutrient pollution.* National Academy Press. Washington, D.C. 375 pp.

Nicholson, S., S. C. Clarke, J. Q. Word, R. Kennish, K. L. Barlow, and C. A. Reid. 2000. "Quality Assurance in the Toxicological Assessment of Hong Kong Dredged Sediments: The Potential Influence of Confounding Factors on Bioassay Results." *Proceedings of the ISWA International Symposium & Exhibition on Waste Management in Asian Cities,* **2**, 196–203 C. S. Poon and P. C. K. Lei (eds.). The Hong Kong Polytechnic University Press.

PIANC. 1996. *Handling and Treatment of Contaminated Dredged Material from Ports and Inland Waterways, "CDM."* Vol. 1; Supplement to Bulletin No. 89. Brussels, Belgium.

Peddicord, R. K. 1988. "Summary of the U.S. Army Corps of Engineers/US Environmental Protection Agency Field Verification Program." U.S. Army Engineer Waterways Experiment Station, tech report D-88-6. 46 pp.

Reish, D. J., 1955. "The relation of polychaetous annelids to harbor pollution," *U.S. Public Health Reports,* **70**: 1168–1174.

Reish, D. J. 1959. *An Ecological Study of Pollution in Los Angeles-Long Beach Harbors, California.* University of Southern California Press. Los Angeles, 119 pp.

Rhoads, D. C., and J. D. Germano. 1982. "Characterization of Benthic Processes Using Sediment Profile Imaging: An Efficient Method of Remote Ecological Monitoring of the Seafloor (REMOTS System)." *Marine Ecology Progress Series,* **8**:115–128.

San Diego Regional Water Quality Control Board. 1952. *Report on the Extent, Effects and Limitations of Waste Disposal into San Diego Bay.* 95 pp.

Sims, J. G., and D. W. Moore. 1995 "Risk of Pore Water Ammonia Toxicity in Dredged Material Bioassays." U.S. ACE Waterways Experiment Station, Misc. Paper D-95-3, 21 pp.

Tessier, A. and D. R. Turner (eds.). 1995. *Metal Speciation and Bioavailability in Aquatic Systems.* John Wiley & Sons, Inc., New York.

U.S. ACE. 1987. *Confined Dredged Material Disposal.* Engineer Manual 1110-2-2-5027. Washington, D.C.

U.S. Congress. 1987. *Wastes in Marine Environments.* Office of Technology Assessment, OTA-0-334; U.S. GPO, Washington, D.C., 313 pp.

U.S. EPA. 1997. *The Incidence and Severity of Sediment Contamination in Surface Waters of the United States.* Vols. 1–3. EPA 823-R-97-006.

U.S. EPA/U.S. ACE. 1977. *Ecological Evaluation of Proposed Discharge of Dredged Material into Ocean Waters.* Implementation Manual for Section 103 of Public Law 92-532 (Marine Protection, Research, and Sanctuaries Act of 1972. Environmental Effects Laboratory, U.S. Army Engineer Waterways Experiment Station, Vicksburg, Miss.

U.S. EPA/U.S. ACE. 1991. *Evaluation of Dredged Material Proposed for Ocean Disposal, Testing Manual.* U.S. Environmental Protection Agency, Office of Water. Washington, D.C. EPA-503/8-91/001

U.S. EPA/U.S. ACE. 1994. *Evaluation of Dredged Material Proposed for Discharge in Waters of the U.S.* Testing Manual (ITM). U.S. Environmental Protection Agency, Office of Water. Washington, D.C. EPA-823-B-98-004.

USEPA/USACE. 1998. *Evaluation of Dredged Material Proposed for Discharge in Waters of the U.S.* Testing Manual (ITM). Office of Water. Washington, DC. EPA-823-B-98-004.

Web Sites

U.S. EPA: *www.epa.gov*

U.S. EPA Office of Science and Technology/Contaminated Sediments: *www.epa.gov/OST/cs*

U.S. ACE: www.usace.army.mil

International Association of Dredging Companies (IADC): www.iadc-dredging.com

CHAPTER 6

MANAGEMENT OF WASTES FROM NUCLEAR FACILITIES

Marve Hyman

Bechtel National, Inc.
Richland, Washington

Ken Hladek

Duratek Federal Services of Hanford, Inc.
Richland, Washington

6.1 INTRODUCTION

This chapter deals with the management of wastes generated in nuclear plants used for the generation of electric power and for the production of material used in nuclear weapons. Wastes can be categorized into three general categories:

- Radioactive
- Hazardous—corrosive, toxic, or reactive properties but not radioactive
- Mixed—a combination of radioactive and hazardous properties.

Emphasis is on radioactive and mixed wastes. However, because nonradioactive chemical reagents and organic solvents are used in nuclear processing, some examples are included for treating hazardous wastes. The technologies described are mainly proved ones. Wagner (1997) describes innovative technologies and new applications of proven technologies to treatment of mixed wastes.

6.1.1 The Nuclear Fuel Cycle

The *nuclear fuel cycle* refers to the series of activities covering the entire life cycle of fuel used in nuclear reactors. The cycle begins with mining of uranium and ends with the final disposition of the irradiated material and waste. The life cycle processes are depicted in Fig. 6.1, which shows the steps that generate radioactive waste. The term *front end* of the cycle refers to the preparation of uranium for use in power or pro-

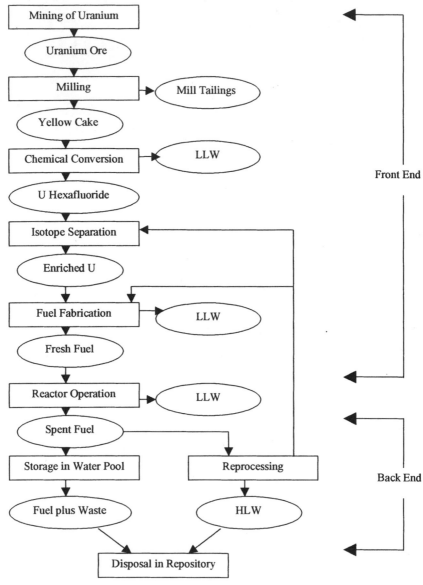

FIGURE 6.1   The nuclear fuel cycle. (*Adapted from the Nuclear Regulatory Commission publication* Regulatory Technical Evaluation Handbook, *NUREG/BR-184, App. C.*)

duction reactors. The term *back end* refers to operations performed on the irradiated, or spent, fuel.

Uranium is mined in several western states of the United States, namely Colorado, Utah, and Wyoming. Natural uranium is extracted from the low-grade ores by a milling process, leaving a large radioactive waste residue called *mill tailings*. A complex uranium compound called *yellow cake* is the useful product that continues in the fuel cycle to be purified by chemical refining.

The yellow cake compound is converted to uranium hexafluoride (UF_6), a gas at ordinary temperatures and pressures. The UF_6 is shipped to the isotope separation plant, where a process of gaseous diffusion separates an enriched or product stream of U-235 content greater than that of natural uranium from a tails stream of depleted uranium. The slightly enriched uranium, still as gaseous UF_6, goes to the fuel fabrication plant, where it is converted to the form of solid pellets of uranium dioxide, UO_2. The pellets are inserted into tubes to form fuel rods, which are sealed, tested, and assembled into bundles for shipment to the reactor site—"fresh fuel."

Each fuel assembly remains in the electric power generator reactor for about 3 years, or a much shorter time in a plutonium production reactor, after which it is removed and placed in a water storage pool for radioactive "cooling." The spent fuel may be temporarily stored in this water-filled pool for eventual disposal in a waste repository if a once-through cycle is selected. Otherwise, reprocessing may be selected, as was done at Hanford in the plutonium production mission during and following World War II. This reprocessing not only produces uranium that could be fabricated into fresh fuel again, but also produces two significant waste streams. The irradiation process results in highly radioactive fission products; these are chemically separated from the uranium and require disposition as radioactive *low-level waste*. The second waste stream, consisting of the chemical dissolution material contaminated with fission products and some small quantity of transuranic (heavier than uranium) elements, is called *high-level waste*. Current policy requires that high-level waste be disposed in a deep geologic repository. Any of the low-level waste that is contaminated with sufficient quantities of transuranic elements is classified as transuranic waste.

Plutonium, separated during the chemical processing, can be blended with slightly enriched uranium to form a fuel called *mixed oxide* (MOX), which is a combination of UO_2 and PuO_2. The uranium in spent fuel has a U-235 content higher than that of natural uranium, and thus is slightly enriched. The uranium that results from the chemical separation process, then, can be returned to the isotope separation plant for reenrichment and reuse.

6.1.2 Processing By-Products

Tank Wastes.   Processing of nuclear fuels (i.e., uranium and plutonium) produces liquid and solid wastes with constituents of principal concern including strontium-90, cesium-137, technetium-99, and transuranics (e.g., americium and plutonium isotopes). A number of other radionuclides in the waste can cause waste handling problems, but are either present in relatively small amounts (e.g., cobalt-60, uranium-233, -235, and -238) or are short-lived (e.g., cesium-134, yttrium-90, barium-137). Tritium is also present, but there is no practical treatment for it, and usually it is allowed to decay, the half-life being 12.26 years. In some instances, iodine-129 and/or carbon-14 are present in large enough amounts to require removal.

The Department of Energy (DOE) keeps these wastes as sludge and liquid in interim storage—mostly in underground tanks. The DOE site with the most of this waste is at Hanford in southeastern Washington state, where over 200 million liters are stored.

In addition to radionuclides, the wastes contain heavy metals and organics, so they are classified as Resource Conservation and Recovery Act (RCRA) mixed wastes. These processing wastes are most often stored in tanks, where undissolved solids are present as sludge, and dissolved solids are present in a floating liquid layer. A semisolid salt crust may form on top of the liquid layer in those instances where the ionic strength is so high that the solubility limit of some salts is exceeded. Some of the main anions present include nitrate, carbonate, chloride, sulfate, chromate, ferrocyanides, aluminate, and hydroxide. Some of the organics form complexes or chelates with heavy metals, and organic complexes of the transuranics dissolve into the liquid phase.

The pH of the liquid phase is usually maintained at 12 or above so that aluminum hydroxide does not precipitate. Such precipitates are gelatinous and form a sludge that would be difficult to handle.

Risks with Storage Tanks and Short-Term Strategy.   Heat from radioactive decay causes evaporation of water. In general, the water vapor is condensed and returned to the tank or the water is replaced. Radiolysis of the organic compounds and water that are present causes evolution of hydrogen and oxygen gases. The tanks are not allowed to go dry, because heat buildup, from radioactive decay, could lead to an explosion. With some tanks, purge air is continually swept through the head space to keep the hydrogen concentration safely below the lower flammable limit and to cool the tank contents. Some storage tanks have a salt crust layer that temporarily traps the evolved gases and intermittently releases them, making control of hydrogen concentrations difficult and dangerous. The need for monitoring and processing the condensate and handling gases leads to potential exposure of workers and the environment to radiation and explosion hazards.

Another risk with storing waste in tanks is leakage. A number of tanks installed over 30 years ago have exceeded their design life and either are leaking or have a high potential to leak. Until the contents of these tanks can be treated, the contents must be transferred to newer, double-walled tanks.

Treatment Options and Long-Term Storage.   The general approach to treatment of tank wastes is to evaporate a substantial portion of the water in order to reduce waste volume and to encapsulate the solids. Micro-encapsulation (e.g., incorporating into cementious grout, encasing in concrete, or vitrifying to form glass) prevents the spread of radionuclides and heavy metals into the environment. Treatment does not include transmutation, so the radioactivity continues after encapsulation. Each container of encapsulated waste must have enough thickness of surrounding metal and/or concrete so that radiation levels near the container are at a safe level. Each container of micro-encapsulated waste must have a limited amount of concentration of radionuclides and must be stored with sufficient surrounding airspace so that decay heat is dissipated.

The long-term disposition of the waste containers depends on the level of radioactivity. Containers with high-level wastes will eventually be stored in underground caverns with a controlled environment such that decay can proceed safely for at least 10,000 years (Johnson, 2000). Some encapsulated wastes are considered to be low in activity because the major constituents of concern have been removed. Such containers can be stored in relatively less secure surroundings than containers with high-level wastes, such as in a ventilated building.

A good strategy is to separate and concentrate high-level and long-lived wastes to the maximum extent possible because of the need for the most stringently secure, long-term storage for the high-level portion. With this approach, relatively few high-level waste containers are produced and many low-activity waste containers are produced. Thus, the volume of high-level waste, which needs the most stringent handling, is minimized. In order to achieve this, the salt cake is dissolved and liquid solution is decanted from the tank, evaporated (not to the extent that solids precipitate), and subjected to removal of high-level constituents. These constituents are evaporated further to form a concentrate that is combined with any sludge that is pumped out of the tank and filtered. The filtrate is combined with liquid solution that is subjected to removal of high-level constituents. (The overall process is shown in Fig. 6.2.) Thus, two products are formed that are micro-encapsulated separately:

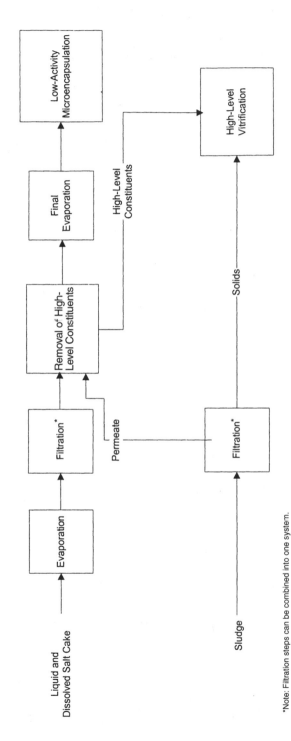

*Note: Filtration steps can be combined into one system.

FIGURE 6.2   Waste feed preparation for low-activity and high-level micro-encapsulation/vitrification.

6.5

- A relatively high volume of treated solution that has low radioactivity, which can be micro-encapsulated and safely stored relatively easily. A storage facility with a design life of a few hundred years is anticipated.

- A small-volume concentrate combined with filtered sludge that has high radioactivity. A vitrified high-level product must be stored in a facility with a design life of thousands of years.

The goal of ultimate storage of low-activity and high-level wastes is to have no free water or extensive void space present. Any water that remains after evaporation must be incorporated in concrete or removed (by further evaporation or by absorption), and voids must be filled with inert material. If vitrification is employed, all water is evaporated in the glass-melting operation, and the melted glass is poured into metal containers such that no void space forms.

Unconfined Wastes—Contaminated Groundwater, Soil. During the preparation of nuclear fuels, whether for commercial electric power plants or for defense activities, radioactive and hazardous constituents have entered soil and in some instances infiltrated down to groundwater. Entry into the soil has occurred from leaks and spills, and at some installations liquid effluents were deliberately discarded into the soil, especially where it was anticipated that soil particles would adsorb the contaminants.

The main contaminants of concern are strontium-90, cesium-137, plutonium, uranium complexes, chlorinated solvents, hydrocarbons, chromate, and nitrate. (The chlorinated solvents, hydrocarbons, chromate, and nitrate are not radioactive, but often are mixed with radionuclides.) Some cleanup standards for groundwater containing the radionuclides, such as drinking water standards, are so stringent that the main remediation technology that is applicable in many instances is ion exchange or adsorption; chemical precipitation and evaporation can also be used. Sometimes, reverse osmosis is used followed by ion exchange; the osmosis process removes almost all of the inorganic contaminants and natural minerals, thereby greatly reducing the load on the ion exchange medium.

Chlorinated solvents and hydrocarbons in groundwater are removed by conventional remediation techniques such as in situ air sparging and pump-and-treat methods including carbon adsorption and air stripping.

Chromate can be controlled in situ by injecting reducing agents or with conventional pump-and-treat methods including ex situ reduction/precipitation and ion exchange.

Nitrate removal from groundwater generally has not been practiced, but both in situ and pump-and-treat methods using anaerobic biodegradation have proved in tests to be effective. The radionuclide concentrations generally are not so high as to sterilize the bacteria. In the event of very high radionuclide concentrations, pump-and-treat techniques that selectively remove the radionuclides followed by biodegradation can be used. Anion exchange could also be used for nitrate removal.

Applicable soil remediation techniques for radionuclides include soil washing, in situ vitrification, and micro-encapsulation with portland cement. Soil vapor extraction and soil heating are used to remove chlorinated solvents and hydrocarbons.

6.1.3 Debris from Deactivation and Decommissioning

Deactivation includes removal of nuclear fuel and preliminary characterization of surfaces and of radiation levels throughout the facility. Decommissioning of obsolete nuclear facilities includes final characterization of contaminants and of radiation levels, removal of hazardous substances, decontamination of surfaces, encapsulation or fixation of certain materials and surfaces, and dismantlement of structures and processing equipment. Various types of debris are generated, including the following:

- Protective clothing used by workers
- Lead bricks and lead sheets formerly used for radiation shielding
- Concrete rubble
- Scrap steel
- Tools and failed or spent equipment

6.2 STORAGE TECHNIQUES

Frequently, wastes at nuclear facilities are stored before final disposal is carried out. Stored wastes must be prevented from container corrosion and leakage, uncontrolled chemical reactions, accumulation of radioactive isotopes that could lead to criticality and runaway fission reactions, uncontrolled expulsion of gaseous decay products (such as hydrogen), uncontrolled heat buildup from chemical reactions and radioactive decay, excessive pressure buildup, and excessive radiation to surroundings. A number of shielding materials are used to shield surroundings from radiation, especially gamma radiation, which can penetrate even very thick shielding. Some common shielding materials are, in the order of increasing shielding effectiveness per unit thickness: water, concrete, steel, and lead.

6.2.1 Water Storage Pools or Basins

Wastes such as spent nuclear fuel are often stored in water basins for long time periods. A typical basin used for storing spent nuclear fuel canisters submerged in water mainly resembles a concrete, flat-bottomed swimming pool at least 16 ft deep. This storage technique is often sufficiently effective for personnel to enter the area surrounding a basin. However, suitable precautions against contact contamination and excessive radiation dosage must be taken. Radiation that escapes from the pool (termed "shine") is significant.

After a number of decades of service, such basins become excessively hazardous from accumulation of products from corrosion of spent fuel components, decay products (notably including strontium-90, cesium-137, and cobalt-60) and deterioration of the concrete. The radioactivity accelerates deterioration of the concrete, leading to cracks and subsequent leakage. At some installations, spent fuel is being removed from the basins and transferred to dry storage.

Smaller pools have been used for short-term storage of spent fuel elements that are expelled periodically from nuclear reactors. This has been the practice at reactors used during World War II and the cold war for producing plutonium for nuclear warheads. A number of such reactors have a graphite core with horizontal tubes that were charged initially via the front vertical reactor face with uranium fuel rods. The water pool just outside of and below the rear face received the spent fuel elements.

6.2.2 Dry Storage/Containment

A variety of dry storage techniques are employed, including casks, vaults, aboveground structures (buildings), and belowground burial, and combinations of these techniques are often used. These techniques provide for retrievable storage; otherwise, the operations amount to disposal as discussed in Sec. 6.4.

Casks, drums, and similar containers are made of a variety of materials, including mainly wood, steel, and stainless steel. The contents must not contain more than 1 percent free water, and void spaces must be eliminated via compaction or filling with inert material. Water is removed to decrease corrosion and to minimize the formation of hydrogen and oxygen gases that are generated by radiolysis of the water. Void spaces are minimized to better utilize storage space or to prevent shifting of waste, and possible container damage, during handling operations.

Removal of the bulk of the free water is done by drainage before containerizing the waste or by pumping water out of the cask or drum via a screen near the bottom. Final removal of free water is done by vacuum drying.

Void spaces are generally filled with an inert substance, such as helium gas or certain foam compounds, depending upon the application. In all cases, the void fill must be compatible with the waste within the container and must meet the requirements for the storage location.

Vaults are typically concrete cells that may be aboveground or belowground. Three primary purposes for using vaults are reducing potential for susbsidence, secondary containment, and shielding to reduce radioactivity exposure levels.

Subsidence can be an important consideration in retrievable storage belowground. Failure of the waste container package under the load of the soil covering or due to heavy equipment moving over the emplaced container could result in a breach that would release contamination to the environment.

Secondary containment may be a significant consideration for longer term storage when it must be shown that release of the waste container contents would result in exceeding limits for worker, public, or environmental protection. This technique is generally employed when specific radionuclides, either those that are mobile in soil columns or are of significance in human ingestion, are identified in the waste in quantities that would be of concern upon release.

In many instances, storage of high-activity radioactive waste in shipping casks is not feasible and some temporary shielding needs to be employed. In these cases, especially for relatively short-term storage, say up to 10 years, use of a vault serves the purpose. Design of the vault for these situations is dependent upon the radioactive decay of concern and the specific application for the storage. Manufacture of the vaults, usually using commercially equivalent concrete, is readily accomplished, either at the storage location or off site, with the vaults then transported to the location.

A building used to house contaminated components or casks, drums, or vaults containing radioactive materials must be monitored and ventilated to prevent buildup of heat and of hydrogen concentrations. Regulatory requirements and worker health and safety concerns are used to establish both ventilation system size and reliability.

Regulations and on-site inspection requirements determine waste container placement configurations within the storage buildings. For routine inspection needs, sufficient aisleways must be left between groups or rows of drums and containers to allow for fulfilling the requirements for visual verification. Also to be considered for placement of the waste containers is the radiation dosage from a group of containers. Routine inspections by operations personnel will result in radiation exposure; placement of the containers should be planned so that this exposure will not be excessive.

Within storage buildings and structures, sufficient operating space for movement of the containers is vital. Two capacities are generally identified for each facility: the design capacity and the operational capacity. In many cases, the operational capacity is no more than 80 percent of the design. Another potential inefficiency in storage facilities is the type and shape of containers. A building loaded with stacked 55-gallon waste drums will have a much different effective storage utilization than one with odd-shaped packages (such as a contaminated waste tank pump in a long storage tube) that cannot be stacked and that may require much more handling space. The waste container "footprint" is significant in determining the amount of waste that might be stored in a given facility.

Generally, within the Department of Energy (DOE) system, retrievably stored burial most likely would be used for transuranic (TRU) waste containers. The TRU waste designation occurred in 1970, and this type of waste is to be retrievable for up to 20 years. As the general waste disposition technique within the Atomic Energy Commission (AEC) defense program in 1970 was direct burial of waste containers, that approach was also applied to TRU waste, with the provision that the waste be retrievable. At several DOE sites, trenches in the existing radioactive waste burial grounds were identified for the TRU waste. Records were maintained on emplacement of containers to establish that the waste met the requirements for TRU waste. Evolution of the record-keeping process resulted in individual container identification and emplacement location, along with specific radionuclide and hazardous constituent inventories.

Several configurations for TRU waste containers were employed: horizontal stacking of drums in trenches with 1 to 2 meters of soil covering, V-trench stacking of drums with a concrete cover (also covered with soil), and vertical drum stacking in trenches with or without an asphalt pad, both with the soil covering.

At the DOE Hanford site, spent fuel assemblies that are stored in water basins are currently being removed, vacuum dried, filled with helium gas, and moved to dry storage where they will be in canisters in 40-foot-deep underground stainless steel vaults below a 5-foot-thick concrete floor in a canister storage building. This building is intended for interim storage up to 40 years, with ultimate disposal at a geologic repository designed for thousands of years' life.

6.3 TREATMENT AND CLEANUP TECHNIQUES

6.3.1 Decontamination and Removal of Nuclear Facilities

Spent Fuel Basins. The steps for removal of an obsolete basin include the following:

- Fuel canisters are removed mechanically.
- Sludge and solid particulates (usually products of corrosion) are removed hydraulically, filtered, and encapsulated for disposal.
- Water is filtered by circulation through filters, followed by pumping out or by filtering as pumping out progresses, and later demineralized. For example, water that was filtered and removed from a fuel basin in eastern Washington was demineralized by reverse osmosis followed by ion exchange in the treatment unit described in Sec. 6.3.2 under "Ion Exchange and Adsorption Case Histories."

- Concrete surfaces are washed, and the concrete structures partially or totally re-moved.

For example, another fuel basin in eastern Washington had its upper walls hydraulically sheared and removed piecemeal with conventional clamshell-type equipment. The concrete rubble is stored in a landfill that is permitted for low-activity nuclear and mixed wastes. The lower basin walls and floor were deep enough to be abandoned in place, and were covered with clean, compacted fill.

Decontamination and Removal of Equipment and Buildings. After surfaces are surveyed for radioactive contamination, and radiation dosages are measured for each room or area in a deactivated nuclear structure, the following activities are undertaken:

- Hazardous materials, including asbestos, PCB transformer oils, mercury ballasts, chemicals, etc. are removed (with appropriate precautions).
- Lubricating oils and liquid fuels are removed from both machinery and from storage areas.
- Contaminated surfaces of rooms and equipment are either sprayed with fixative paint (typically acrylic based) or decontaminated.
- Structures and equipment are segmented (if needed) and removed.

Decontamination is accomplished by detergent washing or by wire brushing, concrete scabbling, grinding, etc. with vacuum capture of dust that is generated.

More details regarding fixative coatings and handling of debris from dismantlement operations are given in "Debris Wastes" in Sec. 6.3.3.

6.3.2 Liquid Wastes

Treatment of Organics

Aqueous-Phase Carbon Adsorption of Organic Contaminants. Carbon adsorption can be applied to both volatile and nonvolatile organics dissolved in water. Hydrocarbons and most chlorinated solvents can be removed to near nondetection levels. Undissolved phases such as oil or grease should be removed first. Although activated carbon adsorbs oil and grease, these can be gravity-separated or filtered at relatively little cost, thereby extending the life of the carbon. The same concept holds for filtration of suspended solids.

If the water contains both organics and metal radionuclides, the carbon adsorption step can be before or after the radionuclides are removed. The disadvantage of using carbon as a first treatment step is that the carbon will pick up more radioactivity than if used last. This makes it more difficult to handle when the carbon needs reactivation or disposal.

In order to make the carbon last as long as possible, it is often used as an organics removal polishing step following removal of the bulk of the organics by air stripping or ultraviolet oxidation.

Aqueous-phase activated carbon is routinely deployed in fixed beds at a number of nuclear facilities for services ranging from treatment of radioactive condensate to groundwater remediation. Figure 6.3 is a simplified flow diagram showing a typical arrangement. Carbon beds are usually designed for at least 10 minutes' fluid residence time, with a hydraulic loading of 2 to 10 gal/min per square foot of bed cross section. Carbon is highly effective for removing most chlorinated solvents and hydro-

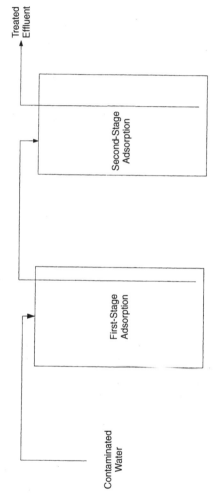

FIGURE 6.3 Two-stage aqueous-phase carbon absorption.

carbons, sometimes removing these contaminants to below the detection level. Low-molecular-weight oxygenated organics, vinyl chloride, and methylene chloride are poorly adsorbed.

The capacity of activated carbon is expressed as the amount of organic contaminant it can adsorb per unit mass of carbon, and is higher at higher organics concentrations in the water being treated. For treating condensate, published capacity values for each organic compound can be used to predict the volume of water that can be treated before the carbon is spent. However, it is important to understand that the capacity values are obtained from equilibrium laboratory tests, and the capacity that is actually obtained under dynamic flow-through conditions is approximately 45 to 55 percent of the equilibrium value (Stenzel and Merz, 1989).

For treating groundwater, laboratory or pilot tests using samples of the groundwater are needed to determine carbon capacity, because naturally occuring organic matter may be present that use some of the capacity. "Bottle isotherm" tests (batch equilibrium tests) with powdered carbon and pilot columns or drums with granular activited carbon determine the amount of organic contaminant that can be adsorbed per unit mass of carbon.

In Situ Air Sparging and ex Situ Air Stripping of Groundwater. In situ air sparging or pump-and-treat air stripping are used to remove volatile organic compounds. Typical air stripping systems remove 90 to 99 percent of volatile organics dissolved in groundwater. Air sparging usually applies where the groundwater table is within 60 ft of the surface and where soil vapor extraction is used simultaneously to remove volatile organics from the soil. Compressed air is injected into the aquifer at depths ranging up to approximately 25 ft below the groundwater table. Either pipes with sparge nozzles near the tip or wells with short screens are used to inject the air. The radius of influence for each sparge point is generally less than 30 ft, so many sparge points must be used in a pattern that encompasses the areal extent of the contaminant plume. In order to make the most efficient use of equipment and energy, the air can be injected into a group of sparge points intermittently and into other groups sequentially. The cleanup time is usually much faster than with pump-and-treat techniques.

The air strips the volatile organics from the aquifer and enters the soil above the aquifer (the vadose zone). Soil vapor extraction is used to clean the soil. Refer to Fig. 6.4. Vadose zone wells or trenches are connected by piping to a water knockout pot

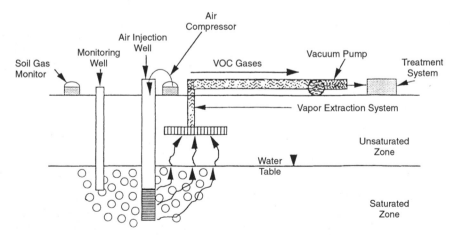

FIGURE 6.4 Cross section of an air sparging/vapor extraction system. [*From U.S. EPA (1992a).*]

(not shown in the figure) in series with a vacuum blower or with an internal combustion engine that creates a vacuum. If a blower is used, it is in conjunction with vapor-phase activated carbon or with an oxidizer (direct thermal or catalytic) on the discharge, to remove or destroy the organics. If an engine is used, it acts as a thermal oxidizer.

The alternative stripping method to in situ air sparging, an ex situ stripper receiving groundwater pumped from wells, is most often a tower with internal packing or sieve trays. The groundwater is injected at the top of the tower from where it flows downward through the internals, while air flows upward. A cooling tower can be used to perform the same function. Air stripping can also be accomplished by diffusing air through a series of chambers containing the water.

When operated near ambient temperatures, air strippers are used to remove only volatile organics. Heated air strippers and steam strippers are used to remove semivolatile as well as volatile organics. The organics in the emissions from strippers are generally not radioactive. Where local, state, and/or federal regulations require emissions abatement, conventional air pollution control systems are deployed.

An air stripper can be circular or rectangular in cross section. Most are cylindrical plastic or metal towers with packing. In recent years, rectangular tray towers have come into widespread use for groundwater remediation. These rectangular tray towers are usually designed with a much larger cross-sectional area than a packed tower would have for the same stripping capacity, and are much shorter. Low-profile tray towers (and diffused-air chambers) are especially advantageous in situations where equipment height is limited. However, low-profile tray towers use much higher air/water ratios than do packed towers.

Where the contaminants in the air emissions from a stripper must be abated, low air/water ratios are desirable. The capital and operating costs for abatement equipment are much higher than for stripping equipment at any given air rate. With a packed tower, the higher the depth of packing, the lower the air/water ratio can be for a given percent removal of a contaminant. The packing depth and corresponding air/water ratio, and pressure drop as well, can be derived from the following sources: (1) for traditional packings (e.g., ceramic Raschig rings, ceramic Berl saddles, metal tellerettes, or Pall rings)—Kavanaugh and Trussell (1980); (2) for modern plastic packings and some traditional packings—the computer program Air Strip from Dave Schoeler, Ames, Iowa.

Traditional packings have higher pressure drop, and the tower cross-sectional area can be determined on the basis of pressure drop, as described by Kavanaugh and Trussell (1980). Many towers with modern packings are designed with a cross-sectional area such that the liquid loading is approximately 20 gal/min per square foot of superficial cross-sectional area. The nominal packing diameter is selected such that the ratio of tower width to packing diameter is at least 12 to 1.

An air blower is used to move air through an air stripper. Most stripping systems use a forced-air blower that feeds the air below the packing or below the bottom tray. However, a vacuum blower on the overhead vapor line can be used instead, by inducing air to enter below the packing or below the bottom tray. A vacuum blower is preferred, because the lower the absolute pressure in the stripper, the better it works. Also, if there is an air leak, with a vacuum blower the leakage is inward, without uncontrolled loss of contaminants to the environment. This is especially important if radioactive contaminants are involved. The blower pressure rise, with either forced air or vacuum, must be enough to overcome air pressure drop through the following equipment: tower internals, ductwork, and any air emissions abatement equipment.

The most common air emissions abatement systems used are vapor-phase carbon and, alternatively, a thermal oxidizer or catalytic oxidizer. At best, a single-stage carbon adsorption system removes 90 percent of volatile organic compounds from typical air stripping offgas. Two stages of carbon adsorption in series may be needed to achieve regulatory air emissions limitations.

Vapor-phase carbon can be regenerated in place with steam by using multiple carbon vessels and alternating between adsorption and regeneration cycles. If off-site reactivation of the carbon is potentially to be used, special consideration must be given to radioactivity concerns.

At any given temperature, vapor-phase carbon has the maximum adsorption capacity when the air relative humidity is below approximately 50 percent, and the colder the better. However, the air exiting from the top of a stripper is at 100 percent relative humidity. The carbon adsorption capacity can be improved by warming the air by 13°C, thereby lowering the relative humidity sufficiently. The maximum capacity can be attained by first refrigerating the air by air conditioning equipment; this causes water vapor to condense, and the water can be removed, thereby lowering the absolute humidity. Then the air is warmed by 13°C, with the final relative humidity being 45 percent at a relatively low temperature, these conditions being the optimum for carbon capacity.

A thermal oxidizer operating above 760°C typically achieves over 99 percent destruction of organics in the offgas. Air stripper offgas fuel value is generally very low, and considerable amounts of auxiliary fuel (e.g., natural gas or propane) are needed. The auxiliary fuel consumption can be cut by more than 50 percent if heat exchange between the hot oxidizer exhaust and the stripper offgas is employed.

The auxiliary fuel consumption can be cut even further by using heat exchange and a catalytic unit. Catalytic oxidizers achieve approximately 95 percent destruction of the organics at 316°C and are operated at up to 371°C to achieve somewhat higher destruction efficiencies. The most widely used catalytic units employ the same platinum catalyst as used in many automobile exhaust system catalytic converters. Such catalyst cannot be used if the stripper is used for groundwater that contains gasoline that contains volatile lead compounds such as tetramethyl lead, or for groundwater that contains medium or large amounts of chlorinated organics. If heavy metals that poison catalysts, such as lead, are present, a thermal oxidizer or carbon adsorption should be used. If chlorinated organics are present, the catalyst manufacturer should be able to estimate how long the catalyst will last before becoming deactivated, depending on chloride concentration and offgas air flow rate. If the catalyst life is intolerably short, a special catalyst that is not affected by halogens can be used.

If radionuclides are present in the offgas, a high-efficiency particulate air (HEPA) filter should be used following the organics abatement system. The relative humidity must be kept well below 100 percent for a HEPA filtration system. If an oxidizer is used, the exhaust temperature must be cooled to below the temperature limit of the filter, and then slightly reheated to the temperature limit so that the filter stays dry.

Bioremediation of Groundwater.   Bioremediation is the use of micro-organisms such as bacteria to convert organic compounds into water and nontoxic carbon compounds. If the organics are mineralized, the end products are water, carbon dioxide, and an increased mass of micro-organisms (typically bacterial cell mass, i.e., bacterial sludge). The process occurs naturally in situ in the aquifer, or an engineered pump-and-treat system can be used with a bioreactor, using added phosphate and fixed-nitrogen nutrients. Most engineered bioremediation systems use aerobic bacteria, with aeration to supply oxygen to aid in the metabolization of the organics.

Pump-and-treat bioremediation systems are generally not used for cleanup of nuclear facilities. The bacterial sludge that is generated could be contaminated with radionuclides, and would be difficult to dispose of safely. In situ natural attenuation (biodegradation of organic compounds) frequently occurs at all types of sites, including nuclear ones.

Other Treatment Techniques for Aqueous-Phase Organics. Organic compounds can be converted to water and carbon dioxide (plus halides in the case of halogenated organics) with oxidizing agents such as hydrogen peroxide and ozone. However, the reaction rates are too slow for practical use unless a catalyst or ultraviolet light is used in conjunction with the oxidizing agent. The most commonly used such advanced oxidation units employ exposure to ultraviolet light with hydrogen peroxide additions. For hydrocarbons and most chlorinated solvents, the destruction efficiency can be over 90 percent, with unsaturated compounds being most readily destroyed. The destruction efficiency depends on the ultraviolet light and peroxide dosages and exposure time.

The liquid being treated must be prefiltered so that suspended solids do not interfere with penetration of the light. Quartz-encased ultraviolet lamps are employed that must be cleaned or replaced as needed, in order to maintain the ultraviolet dosage.

Ozone can be diffused into the fluid instead of injecting hydrogen peroxide solution, or both ozone and peroxide can be used.

In treating liquids that contain transuranics complexed with organics, oxidizing agents are used to destroy the complex. Then the transuranics become insoluble in alkaline solutions and can be removed by ultrafiltration. The solids thus removed are subjected to encapsulation for disposal as may be appropriate.

Chemical Precipitation. Precipitation is carried out either batch-wise or continuously with the following steps: (1) mixing coagulation aides and precipitation agents with the wastes and (2) separating the precipitate by settling and/or filtration. With most chemical precipitation units the sludge precipitate that forms is separated in a clarifier or settler followed by conventional filtering (e.g., a filter press). An alternative separation method for this separation is cross-flow ultrafiltration.

Ultrafiltration units used for separating radioactive solids are often semibatch systems as shown schematically in Fig. 6.5. Most of the liquid circulates through the unit; some goes through the pores in the ultrafiltration membrane and is recovered as permeate. The circulating fluid sweeps the sludge into the sludge collection tank, where the solids concentration increases as the batch processing progresses.

Strontium-90, Carbon-14, and Heavy Metal Isotopes. Chemical precipitation is particularly applicable to removing radionuclides in the following examples:

Isotope	Sr-90	C-14	Heavy metals
Form	Sr^{++}	CO_3^{-}	M^{++} or M^{+++}
Precipitating agent	Sodium carbonate	Barium nitrate	Hydroxide or sulfide
Precipitate	Strontium carbonate	Barium carbonate	Hydroxide or sulfide

Other precipitating agents have been used besides those shown here. Another approach to removing radioactive ionic species by precipitation is with isotopic dilution. An example is the use of nonradioactive strontium nitrate to remove Sr-90 from solutions containing dissolved carbonate, as follows:

$$^{90}Sr^{++} + Sr(NO_3)_2 + 2CO_3^{-} = 2SrCO_3 + 2NO_3^{-}$$

The strontium carbonate shown on the right side of the equation precipitates as a solid containing both nonradioactive strontium and Sr-90 because the addition of nonradioactive strontium shifts the equilibrium to the right.

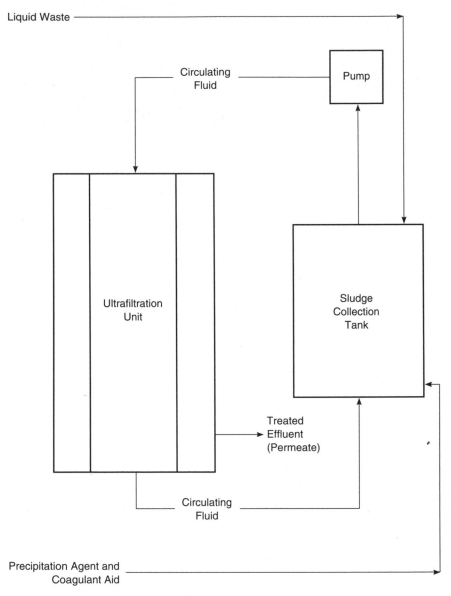

FIGURE 6.5 Semibatch ultrafiltration.

Chromate. Chromate is not radioactive, but is a common water-soluble contaminant at nuclear sites. Usually it is first reduced from the hexavalent form to the trivalent form by using sulfur dioxide, ferrous iron, or other reductants. At pH levels above 7, trivalent chromate forms the insoluble hydroxide.

When chromate is in groundwater, it is generally removed ex situ with a pump-and-treat system. Most groundwater aquifers contain dissolved iron, and chromate can be precipitated by injecting a reducing agent in a buffering solution. One such process is in situ redox manipulation, developed by the Pacific Northwest National Laboratory (Richland, Washington), operated by Battelle Memorial Institute. The buffering agent usually employed is potassium dicarbonate solution. The reducing agent usually employed is sodium dithionite. Dissolved iron naturally in the aquifer is reduced to ferrous ion, which reduces the hexavalent chromate ion to chromic ion, which precipitates as the hydroxide.

Sulfate. In some systems used to vitrify radioactive wastes, sulfate in the waste in medium or high concentrations may interfere with the melting process. With medium-temperature and high-temperature glass melters, sulfate is not a problem. With lower-temperature melters (i.e., 1100°C), a separate sulfate layer forms that does not become incorporated into the glass product. This layer concentrates radioactivity and is corrosive.

Sulfate can be removed from aqueous streams by precipitation with barium. If carbonate is also present, it coprecipitates. The mixed precipitates can be treated with nitric acid if it is desired to remove the carbonate and recover some of the barium for reuse.

Transuranic Metals. Transuranic isotopes (e.g., americium-241, neptunium-237, plutonium-238, -239, -241) may be soluble when complexed with organics. They are precipitated from alkaline solutions when the organics are destroyed with an oxidizing agent.

Reverse Osmosis. Reverse osmosis is a membrane separation process that removes dissolved ions. (This is in contrast to ultrafiltration, which uses membrane separation for removing suspended solids.) The ion removal efficiency is typically in the 80 to 95 percent range. Usually, the permeate must be cleaned up further, and it is sometimes subjected to ion exchange as a polishing treatment step. Approximately 20 percent of the feed stream becomes concentrate that is subjected to evaporation.

Pretreatment steps include lowering of pH, to deter membrane scaling, and filtration.

Ion Exchange and Adsorption. Most ion exchange and adsorption applications for radioactive waste employ semiselective granular solid media that each remove one or a limited number of substances from aqueous streams. Examples include:

Substances removed	Solid medium
Iodine isotopes	Silver nitrate
Plutonium	Bone char
Uranyl complexes	Anion resin
Strontium	Zeolite
Cesium	Zeolite or cation resin or crystalline silicotitanite
Technetium	Resin

Where resins are employed in nuclear applications, they are frequently specific for removing certain ions, unlike media that are commonly used for completely demineralizing (deionizing) water. These resins are regenerable, often by means of a scheme such as the following:

1. Elution with a liquid such as acid solution, followed by water rinsing
2. Treatment with sodium hydroxide solution

The eluate discharged from step 1 contains the isotope of concern and may be treated in an evaporator. The evaporator system concentrates the isotope and recovers the elution stream for reuse by condensing it from the evaporator overhead vapor.

Typically the aqueous stream flows down through the selected solid medium in a column or in columns arranged in series. Elution, if employed, is done with flow in the opposite direction.

Evaporation and Crystallization.   Evaporation can be used to remove virtually all of the water, leaving the contaminants in a concentrated brine or sludge. Or, it can be used in conjunction with other aqueous treatment techniques such as the following:

- *Reverse osmosis.* The concentrate may be further concentrated or dried in a crystallizer.
- *Ion exchange.* The eluate may be evaporated, and condensed for reuse, as described above.

Most evaporators are steam heated with jackets, internal coils, or external reboilers. Alternatives that do not use steam are vapor recompression units and solar ponds.

Jacketed units are generally limited to evaporators with low heating duties, because heat transfer is limited by the relatively small transfer-area/volume ratio available. Internal coils provide for high duties, limited by space available within the evaporation vessel. Very high duties can be obtained by circulation of the process fluid through an external heat exchanger (reboiler). The heat transfer area available with an external reboiler is not limited by the size of the evaporation vessel. Steam is the most commonly used heat transfer medium. Most steam-heated evaporators are operated under vacuum, which lowers the boiling point of the process fluid and thereby reduces corrosion rates.

Vapor recompression units do not need on-site steam because the energy input is from the heat of compression released when the water vapor that is formed is compressed for recirculation. The compressor is usually electric motor driven.

Case Histories of Liquid Waste Treatment.   Case histories are presented here for organics treatment, chemical precipitation, and ion exchange/adsorption.

Organics Treatment Case Histories.   At a nuclear defense processing facility, groundwater is contaminated with chlorinated solvents over an areal extent of 10 km^2. The main contaminants are carbon tetrachloride and chloroform. The carbon tetrachloride had been used prior to 1989 as a diluent in the processing of plutonium. The chloroform is apparently a degradation product derived naturally from the carbon tetrachloride. Used tetrachloride had been dumped into "cribs" that allowed infiltration into soil columns. Over the years, the contaminants have reached an aquifer at depths over 200 ft. The soil has been remediated with soil venting systems (as described in Sec. 6.3.4).

The groundwater contamination plume was advancing in a westerly direction toward a river. A row of groundwater extraction wells has been installed near the western edge of the plume. This strategy has stopped advance of the plume, with the extracted water being remediated in an air stripping unit.

The stripping unit is designed for treating 500 gal/min at a volumetric air/water ratio of 21:1 (Fig. 6.6). The stripping tower, fabricated from fiberglass-reinforced plastic with a solvent-resistant interior gel coat, is 5 ft in diameter with approximately 37 ft of 3½-in nominal diameter plastic Lanpac packing. The allowable carbon tetrachloride concentration limit in the treated effluent is 5 mg/L; up to this level the

FIGURE 6.6   Packed-air stripper for treating 500 gal/min being erected at Hanford.

effluent is returned to the aquifer via five injection wells. These wells are located upgradient of the plume.

Offgas from the stripper is treated in a two-stage vapor-phase activated-carbon adsorption system. The carbon is reactivated off site. The carbon canisters are manifolded with quick-disconnect fittings so they can be readily removed for shipment to the reactivation vendor and replaced with canisters containing freshly reactivated carbon. This replacement cycle is done with the fresh unit placed in the lag position and the former lag unit moved into the lead position.

Because shipping and handling costs are significant with off-site reactivation, it is important to maximize the contaminant loading in the lead canister. This is accomplished by refrigerating the offgas with an industrial air conditioning chiller, which cools it to below 50°F. Moisture in the offgas condenses under these conditions, and the condensate is removed. The chilled offgas is then postheated by approximately 23°F, which reduces the relative humidity to 45 percent. At or below this humidity, water molecules have little effect on the organics adsorption capacity of the carbon.

At another area of the same contamination plume, a field pilot-scale bioremediation unit successfully destroyed the chlorinated hydrocarbons in situ. An extraction well and an injection well were deployed, designed to operate at 100 gal/min. Groundwater from the extraction well was fortified with nitrate additions, and for part of the time with phosphate, to act as nutrients for the native anaerobic bacteria. Sodium acetate was also added to increase the dissolved organic carbon content so that the bacteria had adequate substrate to metabolize. The nitrate addition rate was controlled such that bacterial uptake of the fixed nitrogen resulted in a net reduction of nitrate concentration in the groundwater. The acetate additions were done in pulses. The fortified groundwater was reinjected at approximately the same depth as the extraction well, from where it dispersed into the aquifer. The bacterial destruction of the chlorinated hydrocarbons then took place in the aquifer.

The aquifer hydraulics had been mathematically simulated by a model developed by the Battelle Pacific Northwest National Laboratory (Richland, Washington). As predicted from the model, the injection pressure tended to increase as the bioremediation progressed, because of buildup of bacterial cell mass in the aquifer. To control this buildup so that the aquifer did not become completely plugged, the phosphate additions were discontinued, and the nitrate and acetate fortifications were modulated. A preliminary description of the modeling and cost comparisons to alternative remediation technologies (i.e., air stripping/activated carbon and ex situ bioremediation) are given in Skeen et al. (1993).

Reverse Osmosis Case History. At the Hanford nuclear site in Washington state, the effluent treatment facility treats waste processing water, condensate, and groundwater effluents that are mildly radioactive. Following removal of organics, the wastes flow through a reverse osmosis unit followed by ion exchange. The reverse osmosis unit operates at approximately 400 psig pressure and removes the bulk of the radionuclides. Thin-film composite polyamide membranes are employed. See Fig. 6.7. The ion exchange media are regenerable, and treated effluent from the ion exchange unit is discharged to groundwater via an infiltration gallery.

Ion Exchange and Adsorption Case Histories. At the same site in Washington, ion exchange and adsorption are used in a number of applications besides the effluent treatment facility. An anion exchange unit was used to remove uranyl complexes from groundwater in a 50-gal/min two-stage unit. Clinoptilolite is used in a three-stage groundwater cation exchange unit to remove strontium-90. Bone char was used to adsorb plutonium from groundwater in a unit that processed 28 gal/min of groundwater. In that unit a second-stage vessel contained clinoptilolite to remove cesium-137 and strontium-90 via cation exchange. A third vessel contained a mixed bed of 50 percent bone char and 50 percent clinoptilolite for final removal of the contaminants (Fig. 6.8). None of these groundwater treatment units employed regeneration.

Two new ion exchange/adsorption units are being designed for removal of cesium and technetium from tank-stored wastes. The cesium will be removed with an organic cation exchange resin that will be regenerated by elution with nitric acid. (The nitric acid will be recovered from the eluate by evaporation/condensation. The concentrated stream from the evaporator will be incorporated in glass by vitrification.) The resin is specific for removal of monovalent cations, and does not become loaded with divalent noncontaminants such as calcium and magnesium cations. The technetium will be adsorbed selectively on an organic resin that can be regenerated by elution with hot water.

FIGURE 6.7 End view of a reverse osmosis system installed in eastern Washington.

FIGURE 6.8 An absorption and ion exchange system treating 28 gal/min of groundwater at Hanford.

Evaporation and Crystallization Case History.   The concentrate from the reverse osmosis unit described in "Reverse Osmosis Case History" above, along with ion exchange regeneration effluents, is reduced in volume by evaporation and the brine from the evaporator is dried in a crystallizer. The evaporator is a vapor recompression unit. The crystallizer is a thin-film dryer, where brine slurry from the evaporator is sprayed onto the tank inside walls that are mechanically scraped to produce a powdered product that is drummed.

Reboiled Evaporator Case History.   The 242-A evaporator at the Department of Energy Hanford site is operating at a processing facility originally used for recovery of plutonium. A major function of the evaporator is to reduce the volume of radioactive liquid processing wastes stored in tanks. Figure 6.9 shows the evaporation system. The evaporated water is condensed, and the condensate is disposed. An ion exchange unit was installed to treat the condensate. The product from the bottom of the evaporator is a slurry of crystallized salts. The unit operates under vacuum created by steam-jet ejectors. The source of heating is steam reboiler E-A-1 through which a stream is circulated by pump P-B-1 and returned to the evaporator. The vapor is condensed in shell-and-tube heat exchangers with cooling water.

The design parameters for the system, given in an appendix to Lini, Forehand, and Kelley (1990), are as follows:

Feed rate—70 to 120 gal/min

Reboiler circulation rate—14,000 gal/min

Steam to reboiler—27,000 lb/h at 5 psig

Slurry product rate—70 gal/min

Vapor flow rate to primary condenser—21,000 lb/h

Cooling water flow to primary condenser—3500 gal/min

Steam to first-stage ejector J-EC1-1—680 lb/h at 90 psig

Steam to second-stage ejector J-EC2-1—750 lb/h at 90 psig

6.3.3   Sludges and Solid Wastes

Sludges and granular solid wastes that contain hazardous organics and radioactive isotopes (mixed wastes) are usually treated by first removing the organics and then subjecting the wastes to fixation. Methods of removing organics include solvent extraction and thermal desorption. Some fixation techniques include heating, which often volatilizes the organics, which are then treated in the vapor phase. (Precautions must be taken against accidental combustion of the organics.)

Sludges

Organics Removal.   If hazardous organics must be controlled, removal via solvent extraction can be applied. Hydrocarbon solvents are used for treating hydrocarbon contaminants and halogenated hydrocarbon contaminants. (Water interferes with such extraction processes, and can be removed by conventional dewatering techniques at ambient temperature followed by vacuum drying.) The extraction process is usually carried out batchwise in multiple stages by mixing the sludge with the solvent, followed by settling and draining or pumping out the extract. The extract may be recovered for reuse by distillation or filtered and incinerated. (If high levels of halogenated compounds are involved, an incineration system must include a wet scrubber for removal of acid gases.)

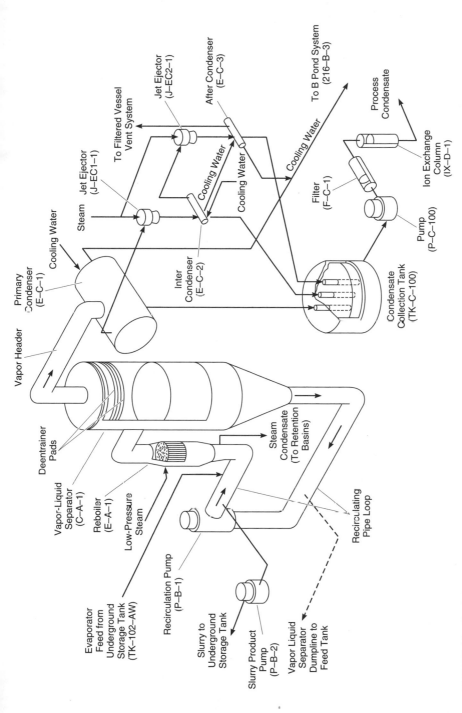

FIGURE 6.9 The 242-A evaporator simplified schematic.

6.23

Thermal desorption vaporizes volatile and semivolatile organics by applying heat at temperatures below the ignition point. One method that minimizes the hazard of accidental ignition is indirect heating. Unlike direct heating that utilizes heated air or burner exhaust gases in contact with the matrix being desorbed, indirect heating employs heat exchange across metal surfaces in contact with the matrix, infrared heating with gas burners behind a metal radiative shield, or electrical heating. The flow rate in the vapor phase is much less than with direct heating, being limited to water vapor and organic vapor that is formed plus injected inert sweep gas.

An emerging technology for removing halogenated organics from mixed waste is molten salt oxidation. A pilot-scale unit has been operating at the Lawrence Livermore National Laboratory (Livermore, California) under sponsorship of the Department of Energy. Commercialization of the technology will be managed by ATG Inc. at its Richland, Washington facility.

Fixation. Fixation of sludges can be accomplished with solidification agents or by subjecting them to vitrification (melting with glass-forming agents). Watery sludges are first reduced in volume by evaporation of part of the water. If it is desired to separate the high-level waste in order to reduce the volume of highly radioactive stabilized or vitrified product, a scheme such as that shown in Fig. 6.2 can be employed. The filtration steps can be carried out with conventional filtering equipment, or with crossflow ultrafiltration as shown in Fig. 6.5.

1. *Solidification.* Solidification techniques as reviewed by Arniella and Blythe (1990) are described here. The waste is mixed with an agent that keeps it insoluble and/or accomplishes micro-encapsulation, thereby minimizing potential leaching of contaminants to the environment. Portland cement is the most common agent used, with cement-to-waste ratios in the 1:5 to 1:1 range. If sulfates are present, general construction-grade portland cement is not used, and special grades are required. If organic substances interfere with the setting and curing, resins (e.g., heated ureaformaldehyde in the presence of a catalyst) or polymers (e.g., heated polyethylene) may be used.

The solidified product must pass certain tests to be acceptable for burial, two of the most important being unconfined compressive strength and resistance to leaching. Compressive strength requirements for non-nuclear solidification applications often depend on how deep the solidified product may be buried in a landfill, and 50 to 100 psi is usually adequate. However, for radioactive wastes, the strength requirement may be 150 psi. Leaching tests are done on a sample that has been subjected to grinding and rinsed with an acidic solution. There are regulatory limits on the amount of specific contaminants in the used rinse solution.

2. *Vitrification.* Sludges can be calcined in a kiln, which removes free liquid, and the calcined product mixed with glass-forming substances and then fed to a melter. Melting is carried out at approximately 1200°C, with some designs at 1100°C. Higher temperatures result in a higher percentage of waste being incorporated in the glass product. However, the lower-temperature designs potentially have a longer melter life. Most of the vitrification in U.S. operation or planned is done with slurry-fed (or liquid-fed) electrically powered glass melters, without the calcining step. Slurry-fed melter technology as reviewed by Chapman and McElroy (1989) is described here.

Glass-forming agents, primarily silicate and borate, and the sludge are melted with a residence time of approximately 40 hours. Because oxygen may form and impede heat transfer, reducing agents such as formic acid or sugar are added to the feed. Where the sludge includes wastes from dissolution of nuclear reactor spent fuel, the sludge composition is dominated by constituents of the fuel rod cladding—zirconium, aluminum, magnesium, stainless steel (iron, nickel, chromium, manganese, molybdenum, etc.). Typical sludges also contain nitric acid wastes some of which have been neutralized with sodium hydroxide, and contain other sodium wastes. The slurry feed must be premixed thoroughly and is fed to the surface of the melt at a rate corresponding to 40 hours residence time.

The percentage of waste that ends up in the glass product is limited by the concentrations of chromium, manganese, nickel, molybdenum, and iron. Excessive concentrations of these transition elements can result in a separate sludge layer to form that does not become incorporated in the glass. Excessive sulfate in the waste can also cause this problem in some melter systems.

The glass product typically contains 10 to 22 percent B_2O_3, 45 to 53 percent SiO_2, up to 14 percent Fe_2O_3, and 8 to 12 percent sodium reported as Na_2O, with significant amounts of aluminum, calcium, lithium, magnesium, manganese, and zirconium. Other constituents include chromic oxide, potassium oxide, nickel oxide, phosphorus pentoxide, sulfate, and titanium dioxide. The product must pass leachate tests.

A slurry-fed melter is constructed with two to four layers of refractory material in a steel box, with offgases withdrawn under a moderate negative draft. The inner refractory must resist glass corrosion as well as provide thermal insulation. The outside is cooled either with water or natural convection. Joule heating is attained with alternating current power, with resistivity of the molten glass typically being in the 2 to 10 Ω-cm range. Up to 0.5 wt% of the feed (0.3 wt% of the glass production rate) is lost to offgases, which are controlled with wet scrubbing.

With a design life ranging down to 2 years, the melters themselves become a disposal problem. They can be segmented and segregated into high-activity and low-activity fractions. Calculations indicate that the high-activity fraction will typically be less than 2 percent of the mass of glass produced.

More information on vitrification processes is given in U.S. EPA (1992b).

Debris Wastes. In the waste management world, the word *debris* can be applied two specific ways: (1) generally describing the waste that results from demolition or cleanup activities and (2) a prescriptive definition found in application of dangerous waste regulations. For this latter definition, "Debris means solid material exceeding a 60 mm particle size that is intended for disposal and that is: a manufactured object; or plant or animal matter; or natural geologic material. However, the following materials are not debris: Any material for which a specific treatment standard is provided in Subpart D, Part 268, namely lead acid batteries, cadmium batteries, and radioactive solids; process residuals such as melter slag and residues from the treatment of waste, wastewater, sludges, or air emission residues; and intact containers of hazardous waste that are not ruptured and that retain at least 75% of their original volume. . . ." (*RCRA Regulations and Keyword Index,* ISSN 1074-1364, 2000 edition, Art. 268.2, p. 716.) More information regarding the detailed description and application of *debris* in the RCRA regulations can be obtained in the referenced document. For the purposes of this section, the general description of debris will be used.

Where practical, facilities undergoing demolition are first decontaminated as much as possible before demolition commences. In some instances, contaminated large steel or concrete structures have a fixative paint coating applied first, to minimize the spread of contamination.

Equipment components are removed. Large equipment, piping, structures, etc. are segmented by various cutting, shearing, and ramming techniques. The debris is mainly metals and concrete rubble, along with hazardous substances such as asbestos-containing insulation and transite wall panels and equipment containing mercury or PCBs. Potential spreading of dust during demolition is controlled by air vacuum and water spraying techniques. Water washing is the most-used technique for decontamination of debris.

Some metals that cannot be decontaminated readily are subjected to remelting and recovery for use within nuclear facilities. Scrap metal that is not radioactive is segregated and recycled via general commercial dealers. Nonradioactive concrete rubble could be recycled, but generally this is not cost-effective. The rubble and most other debris is disposed in permitted landfills as low-activity waste. Debris that is contaminated with transuranics, such as workers' protective clothing and gloves with plutonium, is drummed and is being disposed in artificial, deep caverns at the Waste Isolation Pilot Plant in New Mexico.

Handling of such debris is usually as follows:

- Clothing and personal equipment worn by workers for protection against contamination is often only lightly contaminated if at all. It is double-bagged in plastic and disposed of by landfilling (if not contaminated with transuranics) in a licensed and/or permitted commercial or government-owned disposal facility.
- Most lead used for shielding does not become radioactive, but the surface often becomes contaminated with radioactive substances. The lead surfaces can be decontaminated and the lead reused. Any lead shapes that are not practical to decontaminate can be encased in grout, made with portland cement, for burial.
- Heavily contaminated concrete structures are first decontaminated. Lightly contaminated and potentially contaminated surfaces are spray painted with acrylic paint prior to dismantling the concrete structures, which is generally done with a battering ram. The paint acts as a surface contamination fixative, but is only partially effective and helps to reduce the spread of contamination. During dismantlement, water sprays are used to control concrete dust. The rubble is disposed of in a secure landfill, consistent with limitations on remaining radioactive and/or RCRA contaminants.
- Steel structures that are determined during characterization to be free of contamination, or are known from history of use to be uncontaminated, are dismantled separately and the steel is recycled. The remaining structural steel can be decontaminated and recycled or disposed of in a secure landfill.
- Tools and failed or spent equipment that are difficult to decontaminate are segmented and encased in concrete or bagged or drummed for disposal.

Disposition of large nuclear reactor cores that were used to produce plutonium present a special challenge, and has not yet been accomplished. Presently, after the nuclear fuel is removed, such obsolete reactor cores are surrounded with concrete or steel and left standing pending final disposition. Eventually these reactor cores will likely be buried whole or dismantled and the pieces buried. The dismantling may have to be done with the aid of remotely controlled manipulators or with robotic equipment.

Other Solid Wastes. Most other solid wastes that exhibit radioactivity are encapsulated prior to disposal. Void spaces are removed or filled with inert material such as rigid foam. Free liquids are removed, typically to below 1 wt%. Macro-encapsulation is usually done with metal canisters or concrete.

Case Histories for Treatment of Sludges and Solid Wastes
Treatment of Sludge Wastes

1. *Solvent extraction/thermal desorption for removal of organics from a mixed waste.* Approximately 1485 kg of uranium oxides mixed with oil were treated at a Department of Energy site in Colonie, New York. The oil included halogenated compounds, with a concentration of 16,900 ppm extractable organic halides. The treated waste was to undergo land disposal with a limitation of 1000 ppm. Solvent extraction was used to remove 80 to 90 percent of the organic halides, and thermal desorption was used as a polishing step to meet the limitation. The extract was shipped off-site for incineration. Amrit et al. (1997) describe the solvent extraction treatment as given below.

Preliminary bench-scale treatability studies were conducted with three low-cost candidate solvents—isopropanol, diesel, and kerosene. Kerosene was chosen on the basis of these three criteria:

- Ability to extract a high percentage of the organic halides
- Obtaining a distinct phase separation
- Obtaining a clear extract

A solvent/feed ratio of 1:2 was used in a batch mixer followed by 24 hours of settling. The extract was removed by repeated decantation at 8-hour intervals over another 24-hour period and pumped through a 200-mesh screen. Solids collected on the screen were returned to the solid phase.

The solid phase was subjected to repeated contact with the solvent in the batch mixer. Organic halide removal efficiency was approximately 80 wt% with one contact and up to approximately 90 wt% with double contact.

2. *Heat-enhanced vapor extraction of volatile organics from mixed waste.* Drummed electroplating sump sludges were treated at a Department of Energy remediation site that was formerly used for fabricating radiation shielding components, ballast weights, and projectiles from depleted uranium. The sludges contained low-level radionuclides, volatile halogenated hydrocarbons, and metals contaminants. The volatile organics, principally tetrachloroethylene (perchloroethylene, PCE) and trichloroethylene (TCE) were removed by a drying process, and the treated matrix fixed with cement. The waste analysis indicated concentrations of 7 percent PCE, 0.6 percent TCE, and 8.9 ppm cadmium. Treatment criteria limited PCE to 5.6 ppm, TCE to 5.6 ppm, and total organic halides to 1000 ppm prior to fixation of metals.

The volatile organics were removed batchwise by using drums fitted with a drying apparatus. Each batch was a portion of a drum load. Nineteen loaded drums were treated in 42 batches. The drying was carried out by heating a partially filled drum to approximately 270°F and sparging compressed air through a slowly rotating paddle mixer. The average treatment time was 11 hours per batch.

The vapor emissions were controlled by operating under negative pressure—the vapors were drawn through a high-efficiency particulate air filter (HEPA filter), followed by three activated carbon drums in series connected to vacuum blowers.

The highest chlorinated hydrocarbon concentration in any treated drum was 3.3 ppm PCE.

Decontamination of Used Soil Drilling Pipe.   Steel pipe sections, each up to 20 ft long, became contaminated with soil containing radionuclides at the DOE Hanford site. Bechtel Hanford, Inc. has decontaminated such pipe sections by using a Kelly washer manufactured by Container Products (North Carolina). This unit is a portable container that employs steam cleaning with hot water pressurized to a few hundred psig. Detergents can be added to this process.

Macro-encapsulation of Mixed Waste Debris at the Hanford DOE Site

1. *Overview.* During fiscal year 1997, a pilot project to macro-encapsulate radioactive mixed waste was undertaken at the Hanford site in eastern Washington state

(Fluor Hanford, 1998). The project consisted of compacting drums containing the mixed waste debris, placing the compacted waste inside polyethylene tubes, and seal-welding polyethylene end caps onto the tubes. The project was completed in September 1997 with the compaction and macro-encapsulation of 880 drums (185 m³) of mixed waste debris. The purpose of the project was to demonstrate the macro-encapsulation technology to meet 40 CFR Part 268.45. It also was to conduct an appropriate demonstration of the ability of a container to safely and efficiently encapsulate radioactive mixed waste debris in order to isolate the contents of the container from the environment to meet U.S. EPA treatment and disposal requirements.

The treatment units used on this project were constructed of 30-in outer diameter, 0.923-in wall thickness, high-density polyethylene (HDPE) pipe approximately 21 ft long, with 1.375-in-thick HDPE caps welded to the ends. The resin used in the manufacture of the sleeves and end caps was previously tested by the DOE, and shown to be resistant to leaks and ultraviolet degradation, with an outdoor storage life expectancy of between 100 and 300 years.

2. *Waste feed stock.* The waste feed stock for this project consisted of 163 eighty-five-gallon and 717 fifty-five-gallon drums of mixed waste debris, 880 drums total. The 85-gal drums contained 55-gal drums that required, for various reasons, an overpack drum. The 55-gal drums were removed from the 85-gal overpacks before processing. The 85-gal drums were returned to the storage facility as RCRA empty drums for reuse. The 880 drums contained or possessed the following:

- A total of 0.761 curies of the following isotopes: Am-241, Co-60, Cs-137, Eu-154, Pu-238, Pu-239, Pu-240, Pu-241, Pu-242, and Sr-90
- The following RCRA and Washington state codes for waste: F001–F005, D007–D008, WC02, WP02, WT01, and WT02, where F001–F005 are spent halogenated and nonhalogenated solvent waste codes, D007 represents the waste code for chromium, D008 represents the waste code for lead, WC02 represents the Washington state code for carcinogenic substances, WP02 represents the Washington state code for persistent dangerous halogenated hydrocarbon waste, and WT01 and WT02 represent Washington state codes for toxic dangerous and extremely hazardous toxic waste, respectively.

3. *Compaction step.* U.S. EPA regulation 40 CFR 265.315 requires that containers be at least 90 percent full prior to placement in a landfill. For this project, compaction of the waste containers (drums) by a commercial facility was the chosen method to meet this requirement. During the planning stages of the project, it was estimated that a compaction ratio of 4 to 1 could be achieved; a 1500-ton supercompactor was used by the commercial facility.

The 880 fifty-five-gallon drums of waste were each compacted into a "puck" approximately 6 in thick. These pucks were then loaded into a 70-gal (9.6-ft³) drum. The actual compaction ratio achieved in this project was about 4.8 to 1, with the compaction being accomplished over a period of approximately 3 weeks. A total of 149 of the 70-gal drums were loaded, each with six of the compacted waste drums.

4. *Macro-encapsulation step.* This radioactive mixed-waste treatment demonstration was conducted on the mixed-waste treatment pad at the Hanford T-Plant facility. T Plant is located within a designated radiologically controlled area. A total area of approximately 29,000 ft³ was used in this deployment, that included empty tube storage, the tube loading area, the fusion area, and the treated unit storage area. The 149 seventy-gallon drums were loaded into a total of 22 of the polyethylene tubes over a period of 4 days. Twenty-one of the tubes were the standard 21-ft units, each containing seven of the 70-gal drums; one unit was customized for two of the 70-gal drums.

The loading was accomplished with a crane and rigging crew. It took an average of 8 min to load seven of the 70-gal drums into the polyethylene tubes. The first two tubes required about 12 min each to load. The smooth inside surface of the tubes combined with a well-designed loading rack and a simple ramming device made for a rapid loading process.

Fusion of the end caps on the sleeves was accomplished in two phases. The first phase, fusing of one end cap on empty tubes, was accomplished prior to delivering the tubes to the Hanford site. A total of 34 end caps were fused in approximately 6 days. Trained fusion specialists were used to undertake this operation for both phases.

The second phase, fusion of the second end cap on loaded tubes, was accomplished on the Hanford site at the T Plant. Only 22 tubes were needed; these fusion operations were accomplished over a period of five working days.

The gross weight of all 22 waste-loaded tubes was 143,216 lb; net weight (i.e., less the weight of the HDPE material) was 125,326 lb.

5. *Cost data.* This project was undertaken as a firm fixed-price contract. Using the base scope of the contract, 1000 mixed-waste drums and a compaction ratio of 4 to 1, the following firm-fixed price unit price quotation was received:

Description	Quantity	Unit price	Amount
Compaction (on site)			
Mobilization/demobilization			$150,000
Permitting and regulatory compliance			50,000
Employee training			30,000
Compaction,	1000 drums	at $175	175,000
Total compaction			$405,000
Macro-encapsulation			
Mobilization/demobilization			$50,000
Permitting and regulatory compliance			50,000
Employee training			30,000
Macro-encapsulation,	250 drums	at $1350	337,500
Total macro-encapsulation			$467,500
Estimated project totals			
Total compaction			$405,000
Total macro-encapsulation			467,500
Subtotal			$872,500
20% general and administrative costs			174,500
Total estimated cost			$1,047,000

6.3.4 Soils

Soils at nuclear installations can be contaminated with organics and metals. Generally, the organics are not radioactive. Some metal contaminants are not radioactive (e.g., hexavalent chromium), but a number of light and heavy metals are. In some instances, a site contaminated with radionuclides may be capped with relatively impervious layers so that water infiltration does not carry contamination down into an aquifer. Where remediation via treatment is used, in situ technologies are preferred, especially where cost is less than ex situ technologies; also, in situ approaches are safer from the viewpoint of less exposure to radioactive dust when excavation, hauling, and conveying or dumping of soil are avoided. The in situ remediation technologies that might be considered at a specific site include:

- Soil venting (soil vapor extraction)
- Soil heating and heat-enhanced soil venting
- In situ vitrification
- Fixation with portland cement injected via a hollow auger
- Electrokinetics

This discussion will cover both in situ and ex situ soil treatment processes.

Soil Venting.   Conventional soil venting is practiced at nuclear facilities where volatile organics have contaminated the soil. The process uses ambient air that infiltrates from the ground surface into the soil and is drawn under vacuum through the unsaturated soil layers to wells or trenches for extraction. The most widely used extraction devices, which draw a vacuum, are vacuum blowers. Generally, the radioactive soil contaminants that may be extracted in addition to the volatile organics are adsorbed on soil particulate matter that is filtered out of the extracted airstream above ground. If the discharge must be abated to conform to air pollution control limits on organics discharges, a multistage activated carbon system or an oxidizer is employed.

Figure 6.10 shows a typical extraction well. The header pipe to the blower, shown above the typical well, would connect to a number of wells that are screened within the soil contaminant plume.

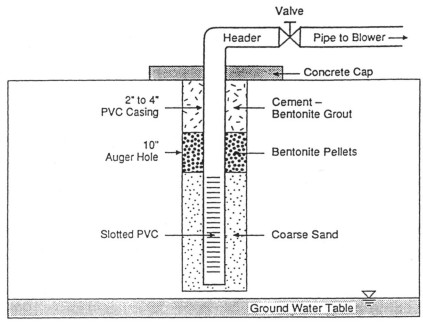

FIGURE 6.10   Typical extraction well schematic for SVE. [*From U.S. EPA (1991).*]

The concept can also be applied ex situ. Piles of excavated soil are imbedded with perforated pipes that are manifolded to a moisture separator (water knockout pot) and a vacuum device.

The air flowing through the soil pores strips volatile organics that may be in the liquid phase between soil particles or adsorbed on the particles. The organics transfer into the vapor phase and are not radioactive. Soil venting removes over 90 percent of hydrocarbons and chlorinated hydrocarbons.

Soil Heating and Heat-Enhanced Soil Venting.   If the soil is heated or if soil-venting air is heated, some semivolatile organics can be removed, as well as volatile organics. If only volatile organics are of concern, the remediation can be accomplished faster than with ambient air.

Various techniques can be used, with either in situ or ex situ applications, some with heated air and some with other media. The exhaust from a vapor abatement system can be heated and recycled by injection into the soil. For in situ systems, steam can be injected into vadose-zone wells or via the type of hollow auger described in "Fixation of Soil Contaminants," below, for in situ soil encapsulation. Electric heating can be applied to the soil with radio-frequency power or with resistance heating.

Soil heating removes over 99 percent of volatile organics and lesser percentages of semivolatile organics.

Soil Washing and Solvent Extraction.   Water-based soil washing is used for removal of inorganics (including radionuclides) and oils. Solvent extraction can be used for removal of a wide variety of organics.

Washing with water, or with water and additives, is effective on soil particles over 63 μm average diameter. Some water-based washing techniques separate the coarse particles from fines. The fines fraction contains almost all of the contaminants and is subjected to further treatment, such as fixation (micro-encapsulation) for radionuclides or incineration for organics. Thus, washing becomes a volume-reduction type of treatment, working best on coarse, sandy soils. For example, at the Hanford nuclear site in eastern Washington, 85 percent of soil contaminated with radionuclides was separated from the remaining fines fraction during large-scale testing.

For removal of cationic metals, water or acid solution are used. Common additives, including those used for removal of oils, are detergents and alkaline agents such as sodium hydroxide.

Solvent extraction is applicable to all soil particulate size fractions. Polar organics can be removed with liquid carbon dioxide. Polar and some nonpolar organics can be removed by solvent extraction with a hydrocarbon or other organic solvent. The solvent is usually recovered by distillation and recycled through the extraction process. For highly efficient contaminant removal, either multiple extraction stages or repeated extraction steps are usually needed.

Micro-encapsulation and in Situ Vitrification

Fixation of Soil Contaminants.   Portland cement is the most commonly used contaminant fixation agent for micro-encapsulation. The ex situ process is usually conducted batchwise in a slurry-mixing device that can handle very high solids percentages, such as a pugmill. Water and cement are mixed with the soil in proportions that form a weak concrete (compressive strength of only up to 150 psi, suitable for deep landfilling). The product is granular—not monolithic concrete. If a batch of product does not pass a leachate procedure when tested, that batch is recycled through the process.

An in situ process that works similarly forms a column of fixed soil up to 12 ft in diameter and up to 120 ft deep. Multiple, overlapping columns are treated sequentially by injecting portland cement as a slurry via a hollow auger.

Such micro-encapsulation processes produce a larger mass and volume of fixed soil than the original amount treated.

In Situ Vitrification. The process marketed by AMEC Earth and Environmental Inc. (Richland, Washington) uses electric-resistance heating between two pairs of graphite electrodes. A block of vitrified soil up to several meters deep forms, resembling black obsidian glass. As the soil heats up, volatilized organics and water vapor are collected under a hood and conveyed at negative pressure to a treatment train that removes particulates and organics. Electric power at over 12,000 V and up to 4600 kW is used. The electrodes and hood are moved for each successive block of soil that is vitrified. The nonvolatile radioactive metals are immobilized, and radiation is shielded by surrounding soil. As soil vitrifies, it shrinks and subsides, so clean fill or conrete is placed over the vitrified blocks, shielding the atmosphere from radiation.

In the plasma-arc process developed at Georgia Tech (Circeo et al., 1994), bore holes are predrilled at a spacing of approximately 1.5 m, and an electrically energized plasma torch is inserted to the bottom of each hole, in turn. The torch is gradually raised, forming a vitrified soil column. The technique is not depth-limited, as is electric-resistance heating described immediately above.

Case Histories for Soil Remediation at Nuclear Facilities

Capping. The Hanford Barrier in eastern Washington is a full-scale test prototype of a maintenance-free capping system designed to contain radionuclides for 1000 years. As described in the U.S. Department of Energy (1999) report DOE/RL-99-11, this barrier that capped a crib (pond) was tested over a 4-year period ending in 1998. The site had been used to dispose of low-level radioactive liquid waste associated with uranium reclamation operations. However, the soil contains high-activity (>1,000,000 picocuries/g) contamination, mainly between 5 and 15 m below the ground surface, that includes strontium-90, cesium-137, plutonium-238, plutonium 239/240, and uranium. The barrier prevents rainwater from infiltrating the soil and transporting the contaminants down to an aquifer. Testing was concerned with structural stability and resistance against water intrusion, wind erosion, and animal intrusion. A surface irrigation system was employed to simulate rainfall up to 3 times the long-term average and to simulate the strongest rainstorm expected over a 1000-year period.

This cap was constructed with the following layers from the top down: 1 m of silt loam/pea gravel with native vegetation, with a 2 percent surface slope; 1 m of silt loam; 0.15 m of sand with geotextile; 0.3 m of gravel; 1.5 m of basalt riprap; 0.3 m of gravel; 0.15 m of asphalt. The sand and gravel layers below the silt loam serve as a capillary break that inhibits downward percolation to layers below and prevents fine soil from filtering downward into the riprap. The riprap and gravel layers extend over the 2:1 side slopes. The riprap deters root penetration and animal burrowing. The asphalt is a hydraulic barrier and redundant biointrusion layer and extends under the riprap side slopes.

Asphalt core samples exhibited comparable or lower permeability than the standard of 1×10^{-7} cm/s for RCRA low-permeability soil for capping. Monitoring indicated no drainage through the silt loam layers. As expected, drainage occurred through the riprap side slopes, but was diverted by the sloped asphalt layer. Surface water runoff under extreme precipitation conditions was minimal except at times that the soil froze, and surface erosion was not significant.

The unit cost, excluding testing and monitoring costs, was $320 per square meter. *In Situ Vitrification.* The AMEC Earth and Environmental Inc. (Richland, Washington) Maralinga project in the Great Victoria Desert in Australia remediated a number of mixed waste underground pits at Taranaki that included plutonium and uranium contamination. Information obtained at www.geomelt.com included the following.

The soil at Taranaki is generally naturally cemented limestone/dolomite with some interspersed sand. It melts at the relatively low temperature of 1200°C. An initial step is the application of a layer of high-silica sand, 0.3 to 0.6 m thick. (The mix of this cover soil and native soil melts at 1500 to 1600°C.)

The pits contained debris and waste drums that were first probed with a hydraulic hammer to collapse or fill large voids and to disrupt the integrity of the drums. Later testing of vitrified product indicated that the plutonium and uranium was uniformly distributed.

After probing, melts were initiated in the high-melting-point sand layer. Melting progressed downward through each pit. A volume reduction of 50 percent occurred as a result the removal of voids and calcination of the limestone/dolomite.

6.3.5 Incineration of Radioactive Mixed Waste

Incinerators burn waste at high temperatures. The main purpose of incinerating radioactive waste is to reduce waste volume, since a large proportion consists of bulky items such as contaminated clothes, lumber, and plastic. Incineration of waste that is a mixture of chemically hazardous and radioactive materials, known as *mixed waste,* has two principal goals: to reduce the volume and the total chemical toxicity of the waste.

Incineration does not destroy metals or reduce radioactivity of wastes. Radioactive waste incinerators, when equipped with well-maintained, high efficiency filters, can capture all but a small fraction of the radioactive isotopes and metals fed into them. The fraction that does escape, however, tends to be in the form of small particles that are more readily absorbed by living organisms than larger particles.

Incinerators, like many combustion devices such as automobile engines, convert combustible materials mainly to carbon dioxide and water (steam). But they generally also create toxic by-products, known as *products of incomplete combustion* (PICs). PICs can be more toxic per unit weight than the original wastes. The total quantity and toxicity of PICs from incinerators is highly uncertain. The most widely studied toxic PICs are known as dioxins.

Dioxins and similar toxic chemical compounds accumulate in fatty tissue, increasing in concentration at each successive level of the food chain. Until 1993, regulations did not factor in food chain exposure. Although special filters can reduce toxic emissions to well below legal limits, they also concentrate toxins in ash. Landfilled ash and contaminated filters present greater threats to groundwater than the original wastes in some cases. Permanent storage of ash in well-monitored structures can minimize the risk of groundwater contamination.

Alternatives can present their own environmental problems. Landfilling liquid wastes can contaminate groundwater, while storing them in tanks can lead to explosions (See "Risks with Storage Tanks and Short-Term Strategy" in Sec. 6.1.2). Emerging techniques for destroying toxic compounds such as supercritical water oxidation and plasma arc pyrolysis may prove preferable to incineration. However, some wastes may not be treatable by a single system, requiring separation (for example, to remove metals) before treatment.

For some existing wastes, it may be impossible to keep risks low for both current and future generations. Reducing the production of waste is therefore the surest way to minimize future health and environmental hazards.

6.4 DISPOSAL TECHNIQUES

This section will describe techniques for disposal of radioactive debris and solid wastes (or solidified liquids) by giving case histories and descriptions of planned facilities. Included are the following:

- Low-level waste disposal in trenches, landfills, and pits
- Geologic depositories

6.4.1 Low-Level Waste Disposal in Trenches, Landfills, and Pits

Hanford Site Shallow-Trench Burial Grounds. The final step in the waste management process for radioactive wastes is disposal, or isolation, of waste until its radioactivity has greatly decreased via decay. Some types of waste can be managed and disposed of safely in facilities using "conventional" methods, such as shallow land burial.

The Hanford site low-level waste disposal facilities are located in an arid (less than 7 in of precipitation per year) region of the State of Washington. The disposal facilities provide for final disposition of both low-level and mixed low-level radioactive wastes. For over 50 years, the Hanford site burial grounds have received radioactive solid wastes from the various missions at the Hanford site as well as from off-site waste generators. From 1944 until April 1970, all of the radioactive solid waste on site, regardless of radionuclide content or hazardous constituents, was buried in shallow trenches. Beginning in 1970, the radioactive waste categorized as transuranic was segregated from low-level waste and was buried in a retrievable configuration. In 1999 mixed low-level waste disposal began in a permitted mixed-waste trench.

The Hanford site has six trench-type burial grounds with over 200 trenches. The general configuration for the disposal of radioactive low-level wastes is a slope-sided trench with a bottom width of 20 to 30 ft and a length of up to approximately 1000 ft. As the disposal facility is in sandy soil, gradual sloping sides are necessary to preclude unwanted backfilling. Containers, drums, or boxes, of low-level waste are placed directly into the trenches. Some bulk wastes, such as soil or debris, are placed in the trenches without containers, and the waste is then covered with a layer of clean soil to keep the waste from dispersing. Some radioactively contaminated equipment, even a railroad flatcar, has been direct buried.

Low-level waste with higher concentrations of radionuclides is overpacked within the disposal facility. The current practice at Hanford is to place this type of waste in small concrete vaults or high-integrity containers, or to encase the waste in cement in place (a process referred to as *trench grouting*).

For final disposal, the low-level waste trenches will be completely filled with soil. The entire burial ground will then be covered with a layer of relatively impermeable material that is sloped to prevent infiltration of water. A second layer of soil over the infiltration barrier will be added and planted with vegetation to stabilize the surface of the facility after closure. This process is generally referred to as *capping*.

One of the Hanford site burial grounds contains two trenches that are permitted for disposal of mixed low-level waste that has been treated to comply with land disposal restrictions under RCRA Subtitle C. Each trench measures approximately 450 by 300 ft, with the excavated depth ranging between 25 and 35 ft.

The trenches are double-lined with leachate collection. The system consists of the following layers from top to bottom:

• Primary leachate collection system
• Primary high-density polyethylene liner
• Secondary leachate collection system
• Secondary high-density polyethylene liner

The primary and secondary liners with leachate collection are designed to prevent any liquid that leaks into the disposed waste from reaching the surrounding environment. The liner material, high-density polyethylene, was selected primarily because of high chemical resistance. The leachate collection and removal system consists of a sump in the polyethylene liner that will collect any liquid that drains into the facility. The system was designed for the "24-hour, 25-year" storm. The secondary collection system will remove any leachate that might penetrate through the primary liner. The trenches are designed for a functional life of 50 years, with 20 years for the operational phase and 30 years for the postclosure monitoring phase.

Landfill Example.   The DOE Hanford Site has a large cell-type landfill called the Environmental Restoration Disposal Facility (ERDF). The ERDF handles low-level soils and a variety of radioactive and mixed wastes derived from cleanup of facilities formerly used for weapons-grade plutonium production.

As described in Eacker and Dronen (1998), the facilities included nuclear reactor complexes, chemical treatment plants, liquid-waste disposal sites, solid-waste disposal sites, research laboratories, and various types of cleanup facilities. For example, a reactor complex includes a reactor structure, primary and secondary cooling piping systems, spent-fuel storage basins, laboratories, and ancillary equipment needed to operate the reactor. Wastes from a decommissioned reactor complex include bulk soils, concrete rubble, contaminated equipment, stabilized/treated sludge, irradiated hardware, scrap piping and structural steel, and other miscellaneous materials. The vast majority of the wastes being landfilled is bulk soils.

Contaminants include radioactive isotopes, RCRA-listed wastes, RCRA characteristic wastes, and other toxic substances. The facility is authorized under a Superfund (CERCLA) Record of Decision. The facility started up in July 1996 at 100 tons/day of soils.

As of 30 January 1998, 730,727 tons had been disposed, of which 94.7 percent was soils. Other wastes included 82,259 tons of concrete rubble, 736 tons with debris containers, 4123 tons of tanks, pumps, and metal containers, and 400 tons of monoliths. Peak handling rates exceed 3500 tons/day. (As of late 2000, 2.5 million tons have been disposed.) Four cells have been constructed, of which two are closed. Two more cells are planned, which will make the total capacity 10 million tons, with 55 ft below ground and 32 ft above ground, and a ground-surface footprint of 1900 × 2400 ft. Each cell is 500 × 500 ft at the base.

Wastes are compacted in two 35-ft layers separated by 2 ft of clean fill. The final 15-ft-thick cover will include clay, compacted fill, and a plastic liner system. Debris and contaminated soil are transported to the facility in 20-ton containers at an average rate of 125 containers per day.

Most of the bulk wastes are transported to the landfill in roll-off containers similar to large-scale municipal solid waste operations. Stabilized and/or containerized waste forms are used where protection against radiation is needed.

Dumped soils are placed with bulldozers and compacted with recovered leachate. Large equipment, including fans, piping, and tanks, requires some form of stabilization to ensure compaction in the landfill. Grout, contaminated fill, and sand have been used for stabilization.

Life-cycle costs have been under $3 per cubic foot (of which 35 percent is for transportation)—much less than with low-level burial ground trenches at over $20 per cubic foot. The DOE paid the capital cost and owns the facility, and a private firm performs the operations.

Pit 9 at Idaho Falls Area.   Pit 9 is part of the Idaho National Engineering and Environmental Laboratory (INEEL) facility in Idaho Falls. The site is an inactive waste disposal pit that covers slightly more than an acre of ground surface. The site is part of an 88-acre landfill at the INEEL contaminated with transuranic waste and organic solvents. Pit 9 measures 127 ft in length, 379 ft in width, and approximately 18 ft in depth. An approximately 8-ft-thick layer of waste is thought to be 4 to 6 ft below the surface.

From November 1967 to June 1969, various wastes ranging from contaminated rags to storage drums with hazardous chemicals, organic solvents, and plutonium-contaminated sludge, many of which derived from DOE's Rocky Flats Plant in Colorado, were dumped and covered with a layer of soil. The DOE estimates that Pit 9 contains approximately 250,000 ft^3 of transuranic wastes, hazardous wastes, and contaminated soil needing treatment. Transuranic wastes are man-made radioactive elements, produced from uranium during a nuclear reactor's operations, that emit alpha particles. The pit contents of primary concern include plutonium and americium (from weapons production), and volatile organics such as trichloroethylene and carbon tetrachloride.

The site was scraped down to the bedrock, materials were buried in boxes and barrel drums, and the pit was then overfilled. No barriers such as those used in current waste disposal were employed. According to Idaho officials, with the regulations that are now in place, Pit 9 would never have been selected as a waste disposal site because of its location 580 ft above the sole-source Snake River aquifer. Although there appears to be no immediate threat to water quality, contamination could affect future generations if the landfill remains untreated.

DOE and its regulators agreed to clean up Pit 9 as an interim action under Superfund by retrieving soil and wastes from the pit, separating those materials that could be returned to the pit without treatment, treating the remaining soil and wastes to achieve at least a 90 percent reduction in volume, and packaging the remaining concentrated materials for on-site storage until final disposal. This Pit 9 cleanup effort was conceived to demonstrate cleanup technologies over a relatively small area prior to selecting a treatment system for the entire INEEL landfill. Of all the sites within the 88 acres, researchers know the most about Pit 9.

6.4.2   Geologic Repositories

Yucca Mountain—A Repository Planned for Nevada.   Yucca Mountain is the Department of Energy's potential geologic repository designed to accept spent nuclear fuel and high-level radioactive waste. If approved, the site would be the nation's first geological repository for permanent disposal of this type of radioactive waste.

It is located in Nye County, Nevada, about 100 miles northwest of Las Vegas, on federally owned land on the western edge of the Department of Energy's Nevada Test Site. If approved, the repository will be built approximately 1000 ft below the top of the mountain and 1000 ft above the groundwater.

Yucca Mountain is a 1200-ft-high flat-topped volcanic ridge extending 6 miles from north to south. The mountain is composed of *tuff*, a rock made from compacted volcanic ash formed more than 13 million years ago. Yucca Mountain has a desert climate and receives about 6 to 7 in of rain and snow per year. The mountain has a deep water table.

As early as 1957, the National Academy of Sciences recommended burying radioactive waste in geologic formations. After more than 2 decades of additional study, the Department of Energy concluded that disposal in an underground mined geologic repository is the preferred approach. Key to a final decision will be the existence of long-stable geological formations and long-lived engineered barriers to isolate wastes. Optimum characteristics of a site would be high stability, no circulating groundwater, location where severe earthquakes or volcanic eruptions are highly unlikely, and deep enough to allow for buffers of the same rock above and below storage.

Spent nuclear fuel and high level radioactive waste make up most of the material to be disposed at Yucca Mountain. Approximately 90 percent of the waste proposed for disposal is from commercial nuclear power plants, with the remainder coming from defense programs.

Spent nuclear fuel and high-level radioactive waste contain short- and long-lived radionuclides. Most radionuclides in this waste will decay to insignificant levels within several hundred years. Some radionuclides will take many thousands of years to decay to nonthreatening levels.

The wastes are currently stored at commercial nuclear power plants and Department of Energy facilities throughout the United States. Spent nuclear fuel is stored in specially designed water-filled pools and aboveground dry storage facilities. Liquid high-level wastes are stored in large underground tanks made of stainless steel or carbon steel. However, storage pools are reaching capacity at some nuclear power plants. Although these sites were designed for temporary storage, new dry storage technology is available to permit extended at-reactor or on-site storage until a repository is eventually established.

DOE would be responsible for operating the facility. Under current plans, the waste would be repackaged and placed into disposal canisters. A remotely operated railcar would carry the canisters down a ramp into a 100-mile network of tunnels. Because of the excessive heat and the high level of radiation, robots would position the canisters.

The goal for the potential repository at Yucca Mountain is to isolate the waste from the environment in these ways:

1. Position the waste above the water table where the relative dryness of rocks would minimize exposure to groundwater.

2. Contain the waste in extremely thick, corrosion-resistant packages.

3. Bury the waste deep—approximately 1000 ft below the land surface—to prevent most kinds of accidental contact with the waste from natural causes such as severe weather.

When an estimated 70,000 tons of waste has been disposed, the repository would be closed.

The site needs to undergo a complex NRC licensing process to determine whether it can safely contain the waste. DOE is planning to complete the process and begin placing the waste in the repository in 2010.

DOE's current plan is to transport the waste by truck and rail to Nevada. The waste will be shipped in casks that are heavily shielded to contain the radioactive material and are certified to withstand extreme accidents, impacts, puncture, and exposure to fire and water. In addition, NRC and Department of Transportation regulations must be met before any waste is shipped to the site.

The transportation routes go through 43 states. The federal government will be working with states, local governments, and tribes in developing emergency response plans.

There is ongoing debate over whether the geologic features and proposed engineered barriers at Yucca Mountain will provide sufficient isolation for permanent disposal. A number of interested parties believe Yucca Mountain has certain characteristics that pose a concern for long-term isolation of highly radioactive materials. The State of Nevada's Nuclear Waste Project Office points to Yucca's location in an active seismic (earthquake) region; the presence of numerous earthquake faults (at least 33 in and around the site) and volcanic cinder cones near the site; evidence of hydrothermal activity within the proposed repository block; and the presence of pathways (numerous interconnecting faults and fractures) that could move groundwater (and any escaping radioactive materials) rapidly through the site to the aquifer beneath and from there to the accessible environment.

The Waste Isolation Pilot Plant in New Mexico.   The Waste Isolation Pilot Plant (WIPP) near Carlsbad is an operating repository licensed to dispose of radioactive transuranic (TRU) waste left from research and production of nuclear weapons. Transuranic elements (heavier than uranium) include isotopes such as americium-241, neptunium-237, and plutonium-238, -239, and -241. Half-lives as much as 24,000 years are involved. An example of TRU waste is protective clothing and equipment used by personnel working in nuclear weapons facilities and laboratories. Radiation from TRU waste consists mostly of alpha particles that travel only a short distance in air. Most of the wastes that have been received at WIPP is in containers transported by truck from Department of Energy nuclear facilities in other states.

Information obtained from a WIPP Web site (www.infocntr@wipp.carlsbad.nm.us) (phone 1-800-336-9477) includes the following:

WIPP is located in the remote Chihuahuan Desert of southeastern New Mexico, and began operations on 26 March 1999. Project facilities include 56 disposal rooms mined 2150 feet underground in a 2000-ft-thick salt formation that has been stable for more than 200 million years. Each room is 33 ft wide × 13 ft high. Rooms are accessed by a series of tunnels 8 miles long. Unlike the planned Yucca Mountain repository, wastes deposited at WIPP will not be retrievable.

Salt formations are favored for such repositories because of the following advantages:

* Most salt deposits are geologically stable, with very little earthquake activity.
* Salt deposits demonstrate the absence of flowing freshwater.
* Salt is relatively easy to mine.
* Rock salt heals its own fractures because of its plastic quality.

As required by the U.S. EPA, warnings about the location and purpose of such repositories must last at least 10,000 years. Included at WIPP are: a large berm; perimeter monuments; an information center; two information storage rooms; buried warning markers; and archives stored in various worldwide locations, with a distinctively bound summary available in six recognized United Nations languages on archival-quality paper.

Over the next 35 years, WIPP is expected to receive approximately 37,000 shipments of TRU waste from up to 23 locations nationwide. Seventeen shipments per week are anticipated. The wastes are packaged for shipment at the sources in vacuum-sealed 10-ft-high domed cylinders, 6 ft in diameter and weighing up to 19,265 lb. Each cylinder is composed of two stainless steel containers, one inside the other, with insulation between them.

REFERENCES

Amrit, S. K., L. L. Baldy, E. T. Newberry, and B. N. Kapoor. 1997. "Developing Mass Balance Nomographs to Assess Solvent Extraction Performance," *Waste Management Conference,* Tucson, Arizona, March.

Arniella, E. F., and L. Blythe. 1990. "Hazardous Wastes," *Chemical Engineering,* pp. 93–102, February.

Chapman, C. C., and J. L. McElroy. 1989. "Slurry-Fed Ceramic Melter—A Broadly Accepted System to Vitrify High-Level Waste," in *High Level Radioactive Waste and Spent Fuel Management,* Vol. II, S. C. Slate, R. Kohout, and A. Suzuki (eds.). Book no. 10292B, American Society of Mechanical Engineers, New York.

Circeo, L., S. L. Camacho, G. K. Jacobs, and J. S. Tixier. 1994. "Plasma Remediation of In Situ Materials—The Prism Concept," in G. W. Gee and N. R. Wing (eds.), 33d Hanford Symposium on Health and the Environment, *In Situ Remediation,* November 7–11, Pasco, Washington, part 2, Batelle Press, Columbus, Ohio.

Eacker, J. A., and V. Dronen. 1998. "Hanford Environmental Restoration Disposal Facility: An Operation and Privatization Success," paper presented at Waste Management Conference, Tucson, Arizona.

Fluor Hanford. 1988, "Macroencapsulation of Mixed Waste Debris at the Hanford Nuclear Reservation," Final Project Report by AST Enviromental Services, LLC, HNF-1846. Prepared for the U.S. Department of Energy.

Johnson, Jeff. 2000. "Brief Life for Waste Bill?" *Chemical and Engineering News,* pp. 27–28, April.

Kavanaugh, M. C., and R. R. Trussell. 1980. "Design of Aeration Towers to Strip Volatile Contaminants from Drinking Water," *J. Amer. Water Works Assoc.,* **72,** 12: 684–692, December.

Lini, D. C., G. D. Forehand, and D. E. Kelley. 1990. "Restart of the 242-A Evaporator," WHC-EP-0342, prepared for U.S. Department of Energy by Westinghouse Hanford Co., Richland, Wash., April.

Skcen, R. S., et al. 1993. "In-Situ Bioremediation of Hanford Groundwater," *Remediation,* pp. 353–367, Summer.

Stenzel, M. H., and W. J. Merz. 1989. "Use of Carbon Adsorption Processes in Groundwater Treatment," *Environmental Progress,* **8,** 4: 257–264, November.

U.S. Department of Energy. 1999. "200-BP-1 Prototype Barrier Treatability Test Report," DOE/RL-99-11 Rev. 0, Richland, Washington, August.

U.S. EPA. 1991. "Soil Vapor Extraction Technology," reference handbook, EPA/540/2-91/003, February.

U.S. EPA. 1992a. "A Citizen's Guide to Air Sparging," EPA/542/F-92/010, March.

U.S. EPA. 1992b. "Vitrification Technologies for Treatment of Hazardous and Radioactive Waste," EPA/625/R-92/002, May.

Wagner, Julie. 1997. "New and Innovative Technologies for Mixed Waste Treatment," prepared for U.S. EPA Office of Solid Waste, Permits and State Programs Division. Available at www.epa.gov/radiation/mixed-waste/docs/innotech.

CHAPTER 7

INNOVATIVE STRATEGIES IN REMEDIATING MINING WASTES

Andy Davis

Geomega, Boulder, Colorado

Cynthia Moomaw

Geomega, Boulder, Colorado

George Fennemore

Geomega, Boulder, Colorado

Randy Buffington

Placer Dome, Bald Mountain Mine
Elko, Nevada

Approximately 2670 million metric tons of metal ore (e.g., copper, gold, iron, zinc) was handled at surface and underground mines in the United States in 1998, with surface mines accounting for 98 percent of the total (USGS, 2000a). Production in the metal industry accounted for 9790 million dollars in 1998, which was 0.1 percent of the gross domestic product (USGS, 2000b). The United States is second only to Africa as the world's largest producer of gold and second to Chile as the world's largest producer of copper (USGS, 2000b). Gold production in 1999 was dominated by mines located in Nevada and California, with a combined production accounting for nearly 80 percent of the U.S. total. The value of mine production in 1999 was approximately $3.1 billion.

Mining and mineral recovery generates many wastes. Mining overburden (material removed to gain access to the ore body) is excluded by EPA as a waste because it is not considered a discarded material within the scope of the Resource Conservation and Recovery Act (RCRA), and is thus exempt from Subtitle C [261.4(b)(3)].

Mining wastes fall into three categories, including extraction, beneficiation, and mineral processing. Extraction, beneficiation, and 20 specific mineral processing wastes are exempt from RCRA Subtitle C regulation [261.4(b)(7)] according to the Bevill Amendment. Other mineral processing wastes are regulated as hazardous waste if they exhibit one or more RCRA characteristics.

The scope of RCRA as applied to mining waste was amended in 1980 by the Bevill Amendment, Sec. 3001(b)(3)(A), which states that "solid waste from the extraction, beneficiation, and processing of ores and minerals" is excluded from the definition of hazardous waste under Subtitle C of RCRA [40 CFR 261.4(b)(7)]. After studying the wastes in accordance with RCRA Sec. 8002(f) and (p), the EPA concluded in 1985 that regulation of extraction and beneficiation wastes under Subtitle C was not appropriate, primarily because of the large volumes (U.S. EPA 1994, U.S. EPA 1999a). They further concluded that a wide variety of existing federal and state programs already addressed many of the risks posed by extraction and beneficiation of wastes.

A variety of wastes and other materials are generated and managed in mining operations, including waste rock piles or dumps, tailings ponds, spent ore piles, and various mine waters generated from dewatering activities or leach solutions. Although these are managed as wastes, some may be used for other purposes such as construction and foundation or cover materials. Solutions used to leach ore may be reused within the mill circuit or other location, and recycling may be conducted to recover additional mineral value. Recycling of secondary material is excluded from Subtitle C [261.4(a)(17)] if these wastes are managed in containers, tanks, containment buildings, or on approved pads.

From 1998 to 1999, 12 gold mines were closed, 2 new gold mines were opened, and 1 gold mine was reopened in the United States (USGS, 2000b). Successful closure and remediation of mining facilities, some of which are over 100 years old, can be a daunting task in the twenty-first century regulatory arena. Indeed the art of the exercise is to optimize closure such that a long-term benefit accrues, while controlling expenditures, and ideally incurring costs during mining. In theory, such a management approach would result in the last ton of ore being milled concurrent with environmental restitution.

Clearly, there are practical impediments to this Orwellian prospect; nevertheless, closure has become an integral element of mining that more companies proactively plan during the mine life itself. The goal of this chapter is to describe recent advances in closure, and in particular the incorporation of risk-based assessments, already well-accepted in developing remedial alternatives at other industrial sites, to mining facilities.

There are many issues that affect cleanup standards, including derivation of an appropriate background in local soils and groundwater around the mine. Indeed, the very nature of a metal-based mineral deposit suggests that natural background metal concentrations in mine area media will be higher than in a nonmineralized area. Identification of the appropriate site baseline for impacted media allows delineation of the area that requires attention, and helps selection of the remedial goals and realistic standards to be achieved by the remedial activity.

Within a mine site, there are multiple facilities with unique characteristics that require tailored approaches to closure. For example, there are always waste rock dumps, some of which may be oxide and as such easily reclaimed, while others may be acid generating. At older facilities, some dumps straddle drainage features, resulting in the generation of low-pH leachate ponds at the dump toe. In groundwater-impacted systems, natural attenuation of a variety of solutes released from impoundments may be a viable strategy if there is a substantial depth (e.g., several hundred feet between ground surface and the first usable water). This option is enhanced if there are subjacent limestone units that neutralize low-acid mine drainage emanating from pyritiferous waste rock dumps. As an example, we describe steps taken to successfully close the low-pH Intera Pond in the Robinson Mining District that incorporated active remediation in conjunction with natural attenuation.

At some mines, the ore may have been milled, generating an alkaline sand-sized slurry that was sent to tailings ponds, while at other mines low-grade ore has been leached with cyanide solutions (generally on containment) to recover the gold. At the conclusion of economic leaching activity, it is necessary to close these facilities by rinsing the heap with fresh water, applying the draindown leachate in a land application cell, then crafting a final solution, e.g., a passive wetland to provide a final polishing step. The Bald Mountain No. 1 Pad, in conjunction with the Red Springs and Buckhorn bioreactor provide examples of these closure strategies.

Recent improvements in mining technology have allowed economic beneficiation of low-grade disseminated deposits (<0.2 oz Au/ton) resulting in excavation of large pits up to 1.5 miles in diameter. In Nevada, it is estimated that there will be up to 35 pits that will result in formation of pit lakes of varying water quality as the ground watertable rebounds following cessation of pumping (Shevenell et al., 1999; Davis and Eary, 1997; Miller et al., 1996). Of the pits that have formed pit lakes over the last 5 years, a few have generally good water quality, although most have at least one or two parameters exceeding some numerical water quality standards and a few have poor water quality (Davis and Ashenburg, 1989).

To date there have been no closures of the large pit lakes (i.e., those with more than 1 billion gallons of water). Recently, in-pit translocation of waste rock has been proposed to reduce the size of the surface impoundments with the environmental benefits of reduced surface disturbance and a reduction in size of the pit lake. However, it is necessary to evaluate the effect of waste rock on the pit lake, using a method akin to that described here for a pit lake in the Western United States.

7.1 DELINEATION OF BACKGROUND SOIL METAL CONCENTRATIONS

Definition of natural background metal concentrations is imperative in understanding mining-related impacts. There have been many studies that have used the naturally enriched soil-metal concentrations found in sediments and soils around mining targets to focus further investigations into the nature and extent of the mineralized deposit (Friedrich et al., 1984; Learned et al., 1985; Ribeiro, 1979). Mineralized areas in mining districts naturally contain elevated concentrations of many metals, which must be taken into account to assign realistic background levels and hence comparative standards against which these concentrations can be compared.

There have been numerous studies identifying natural background conditions, but fewer dealing explicitly with potential or active mining properties, and to our knowledge, even fewer where a database consisting of thousands of samples has been assembled. At the Bald Mountain Mine in Nevada, antimony (Sb) and arsenic (As) have been analyzed in over 10,000 shallow (0 to 25 cm) soil samples (Fig. 7.1). Because of the number of samples, the greater sample population can be subdivided

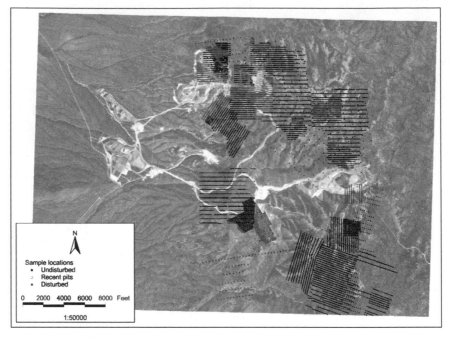

FIGURE 7.1   Soil samples in the vicinity of the Bald Mountain Mine.

into subareas, including (1) undisturbed by anthropogenic activities (referred to as *natural background*); (2) disturbed, nonexcavated areas, primarily comprising waste rock dumps (referred to as *ambient background*); and (3) mineralized areas, comprising recent mine pits (referred to as the *mineralized background population*). Within each class, there are sufficient samples to assess the regional background concentrations of As and Sb in the vicinity of the mine.

The data set demonstrates that As and Sb concentrations increase from undisturbed to mineralized populations for both metals (Figs. 7.2 and 7.3), and that the natural background levels of As and Sb in undisturbed soils at Bald Mountain are higher than from unmineralized locations in Nevada. For example, average Sb in 5829 undisturbed soil samples was 16 mg/kg compared to the normal range of Nevada soils of <1 to 10 mg/kg (Shacklette et al., 1984). Similarly, average As in 3490 undisturbed soil samples was 120 mg/kg compared to the normal range of Nevada soils of 0.1 to 100 mg/kg (Shacklette et al., 1984). On-site soil cover material suitable for closure of a heap leach pad contained average As and Sb (67 mg/kg and 3.3 mg/kg, respectively) above regional background soil conditions. However, these concentrations are well within the range of these metals occurring naturally in undisturbed soils in the Bald Mountain area. Thus, despite the As and Sb concentrations, the presence and potential use of these soils as closure cover materials are consistent with the ambient local condition.

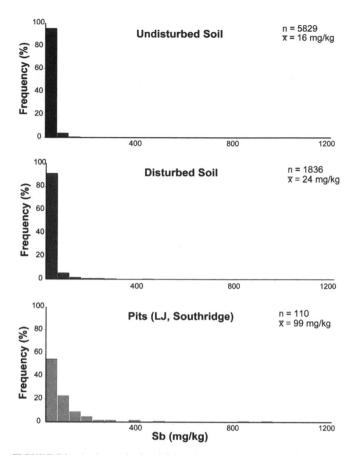

FIGURE 7.2   Antimony in the vicinity of the Bald Mountain Mine.

7.2   IDENTIFYING BACKGROUND GROUNDWATER METAL CONCENTRATIONS

Several studies have shown that groundwater in mining areas may have naturally elevated metal concentrations above nonmineralized areas (Bowers, 1996; Gorett, 1983; Miller and McHugh, 1999; Rose et al., 1979; Runnells et al., 1998). Recently a detailed investigation was conducted to determine local background groundwater conditions in the Robinson Mining District, a historical mining area operated for over 100 years.

Water samples were collected from nine surface water locations, including six waste rock seep/pond locations, three pit lakes (Ruth, Kimbley, and Liberty), and 24 groundwater monitoring wells. The surface water and groundwater sample locations

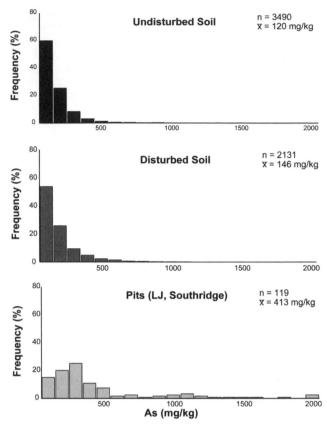

FIGURE 7.3   Arsenic in the vicinity of the Bald Mountain Mine.

were selected because they represented each major hydrolithologic province in the Robinson Mining District, as well as strategic locations with respect to potential sources of mining-related impacts (e.g., waste rock seeps and pit lakes) and potential groundwater migration pathways. Traditional major cations, metals and anions, trace and precious metals, rare earth elements (REEs), and selected stable isotopes ($^{18}O/^{16}O$, ^2H-deuterium, $^{13}C/^{12}C$, $^{34}S/^{32}S$, and tritium ^3H) were analyzed. Groundwater from Murray Spring (the town of Ely's water supply) and two temporary water supply wells (GQ-1 and GQ-2) located in a different lithologic setting and over 2 miles south of the Robinson Mining District are representative of background conditions for non-mineralized groundwater in the area.

7.2.1   Geochemical Characterization

Geochemical evaluation of the groundwater and surface water chemical data was undertaken to characterize site water quality, and to identify trace metal and rare earth elements (REEs) that could be used as indicator elements or tracers for evaluating the potential influence of mine-impacted waters along known and suspected groundwater flow pathways.

Several geochemical and statistical approaches were utilized in the evaluation, including graphical techniques (Piper diagrams, ternary plots, and quaternary plots) that provided three-component and four-component characterization of site constituents. In addition, statistical evaluations (cluster analyses) were used to determine tracer and indicator elements for the Robinson Mining District. The results of the geochemical data analysis were integrated with site hydrological data to elucidate groundwater flow paths and potential mine-related impacts (Fig. 7.4a).

Stable Isotopes (^2H and ^{18}O).   The stable isotopes of hydrogen (^2H or deuterium) and oxygen (^{18}O) were analyzed because of their propensity for fractionation as a result of hydrologic processes operating at the earth's surface. Fractionation between the isotopes of hydrogen (^1H and ^2H) and oxygen (^{16}O and ^{18}O) results from mass differentiation between each isotopic pair by physical processes, such as evaporation, that act to preferentially enrich a medium in one isotope relative to the other isotope. The difference in ratios between isotopes of a sample relative to a standard is expressed as the delta (δ) notation in units of parts per thousand, or per mil ($^0/_{00}$), so that the deuterium content of a sample is expressed as δD, and the ^{18}O content is expressed as $\delta^{18}O$.

The groundwater and surface water δD versus $\delta^{18}O$ data were plotted relative to the world meteoric water line (MWL) that establishes the ratio of δD to $\delta^{18}O$ for waters of meteoric origin (Craig, 1961). Precipitation at Robinson Mining District (labeled RMD meteoric) plots directly on the MWL (Fig. 7.4), thus tying the analyses at Robinson to other investigations performed around the world. Waters that plot to the right of the MWL indicate evaporative loss, resulting in enrichment of the remaining water with the heavier isotope of oxygen (^{18}O) relative to the lighter isotope. Evaporation also causes a slight upward shift in the δD value (indicating enrichment in D) relative to the meteoric water sample, as the lighter isotope of hydrogen is preferentially removed with water vapor.

The δD-$\delta^{18}O$ plot identified several obvious groupings of water based on similar δD and $\delta^{18}O$ ratios, and in doing so identified which water bodies are undergoing similar hydrologic evolution. For example, water samples that plot close to the Robinson District meteoric water point are derived from local meteoric precipitation, and are grouped as *meteoric recharge* (Fig. 7.4b). These include all groundwater samples, Riepetown seep, and Murray Spring water. The meteoric origin of these waters is consistent with their water quality, characterized by generally low total dissolved solids content, and indicates that background or near-background water quality conditions exist at these locations.

In contrast to the meteoric-derived δD-$\delta^{18}O$ relationships observed for background locations, waters that may have been impacted by Robinson District mineralization and/or anthropogenic activities are enriched in δD and $\delta^{18}O$ to a degree that clearly separates them from other water sources. These include all pit lake samples and waste rock seeps. For example, Mollie Gibson Seep (see the *moderate evaporation* group) appears to represent water that is intermediate between the upgradient Ruth Pit water and meteoric recharge.

Indicator Elements and Tracers.   Ideally, analytes used for source discrimination (tracers) must be above analytical method detection limits, and migrate nonreactively in the subsurface. In addition, indicator elements may be unique to a particular water type, but may not be useful tracers because of their immobility downgradient from a

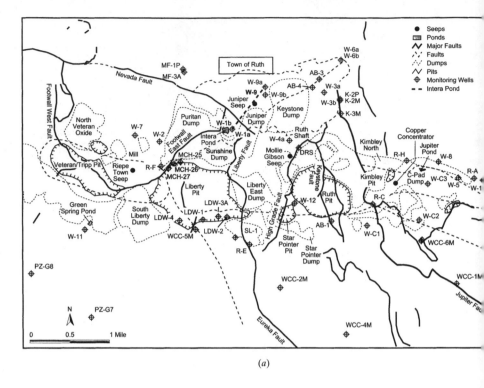

(a)

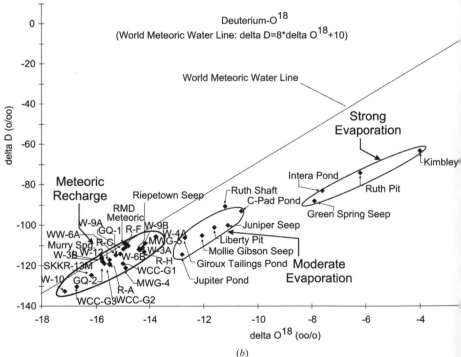

(b)

FIGURE 7.4 Robinson Mining District. (a) Sample locations and the Intera Pond. (b) Isotopic groupi[?] Robinson District waters.

source due to geochemical constraints (e.g., solubility). For example, evaluation of REE geochemistry determined that, while the REEs are diagnostic indicators, they are of less use as tracers because reactivity in the subsurface reduces their solubility through precipitation as REE-carbonate phases. However, such elements are useful for "fingerprinting" sources of mining-related impacts, because individual seeps and pit lakes have distinct REE concentrations depending on the local lithology and mineralogy. On the other hand, sulfate is a conservative tracer, but it is common to most mineralization, and, therefore, its source at Robinson is nonunique. However, viable indicator (fingerprint) compounds were identified at Robinson in reference to sulfate concentrations along potential mine-impacted hydrologic and hydrogeologic flow pathways.

Three indicator elements, rhenium (Re), scandium (Sc), and rubidium (Rb), that are both soluble and consistently elevated above method detection limits were compared with sulfate concentrations (a conservative but ubiquitous tracer in the Robinson District) downgradient from a mine-impacted source area (Intera Pond) to assess their relative transport characteristics and use as indicator elements.

Re, Sc, and Rb form stable aqueous complexes in the presence of sulfate. Therefore, the extent of mine-related impacts can be determined by measuring the concentrations of these elements compared to background, and specific source areas can be identified by comparing their relative concentrations to established source fingerprints.

A fourth element, barium (Ba), was also utilized in the analyses because it is insoluble in the presence of sulfate. Barium precipitates in the presence of sulfate to form barite (Fig. 7.5), which is extremely insoluble (e.g., 0.02 mg/L Ba^{2+} at SO_4^{2-} concentrations of 100 mg/L). The precipitation of barite controls the concentrations of barium, because on a molar basis, sulfate concentrations typically exceed barium in mine-impacted groundwater. Hence the absence of barium, together with excess sulfate, may be indicative of anthropogenic impacts. Conversely, background groundwater locations tend to have detectable barium because of the relatively low concentration of sulfate relative to mine-impacted locations.

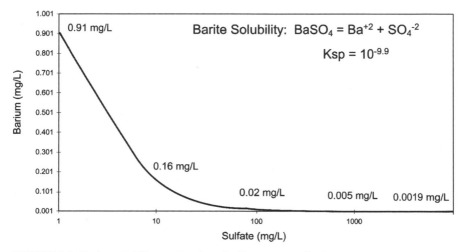

FIGURE 7.5 Barium solubility as a function of sulfate concentration in pure water.

7.2.2 Piper Diagrams

The spatial relationship between selected surface water and groundwater chemistry is typically interpreted on the basis of major cation and anion ratios (Fig. 7.6). At Robinson, Piper diagrams failed to provide the explicit resolution necessary to fingerprint individual water sources at the site. However, they do exhibit broad trends in major element chemistry that contribute to the geochemical evaluation. For example, mine-impacted waters (Group 2 including Intera Pond, Green Springs Pond, Ruth Pit, Jupiter Seep, Kimbley Pit, Mollie Gibson Seep, and Liberty Pit) have similar ratios of major cations and anions with subtle variations (Fig. 7.6). In general, these surface waters may be characterized either as Mg-SO$_4$ waters (Intera Seep, Jupiter Seep), Ca-SO$_4$ waters (Liberty Pit, Ruth Pit, Kimbley Pit), or Mg-Cl waters

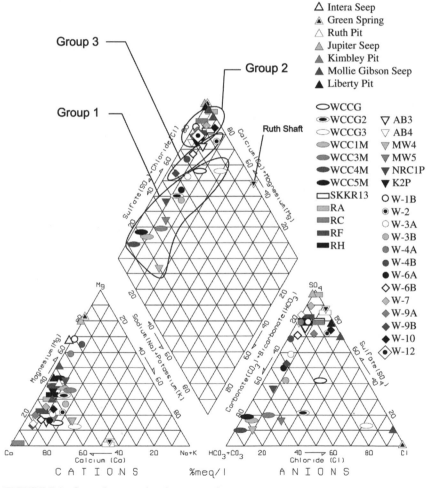

FIGURE 7.6 Groundwater major element results.

(Green Spring and Mollie Gibson Seep). This diagram shows that these waters can be differentiated on the basis of variations in the cation ratio (Ca versus Mg) and anion ratio (Cl versus sulfate), but specific differentiation between potential mine-impacted water is not possible with Piper diagrams alone.

7.2.3 Ternary and Quaternary Diagrams

Ternary diagrams were used to graphically evaluate the relationship between selected trace metals and REE in site surface waters and groundwaters at the Robinson Mining District (Fig. 7.7). Barium, Sc, Re, and Rb were chosen on the basis of statistical criteria, as elements useful to fingerprint the potential source relationships for site waters. Background groundwaters are generally associated with high proportions of Ba; waste rock seeps and ponds with high proportions of Sc; and pit lakes with high proportions of Rb and Re. To better distinguish affiliations of waters in the center of the ternary diagram, a quaternary plot was constructed (Fig. 7.8) using the Rb, Ba, Sc ternary as the base, and the Re molal percentage as the vertical axis. This four-component representation allowed further discrimination between waters.

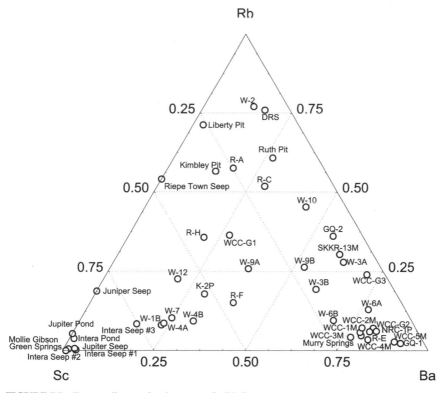

FIGURE 7.7 Ternary diagram for the system Sc-Rb-Ba.

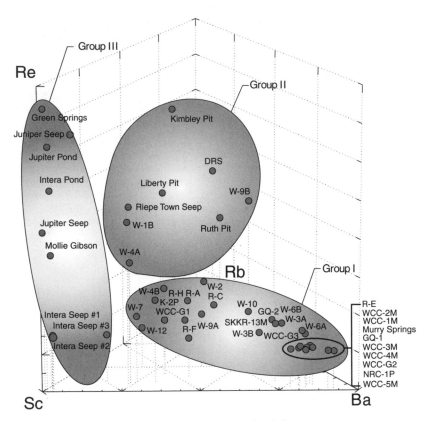

FIGURE 7.8   Quaternary diagram for the system Re-Sc-Rb-Ba.

7.2.4 Distribution Analysis

Solute concentrations in the Robinson Mining District groundwater and surface water were analyzed graphically to determine their underlying statistical distributions. On the basis of this analysis, the underlying distribution of the concentrations is best characterized as lognormal, superimposed with pronounced bimodal or trimodal populations that are indicative of potentially multiple background and mine-impacted water chemistries. For example, the sulfate distribution shows three distinct groups (Fig. 7.9). There is one background population with natural log concentrations from 0 to 3.5 ($\approx$ 33 mg/L), one from 3.5 to 7.2 (33 to 1300 mg/L) in mineralized areas, and one group >7.2 (>1300 mg/L) corresponding to pit lake and waste rock related sources.

In the mineralized zone there are naturally occurring sulfate concentrations. Of particular interest is monitoring well W-12, located in the mineralized zone of Ruth

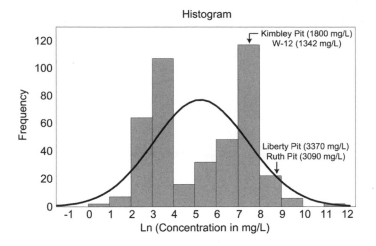

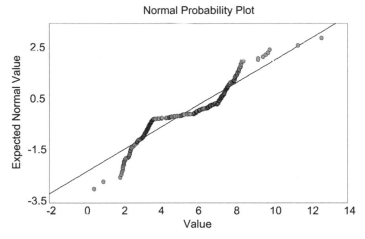

FIGURE 7.9 Natural logarithm distribution of sulfate concentrations in RMD surface and groundwaters.

Pit. The concentrations of calcium and sulfate in W-12 and Ruth Pit are similar, with concentrations from W-12 falling within the pit lake group. However, W-12 has a copper log concentration of -5.74 (≈ 0.007 mg/L) consistent with background, whereas Ruth Pit Lake contains 16 mg/L (Fig. 7.10). These data, together with ambient sulfide mineralization and the groundwater flow direction toward the pit lake, suggest that sulfate concentrations in W-12 are due to the formation mineralogy, rather than to migration of Ruth Pit Lake into the surrounding groundwater.

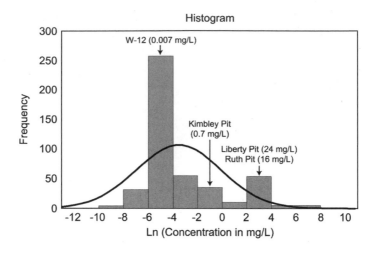

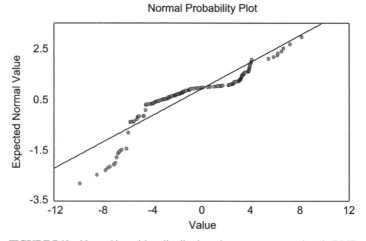

FIGURE 7.10   Natural logarithm distribution of copper concentrations in RMD surface and groundwaters.

7.3   *CLOSURE AND RECLAMATION OF HISTORIC WASTE DUMPS AND SEEPAGE POND*

Two years prior to ceasing active mining operations of a process unit, the operator is requested to develop a closure plan to initiate final closure and reclamation of the site to natural environmental conditions that are consistent with the site's premining land use. The objective of the closure is to comply with state and federal regulations and guidelines that include water management to "prevent the degradation of waters of the state," to provide "stabilization" of spent ore and tailings, and to prevent any "unnecessary or undue degradation" of public lands and associated area resources. In essence, the goal of the closure plan is to minimize or eliminate all potential contaminant pathways to area resources, in a manner that will minimize long-term obligations for active management or maintenance of facilities, equipment, or restricted land uses.

The following case study summarizes the technical work conducted to support closure of a copper oxide leaching operation in Nevada.

7.3.1 Case Study Background—Intera Pond

The Intera Facility, located ¾ mile south of Ruth, Nevada (Fig. 7.4a), consists of three leach dumps (Puritan, Sunshine, and Juniper), Intera Pond, and a solution pumping station. The dumps form a U shape that lies across the largest premining drainage in the Robinson Mining District. The combined Puritan, Sunshine, and Juniper waste rock dumps contain approximately 19 million yd^3 of waste rock, and cover approximately 78 acres to a height of 150 ft. Historically, acid leaching for recovery of copper values was performed on the Puritan and the Sunshine waste rock dumps, which consisted of run-of-mine, end-dumped material containing a significant amount of silt and clay. Infiltration galleries were constructed on top of the dumps to enhance infiltration of acid leach solutions. After leaching was discontinued, these galleries continued to facilitate infiltration of stormwater.

It is not clear when Intera Pond was commissioned for in-dump leaching of oxide copper; however, it is apparent that significant leaching activities did occur in the 1950s through 1970s. A copper precipitation plant operated at the current Intera Pond site from the early 1950s through the late 1960s. The 2½- to 3-acre Intera Pond was constructed to collect leachate from the Puritan and Sunshine dumps, retaining approximately 8 to 10 million gallons. The surface leach dump catchment area of approximately 280 acres contributed recharge to Intera Pond. Leachate draining from the topographic low ephemeral stream valley collected in the pond, where it was subsequently transferred to a solvent extraction/electrowinning recovery plant. Apparently because of uneconomic recovery conditions, operations at the site were discontinued. The pond was constructed by placement of a fill embankment across the existing topographic drainage. A compacted clay layer was placed in the pond, which is underlain by upturned middle Devonian Guillemette limestone with zones of perched water at a depth below 60 ft and groundwater at a depth of 700 ft.

The objective of the closure was to minimize infiltration of precipitation through the dumps and subsequent generation of acidic leachate, thus preventing potential adverse impacts to waters of the state. This objective was to be achieved in the short term by installing a system to actively measure, control, and contain seepage from the leach dumps in the Intera Pond area. The low pH solutions from Intera Pond would be incorporated into the mill circuit at a rate such that discharge limits to the tailings facility would be met. Over the long term, the objective will be to mitigate seepage by a waste rock management program that would incorporate stormwater run-on/runoff provisions. With the run-on/runoff controls, seepage from the leach dumps would eventually discontinue so that recovery and disposal of acidic Intera Pond water would no longer be necessary. Additionally, an evaluation of potential past and future impacts to groundwaters of the state was required for the closure.

The closure plan included placement of 30 to 50 ft of nonreactive rhyolite over the pond area to inhibit infiltration into the pond area, followed by placement of waste rock from active operations until Intera Pond was incorporated into an integrated Puritan, Sunshine, and Juniper waste rock disposal area. The rhyolite provides alkalinity to neutralize infiltrating waters to the pond. The integrated waste rock dump area would be graded to promote runoff and minimize infiltration, percolation, and seepage. Eventually, the dumps would be revegetated to stabilize the soil and promote evapotranspiration of infiltrating meteoric water. Upgradient surface water in the historic drainage covered by the waste rock was diverted around the dumps by a drainage ditch.

7.3.2 Intera Pond Sediment Leachability

An estimated 2300 yd³ of sediment accumulated in Intera Pond. To ascertain both past and future potential for solute leaching from the sediments via meteoric water percolation, meteoric water mobility procedure (MWMP) tests were conducted. Three MWMP tests were performed, Column 1 with only sediment (solute generation only), Column 2 with sediments overlying limestone (current conditions), and Column 3 with sediments between rhyolite and limestone (future mitigation scenario).

The MWMP is a 24-hour test, with fluid throughput volume equivalent to the dry mass of solids in the column. The run times for Column 2 (current conditions) and Column 3 (future conditions) were extended to better represent long-term potential leaching of Intera sediments and performance of the underlying bedrock and overlying rhyolite. These two column tests were continued until geochemical equilibrium was achieved.

The MWMP test on sediments indicated that, although contaminants may be leached from the sediment at concentrations above the MCLs (Fig. 7.11), addition of the underlying limestone and superimposed rhyolite cover would mitigate any acid and solutes generated by the sediments. This form of the test was deemed acceptable by the Nevada Department of Environmental Protection.

7.3.3 Demonstration of Monitored Natural Attenuation

Further evaluation of potential groundwater impacts and solute attenuation capabilities of the Devonian Age Guillemette limestone underlying Intera Pond was performed by installing two temporary monitoring wells through the embankment adjacent to the eastern edge of the pond. Well 1a was screened at the first encounter with subsurface water, at a depth of approximately 60 ft below ground surface (bgs), while Well 1b was completed at approximately 140 ft bgs within a lower zone of subsurface water.

Comparing analytical water quality results from Wells 1a and 1b with Intera Pond water chemistry (Table 7.1) suggests that historical vertical seepage from the base of Intera Pond has occurred, and that significant solute attenuation has occurred via water-rock interactions with the Guillemette limestone bedrock beneath the pond. This hypothesis is supported by the increase in pH and decrease in reactive solute concentrations (e.g., copper, iron, and sulfate) observed from Intera Pond to the perched subsurface water zone.

Attenuation of inorganic solutes (e.g., base metal cations) in subsurface environments occurs principally by precipitation, coprecipitation, and/or adsorption processes (Langmuir, 1997). Solid phases precipitate in response to a change in pH that occurs when an acidic solution is neutralized by an alkaline solution (high pH) or by a neutralizing solid phase such as calcium carbonate, which is abundant in the limestone aquifers at the Robinson Mining District. Neutralization of the acidic solution has caused precipitation of secondary solid phases (hydroxides and carbonates), reduced

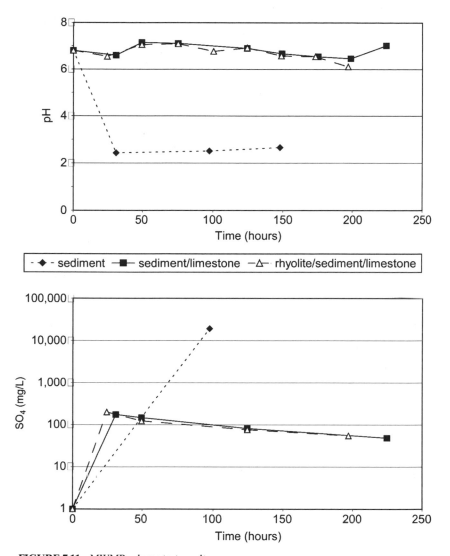

FIGURE 7.11   MWMP column test results.

metal solubility, and thus decreased solute concentrations. In addition, the presence of highly reactive minerals with large surface area per unit mass such as amorphous ferric hydroxide [$Fe(OH)_3$], and oxyhydroxides of aluminum and manganese also lower solute concentrations by providing reactive surfaces on which dissolved solutes (e.g., copper) are rapidly sorbed (Parkhurst, 1995; Dzombak and Morel, 1990). Removal of solutes by solid solution and replacement reactions (e.g., replacement of Cd^{2+} for Ca^{2+} in calcite) are also common attenuation reactions when reactive carbonate and hydroxide solids are present (McBride, 1994).

TABLE 7.1 Water Chemistry in Intera Pond and Wells 1a and 1b
(Concentrations in mg/L unless otherwise noted; pH in standard units)

Parameter	MCL	Intera Pond	Well 1a	Well 1b
Aluminum	0.05–0.2	918	0.299	6.85
Arsenic	0.05	0.014	<0.005	<0.005
Barium	1	<0.009	0.042	0.085
Cadmium	0.005	0.454	0.005	0.123
Calcium	N/A	580	669	865
Chromium	0.1	0.263	<0.005	0.074
Copper	1.0	467	0.48	8.75
Fluoride	2	52.8	1.25	15.89
Iron	0.3–0.6	582	0.03	0.12
Lead	0.015–0.05	0.024	0.14	0.1
Magnesium	125–150	780	1200	760
Manganese	0.05–0.1	231	45.3	143
Mercury	0.002	N/M	N/M	N/M
Potassium	N/A	<2.3	8.82	3.76
Selenium	0.05	0.083	0.02	0.016
Silver	0.05	0.042	0.01	0.009
Sodium	N/A	43.8	57	39
Zinc	5.0	103	3.23	51.9
Alkalinity	N/A	N/M	1244	613
Chloride	250–400	52.7	35.2	32.9
$NO_2 + NO_3$ (as N)	10.0	<2.5	N/M	N/M
pH	6.5–8.5	2.65	6.52	6.34
Sulfate	250–500	11,600	4720	4364
TDS	500–1000	19,800	6316	6356
WAD CN	0.2	N/M	0.017	0.008

N/A = not applicable.
N/M = not measured.
TDS = total dissolved solids.
WAD CN = weak acid dissociable.

Core material was obtained from Well 1b and examined by electron microprobe analysis (EMPA). EMPA was used to determine if secondary metal-bearing precipitates were present, the existence of which would indicate the occurrence of attenuation reactions. Geochemical modeling of the solution chemistry of Intera Pond and the groundwater chemistry at downgradient Wells 1b, 9a, and 9b was also performed by using the established geochemical code PHREEQC (Parkhurst, 1995) to determine if the precipitation of secondary solid phases was thermodynamically reasonable. The precipitation of secondary solids is favored when the saturation index for a mineral is greater than 0.0 (Parkhurst, 1995).

Saturation index values indicate that chemical conditions are favorable for the precipitation of several metal phases in Intera Pond and in the aquifer downgradient of the pond, including phases containing Al, Ba, Ca, Cd, Cr, Fe, K, Mn, Na, Sb, and Sr (Table 7.2). For example, geochemical model results indicate that aluminum concentrations in Intera Pond are controlled by the precipitation of aluminum sulfate ($AlSO_4$), demonstrated to precipitate in void space as limestone dissolves in response to acidic percolation (Fig. 7.12a). The absence of aluminum in groundwater downgradient of Intera Pond supports the model prediction, indicating that aluminum is actively being removed and is not migrating downgradient in groundwater. EMPA photomicrographs of core material downgradient of Intera Pond (Figs. 7.12b and c) demonstrate the presence of a mixed Al-Fe-Cu sulfate precipitate that supports the model prediction. The presence of copper could be the result of adsorption or coprecipitation to Fe and Al oxide surfaces. The net result is a decrease in copper concentration from 470 mg/L in Intera Pond to <0.2 mg/L in Well 1b.

TABLE 7.2 Mineral Saturation Indices for Intera Pond
(Computed by using PHREEQC)

Mineral	Intera Pond	Well 1a (60 ft)	Well 1b (140 ft)
$Al_1(OH)_{10}SO_4$	−10.7	4.3	3.6
$Al(OH)_3$	−5.6	0.15	−0.12
Gypsum	0.36	0.32	0.45
Ferrihydrite	−8.7	−1.3	−1.2
Fluorite	−6.3	−1.1	0.2
Calcite		0.36	0.01

Similar results were obtained for gypsum, goethite, and the H-, K-, and Na-jarosites (hydrated iron sulfates). Model results predict that Intera Pond and Well W 1b are supersaturated with respect to both solid phases. EMPA photomicrographs show the presence of abundant secondary gypsum mixed in with Al-Fe-Cu sulfate and FeOOH (goethite) phases (Fig. 7.12a) and associated with calcite, Al-SO$_4$, and Fe-bearing aluminum hydroxide [$Al(OH)_3$] (Fig. 7.12c). Model results also indicate supersaturation with respect to strontium sulfate ($SrSO_4$) and barite ($BaSO_4$), in addition to gypsum (based on the evident growth of euhedral crystals shown in Fig. 7.12d).

These geochemical controls contribute to decreases in sulfate concentrations along the potential flow path from Intera Pond (12,800 mg/L) to Well 1b (3200 mg/L). Sulfate is also weakly adsorbed by ferric hydroxides [e.g., $Fe(OH)_3$ and FeOOH] which may also contribute to sulfate attenuation (Dzombak and Morel, 1990). The EMPA provided important verification of phases identified as solubility controls in the geochemical modeling. Use of these phases provided a reasonable explanation for the rapid decreases in solute concentrations along the flow path from Intera Pond to Wells 1a and 1b. The collected data and modeling support the use of natural attenuation as a viable mechanism in preventing degradation of waters of the state and in closing the pond and associated dumps.

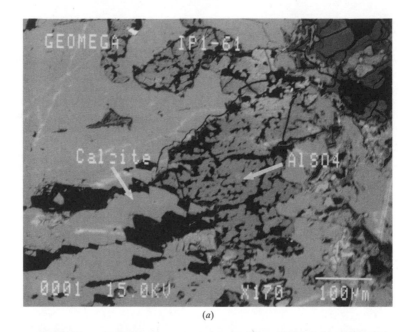

(a)

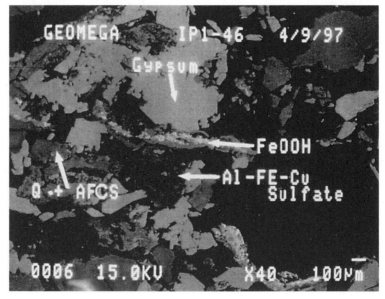

FIGURE 7.12 Photomicrographs of Guillemette limestone core drilled near Intera Pond, showing the presence of secondary precipitates of Al-Fe-Cu sulfate, gypsum, and hydroxide minerals.

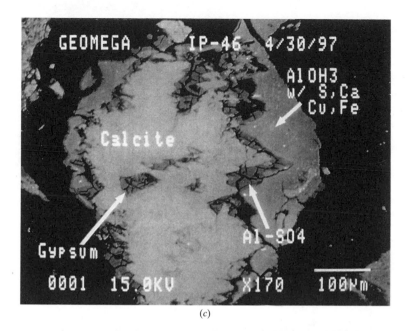

(c)

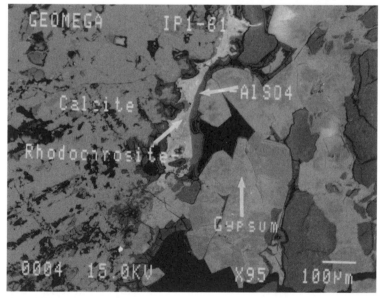

FIGURE 7.12 (*Continued*)   Photomicrographs of Guillemette limestone core drilled near Intera Pond, showing the presence of secondary precipitates of Al-Fe-Cu sulfate, gypsum, and hydroxide minerals.

7.4   CLOSURE AND RECLAMATION OF A HEAP LEACH FACILITY

The regulations and objectives for closure of a heap leach facility are the same as that described for the Intera Facility, namely to prevent degradation of waters of the state and prevent unnecessary or undue degradation of public lands and associated area resources. The key issues related to closure of heap leach pads are managing pad draindown, which may contain metals and cyanide in solution, minimizing future infiltration through heap leach pad materials, and preventing the exposure of heap material to human and ecological receptors.

Closure activities for a heap leach facility usually start with rinsing the pad to reduce cyanide and pH to prescribed levels. To satisfy regulatory requirements, it must be demonstrated that contaminants in the heap effluent will not degrade surface water or groundwater through modeling or similar studies. Field activities typically include regrading the heap to stabilize slopes against erosion and promoting runoff and covering the heap with a vegetated soil cover to reduce generation of heap effluent by minimizing infiltration of precipitation and maximizing evapotranspiration. Over the long term, the heap draindown is managed by methods such as land application and/or passive wetlands.

The following is a case study illustrating the scientific studies conducted to support the closure plan for a heap leach pad in Nevada.

7.4.1   Case Study Background—Bald Mountain

The Bald Mountain Mine is located approximately 100 miles northwest of Ely, Nevada, and directly south of the structurally continuous Ruby Range. The Bald Mountain mining district is the most northwestern component of the larger Alligator Ridge–Bald Mountain mining district, typified by disseminated gold deposits hosted by intercalated carbonate and siliciclastic strata of Lower to Middle Paleozoic Age (Hitchborn et al., 1996).

Processing at one of the heap leach pads, Pad 1, was discontinued and closure activities initiated. Pad 1 consists of 12 million tons of oxide ore and covers approximately 55 acres to a height of approximately 100 ft above the pre-existing ground surface (Fig. 7.13). The heap is composed of rock from the ore body, originating from fault-controlled oreshoots and disseminated and stockwork ores in Paleozoic sedimentary and late Jurassic intrusive rocks. The Bald Mountain mining district defines a southeastern continuation of the Carlin trend. All of the gold deposits in the Bald Mountain mining district have elevated arsenic and antimony, similar to those of Carlin and other sedimentary rock-hosted deposits of northern Nevada. This is illustrated in Table 7.3 by a comparison of Bald Mountain soils with typical Nevada soils, Western U.S. soils, and U.S. soils. Groundwater in the vicinity of Pad 1 is approximately 500 ft below the surface.

A geotechnical liner prevents direct contact of the pad and pad effluent with the subjacent soil material. Surface runoff and effluent from the pad is contained on the liner and collected by five ponds (Pond 1, Pond 2, Pond 3, the settling pond, and the barren pond).

7.4.2   Pad Rinsing

Subsequent to cyanide leaching to remove gold, the leach pad was rinsed from 1996 through 1999 with a solution containing heap effluent augmented with makeup water

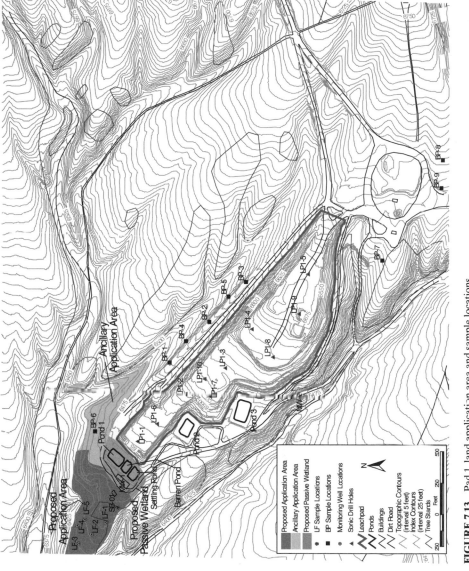

FIGURE 7.13 Pad 1, land application area and sample locations.

7.23

TABLE 7.3   Ambient Soil Chemistry in Pad 1 Area

(Concentrations in mg/kg)

Analyte	Bald Mountain soils	Nevada*	Western United States[†]	U.S. soils[‡]
Calcium	60,500	0.013–33	18,000	
Chloride	2.6			
WAD Cyanide	<0.5			
Fluoride	1.9			
Potassium	3,230	0.22–6.5	18,000	
Magnesium	5,530		7,400	
Sodium	451	0.3–10	9,700	
Nitrate + Nitrite as N	12.1			
Sulfate	12.1			
Silver	1.9			0.7
Aluminum	12,500	0.07->10	58,000	
Arsenic	69	0.1–100	5.5	7
Boron	20	<20–300	23	45
Barium	167	10–5,000	580	560
Beryllium	0.8	<1–15	0.68	1.6
Cadmium	<0.24			0.41–0.57
Chromium	10.1	1–2,000	41	50
Copper	16	<1–700	21	26
Iron	10,800	0.01->10	21,000	
Mercury	<0.1	<0.01–5.1	0.046	0.17
Manganese	303	<2–7000	380	490
Nickel	12.2	<5–700	15	18.5
Lead	15	<10–700	17	26
Antimony	0.4	<1–10	0.47	0.25–0.6
Selenium	<0.5	<0.1–5	0.23	0.31
Thallium	0.1		9.1	0.02–2.8
Zinc	36	28–3500	1.79	73.5

* Range of concentrations from Nevada soils. Values summarized from analyte maps in Shacklette and Boerngen, 1984.
[†] Shacklette and Boerngen, 1984. Values are from Table 2; they represent the geometric mean of samples collected west of the 96th meridian.
[‡] Kapata-Pendias, A., and H. Pendias. 1984, *Trace Elements in Soils and Plants,* CRC Press, Boca Raton, Fla. Values summarized from analyte-specific tables.

derived from groundwater. The rinsing was conducted to degrade cyanide to an acceptable level (<0.2 mg/L). The degradation results in a concomitant reduction in metals that are complexed by cyanide (e.g., Ni) while nitrate increases as a result of the cyanide breakdown. Other metals, such as arsenic, tend to stay the same or slightly increase. For Bald Mountain, cyanide, mercury, and nickel were initially above the Nevada DEP standard in June of 1998, but decreased to near or below their respective standards in the summer of 1999, and did not change significantly after (Fig. 7.14). Other analytes, such as antimony, arsenic, nitrate, selenium, sulfate, and pH, remained above the standards and did not change significantly after June of 1998 (Figs. 7.15 and 7.16). Therefore, the temporal solute data indicate that the effluent chemistry has generally reached steady state and that further rinsing would provide little incremental benefit. The chemistry of the pad effluent in the future was also determined, as discussed later.

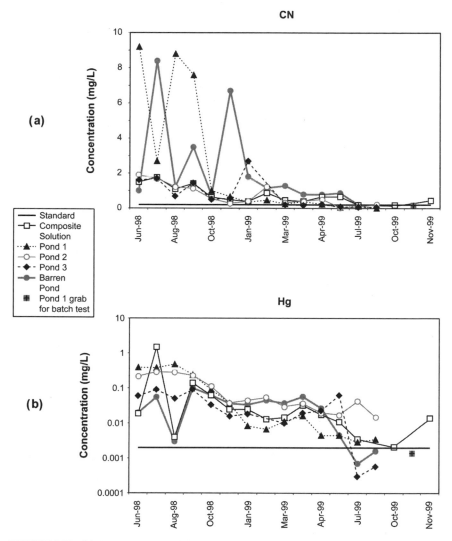

FIGURE 7.14 (*a*) Aqueous cyanide; (*b*) mercury concentrations in ponds. (Refer to Fig. 7.13 for locations.)

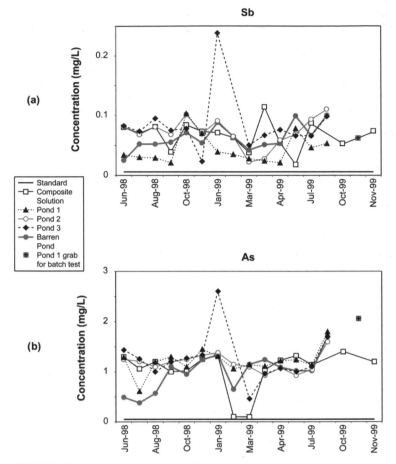

FIGURE 7.15 (*a*) Aqueous antimony; (*b*) arsenic concentrations in ponds.

7.4.3 Leach Pad Characteristics

The leach pad must be characterized to determine the volume of water that will drain from the heap over time and the mass of solutes available for leaching so that disposal options can be investigated. The most important factors influencing draindown and chemistry are the pad lithology and the bulk chemistry and acid-generating potential of the pad material.

Lithology. The pad lithology, determined from borehole samples, is characterized by interbedded layers of gravel, loamy sand, sandy loam, and sandy clay loam (Table 7.4). In general, the pad is coarse-grained and permeable. The moisture content of the material was generally low, from 2.8 to 12.4 percent, with perceptible interstitial water saturation observed at only 5 of the 48 sample locations. Free water that could

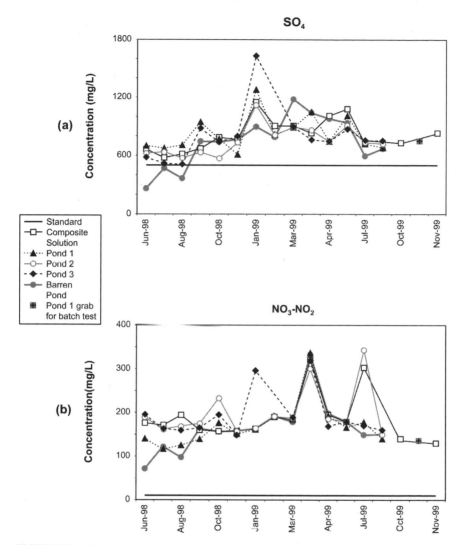

FIGURE 7.16 (*a*) Aqueous sulfate; (*b*) nitrate-nitrite concentrations in ponds.

be readily drained from the heap material was observed at one location and was sampled to determine the partitioning between the solid and aqueous phases.

The relatively coarse grain sizes and low moisture content indicates that pad material is well drained, which is consistent with operational records that show pad influent and effluent rates varying synchronously during the heap rinsing.

Bulk Chemistry. The bulk solids chemistry of the pad was quantified for total metals and major ions. The bulk chemistry of the heap materials (Table 7.5) was mea-

TABLE 7.4 Grain Size of Heap Leach Material

Sieve number	Grain size, mm	Percent retained		Description
		LP1-1-20	LP1-8-30	
5	>4	87	58	Gravel
10	2–4	6	6	Coarse sand
18	1–2	3	6	Medium sand
35	0.5–1	2	7	Medium sand
120	0.125–0.5	1	22	Fine sand
230	0.063–0.0125	0	1	Fine sand
	<0.063	0	0	Silt and clay

TABLE 7.5 Summary of Bulk Chemistry Analytical Results for Pad 1 Material*
(Concentrations in mg/kg)

Analyte	Average	Standard deviation	Minimum	Maximum
% solids	91.7	3.4	85.1	98.9
TOM[†]	0.2	0.3	0.1	1.9
Ca	49,613	53,386	1,100	226,000
Cl	13.2	5.0	5.5	22.9
CN	1.5	2.1	0.5	10.5
F	2.8	2.3	1.0	10.0
K	1,798	1,510	300	6,330
Mg	6,075	8,086	250	52,400
Na	396.2	386.2	73.4	1,700.0
NH_3-N	2.2	2.5	0.6	14.0
SO_4	765.4	828.8	104.0	3,580.0
Ag	2.6	2.2	0.6	16.0
Al	5,928	4,358	1,200	21,500
As	1,367	1,212	170	5,440
B	8.0	2.5	3.6	15.5
Ba	128.2	124.3	40.4	746.0
Be	0.4	0.3	0.2	1.3
Cd	0.4	0.5	0.2	2.8
Cr	15.7	11.0	3.1	42.4
Cu	129.0	203.2	16.4	1,060.0
Fe	24,655	9,068	7,960	49,900
Hg	2.9	4.3	0.2	26.3
Mn	331.2	185.7	6.2	682.0
Ni	14.2	10.2	2.4	39.7
Pb	43.4	46.8	5.4	299.0
Sb	46.1	62.6	3.2	335.0
Se	4.4	1.9	4.0	14.4
Tl	42.8	55.5	10.0	250.0
Zn	148.4	177.9	12.4	732.0

* Summary statistics include results for all locations and samples.
† Total organic matter.

sured to determine future effluent water quality by characterizing the masses of potential solutes remaining within the pad. High concentrations of several constituents of potential concern were identified in the pad material, in particular arsenic at an average concentration of 1367 mg/kg.

Although individual locations within the pad vary in concentration by more than an order of magnitude, depth-averaged concentrations do not vary perceptibly. This is shown for antimony and arsenic in Fig. 7.17. The only exception is sulfate, with depth-averaged concentrations decreasing by approximately 75 percent between surficial material and material at the base of the pad. The heterogeneous distributions of constituents indicate that effluent water quality derived from individual segments of the heap material may vary by an order of magnitude. However, the net effluent from combinations of these segments will be relatively consistent throughout the pad because average bulk chemistries do not vary with depth with the exception of sulfate.

Acid-Generating Potential.   Leaching of metals from the heap material is increased when the leaching solution is acidic. The circumneutral to alkaline pH values measured by paste pH of heap materials indicates that acid-generating reactions that release solutes when leached by infiltrating water are not consequential to pad effluent. Low specific conductivity measurements indicate that the total reactivity of the material contributes <200 mg/L of solutes to solution when leached.

A more rigorous testing that included acid-base accounting analysis on samples with high metals concentrations was conducted to verify this result (Table 7.6). The primary factor influencing leachate chemistry is the propensity for sulfide minerals in the rock to oxidize, releasing solutes when leached, together with the neutraliza-

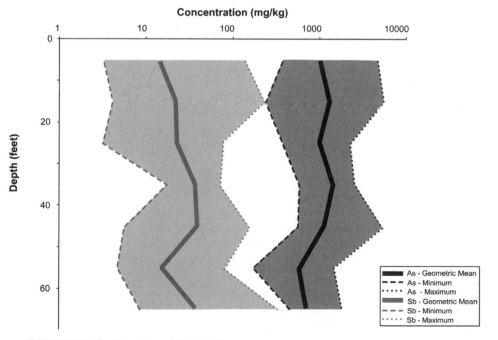

FIGURE 7.17   Profiles of As and Sb in Pad 1.

TABLE 7.6   Acid-Base Accounting Analytical Results and Net Carbonate Values (NCV)

Sample ID	Total S, %	Sulfate, %	Nonextract S, %	Pyritic S, %	AGP, ppt CaCO₃	ANP, ppt CaCO₃	NCV, %CO₂
LP1-1-3	0.03	<0.01	<0.01	0.03	0.9	<0.5	0.0
LP1-1-3-DUP	0.03	<0.01	<0.01	0.03	0.9	<0.5	0.0
LP1-2-2.5	0.04	<0.01	<0.01	0.04	1.3	70.1	3.0
LP1-2-16	0.05	0.01	0.01	0.03	0.9	77.2	3.4
LP1-3-16	0.16	0.14	0.01	0.01	0.3	<0.5	0.0
LP1-4-4.5	<0.01	<0.01	<0.01	<0.01	<0.3	659	29.0
LP1-5-39	0.35	0.04	0.15	0.17	5.31	<0.5	−0.2
LP1-6-25	2.26	0.97	0.29	1	31.3	53.8	1.0
LP1-7-25	0.02	0.02	<0.01	<0.01	<0.3	13.1	0.6
				Totals:	40.3	874	
			Total ANP/AGP:			22	

AGP = acid-generating potential.
ANP = acid-neutralizing potential.

tion capacity of carbonate minerals. The potential for acid generation is quantified by the net carbonate value (NCV), calculated from:

$$NCV = (3.67)(\% \text{ carbonate}) - (1.37)(\% \text{ sulfide})$$

The coefficients in the equation above are specified in terms of the acid-generating potential (AGP) and the acid-neutralizing potential (ANP), reported as $\%CO_2$, i.e.,

$$AGP(\%CO_2) = (1.37)(\% \text{ sulfide}) = 0.044 AGP(\text{ppt } CaCO_3)$$

$$ANP(\%CO_2) = (3.67)(\% \text{ carbonate}) = 0.044 \times \text{calcite(ppt } CaCO_3)$$

A positive NCV indicates that the rock is neutralizing (consumes acid), whereas a negative NCV indicates that the rock has the potential to release acid. However, many samples that have a negative NCV do not, in fact, generate acid because, sulfide minerals are incarcerated in refractory rock that prohibits oxidation reactions and acid generation.

The minimum NCV measured was −0.2, indicating a very limited potential for this sample to generate acidity. However, when tested, the paste pH of this sample was 8.2, indicating that acidity has not been generated under site conditions. Overall, the ANP:AGP ratio was about 22:1. These data together with the paste pH measurements indicate that the pad materials will not generate acidity over the long term.

The mineralogy of samples that contained high concentrations of constituents of environmental interest (e.g., arsenic, antimony, iron, nickel, copper, aluminum, and sulfate) was identified by electron microprobe analysis (EMPA). The EMPA confirmed the acid neutralizing nature of the pad material because abundant calcite precipitates were observed. Pyrite grains in heap material often appeared without signs of oxidation, suggesting that the oxidation reaction is not occurring in the alkaline heap environment. Most metals were present in oxide minerals and included within iron oxides. The EMPA identified Sb-As-Fe replacing rhombohedral carbonate (Fig. 7.18a) also containing antimony (Sb) and vuggy quartz (formed by alkaline dissolution of the siliceous framework) on which has precipitated a veneer of iron hydroxide containing As, Cu, and Sb (Fig. 7.18b).

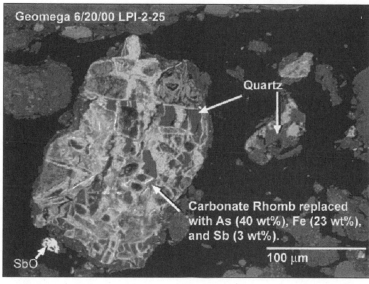

(*a*)

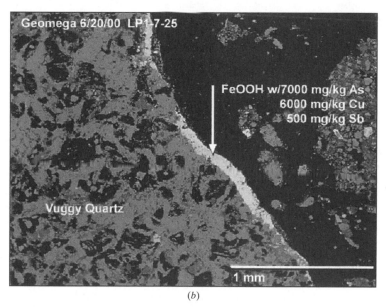

(*b*)

FIGURE 7.18 (*a*) A complex alteration paragenetic sequence consisting of primary rhombohedral carbonate replaced with reprecipitated arsenic, antimony, and iron. (*b*) Vuggy quartz generated by dissolution of primary silica with highly alkaline solutions showing iron oxide rinds with elevated metal concentrations precipitated in situ in Pad 1.

The calcite buffers the pH of the pad effluent, limiting metal solubility. Metals in the pad effluent are a result of dissolution or desorption of those constituents from the metal oxides observed in the heap material in relative concentrations observed in the collocated solid and interstitial water samples.

7.4.4 Evaluation of Leach Pad Cover Designs

Heap leach pad covers are used to minimize infiltration of precipitation that could mobilize constituents from heap materials and transport them from the pad. Covers constructed from soil materials are used for many purposes, including landfills, chemical wastes, radioactive wastes, and heap leach pad facilities (Fayer and Gee, 1997; Waugh et al., 1994; Wing and Gee, 1994).

In the arid and semiarid western United States, landfill cover designs typically include a capillary break feature, where coarse-grained materials underlie a finer-textured layer (Fayer and Gee, 1997). The capillary break promotes the storage of water in the upper layer, where the stored water is then subject to removal by evaporation and transpiration by vegetation. For this case study, the cover material consisted of clayey to silty sands with relatively low permeability (3.2×10^{-6} to 2.9×10^{-5} cm/s) compared to the coarser-grained heap material with a higher permeability (3.5×10^{-4} to 8.1×10^{-3} cm/s).

Cover effectiveness in reducing infiltration of meteoric waters into Pad 1 was determined by using the variably saturated flow model HYDRUS_2D (Simunek et al., 1996). The applicability of numerical models to cover design has been the focus of numerous investigations (Allison et al., 1994; Wing and Gee, 1994; Fayer and Gee, 1997). Recent investigations have focused on validating models by comparison to field data (Gee et al., 1998). Compared to measurements taken from covers, numerical models tend to underestimate storage during periods of soil wetting and overestimate storage during periods of soil drying (Fayer and Gee, 1997). The tendency for a numerical model's inability to resolve seasonal extremes is attributed to uncertainty in characterizing unsaturated hydraulic conductivity, sorption-desorption hysteresis, soil freezing, evapotranspiration, and differential temperature effects on the field scale.

Failure to resolve seasonal extremes potentially results in underestimation of seepage through the cover during exceptional wetting periods (Ward and Gee, 1997). However, the potential underestimation of seepage can be minimized through use of conservative modeling assumptions (e.g., relatively high hydraulic conductivity values, weak hydraulic barriers) and checked by calculating the storage capacity of the cover. The calculation verifies that the cover has sufficient storage to contain the volume of precipitation incident to the cover during wetting periods. By verifying that the cover has sufficient storage capacity to retain ambient infiltration above a capillary barrier and/or low-permeability layer, uncertainty in numerical modeling can be assuaged.

Model results indicated that covers consisting of >2 ft of soil were sufficient to prevent seepage greater than 1 gal/min under observed ambient meteorological conditions. Covers <2 ft thick resulted in seepage rates between 1 and 2 gal/min compared to a seepage rate of approximately 2 gal/min through the heap with no cover.

The temporal change in water storage in the 2-ft cover varies between 2.8 and 4.2 in (Fig. 7.19a) compared to a total storage capacity of approximately 9.7 in. The range of water stored within the cover varies by <1.5 in because most of the rainfall in the site area occurs during months when evapotranspiration rate exceeds precipitation rate (Fig. 7.19b). Therefore, water has little chance to accumulate in the cover before removal by evapotranspiration. Figure 7.20 illustrates the moisture content

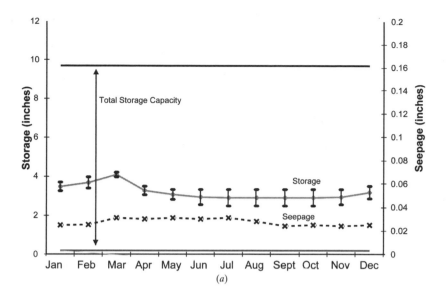

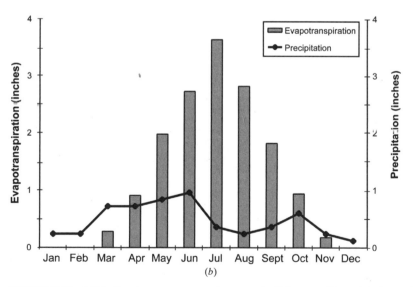

FIGURE 7.19 (*a*) Predicted cover storage and seepage for 2-ft cover; (*b*) precipitation and evapotranspiration.

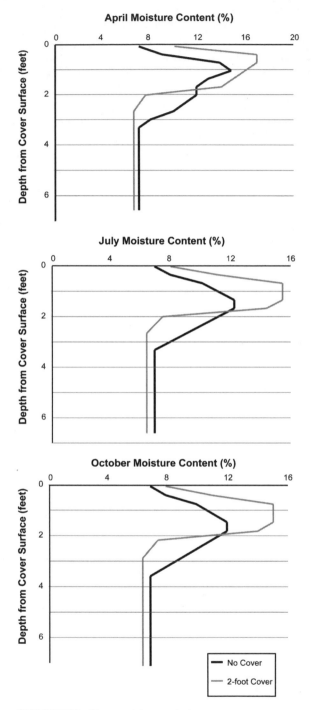

FIGURE 7.20   Cover moisture content curves.

(in percent) with depth for April, July, and October for no cover and a 2-ft cover. In general, more moisture is found at depth with no cover material.

The validity of model predictions was verified through the storage capacity calculations according to the method of Wing and Gee (1994). The method assumes the potential wetting period for the site falls between November and March, inclusive, when evapotranspiration from the cover is minimal. These calculations indicated that a 24-in cover has sufficient storage capacity to contain 100 percent of annual precipitation in the region. Therefore, from a hydrologic standpoint, a 2-ft cover was considered adequate in reducing infiltration. The question that remained was whether a 2-ft cover is sufficient in reducing the ecological risk posed by the heap material, which is discussed later.

7.4.5 Heap Draindown Rate

Future heap draindown effluent volumes were simulated by using HYDRUS-2D and the lithologic and operational data. The infiltration into the heap from the cover, previously described, was used so that the effect of cover thickness on the ultimate heap effluent rate could be determined. The heap effluent rate is influenced by two major components: (1) infiltration from precipitation and (2) drainage of existing moisture contained in the pad from the leaching and rinsing phases of the pad. The operational heap rinsing data was used to calibrate and verify the selection of model parameters for which there was not site-specific data.

With a 2-ft cover in place, the bulk (approximately 63 percent or 120 million gal) of the residual moisture from Pad 1 was predicted to drain within the first 18 months following cessation of application of rinse water (Fig. 7.21). The draindown rate with

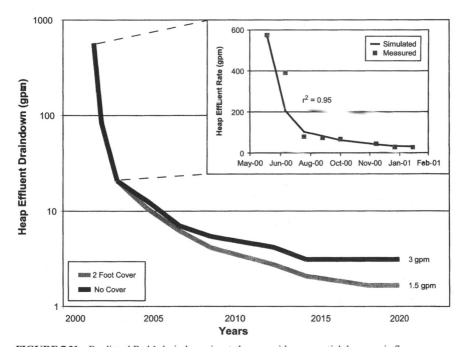

FIGURE 7.21 Predicted Pad 1 draindown, inset shows rapid exponential decrease in flow.

a 2-ft cover is predicted to decline from an initial 575 gal/min to approximately 80 gal/min within 6 months, to 20 gal/min after 1.5 years, to <10 gal/min after 3.5 years, and <2 gal/min after 13.5 years. The presence of the 2-ft cover reduces the predicted volume of effluent from Pad 1 by approximately 1.5 gal/min over the long term.

A comparison of the predicted effluent volume with the total volume of water in Pad 1 (calculated from laboratory water content analyses) verifies the model results. The calculated volume of water in Pad 1, based on the water content analyses, is approximately 180 million gals. From modeling simulations, approximately 160 million gals (90 percent of the total) will drain from the heap during the first 20 years after rinsing, while the remaining 20 million gals residual moisture will drain slowly at <500,000 gals per year.

However, comparing modeled versus subsequent measured draindown curves at three other heap leach facilities demonstrates that the modeled leachate volume invariably overestimates the actual draindown (Fig. 7.22). The implications of these data are that unnecessary costs are expended in either preparing a leach field for surface application of excess water, or applying cover thickness necessary to minimize draindown. The factors causing overestimates of modeled draindown remain to be determined. However, at one facility underestimation of the saturated hydraulic conductivity of fine-grained material in the heap (3 ft/day versus 6 ft/day) was the source of long-term draindown overestimation. This underestimation of hydraulic conductivity was likely due to formation of preferential flow pathways through fine-grained material by leaching operations, resulting in higher-than-expected conductivities in the material.

7.4.6 Heap Effluent Water Quality

PHREEQC (Dzombak and Morel, 1990) was used to determine the effluent water quality from Pad 1. The equilibrium model approach is appropriate for simulating geochemical reactions within the heap material because the heap effluent chemistry has been consistent over the past 2 years (suggesting soil-water equilibrium) and the kinetic mechanisms that potentially release solutes (e.g., pyrite oxidation and acid generation) are not apparent anywhere in the heap material.

PHREEQC was linked with HYDRUS_2D by tracking the volumetric fluxes of infiltrating waters predicted by the flow model. Infiltrating waters were proportionally mixed with ambient pore waters according to the volumetric fluxes. The leaching and adsorption mechanisms were quantified by the chemistry of a collocated heap soil and aqueous sample.

The two primary geochemical processes dictating effluent chemistry are gas-phase equilibrium and solid-phase (mineral) equilibrium/leaching. Equilibrium phases are pure gases or solids that can react reversibly with the infiltrating solution. Equilibrium phases for selected gas (e.g., carbon dioxide at $10^{-2.8}$ atm) and solid phases were set in the effluent water chemistry model to allow the infiltrating solution to leach and equilibrate with solid phases known to dissolve or precipitate within the heap material.

The predicted effluent water quality shown in Table 7.7 is generally consistent with the current chemistry. Significant reductions in concentrations will not be realized for many years because the infiltrating precipitation rate is very low (<1 gal/min). In addition, regrading and covering the heap may influence the immediate effluent water quality because of perturbation of the top and sides of the pad and exposure of minerals previously incarcerated within rock particles to air and/or percolating water. Reactions between the freshly exposed mineral surfaces and effluent water may result in a transitory increase in some constituent concentrations.

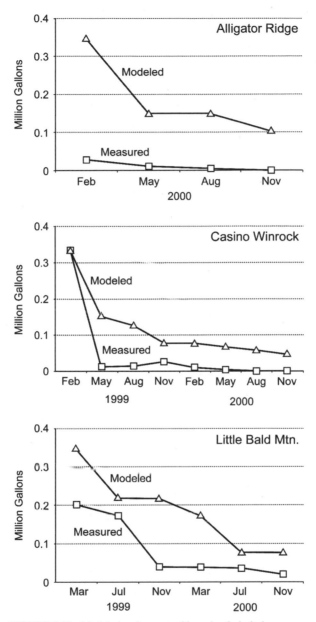

FIGURE 7.22   Modeled and measured heap leach draindown.

TABLE 7.7   Pad 1 Effluent Water Quality

(Concentrations in mg/L; pH in standard units)

Analyte	Drinking water standard	Livestock standard	Irrigation standard	Predicted heap effluent
pH	6.5–8.5			7.8
TDS	500 (1000)			2,120
Alkalinity, bicarbonate				140
WAD cyanide	0.2			0.11
Nitrate + nitrite as N	10			200
Sulfate	250 (500)			960
Arsenic	0.05	0.2	0.1	1
Copper	1.3	0.5	0.2	1.1
Mercury	0.002	0.01		0.0016
Manganese	0.05 (0.1)		0.2	0.143
Antimony	0.146			0.07
Selenium	0.05	0.05	0.02	0.09
Zinc	5	25	2	0.026

7.4.7   Land Application and Natural Attenuation

One potential means to manage leach pad draindown is to land-apply the solution. Land application involves discharging solution over an area so that the fluids are lost to evaporation and/or infiltration and solutes are safely stabilized in the subsurface by sorption mechanisms. The key concerns in land-applying heap leach effluent are prevention of infiltration to groundwater and increase of metals concentrations in the soils that will pose a risk to ecological receptors. Soil is typically excavated from the land application area and fluids applied within the excavated area. This is operationally easier as fluids can be contained within the excavation, runoff is eliminated, and ponding (excessive application rates) can be easily monitored. The excavated soil is then replaced at the end of the application period, reducing the ecological risk posed by the land application. Additionally, the land application can easily be converted to a subsurface leach field if required.

Land Application Rates.   The lithology and geotechnical properties of the land application area define the ability of the soil to accept infiltrating solutions. The selected land application area for Bald Mountain covers an area of about 11 acres and consists of a relatively homogenous gravel and cobble mix in a loamy sand matrix. The measured hydraulic conductivity of the soil, measured using a Guelph permeameter, ranged from 0.03 to 5.8 ft/day with an average of 2.8 ft/day, indicating that this area will be suitable for land application.

Several model iterations were completed to determine the maximum rates that effluent could be land-applied while not completely saturating soil or infiltrating to groundwater. Model results indicate that the rate of application during the first 6 months will be limited to a maximum of 160 gal/min (0.48 gal/ft^2-day) or 28 percent of the total effluent. After 6 months, the full amount of effluent could be applied at rates starting at 80 gal/min (0.25 gal/ft^2-day) and decreasing to <10 gal/min (0.02

gal/ft^2-day) within 5 years. A long-term steady-state rate of 1.7 gal/min could be applied through 20 years; however, a passive wetland will ultimately be used to attain a zero-discharge system.

At the application rates, the wetting front that resulted from land application of Pad 1 effluent remained >160 ft from the water table at steady state. The moisture content prior to land application is 2 to 15 percent of the total capacity of the soil. After land application for 20 years, the moisture content of the soils will be at approximately 34 percent of their capacity. Once land application is ceased, the moisture content will decrease very slowly over several decades back toward the moisture content prior to land application.

Soil Chemistry Impacts. The potential for attenuation (adsorption) of metals and other constituents applied to the soil in the land application area was investigated by using simple adsorption batch tests, following a U.S. EPA (1987) protocol. This approach is in contrast to typical column tests and sequential extractions in that it allows evaluation of attenuation coefficients K_d over a range of solute concentrations, thus addressing the increased partitioning that occurs as concentrations decrease during attenuation through the subsurface.

The batch adsorption technique involves mixing and equilibrating an aqueous solution of known concentrations of solutes with a known amount of soil. The solution is separated from the soil after equilibrium of the system is reached, and analyzed. The difference between the initial solute concentration and the final solute concentration after mixing represents the adsorbed concentration.

Arsenic and copper have the greatest potential to adsorb to land application soils, as indicated by their measured K_d from the batch tests (Table 7.8). Soil concentrations resulting from adsorption of metals in the effluent to soils in the land application area were estimated conservatively by assuming that 100 percent of the metals in the effluent will adsorb to the top meter below the application point. Under this worst case, arsenic increases from 57 to 78 mg/kg, copper from 17 to 19 mg/kg, manganese from 189 to 190 mg/kg, and antimony from <1 to 1 mg/kg. These increases do not result in soil concentrations above the ranges observed in typical Nevada soils (Table 7.3).

TABLE 7.8 Calculated Partition Coefficients (K_d)

Analyte	K_d L/kg
As	140
Sb	0.61
Mn	0.85
Cu	21

Other constituents do not adsorb significantly to soils and will remain in solution. These constituents, however, will not influence groundwater quality because infiltrating land application waters will not recharge groundwater at the prescribed application rates. Ecological risk due to solute loading in the soils will largely be mitigated by removal of the topsoil, followed by its replacement and revegetation at the end of the application process. The increase in metals concentrations from land application will not increase ecological risk over what might already be posed by the naturally high metals concentrations in the ambient soils (Sec. 7.4.8).

7.4.8 Ecological Risk Assessment of Cover Design

Ecological risk assessment is a process that evaluates the likelihood that adverse ecological effects may occur or are occurring as a result of exposure to one or more stressors (U.S. EPA, 1998). A complete exposure pathway between a biological receptor and a contaminated medium must be present for ecological risk to occur. For the heap closure scenario where the heap is covered with ambient soils, the heap soils are considered the contaminated medium and terrestrial plants and local wildlife species are the potential receptors. The exposure pathway is plant root uptake of metals from the heap materials and potential transfer to other biological receptors.

A probabilistic risk assessment (PRA) was used to conduct a comprehensive assessment of the incremental reduction in risk by adding cover to the heap. A PRA is a risk assessment that uses probability distributions to characterize variability or uncertainty in risk estimates, instead of the more traditional point estimate method that uses single values. Eco@RISK was developed by using @RISK (Palisade, 1996), a risk analysis and simulation tool that adds Monte Carlo capabilities to spreadsheet analyses.

Eco@RISK relies on EPA methods developed for PRA (U.S. EPA, 1999c) in addition to the fundamental concepts and equations of the traditional point estimate approach (U.S. EPA, 1999b; New Mexico Environment Department, 1999; Texas Natural Resource Conservation Commission, 1999; U.S. Army Corps of Engineers, 1996). A Monte Carlo analysis is the most frequently used method to conduct a PRA. In Eco@RISK, the risk equations are calculated thousands of times by using statistical techniques to randomly select variable exposure parameters from their probability distributions. The result is a distribution that represents the range of exposure doses experienced by the population of concern. The modeling was conducted to verify that a 2-ft cover based on the hydrologic modeling will be protective of plants and wildlife. Variability was incorporated into the Eco@RISK modeling for heap soil concentrations, cover soil concentrations, plant root depths, monthly soil moisture content, and percent of diet obtained from the heap leach pad (Fig. 7.23)

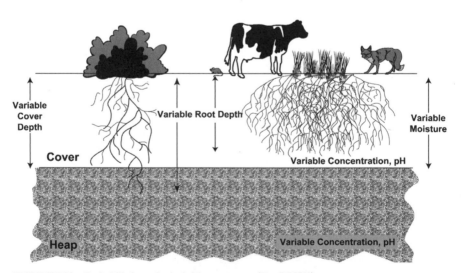

FIGURE 7.23 Probabilistic ecological risk assessment (Eco@RISK).

Plant Risk Evaluation.   The plant species considered for revegetation and consequently as biological receptors comprised eight grass species, three forbs, and two shrubs as follows:

Grasses

- *Pseudoroegneria spicata spp. spicata; Agropyron spicatum* (bluebunch wheatgrass)
- *Agropyron dasystachyum* (thickspike wheatgrass)
- *Agropyron smithii* (Western wheatgrass)
- *Elmus trachycaulus spp. Trachycaulus* (slender wheatgrass)
- *Sitanion hystrix* (squirrel tail)
- *Poa canbyl* (Canby bluegrass)
- *Oryzopsis hymenoides* (Indian ricegrass)

Forbs

- *Linum lewisii* (Appar blue flax)
- *Penstemon palmeri* (Palmer penstemon)
- *Sanguisorba minor* (Delmar small burnett)

Shrubs

- *Atriplex confertifolia* (shadscale)
- *Purshia tridentata* (antelope bitterbush)

Plants will be exposed to analytes in the heap soils through their roots, and provide an obvious exposure route to secondary receptors. Figure 7.24 illustrates the exposure pathways and model inputs used in the modeling. Because little data are available that quantify risk and analyte uptake for plant species considered in this study, it was assumed that plant uptake is proportional to the depth of the root exposed to the heap materials.

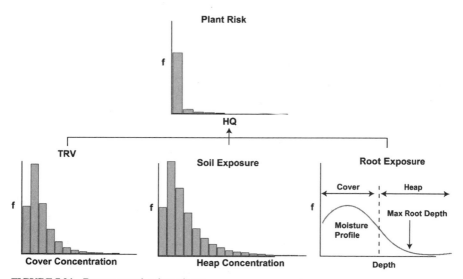

FIGURE 7.24   Receptor evaluation: plants.

Plant root exposure was determined for each plant species by considering rooting depths and monthly soil wetting profiles obtained from the HYDRUS modeling discussed previously. Soil wetting depth, defined as the maximum depth at which soil moisture levels are greater than ambient moisture, is the primary factor that determines the extent of rooting; however, reported and estimated rooting depths were also considered (Klepper et al., 1985; Lee and Lauenroth, 1993; Rundel and Nobel, 1991; USDA, 2000).

Monthly soil moisture profiles for April through October, the primary months of plant growth, were used to define the probability that a root would be found at a particular depth for a particular cover depth. The curve was modified to reflect the probability given the maximum rooting depth.

The risk to each of the plant species was determined from the hazard quotient (HQ), calculated by:

$$HQ = \frac{exposure}{TRV}$$

where the exposure is the soil concentration in mg/kg that the root is exposed to and TRV is the toxicity reference value or safe benchmark value. Healthy flora were observed at the site in areas where cover soils will be obtained; therefore, to assess plant risk, traditional toxicity reference values were replaced with ambient soil concentrations so that the incremental risk due to the heap soils alone could be simulated. This assumption is reasonable because available TRVs for plant toxicity are derived from studies on plants not included in this analysis. In addition, the plants, soils, and growth conditions used in the studies to develop these benchmarks are not similar to site conditions, in particular the high metals concentrations that occur naturally in the area (Sec. 7.1).

The dose (soil concentration) was calculated from the relative portions of the plant root exposed to the cover and heap soils. Probability distributions were calculated for analytes in the heap and cover soils by using S-PLUS, an environmental statistics package (Millard, 1997). Most analytes were normal or lognormally distributed. However, for some analytes with all or most values below detection limits, a discrete distribution was used to reflect the data. Figure 7.25 shows the distributions used for arsenic in the cover and heap soils.

The goal of this study was to be protective of the plant population, which was estimated from the fiftieth percentile (U.S. EPA, 1999c). For most plant species populations, the arsenic risk is equivalent to that of the cover soils (HQ = 1), except for squirrel tail and antelope bitterbush, the deepest-rooted plants. The incremental risk for these two species ($\Delta HQ > 1$) in comparison to cover soils is still low. An HQ of 1 indicates that the estimated exposure of a chemical is the same as the safe benchmark, and a HQ less than 1 indicates that estimated exposure is less than the safe benchmark. In general, a HQ greater than 1 is interpreted as a level at which adverse ecological effects may occur; however, the HQ should not be viewed as a statistical value. For example, an HQ of 0.01 does not indicate a 1-in-100 probability of an adverse effect, but does indicate that the estimated exposure is 100 times less than the determined safe benchmark. Additionally, the ratio of estimated exposure and safe benchmark does not infer a linear relationship because the safe benchmarks are not concise descriptors of toxicity, and dose-response relationships vary with chemical type.

For all plant species except squirrel tail and antelope bitterbrush, there is a 100 percent reduction in risk with a 2-ft cover compared to the same plants grown directly on heap soils. For squirrel tail and antelope bitterbrush, the reduction in risk

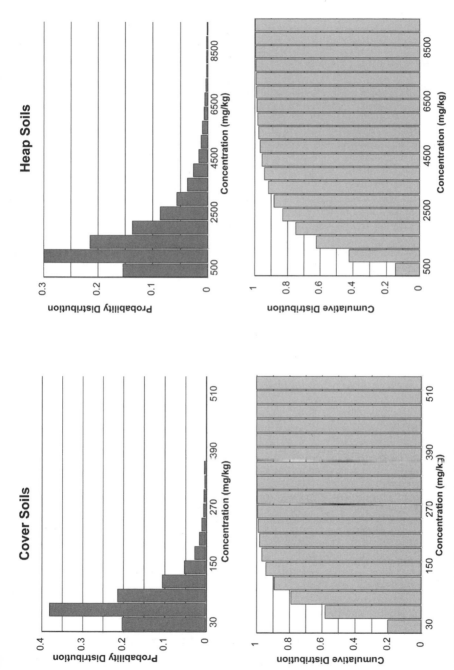

FIGURE 7.25 Probability and cumulative distribution model inputs for arsenic soil concentrations.

compared to heap soils is 94 and 92 percent, respectively. Therefore, to be protective of plants, squirrel tail and antelope bitterbrush would be removed from the seed mix for a cover of 2 ft.

Wildlife Risk Evaluation.   Exposure to upper trophic levels in the food web is through ingestion of plants that have taken up analytes from the heap soils. To assess potential uptake by plants, regression equations and uptake factors were applied. Available single variate (soil concentration) and multivariate (soil concentration and soil pH) regression models were used to calculate plant uptake for As, Cd, Cu, Hg, Ni, Se, and Zn (ORNL, 1998). These models were developed by using published data from soil contaminated in the field to estimate aboveground plant tissue concentrations. Simple uptake factors that represent a fraction of the soil concentration were applied to the remaining analytes (Baes et al., 1984; U.S. EPA, 1999b; NCRP, 1989).

The soil concentration used in both methods for calculating uptake was calculated from the relative portions of the plant root exposed to the cover and heap soils. The plant uptake distributions represent the exposure to wildlife with diets containing plants. Arsenic uptake for a population of plants growing in the cover soil only (no exposure to heap soils) is approximately 1.3 mg/kg compared to 7.3 mg/kg for plants grown directly in the heap. For a 2-ft cover, arsenic uptake for the majority of plant species is similar to that for the cover soils.

Wildlife receptors that ingest plants or prey that have been exposed to heap analytes are termed *secondary receptors*. These species are not directly exposed to the heap soils if a soil cover is constructed for the heap but may be exposed to heap analytes through the food chain. From a survey of species occurrence for Ruby Lake National Wildlife Refuge, located approximately 28 km (17 miles) to the northeast of the Bald Mountain project site, species representative of the significant trophic levels and feeding guilds were selected. No protected terrestrial species (federal or state) are in the immediate vicinity of the Bald Mountain project site.

Figure 7.26 is a diagram of the potential terrestrial food web at Bald Mountain and species selected for the modeling. Several criteria were considered in selecting appropriate wildlife receptors, including: (1) high exposure potential relative to other species, (2) representation of other species, and (3) availability of information (e.g., TRVs). The representative biological receptors selected include terrestrial plant (trophic level 1), deer mouse (trophic level 2, herbivorous), blacktail jackrabbit (trophic level 2, herbivorous), cow (trophic level 2, herbivorous), coyote (trophic level 3, carnivorous), and horned lark (trophic level 3, omnivorous). The evaluation of these species is expected to be inclusive of other species within the local food web because of the diverse exposure potential web represented by these species.

The deer mouse, blacktail jackrabbit, and cow represent trophic level 2 and are characterized as herbivorous. These receptors may be indirectly exposed to heap analytes when they ingest plants that have accumulated heap analytes. The deer mouse and blacktail jackrabbit, in turn, are commonly preyed upon by other upper trophic level predators and thus can transfer tissue-accumulated analytes to animals that would not otherwise be exposed. These receptors are also important because they have different-sized home ranges and therefore represent varying degrees of potential exposure.

The horned lark and coyote represent trophic levels 3 and 4, respectively. The diets of these species are primarily animal tissues. Therefore their only potential exposure to heap analytes is through ingestion of prey tissues that have accumulated heap analytes. The horned lark is primarily an insectivore and the coyote is a carnivore with a varied diet consisting largely of small and medium mammals. Figure 7.27 illustrates the general exposure model used for the wildlife receptors and inputs to the model.

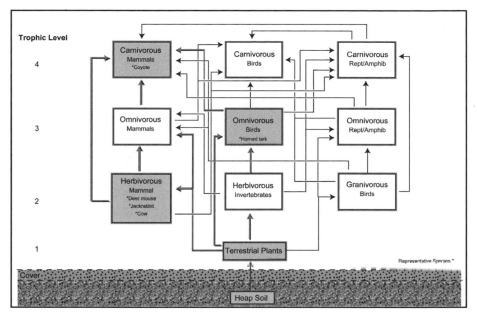

FIGURE 7.26 Bald Mountain terrestrial food web.

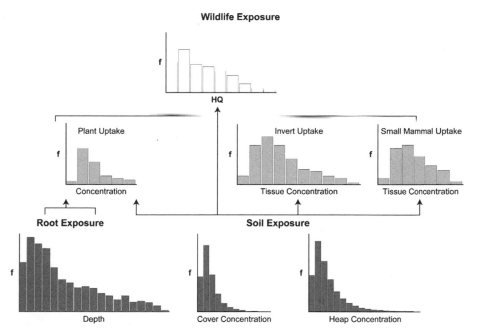

FIGURE 7.27 Receptor evaluation: wildlife.

Dietary exposure is the sum of the dietary components calculated as:

$$\text{Exposure} = \frac{[(C_{\text{soil}} \times \%\text{diet}) + (C_{\text{invert}} \times \%\text{diet}) + (C_{\text{plant}} \times \%\text{diet}) + (C_{\text{mammal}} \times \%\text{diet})] \times \text{FIR} \times \text{EMF}}{\text{body weight}}$$

Concentrations in the invertebrates and small mammals (represented by deer mouse) were estimated by:

$$C_{\text{invert}} = (\text{TF}_{si})(C_s)$$

$$C_{\text{mammal}} = (\text{FtM TF})(C_{\text{plant}})$$

Soil-to-invertebrate transfer factor (TF_{si}) and food-to-muscle transfer factor (FtM TF) for each analyte were taken from the U.S. EPA (1999b) and NCRP (1989), respectively.

Empirical formulas from the *Wildlife Exposure Factors Handbook* (U.S. EPA, 1993) were used to estimate daily food intake rates (FIR) for the various representative receptors, i.e.:

$$\text{Deer mouse (rodent) FIR (kg/day)} = 0.621 \times (\text{body weight, g})^{0.564} / 1000$$

$$\text{Jackrabbit (herbivore) FIR (kg/day)} = 0.577 \times (\text{body weight, g})^{0.727} / 1000$$

$$\text{Coyote (mammal) FIR (kg/day)} = 0.235 \times (\text{body weight, g})^{0.822} / 1000$$

$$\text{Horned lark (passerine) FIR (kg/day)} = 0.398 \times (\text{body weight, g})^{0.850} / 1000$$

Specific FIR values reported in the literature for the cow were used in the analysis to be conservative (U.S. EPA, 1990). In addition to FIR, other receptor parameters such as bioavailability, home range, and sensitive life stage were included as exposure-modifying factors (EMFs) in the estimate of dietary exposure. The percentage of home range that includes the heap leach pad was defined as a distribution. For the deer mouse, whose home range is much less than the area of the pad, the percent of diet derived from the heap was varied from 0 (home range entirely off the heap) to 100 percent (home range entirely on the heap). Therefore, the EMF was varied from 0 to 1. For receptors with a larger home range, the percent of their diet (EMF) was varied from 0 (home range entirely off the heap) to the maximum percent of the home range area that includes the pad.

Risk (HQ) was calculated for each analyte and wildlife receptor by the same formula used for plant risk (HQ = exposure/TRV). Risk was calculated on the assumption that each receptor's diet consisted of equal portions of each plant species (except squirrel tail and antelope bitterbush, which will not be included in the seed mix because of their potential for low risk). For wildlife, the TRV represents a no observable adverse effects level (NOAEL) dose from oral exposure. The studies used to derive TRVs were based on common laboratory species (i.e., mouse and rat) and not on wildlife species. Therefore, it was necessary to adjust the benchmarks to reflect the body weight of the wildlife receptors. The following equation was used to convert NOAEL TRVs from test species to mammalian wildlife (Sample et al., 1996).

$$\text{NOAEL}_w = \text{NOAEL}_t (\text{BW}_t / \text{BW}_w)^{0.25}$$

where subscripts w and t represent wildlife species and test species, respectively, and BW = body weight. Adjustment for plant and avian benchmarks was not done and is not recommended by guidance.

A 2-ft cover reduces the risk to a deer mouse population by 96 percent, jackrabbit by 91 percent, cow by 95 percent, coyote by 100 percent, and horned lark by 95 percent compared to the risk if the diet of receptors were composed of plants and upper trophic levels exposed directly to the heap. Although the potential risk due to the heap is not reduced by 100 percent for all receptors, there is no incremental potential risk relative to the cover material.

Based on the modeling and studies conducted for the Bald Mountain project, a 2-ft cover was selected as the optimal cover thickness from both a hydrologic and ecological standpoint. Additional cover would not significantly reduce the heap effluent and 2 ft is sufficient to protect plants, wildlife, and livestock.

7.5 THE LONG-TERM EFFICACY OF PASSIVE WETLANDS

Passive wetland treatment systems have been shown at the demonstration level as a viable treatment option to manage low flows of acidic pH mine drainage (Wildeman and Laudon, 1989; Howard et al., 1989; Faulkner and Skousen, 1994). However, there appear to have been no reported instances either of the efficacy of this technology in treating acid waters over the long-term (e.g., a decade), or of their use in treating circumneutral pH heap leach drainage.

The prevailing opinion is that wetland systems are best suited to treat moderate water quality and flow rates (Skousen, 2000) because passive treatment for sites with extremely poor water quality and/or high flow rates may be cost prohibitive. However, data collected from the Red Springs wetland (Fig. 7.28) at the former Buckhorn Mine in Nevada (see Sec. 7.5.2) demonstrate that passive marshes represent a viable long-term alternative that can effectively treat poor water quality.

7.5.1 Biogeochemical Processes Controlling Metal Solubility

The natural biogeochemical reactions in the passive wetland will promote nitrate reduction and subsequent off-gassing of ammonia, and sulfate removal by precipitation of sulfates and sulfides, hence reducing the effluent load of these constituents (Fig. 7.29). In addition, subsurface chemical and biological reduction reactions in the anaerobic layer produce and stabilize hydrogen sulfide, which complexes various metals (e.g., Sb, As, and Fe), resulting in the precipitation of metal sulfide minerals. In particular, acute toxicity to benthic organisms and metal bioavailability is significantly reduced by precipitation of metals as insoluble sulfides, as shown in laboratory experiments (Di Toro et al., 1992; Ingersoll et al., 1997). Additionally, metals were not accumulated in invertebrates, plants, or small mammals in studies conducted on wetlands treating acid mine drainage (Albers and Camardese, 1993; Pascoe et al., 1994; Lacki et al., 1992).

7.5.2 The Red Springs Passive Marsh System

The cascading three-stage Red Springs passive wetland has successfully treated low pH waters emanating from a former adit for almost a decade. Prior to treatment with the passive wetland system, the effluent was acidic with a pH of 2.0. Following installation of the wetland system, the pH has increased to between 6.8 and 8.1 from 1993 to the present (Fig. 7.30a). Effluent sulfate concentrations have decreased over the last 6 years (Fig. 7.29b), while As, well above the livestock standard (0.05 mg/L)

FIGURE 7.28   The Red Springs wetland.

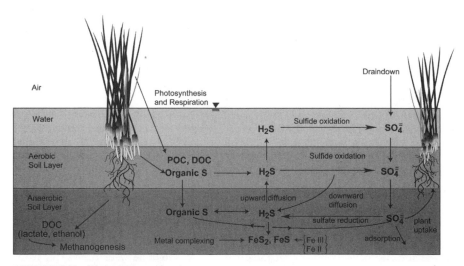

FIGURE 7.29   Wetland biogeochemistry.

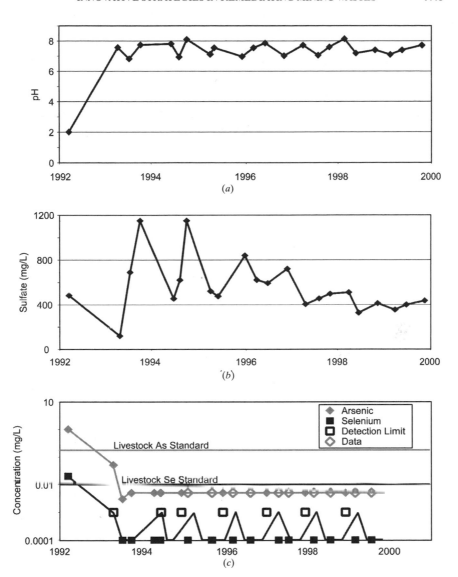

FIGURE 7.30   Red Springs wetland parameters.

in 1992 (1.01 mg/L), has been ≤0.005 mg/L over the last 7 years (Fig. 7.30c). Selenium has decreased from 0.02 mg/L to ≤0.0005 mg/L or less (Fig. 7.30c), and the effluent now meets the Se livestock standard (0.01 mg/L).

The efficiency of metal removal at the Red Springs wetland is further demonstrated by the sediment concentrations in each stage. The upper pond contains the highest metal concentrations, decreasing through the middle to the lower pond (Table 7.9). For example, As decreases from the upper pond (590 mg/kg), through

TABLE 7.9 Metals in the Red Springs Wetland
(All concentrations in mg/kg)

	As	Sb	Cd	Cu	Se
Upper pond	590	56	1.2	29	38
Middle pond	212	21	0.6	30	6.4
Lower pond	71	7	0.4	13	10

the middle pond (212 mg/kg) into the lower pond (71 mg/kg). The sequential decrease in sediment metal concentrations with each stage of ponding demonstrates that metals are being complexed and removed from solution by precipitation of stable phases and/or sorption processes. Furthermore, the low concentrations in the lowest pond indicates that the system is not being oversaturated with heavy metals and acidity.

The addition of precipitates in the ponds manifests itself as ongoing accrual of sediments that continue to build mass through which the vegetation (*Carex sp.*) grows. The residual organic matter decays to provide an ongoing source of carbon to facilitate reduction (Fig. 7.29) as oxygen is consumed by mineralization reactions.

7.5.3 The Buckhorn Mine Bioreactor

The Buckhorn Mine bioreactor is a flow-through system consisting of an underground storage tank filled with straw and gravel. Over the last 18 months, residual heap draindown has been routed through the reactor to improve water quality by taking advantage of natural biogeochemical processes similar to those in a passive wetland system. Influent and effluent waters have both had a circumneutral pH ranging from 7.2 to 8.1 (Fig. 7.31*a*). Reductions in nitrate concentrations have averaged 70 percent (Fig. 7.31*b*) due to either reduction to ammonia or biological uptake, while sulfate removal has ranged from 48 to 99 percent (Fig. 7.31*c*). Particulates analyzed by electron microprobe from a sump downgradient from the bioreactor were predominantly iron phases, e.g., Fe_3O_4, biological FeO, and $Fe(OH)_3$ with authigenic gypsum ($CaSO_4$) (Fig. 7.32). In situ precipitation of gypsum accounts for the decrease in sulfate concentrations.

7.6 BACKFILLING/IN-PIT WASTE ROCK TRANSLOCATION

Depending on the nature of the waste, area groundwater, the wallrock exposed in the ultimate pit surface (UPS), and the nature of the deposit, in-pit translocation may represent an excellent opportunity to facilitate closure while reducing the acres disturbed in the vicinity of the mine. Translocation refers to the controlled in-pit relocation of waste rock in a single haul. It differs from backfilling, which necessitates initial removal of waste rock to a location exterior to the pit, followed by a subsequent haul and dump into the pit.

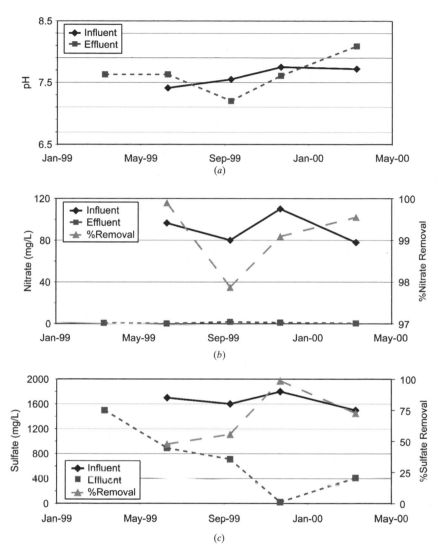

FIGURE 7.31 Buckhorn Mine bioreactor.

   Closure of open pit mines results in the creation of a "pit lake" if the groundwater table is higher than the bottom of the pit when it is decommissioned. Hence, for in-pit translocation to be viable, it is necessary to understand the relative contribution of in-pit waste rock on ultimate pit lake chemistry and/or groundwater chemistry.

   At a mine where translocation was evaluated, the deposit consists of ore hosted in Paleozoic sedimentary rocks that lies beneath several hundred feet of unconsolidated alluvium. The purpose of this study was to assess the effect of proposed in-pit

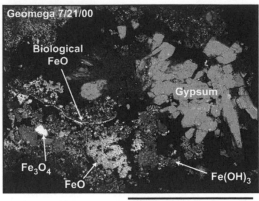

300 µm

FIGURE 7.32   Bioreactor precipitates.

translocation of 360 million tons of mine waste that would result in formation of three smaller pit lakes (A, B, and C), rather than one large pit lake, and specifically to compare the resulting water chemistry with, and without, the translocated waste.

The proposed in-pit waste rock translocation is envisioned to result in Pit Lake A (about 800 ft deep; 3 billion ft³), Pit Lake B (about 500 ft deep; 1 billion ft³), and smaller, higher-elevation Pit Lake C (about 200 ft deep; 200 million ft³) on the pit high wall. In the absence of waste rock translocation, one large pit lake would be created (about 800 ft deep; 8 billion ft³). Under the new scenario, the pit lakes would be dissected by the translocated waste rock that would form a saddle in the center of the pit (Fig. 7.33).

7.6.1   Structure of the Predictive Pit Lake Model

The study used site-specific chemical and hydrologic data in conjunction with laboratory tests and predictive computer modeling to generate a paradigm of pit/aquifer geochemical interactions and evolving pit water quality—i.e., *chemogenesis*—associated with inundation of the three pits following closure. To accomplish this, the geochemical attributes of the wall rock in the ultimate pit surfaces and the exposed waste rock were superimposed on the flow domain to predict the final water quality in the future pit lake.

The groundwater chemistry and wall rock leachate varies with the different rock types occurring in the various segments of the ultimate pit surface. The processes that determine the future pit lake chemistry were modeled by integrating three distinct analyses, the quality and quantity of temporal groundwater inflow, pyrite oxidation rates, and geochemical mixing (Fig. 7.34).

The *groundwater flow model* predicts the relative volumes of groundwater that will flow through the wall and waste rock on a temporal basis, leaching solutes and forming the pit lake. These data provide the rate of pit filling, hence the duration of wall rock exposure and time available for pyrite oxidation in the wall rock, and the relative volumes of wall rock runoff and evaporation from the pit. The *pyrite oxida-*

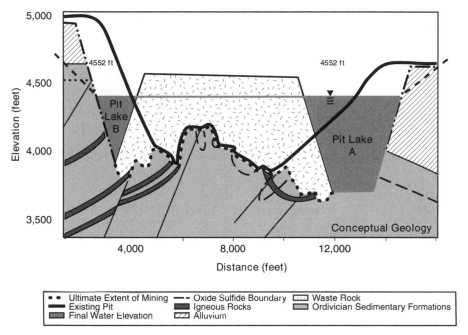

FIGURE 7.33   Cross section of the in-pit translocation.

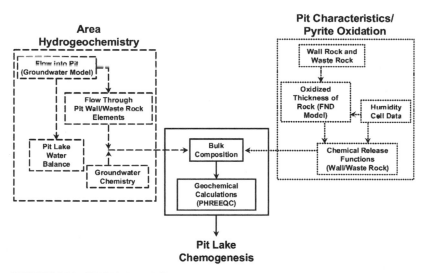

FIGURE 7.34   PITQUAL modeling components.

tion model predicts the total masses of solutes that are available for transport into the pit on the basis of the results of field and laboratory testing. The *geochemical mixing model* uses the groundwater flow and bulk chemistry results from the other models and predicts a final chemistry determined by the chemical reactions that take place between mixing waters (i.e., acid neutralization, precipitation, chemical adsorption). Together these elements comprise the pit lake water quality (PITQUAL) model.

7.6.2 Area Hydrogeology

After the pit closes and pumping ceases, the groundwater will flow into the pit through the ultimate pit surface (UPS) wall rock. The quantity and direction of groundwater inflows were predicted using a numerical flow model. Once the lake has reached full recovery, flow will be from north to south such that the Pits B and C will be composed solely of groundwater flow through the UPS, while Pit A will receive components from waste rock and its high wall UPS (Table 7.10).

TABLE 7.10 Pit Inflows by Rock Type

	% of total inflow		
Rock type	Pit C	Pit B	Pit A
Alluvium			4
Oxide	11		7
Upper Valmy (NCV > 0)	6	4	9
Upper Valmy (NCV < 0)		95	
Lower Valmy	83		19
Fill		1	61

Background groundwater quality was available for all lithologic units in the UPS, including oxide, alluvium, and sulfide zones. Under ambient hydrologic conditions, area groundwaters generally meet drinking water standards (Table 7.11) with the exception of arsenic (up to 0.78 mg/L in a portion of the sulfide zone) and manganese (0.45 mg/L in the alluvial aquifer).

Temporally varying results of the groundwater flow model were incorporated into the pit lake model by determining the volume of flow through each 400×400-ft segment of the pit surface on an annual basis. In this manner, the volume and chemistry of water entering the pit were predicted for 140 years (representing >95 percent pit full capacity) for each pit on an annual basis after the pit begins to fill.

7.5.3 Wall and Waste Rock Geochemistry

The three pits have distinct ultimate pit surface (UPS) chemistries. For example, Pit B UPS has high (about 2000 mg/kg) As, 2 to 5 percent sulfide, with 40 percent of the UPS NCV < 0. In contrast, Pit A UPS has low (about 500 mg/kg) As, <2 percent sulfide (representing only a small proportion of the UPS), and 100 percent NCV > 0,

TABLE 7.11 Background Groundwater Composition
(mg/L unless otherwise noted)

Analyte	Alluvium	Oxide	Sulfide	Sulfide (fault zone)
pH (s.u.)	7.62	7.87	7.53	7.52
TDS	300	360	370	370
Alkalinity	135	191	200	214
Antimony	0.012	0.036	0.051	0.022
Arsenic	0.014	0.143*	0.272*	0.778*
Calcium	40	40	45	46
Chloride	39	16	18	12
Iron	0.043	0.003	0.085	0.133
Manganese	0.45*	0.001	0.024	0.041
Mercury	0.0001	0.0001	0.0002	0.0001
Sodium	35	39	26	27
Sulfate	29	43	48	50
Zinc	0.003	0.003	0.002	0.01

* Numbers exceed drinking water standards.

indicating that the acid-neutralizing potential exceeds acid-generating potential in the pit surface. Pit C also has relatively high As (about 2000 mg/kg), <2 percent sulfide, and 100 percent NCV > 0. Hence it is reasonable to expect the pit lake chemistry in each pit to vary with the individual UPS chemistry, modified by any waste rock leachate entering the pit lake.

As the UPS wall rock is exposed during mining, pyrite will weather and oxidize because of direct contact with seepage, precipitation, and atmospheric oxygen, facilitating release of solutes from the rock matrix. The first influx of groundwater will interact with the wall rock, leaching salts from pyrite oxidation in the UPS into the pit.

The propensity for solute release from each rock type in the pit wall is directly related to the NCV of that rock, which ranges from −18 to +16 in the pit walls. In this study, site-specific laboratory data were collected (humidity cell tests) to determine the solute concentrations that would be leached from wall rock with progressive flushing.

7.6.4 Determination of Oxidized Wall and Waste Rock Volume

After exposure to the atmosphere, sulfides in the wall rock oxidize, becoming soluble and prone to leaching by incident precipitation on the high wall and groundwater. Pyrite oxidation requires both oxygen and water, i.e.,

$$FeS_2 + \tfrac{15}{4} O_2 + \tfrac{1}{2} H_2O = Fe^{3+} + 2SO_4^{2-} + H^+$$

Hence, accurate simulation of this reaction is important to determine the volume of wall rock associated with the ultimate pit surface from which oxidation products are available for leaching. The proclivity of waste rock to leach was delineated in the same manner. The resulting volume depends on several factors, including diffusion of atmospheric oxygen into the wall rock matrix, diffusion of atmospheric oxygen into individual rock particles, fracture density of the wall rock, moisture content of the wall rock, dual air-filled and water-filled porosity of the wall rock, and chemical kinetics of pyrite oxidation.

The FND pyrite oxidation model (Fennemore et al., 1999) was used to calculate the volume of the wall rock zone that forms a rind around the pit. The FND model predicts oxidized wall rock volumes, oxygen concentration profiles, and sulfate releases from the wall rock. Predicted sulfate releases were matched to field results to calibrate the model, and the oxidized rind thickness calculated for the time periods over which segments of the UPS will be exposed to atmospheric oxygen.

7.6.5 Pit Lake Geochemistry

As water enters the pit from various wall rock source regions, in situ mixing, precipitation, sorption, and evaporation occur, determining pit lake chemogenesis. The bulk wall rock leachates for the pits are alkaline. Therefore, geochemical mixing results in precipitation of amorphous ferric hydroxide (AFH), calcite, and other metal hydroxides (Mn, Al, and Cr). These geochemically active solids remove solutes as the floc settles through the water column, accumulating as a sludge on the bottom of the pit lake. Chemical mixing, precipitation, sorption, and evapoconcentration were modeled by using PHREEQC (Parkhurst, 1995).

The long-term nature of the pit lake chemogenetic analysis (>100 years) mandates inclusion of pit water evaporation in the pit water chemistry, resulting in a loss of the total influent volume to evaporation. As pit water evaporates, solutes are retained in the remaining pit water, resulting in evapoconcentration of the solutes. The resulting pH shift can influence solute solubility and hence the solution chemistry.

7.6.6 Results

The pit lake is projected to be a terminal lake (no pit outflow) that will not be used for drinking water or recreation, nor will it be stocked with fish. Therefore, wildlife represents the only potential receptors. Compared to the single pit lake scenario, water quality in Pit Lakes A and B (Tables 7.12 and 7.13) is generally improved because the sulfide zone (the worst background groundwater quality) does not discharge into either pit lake. In addition, the Pit C area (the worst UPS chemistry) will no longer contribute water and solutes to Pit Lakes A and B.

For Pit Lake A, the predicted arsenic and antimony concentrations were similar to the existing predictions for the pit lake that would form without any translocation of waste rock. The water quality influenced by waste rock addition to the pit lake is evident in increased major ion concentrations (e.g., sulfate, chloride, sodium), constituents that leach preferentially from the waste rock (Table 7.12).

For Pit B, predicted arsenic and antimony concentrations were less than existing predictions for the single pit lake scenario because of the precipitation of AFH from solution (Table 7.13). The water quality influence of waste rock to the pit lake chemistry is evident in increased major ion concentrations (e.g., sulfate, chloride, sodium, etc.), primarily due to sulfate generation by pyrite oxidation in the pit walls.

The Pit lake C water quality will be worse than that predicted for the unfilled pit because leachable wall rock solute concentrations will no longer be ameliorated by

TABLE 7.12 Comparison of Predicted Pit Lake Water Quality (mg/L except where noted) with and without Waste Rock Translocation (Pit A)

	Pit A (140 years of refilling)	Pit without translocated waste rock (130 years of refilling)
pH	8.26	8.6
Alkalinity	197	97
Al	0.035	0.17
Sb	0.216	0.48
As	0.277	0.31
Ba	0.228	0.027
Cd	0.0038	0.0014
Ca	41	8.1
Cl	75	29
Cr	0.0007	0.0007
Cu	0.012	0.0007
F	2.55	1.1
Fe	0.0004	0.0006
Pb	0.0000	0.0001
Mg	65	41
Mn	0.0000	0.0001
Hg	0.0009	0.0002
Ni	0.063	0.025
K	20	10
Se	0.016	0.0094
Ag	0.005	0.0034
Na	158	59
SO_4	446	130
Tl	0.0013	0.012
Zn	0.029	0.0073

the lower concentrations in the rest of the system. Specifically, arsenic in Pit Lake C would exceed the concentration predicted for the unfilled pit (Table 7.14).

This information led to the important recognition that a management strategy could be designed to ameliorate the issue of potential poor surface water quality in Pit Lake C occurring upon closure of the facility. Specifically, the mine plan was modified to preclude formation of Pit Lake C by changing the shape of the ultimate pit surface. Clearly, responsible stewardship of land and water quality issues was exemplified resulting in less land disturbance because of in-pit translocation (due to a reduced waste rock dump footprint), and interdiction of a potentially poor water quality (in Pit C). There will also be a concomitant improvement in the water quality of Pits A and B because the translocated waste rock will preclude leaching of high wall sulfide into the two smaller pit lakes.

TABLE 7.13 Comparison of Predicted Pit Lake Water Quality (mg/L except where noted) with and without Waste Rock Translocation (Pit B)

	Pit B (140 years of refilling)	Pit without translocated waste rock (130 years of refilling)
pH	8.1	8.6
Alkalinity	143	97
Al	0.03	0.17
Sb	0.15	0.48
As	0.04	0.31
Ba	0.199	0.027
Cd	0.002	0.0014
Ca	137	8.1
Cl	55	29
Cr	0.0008	0.0007
Cu	0.0037	0.0007
F	5.3	1.1
Fe	0.0004	0.0006
Pb	<0.0001	0.0001
Mg	48	41
Mn	<0.0001	0.0001
Hg	<0.0001	0.0002
Ni	0.286	0.025
K	14	10
Se	0.017	0.0094
Ag	0.0029	0.0034
Na	748	59
SO_4	1900	130
Tl	0.003	0.012
Zn	0.06	0.0073

7.7 SUMMARY

Mine remediation and closure is, by and large, a site-specific exercise that requires consideration of geology, hydrology, hydrogeology, and geochemistry in concert with mine excavation, hauling, and beneficiation operations. The large scale of mining operations requires that optimal decision making be an essential part of cost-effective closure and remediation. In all cases, this decision making is facilitated by thorough site characterization and prospective modeling analyses that allow quantification of not only human health and ecological risks but also the incremental benefits and costs of remedial and closure activities. In addition, because of the large-scale nature of mining

TABLE 7.14 Comparison of Predicted Pit Lake Water Quality (mg/L except where noted) with and without Waste Rock Translocation (Pit C)

	Pit C (140 years of refilling)	Pit without translocated waste rock (130 years of refilling)
pH	8.4	8.6
Alkalinity	257	97
Al	0.048	0.17
Sb	0.075	0.48
As	1.98	0.31
Ba	0.218	0.027
Cd	0.0034	0.0014
Ca	19	8.1
Cl	48	29
Cr	0.0068	0.0007
Cu	0.007	0.0007
F	1.9	1.1
Fe	0.0004	0.0006
Pb	<0.0001	0.0001
Mg	64	41
Mn	<0.0001	0.0001
Hg	0.0004	0.0002
Ni	0.018	0.025
K	14	10
Se	0.006	0.0094
Ag	0.0034	0.0034
Na	63	59
SO_4	172	130
Tl	0.002	0.012
Zn	0.034	0.0073

sites, natural attenuation and intrinsic remediation play larger roles in their cost-effective closure and remediation than at smaller industrial facilities.

REFERENCES

Albers, P. H., and M. B. Camardese (1993), Effects of acidification on metal accumulation by aquatic plants and invertebrates, 1: Constructed wetlands," *Environ. Toxicol. Chem.*, vol. 12, pp. 959–967.

Allison, G. B., G. W. Gee, and S. W. Tyler (1994), "Vadose-zone techniques for estimating groundwater recharge in arid and semiarid regions," *Soil Science Society of America Journal,* vol. 58, pp. 6–14.

Baes, C. F. I., R. D. Sharp, A. L. Sjoreen, and R. W. Shor (1984), "A review and analysis of parameters for assessing transport of environmentally released radionculides through agriculture," ORNL-5786. U.S. Dept. of Energy, Oak Ridge, Tenn.

BLM (1996), "Twin Creeks Mine—Final Environmental Impact Statement," United States Department of the Interior, Bureau of Land Management, Winnemucca District Office, Winnemucca, Nev., December.

Bowers, T. S. (1996), "Distinguishing the Impacts of Mining from Natural Background Levels of Metals," *Geologic Society of America, Abstracts with Programs, 1996 Annual Meeting.*

Craig (1961), "Standard for reporting concentrations of deuterium and oxygen-18 in natural waters," *Science,* vol. 133, pp. 1833–1934.

Davis, A., and D. Ashenburg (1989), "The aqueous geochemistry of the Berkeley Pit, Butte, Montana," *Applied Geochemistry,* vol. 4, pp. 123–136.

Davis, A., and L. E. Eary (1997), "Pit lake water quality in the Western U.S.: An analysis of chemogenetic trends," *Mining Engineering,* June, pp. 98–102.

Di Toro, D. M., J. D. Mahony, D. J. Hansen, K. J. Scott, A. R. Carlson, and G. T. Ankley (1992), "Acid volatile sulfide predicts the acute toxicity of cadmium and nickel in sediments," *Environmental Science and Technology,* vol. 26, pp. 96–101.

Dzombak, D. A., and F. M. M. Morel (1990), *Surface Complexation Modeling: Hydrous Ferric Oxide,* John Wiley and Sons, New York.

Faulkner, B. B., and J. G. Skousen (1995), "Effects of land reclamation and passive treatment systems on improving water quality," *Green Lands,* vol. 25, no. 4, pp. 34–40.

Fayer, M. J., and G. W. Gee (1997), "Hydrologic model tests for landfill covers using field data," in *Landfill Capping Symposium Proceedings,* ESRF-019.

Fennemore, G. G., W. C. Neller, and A. Davis (1999), "Modeling pyrite oxidation in arid environments," *Environmental Science & Technology,* vol. 32, pp. 2680–2687.

Fennemore, G., A. Davis, L. Goss, and A. Warrick (2000), "A rapid screening-level method to optimize location of infiltration ponds," *Ground Water,* vol. 39, pp. 230–238.

Friedrich, G., Herzig, P., Keyssner, S., and Maliotis, G. (1984), "The distribution of Hg, Ba, Cu and Zn in the vicinity of cupriferous sulfide deposits, Troodos Complex, Cyprus," *Journal of Geochemical Exploration,* vol. 21, pp. 167–174.

Gee, G. W., A. L. Ward, and R. R. Kirkham (1998), "Long-term performance of surface covers estimated with short-term testing," U.S. Department of Energy, CONF-980652, pp. 67–81.

Gorett, G., ed. (1983), *Rock Geochemistry in Mineral Exploration, Handbook of Exploration Geochemistry,* vol. 3, Elsevier Scientific, Amsterdam.

Hitchborn, A. D., D. G. Arbonies, S. G. Peters, K. A. Connors, D. C. Noble, L. R. Larson, J. S. Beebe, and E. H. McKee (1996), "Geology and gold deposits of the Bald Mountain Mining District, White Pine County, Nevada," in *Geological Society of Nevada 1996 Spring Fieldtrip, Geology and Gold Deposits of Eastern Nevada, Bald Mountain Mine, Alligator Ridge, Robinson Project, Mount Hamilton,* May 3–5, Special Publication No. 23.

Howard, E. A., J. C. Emerick, and T. R. Wildeman (1989), "Design and construction of a research site for passive mine drainage treatment in Idaho Springs, Colorado," in *Constructed Wetlands for Wastewater Treatment: Municipal, Industrial and Agricultural.,* D. A. Hammer (ed.), Lewis Publishers, pp. 761–764.

Ingersoll, C. G., T. Dillon, and G. R. Biddinger (1997), *Ecological Risk Assessment of Contaminated Sediments,* Setac Press, Pensacola, Fla.

Klepper, E. L., K. A. Gano, and L. L. Cadwell (1985), "Rooting Depth and Distributions of Deep-rooted Plants in the 200 Area Control Zone of the Hanford Site," Pacific Northwest Laboratory, PNL-5247.

Lacki, M. J., J. W. Hummer, and H. J. Webster (1992), "Mine drainage treating wetland as habitat for herpetofaunal wildlife," *Env. Manag.,* vol. 16, pp. 513–520.

Langmuir, D. (1997), *Aqueous Environmental Geochemistry*, Prentice Hall, Upper Saddle River, N.J.

Learned, R. E., Chao, T. T., and Sanzolone, R. F. (1985), "A comparative study of stream water and stream sediment as geochemical exploration media in the Rio Tanama Porphyry Copper District, Puerto Rico," *Journal of Geochemical Exploration*, vol. 24, pp. 175–195.

Lee, G. A., and W. K. Lauenroth (1993), "Spatial Distributions of Grass and Shrub Root Systems in the Shortgrass Steppe," *The American Midland Naturalist*, vol. 132, no. 1, pp. 117–123.

McBride, M. D. (1994), *Environmental Chemistry of Soils*, Oxford University Press, New York, 406 pp.

Millard, S. P. (1997), "EnvironmentalStats for S-PLUS Help," version 1.1. Probability, Statistics & Information, Seattle.

Miller, G. C., W. B. Lyons, and A. Davis (1996), "Understanding the water quality of mining pit lakes," *Environmental Science & Technology*, vol. 30, pp. 118A–123A.

Miller, W. R., and J. B. McHugh (1999), "Natural Acid Drainage from Altered Areas Within and Adjacent to the Upper Alamosa River Basin, Colorado," U.S. Geologic Survey Open File Report 94-144.

NCRP (1989), "Screening techniques for determining compliance with environmental standards: releases of radionuclides to the atmosphere," National Council on Radiation Protection and Measurements Commentary No. 3, revised Jan. 1989. Bethesda, Md.

New Mexico Environment Department (1999), "Guidance for Assessing Ecological Risks Posed by Chemicals: Screening-Level Ecological Risk Assessment," Final Draft.

ORNL (1998), "Empirical models for the uptake of inorganic chemicals from soil by plants," September, BJC/OR-133.

Palisade (1996), "@RISK, Advanced Risk Analysis for Spreadsheets," Palisade Corp., New York.

Parkhurst, D. L. (1995), "User's guide to PHREEQC—A computer program for speciation, reduction-path, advective transport, and inverse geochemical calculations," U.S. Geological Survey Water Resources Investigation Report, 95-4227.

Pascoe, G. A., R. J. Blanchet, and G. Linder (1994), "Bioavailability of metals and arsenic to small mammals at a mining waste-water contaminated wetland," *Arch. Environ. Contam. Toxicol.*, vol. 27, pp. 44–50.

Ribeiro, M. J., M. M. Dos Santos, and S. R. Bressan (1979), "Geochemical exploration over a mafic-ultramafic complex, Americano do Brasil, Goias, Brasil." *Journal of Geochemical Exploration*, vol. 12, pp. 9–19.

Rose, A. W., H. E. Hankes, and J. S. Welb (1979), *Geochemistry in Mineral Exploration*, 2d ed., Academic Press, London.

Rundel, P. W., and P. S. Nobel (1991), "Plant Root Growth, an Ecological Perspective," Special Publication Number 10 of the British Ecological Society, Blackwell Scientific Publications, Oxford.

Runnells, D. D., D. P. Dupon, R. L. Jones, and D. J. Cline (1998), "Determination of Natural Background Concentrations of Dissolved Components in Water of Mining, Milling, and Smelting Sites," *Mining Engineering*, February, pp. 65–71.

Sample, B. E., D. M. Opresko, and G. W. Suter II (1996), "Toxicological Benchmarks for Wildlife: 1996 Revision," ES/ER/TM-86/R3.

Shacklette, H. T., and J. G. Boerngen (1984), Element Concentrations in Soils and Other Surficial Materials on the Conterminous United States. USGS Professional Paper 1270.

Shevenell, L., K. A. Connors, and C. D. Henry (1999), "Controls on pit lake water quality at sixteen open-pit mines in Nevada," *Applied Geochemistry*, vol. 14, pp. 669–687.

Simunek, J., M. Sejna, and M. Th. van Genuchten (1996), "HYDRUS_2D: Simulating Water Flow and Solute Transport in Two-Dimensional Variably Saturated Media," U.S. Department of Agriculture.

Skousen (2000), www.wvu.edu/~agexten/landrec/passtrt/passtrt.htm.

Texas Natural Resource Conservation Commission (1999), "Guidance for Conducting Ecological Risk Assessments at Corrective Action Sites in Texas," Interim Guidance.

U.S. Army Corps of Engineers (1996), *Risk Assessment Handbook,* vol. II, *Environmental Evaluation,* EM 200-1-4.

USDA (2000), http://plants.usda.gov/plants.

U.S. EPA (1987), "Batch-type adsorption procedure for estimating soil attenuation of chemicals," EPA/530-SW-87-006.

U.S. EPA (1990), "Methodology for Assessing Health Risks Associated with Indirect Exposure to Combustor Emissions, Interim Final," EPA/600/6-90/003.

U.S. EPA (1993), *Wildlife Exposures Factors Handbook,* Office of Research and Development, 2 vols., EPA-600-R93-187A&B.

U.S. EPA (1994), *Technical Resource Document, Extraction and Beneficiation of Ores and Minerals,* vols. II and IV, July.

U.S. EPA (1998), "Guidelines for Ecological Risk Assessment, Final," EPA-630-R-95-002F.

U.S. EPA (1999a), "RCRA, Superfund & EPCRA Hotline Training Module: Introduction to Solid and Hazardous Waste Exclusions (40 CFR 261.4)," October.

U.S. EPA (1999b), *Screening-Level Ecological Risk Assessment Protocol for Hazardous Waste Combustion Facilities,* vol. 1, EPA530-D-99-001A.

U.S. EPA (1999c), *Risk Assessment Guidance for Superfund,* vol. 3, Part A, "Process for Conducting Probabilistic Risk Assessment," EPA000-0-99-000.

USGS (2000a), *Minerals Yearbook,* vol. I, *Metals and Minerals,* United States Geologic Survey.

USGS (2000b), *Mineral Commodity Summaries,* United States Geologic Survey.

Ward, A. L., and G. W. Gee (1997), "Performance evaluation of a field-scale surface barrier," *Journal of Environmental Quality,* vol. 26, pp. 694–705.

Waugh, W. J., M. E. Thiede, D. J. Bates, L. L. Caldwell, G. W. Gee, and C. J. Kemp (1994), "Plant and Environment Interactions, Plant Cover and Water Balance in Gravel Admixtures at an Arid Waste-Burial Site," *Journal of Environmental Quality,* vol. 23, pp. 676–685.

Wing, N. R., and G. W. Gee (1994), "Quest of the perfect cap," *Civil Engineering,* vol. 64, no. 10, pp. 38–41.

Wildeman, T. R., and L. S. Laudon (1989), "The use of wetlands for treatment of environmental problems in mining: non-coal mining applications," in *Constructed Wetlands for Wastewater Treatment: Municipal, Industrial and Agricultural,* D. A. Hammer (ed.), Lewis Publishers, pp. 221–231.

CHAPTER 8
THE REMEDIATION OF HAZARDOUS WASTES FROM OIL WELL DRILLING

Stephen M. Testa
Testa Environmental Corp.
Mokelumne Hill, California

James A. Jacobs
Fast-Tek Engineering
Redwood City, California

8.1 INTRODUCTION

8.1.1 History of Oil and Gas Exploration

Petroleum has been part of human history for thousands of years, although not in the refined state as we know it, but as bitumen, an asphalt like form that extrudes from the surface as natural seeps. It has been used for a variety of purposes through the ages. The Chinese, however, were the first to drill for crude oil and natural gas. The first such wells were drilled around A.D. 347 to a depth of about 800 ft with primitive bits attached to bamboo poles much like modern cable tool rigs. It was 15 centuries later that this technology was reinvented. On August 27, 1859, Edwin Drake struck oil near some surface oil seeps at a depth of 69.5 ft below the ground surface at Oil Creek near Titusville, Pennsylvania (Fig. 8.1a). Drake used a rig that essentially punched or pounded a hole, pulverizing the rock and soil. The broken drill cuttings were removed by flushing the borehole with water. It took 15 days to reach 69.5 ft in depth. By 1865, the first oil pipeline, 2 in in diameter and 32,000 ft long, was laid to transport oil from Oil Creek to the Oil Creek Railroad. Many of the first tanks, barrels, and even pipelines in the western Pennsylvania oil boom in the 1860s were constructed of wood. In 1896, the first "offshore" wells were drilled from piers extending into southern California waters in the Santa Barbara channel.

Historic photos from the Bakersfield area in southern California show the technology of the early days (Fig. 8.1b). Since these early days, the petroleum industry

(a)

FIGURE 8.1   (a) Historic photograph showing Drake's drilling rig in 1859 and (b) historic photograph of blowout in Bakersfield, California, area.

has chronicled a continuous stream of technological development. In 1974 a record-depth exploratory gas well was drilled to 31,441 ft in Oklahoma. Instead of hitting oil or gas, this well encountered molten sulfur. In 1979, the world's tallest fixed-leg platform, 1265 ft tall and weighing 59,000 tons, was installed in 1025 ft of water in the Gulf of Mexico, and, in 1984, an exploratory well was drilled at a new world's record water depth of 6942 ft off the coast of New England. The United States is the most thoroughly oil-explored and drilled nation in the world. With roughly 4,600,000 oil and gas wells drilled in the world, about 3,400,000, or 74 percent, have been drilled in the United States. There were 916 rigs in operation in the United States in 2000 and 533,550 producing oil wells and 322,932 producing gas wells in the country (World Oil Exploration Drilling and Production, 2001).

Today, the most commonly used drilling equipment is the rotary drill. A rotary drilling rig consists of a power source, derrick with lifting and lowering devices, and a bit attached to a length of tubular high-tensile steel referred to as a *string* (Fig. 8.2). The drill string passes through a rotary table that turns. As the rotary table turns, it

(*b*)

FIGURE 8.1 (*Continued*)   (*a*) Historic photograph showing Drake's drilling rig in 1859 and (*b*) historic photograph of blowout in Bakersfield, California, area.

provides the torque needed to turn the drill string and drilling bit. As the drill bit is lowered into the earth, additional drilling pipe is added to the top. Average well depths today extend about a mile deep.

Most onshore rigs are portable and include tall derricks that handle the tools and equipment that descend into the hole or well. The modular drilling equipment is transported to the drill site by trucks or barges. Offshore drilling can be performed from bottom-based platforms, drill ships, or submersible platforms. Each is self-contained with its own set of equipment. The average cost in 1992 dollars was $442,547 for

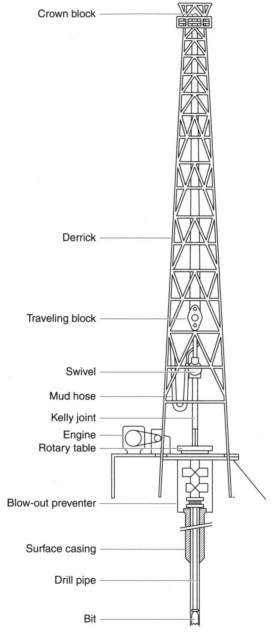

Crown block

Derrick

Traveling block

Swivel

Mud hose

Kelly joint

Engine

Rotary table

Blow-out preventer

Surface casing

Drill pipe

Bit

FIGURE 8.2 Schematic of the various components of an oil-drilling rig.

drilling an onshore exploration well and $4.2 million for an offshore exploration well. Eight out of ten exploration wells do not produce economic quantities of oil or gas, and only one out of ten is commercially productive.

8.1.2   Location of Major Oil and Gas Fields

Oil and gas fields come in all sizes, but they are classified as a giant field if they originally contain at least 100 million barrels of economically recoverable oil and gas, or a supergiant field if they contain at least 5 billion barrels of economically recoverable oil and gas. The size of the individual fields is a reflection of the amount of regional folding, faulting, and tectonics. In general, the larger the amount of earth movement and faulting, the higher likelihood that the traps that accumulate oil and gas will be smaller in size. For example, in the Saudi Arabia oil province, large homoclines create some of the world's largest reservoirs. By contrast, the highly faulted and folded Los Angeles Basin in southern California is one of the world's deepest and most prolific oil producing basins. Because of the active geologic setting, including many reverse faults, the Los Angeles Basin has numerous stacked reservoirs at various levels, creating a large number of smaller oil fields with only an occasional giant field.

Approximately 40,000 fields have been discovered worldwide, of which (as of 1994) 34,067 are located in the United States (Youngquist, 1997). Less than 9 percent (288) of the United States fields are giants, but these contribute nearly 60 percent of the total United States production, and contain more than 61 percent of the remaining United States reserves. Worldwide there are about 40 supergiants. Twenty-six of these are located in the Persian Gulf. The remaining supergiant fields are located in China, Kazakhstan, Libya, Mexico, United States, and Venezuela. Supergiant fields can have very significant economic and political importance. Oil deposits can vary greatly in depth. They can be at the surface, such as the vast Athabasca tar sands in Alberta, or can be a few hundred to over 20,000 ft deep. Regardless of depth, once a field is discovered, certain equipment is necessary in order for the petroleum to be produced.

Both oil and gas wells require that a wellbore be drilled, and production casing positioned inside the wellbore. Many wells can produce oil and gas. About three-quarters of the wells within the United States are referred to as stripper wells, those producing less than 10 barrels per day. Of about 589,000 producing wells in the United States as of January, 1994, about 77 percent, or 452,248 fall in this category, with an average production of about 2.2 barrels per day (American Petroleum Institute, 1995). The worldwide distribution of oil and gas fields indicates spatial variability related to field geology and source rock type, as well as depth and temperature of source rock and reservoirs (Fig. 8.3).

During primary production, gas wells have enough reservoir pressure that natural gas flows to the surface, whereas oil wells usually require a pump to be installed over the wellbore (Fig. 8.4). Gas wells are connected to a pipeline gathering system that delivers natural gas directly to its end-use applications such as a petrochemical plant, an electrical power generating station, or a residential home.

Oil must be transported via truck, barge, tanker vessel, or pipeline to a refinery where it undergoes needed processing that converts it to useful products. Many environmental concerns derive from unwanted releases of petroleum during its transport before and after refining.

The known recoverable world's oil resources have been increasingly abundant. This trend will increase because of technological advances that allow us to more ef-

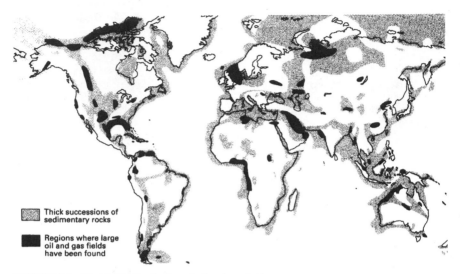

Thick successions of
sedimentary rocks

Regions where large
oil and gas fields
have been found

FIGURE 8.3 Worldwide distribution of oil and gas fields.

FIGURE 8.4 Series of conventional pumping wells operating from a very productive oil-bearing zone, Long Beach, California.

fectively locate new subsurface deposits, and more cost-efficiently recover petroleum from them. However, there are factors that also inhibit exploration for and recovery of oil. Prospective areas for new discoveries may be environmentally sensitive and closed to exploration. Barriers to petroleum exploration and development also come from international political turmoil and unrest, institutional barriers, uncertain property rights and territorial disputes, and urbanization. The location of new petroleum resources in remote regions without transportation facilities can also inhibit their development.

8.1.3   Environmental Concerns during Exploration and Production

Environmental impact to soil and groundwater resources from the drilling of wells in oil and gas fields is an important issue in many parts of the world. Oil fields can cover tens to hundreds of square miles, and have hundreds of production wells. Oil field facilities include production wells as well as sumps for the storage of waste fluids (mostly water), injection wells for subsurface disposal of waste fluids, pumping facilities, storage tanks for recovered oil, and pipelines. In the United States, contamination of water wells and streams by petroleum hydrocarbons extends back to the latter half of the nineteenth. In Marion, Indiana, local streams and the single source of groundwater involving 200 to 300 surface and rock wells were contaminated by adjacent petroleum production activities (Sackett and Bowman, 1905). Significant releases of petroleum hydrocarbons from unlined surface impoundments in oil fields have also been reported as far back as the early 1900s (Bowie, 1918). One unlined surface oil reservoir located in the Kern River field, southern California, had a reported fluid loss on the order of 500,000 barrels. Excavated pits showed oil penetration to depths exceeding 20 ft. Another loss of one million barrels over a period of 6 years occurred from another unlined reservoir in the same field, although some of this loss was through evaporation.

In addition, there are an estimated 1,200,000 abandoned wells in depleted oil and gas fields in the United States. Many old oil and gas wells were not adequately plugged upon abandonment, and have potential to leak well fluids (oily brines primarily) to local groundwater and ground surface (Fig. 8.5).

With about 17 percent of the United States oil production derived from offshore wells, marine settings for the exploration, discovery, and production of oil presents many technical and environmental challenges. Producing oil and gas fields in marine environments present unique challenges due to the proximity of sensitive ecosystems such as fisheries, breeding grounds, coral reefs, wetlands, and salt marshes. Oil spills to the open waters from discovery and recovery operations typically account for less than 5 percent of the total volume of all oil discharges to the marine environment. However, there have been several large oil spills from offshore blowouts at production wells (Table 8.1). Natural seeps release much more petroleum to the marine environment than production facilities. In fact, natural seeps in the North Sea are estimated to contribute 4 times as much oil to the marine environment there as all spills from discovery and recovery activities worldwide.

Exploration and production of oil and gas is now stringently controlled through a myriad of regulations, with many changes in the manner such activities are currently conducted. Technological advances and new operational procedures have enabled such activities to be conducted in an environmentally sound and responsible manner. In addition, there has been much progress toward understanding and mitigating the environmental impacts of petroleum in the environment.

FIGURE 8.5 Petroleum and brine outflows induced by waterflooding activities from an abandoned oil well, Martha Oil Field, Kentucky. (*From Eger and Vargo, 1989.*)

This chapter provides a synopsis of environmental issues associated with the oil and gas industry. Discussed is the history of oil and gas exploration and production, location of major oil and gas fields, and characterization of crude oil and petroleum. These topics are subsequently followed by discussion of primary sources of environmental concern, behavior of crude oil in the subsurface, techniques for mitigation of environmental impact, preventive measures, and case histories, both within the United States and international.

TABLE 8.1   Largest Oil Spills as a Result of Well Blowouts*

Incident	Country	Location	Year month/day	Tons, thousands
IXTOC blowout	Mexico	Gulf of Mexico	1979, June 5	450
Oil well blowout	Uzbekistan	Fergana Valley	1992, March 2	330
Oil platform blowout	Iran	Nowruz field	1993, February 4	300
Oil well blowout	Libya	Inland	1980, August 11	140
Abkatun Production well blowout	Mexico	Bay of Campeche	1986, October 3	33
Funiwa no. 5 well blowout	Nigeria	Off Forcados	1980, January 17	25
Ray Richley well blowout	United States	Ranger, Texas	1985, November 6	15
Laban Island well blowout	Iran	Persian Gulf	1971, December 2	13
Oilwell blowout, Chevron Main Pass	United States	Louisiana	1970, February 10	9
Oil well blowout	United States	Santa Barbara, Calif.	1969, January 28	9
Corpoven well blowout	Venezuela	El Tigre	1979, January 1	8
Oil well blowout	Uzbekistan	Fergana Valley	1994, March 2	8
Trinimar 327 well blowout	Venezuela	Guiria	1973, August 8	7

* Compiled from data provided by American Petroleum Institute, *Environment Canada, Oil Spill Intelligence Report,* United States Coast Guard, and others.

8.2 NATURAL SOURCES OF CRUDE OIL AND PETROLEUM IN THE ENVIRONMENT

8.2.1   Natural Formation of Petroleum Deposits

Petroleum is a naturally occurring mixture that usually exists in a gaseous phase (natural gas) or in a liquid form (crude oil), but can also exist as a solid (waxes and asphalt). Primarily composed of hydrocarbons, which are compounds that contain only hydrogen and carbon, petroleum varies widely in chemical complexity and molecular weight.

In order for significant accumulations of oil and gas to form, four geologic elements are needed:

- Source rock that generates hydrocarbons
- Reservoir rock that stores hydrocarbons
- Geologic trap that forces migration of hydrocarbons
- Seal that inhibits dispersal of hydrocarbons from the trap

Since these four elements are not always present, only about 2 percent of the organic material in rock is actually transformed into petroleum.

Source Rock.   Petroleum source rock is rich in organic matter. This organic matter is most commonly derived from microscopic plants (phytoplankton) and animals (zooplankton) that lived in the world's oceans. Because phytoplankton require sunlight to live, this part of the ocean's biomass is produced where sunlight can penetrate (the photic zone). Compounds produced by marine organisms fall into three major groups of biochemicals: carbohydrates, proteins, and lipids. The elemental compositions of proteins and carbohydrates are not favorable for the generation of petroleum. The lipids, however, have carbon and hydrogen compositions that are very similar to those of petroleum and, in fact, this class of biochemical includes hydrocarbons produced directly from organisms as metabolic products. Several such hydrocarbons are known to exist both in organic-rich sediments and in petroleum mixtures.

Much of the ocean organic matter is consumed in the food chain. In addition, as biomass settles through well-oxidized marine water, much of the carbon and hydrogen—the two elements required to form hydrocarbons—is lost as a result of oxidation. The carbon reacts with oxygen to form carbon dioxide, whereas hydrogen combines with oxygen to yield water. Thus, hydrogen and carbon hydrocarbons are significantly depleted in open marine environments. In some settings, biologic productivity is high enough that the highly oxidizing conditions cannot be maintained. These areas become stagnant and give rise to conditions that are optimal for preserving organic matter. Swamps and marshes are well known for having high levels of biologic productivity, and in these cases, the organic matter commonly becomes peat, or coal. Under marine conditions, if organic matter is not destroyed, sediments on the ocean floor become enriched and organic-rich shales are formed. It is these organic-rich layers that are the most common source rocks that generate petroleum.

Source rocks must be exposed to elevated temperatures for a period of time in order for petroleum to be generated. The level of temperature and the length of time vary greatly, depending on the kind of organic matter in the source rocks. Temperatures rise with depth in the earth, and it is the earth's thermal energy that converts organic material in source rocks to petroleum. The level of temperature and the amount of time required, however, vary considerably and are somewhat interchangeable. For example, certain sedimentary basins in South America have existed for over 100 million years, but because temperatures are low, the source rocks have generated little petroleum. By contrast, some 10 to 15 million-year-old deposits in the southern California Los Angeles Basin have high temperatures due to their rapid burial depth and high isotherm, and these source rocks have generated tremendous quantities of petroleum. If temperature conditions are not sufficiently high for the organic matter in rocks to be converted to petroleum, a special type of rock called *oil shale* is formed.

Many parts of the western United States have oil shale deposits either at the surface or at very shallow subsurface depths. Such rocks can be artificially heated to produce petroleum. Several attempts have been made to economically convert oil shales to petroleum. This has taken place most notably in the states of Colorado, Wyoming, and Utah. Oil yields from oil shale situated in the western states are estimated to be greater than yields in the entire Middle East. However, economics associated with the oil shale conversion process in relation to the price of oil, and stringent environmental regulations and constraints, prohibit further development of these resources at this time.

Reservoir Rocks.   The petroleum generated by the heating of source rocks is buoyant and will slowly migrate where it can in the subsurface. Reservoir rocks provide subsurface space for migrating petroleum to accumulate in reservoirs. Reservoir rocks have interconnected voids that allow petroleum to accumulate in them. Some reservoirs can be similar to beach sand. The sand grains are made up of small mineral or rock particles that are packed closely together. The openings between sand grains are occupied by air. When these sand grains are consolidated together to form rock, some voids remain and are filled with air if the rock is shallow and above the water table, or with a fluid such as water or petroleum if it is deeper.

Geologic Traps.   Geologic traps force the subsurface migration of petroleum and enable reservoir rocks to become saturated with petroleum. Traps can take several forms or combinations of forms. The two main types are structural and stratigraphic. Structural traps are formed when reservoir rocks are folded or become displaced relative to one another (Fig. 8.6). They can occur where underground rock layers have been warped or arched upward, creating a dome (anticline). Anticlines are some of the most important traps in the world. For example, the Spindletop field of eastern Texas was discovered in 1901, and was structurally only 12 ft high.

Fractures of rock formations, often with displacement by faulting, can also inhibit the migration of petroleum. Stratigraphic traps as found in Prudhoe Bay, Alaska, are created where there was a change in the character or extent of the reservoir rock, such that hydrocarbons cannot migrate past it. In all traps, the petroleum will rise to the shallower parts. Being lighter than water, oil and gas will rest above water within these traps.

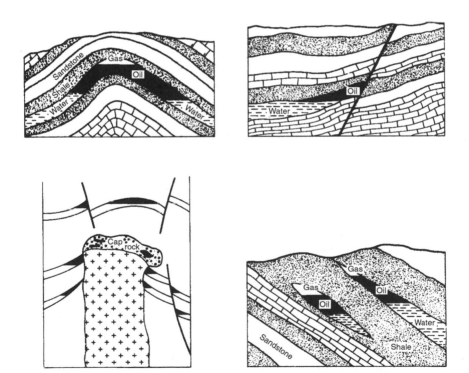

FIGURE 8.6   Schematic showing reservoir rocks and oil and gas accumulations within a stratigraphic trap.

Seals. The fourth geologic component required for the appreciable accumulation of hydrocarbons is a seal. A seal is an impermeable layer that acts as a stopper or cap, and prevents the hydrocarbons from rising through or around a trap. Any rock that has low permeability, such as shale, can serve this purpose.

8.2.2 Natural Seeps

If geologic traps and their seals are not effectively developed, they leak. The resulting petroleum releases are natural seeps to the environment. Natural seeps occur on land and on the ocean floor.

Seeps on Land. Naturally occurring oil seeps may provide evidence of large petroleum deposits at depth, and many of the giant oil fields in the Middle East were discovered by drilling at such locations. Under some circumstances, surface deposits can be quite large. The tar sand deposits that occur naturally in Utah hold an enormous amount of asphalt-like bitumen. Although current economic and environmental pressures make the Utah deposits too expensive to exploit as sources of petroleum, the surface tar sand accumulations in northern Alberta, Canada, have been commercially surface-mined for over 20 years. A large part of the world's remaining oil resources is found in the form of oil shales, heavy and extra heavy oils, and bitumens. These unconventional resources are 10 times greater in volume than the recoverable oil resources that remain. These resources are currently not economically feasible for recovery.

Seeps in the Oceans. Geologists estimate that about 250,000 to 600,000 metric tons of oil per year is derived from natural seeps in oceans. This represents an estimated 11 percent of the oil that ends up in the ocean waters. Marine seeps are common offshore California, in the northern Mediterranean Sea, and offshore Gulf of Mexico. Seeps along the margins of the Pacific Rim contribute about 40 percent of the world's total natural seepage to the ocean environment. This is not surprising, since this area is characterized by complex geology, extensive and complex fault systems that extend to the ocean floor, high tectonic activity, and high volcanic and earthquake activity, allowing for oil to escape.

In southern California, within a 1000 mi^2 area from Point Conception to Point Fermin, about 50 seeps seasonally can vent upward of 900 bbl/day (Fig. 8.7). In the Point Conception area, one oil company uses a concrete structure in the shape of an inverted funnel to trap oil that is seeping naturally from the seafloor. Seepages often decrease in time, reflecting less available oil underground and oil removal by commercial drilling. In the Gulf of Mexico, an individual oil slick on the ocean's surface is estimated to be about 100 m wide, 0.1 µm thick, and over 100 km in aerial extent. Such a slick contains about 100 L of oil. A slick of this size, unless replenished at 100 L/day, is estimated to have a life span of about 24 hours before it dissipates. Since there are an estimated 100 such seeps in the Gulf at any time, an estimated 40,000,000 L of oil seeps into the Gulf every decade or 25,200 bbl/yr, a small fraction of the annual Gulf oil extraction. What these estimates tell us is that the quantity of oil derived from natural seeps over time approaches the amount extracted by the petroleum industry.

Oil seeps in marine environments have occurred for millions of years, and are a natural part of the ecosystem in many areas. Some marine species thrive in marine environments where oil seeps are present, notably in close proximity to a vent. The resulting biological activity produces metabolic by-products of microbes, which tend to plug up pores and fissures by the precipitation of calcium carbonate. Offshore southern California, a healthy community of bottom-dwelling marine organisms is

FIGURE 8.7   Naturally occurring oil seep from the Monterey Formation exposed at the beach in Point Arena, California.

associated with these seeps, with the population of certain species being greater in seep areas than other areas. Karo Island in the Persian Gulf region was named for the tar that seeps from the sea bottom near the island. In this region, where several underwater vents are evident, the acclimated micro-organisms actually promote hydrocarbon biodegradation and photo-oxidation of the released petroleum. Natural seepage of oil to the open sea imposes fewer harms than if it impacts more localized and sensitive ecosystems such as wildlife nesting sites or estuaries. Natural seeps also present less impact than spills on the ocean surface. With natural seeps, the oil is dispersed throughout the water column, but surface oil spills commonly form a thick, gooey mousse that is difficult to disperse.

8.3   ENVIRONMENTAL CONCERNS IN OIL AND GAS FIELDS

8.3.1   Physical and Chemical Characteristics of Petroleum

Oil is naturally occurring and is often referred to as *petroleum.* Crude oil, or *crude,* is unrefined oil or petroleum. More specifically, petroleum is a naturally occurring mixture that usually exists in gaseous form (natural gas) or liquid form (crude oil), but can also exist as a solid (waxes and asphalt). Primarily composed of hydrocarbons that are compounds that contain only hydrogen and carbon, petroleum varies widely in chemical complexity and molecular weight.

Petroleum can be any mixture of natural gas, condensate, and crude oil. The term *petroleum* is derived from the Latin *petra* for rock and *oleum* for oil. A petrochemical is a chemical compound or element recovered from petroleum or natural gas, or derived in whole or in part from petroleum or natural gas hydrocarbons, and intended for chemical markets. Petrochemicals and hydrocarbons are simply compounds of hydrogen and carbon that can be distinguished from one another by composition and structure (Fig. 9.6a).

Crude oil (commonly just called *crude*) is the initial oil extracted from the subsurface without any refinement into other liquid forms, or products. It is a naturally occurring heterogeneous liquid consisting almost entirely of the elements hydrogen and carbon. The composition of crude oil can vary significantly, depending on its origin, age, and history. Crude generally ranges from 83 to 87 percent carbon (by weight) and 11 to 14 percent hydrogen, with lesser amounts of sulfur (0.1 to 5.5 percent), nitrogen (0.05 to 0.08 percent), and oxygen (0.1 to 4 percent). Trace constituents of less than 1 percent in total volume include phosphorus and heavy metals such as vanadium and nickel.

Crude is classified according to the relative content of three basic hydrocarbon structural types: paraffins, naphthenes, and aromatics. About 85 percent of all crude oil can be classified as either asphalt base, paraffin base, or mixed base. Sulfur, oxygen, and nitrogen contents are often relatively higher in comparison with paraffin base crude, which contains little to no asphaltic materials. Mixed-base crude oil contains considerable amounts of both wax and asphalt. Chemically, crude oil is composed of methane (normal straight chain paraffins), isoparaffins (branched-chain paraffins), cycloparaffins or naphthenes (ring structures), aromatics (benzene-ring structures), and asphaltics (Fig. 9.7).

8.3.2 Constituents of Environmental Concern

Certain materials generated as part of exploration and production activities are exempt from regulation as waste materials. Exempt status depends on how the material was used or generated as waste, not necessarily whether the material is considered toxic or hazardous. Some exempt materials may be considered hazardous whereas some nonexempt materials may not be as harmful. Essentially, if the material or waste is derived from downhole (i.e., was brought to the surface during oil and gas operations), or has been generated by contact with the oil and gas production stream during the removal of produced water or other contaminants from the product, then the material or waste is likely considered exempt from RCRA Subtitle C regulations (U.S. EPA, 1995). However, this does not preclude regulatory control under state regulations, or federal solid waste regulations or other appropriate federal regulation. A tabulation of exempt and nonexempt RCRA wastes is presented in Table 8.2.

Ninety-eight percent of the waste from producing oil and gas is water, frequently containing high salinity and high dissolved solids. The produced water is called *drilling brine*. The brines are brought up with the oil and are usually collected in tanks, and are either injected back into the well to help recover more oil and gas, or injected into the underground formations in a manner that prevents contamination of surface water and underground drinking water sources. The primary constituents of environmental concern at oil field and gas field sites are:

- Methane
- Crude oil
- Drilling muds
- Refined petroleum products and constituents
- Naturally occurring radioactive material (NORM)

TABLE 8.2 Summary of Exempt and Nonexempt Exploration and Production RCRA Wastes*

Exempt wastes
Produced water
Drilling fluids
Drill cuttings
Rigwash
Drilling fluids and cuttings from offshore operations disposed of onshore
Well completion, treatment, and simulation fluids
Basic sediment and water and other tank bottoms from storage facilities that hold product and exempt waste
Accumulated materials such as hydrocarbons, solids, sand, and emulsions from production separators, fluid treating vessels, and production impoundments
Pit sludges and contaminated bottoms from storage or disposal of exempt wastes
Workover wastes
Gas plant dehydration wastes, including glycol-based compounds, glycol filters, filter media, backwash, and molecular sieves
Gas plant sweetening wastes for sulfur removal, including amine, amine filters, amine filter media, backwash, precipitated amine sludge, iron sponge, and hydrogen sulfide scrubber liquid and sludge
Cooling tower blowdown
Spent filters, filter media, and backwash
Packing fluids
Produced sand
Pipe scale, hydrocarbon solids, hydrates, and other deposits removed from piping and equipment prior to transportation
Hydrocarbon-bearing soil
Pigging wastes from gathering lines
Wastes from subsurface gas storage and retrieval, except for the listed nonexempt wastes
Constituents removed from produced water before it is injected or otherwise disposed of
Liquid hydrocarbon removed from the production stream but not from oil refining
Gases removed from the production stream such as hydrogen sulfide and carbon dioxide, and volatized hydrocarbons
Materials ejected from a producing well during blowdown
Waste crude oil from primary field operations and production
Light organics volatilized from exempt wastes in reserve pits or impoundments or production equipment

Nonexempt but not necessarily hazardous waste
Unused fracturing fluids or acids
Gas plant cooling tower cleaning wastes
Painting wastes
Oil and gas service company wastes such as empty drums, drum rinsate, vacuum truck rinsate, sandblast media, painting wastes, spent solvents, spilled chemicals, and waste acids

TABLE 8.2 Summary of Exempt and Nonexempt Exploration and Production RCRA Wastes* (*Continued*)

Nonexempt but not necessarily hazardous waste
Vacuum truck and drum rinsate from trucks and drums transporting or containing nonexempt waste
Refinery waste
Liquid and solid wastes generated by crude oil and tank bottom reclaimers
Used equipment lubrication oils
Waste compressor oil, filters, and blowdown
Used hydraulic fluids
Waste solvents
Waste in transportation pipeline-related pits
Caustic or acid cleaners
Boiler cleaning waste and refractory bricks
Incinerator ash
Laboratory wastes
Sanitary wastes

American Petroleum Institute intended exempt wastes not listed under RCRA
Excess cement slurries and cement cuttings
Sulfur-contaminated soil and sulfur waste from sulfur recovery units
Gas plant sweetening unit catalyst
Produced water contaminated soil
Wastes from the reclamation of tank bottoms and emulsions when generated at a production location
Production facility sweetening and dehydration wastes
Pigging wastes from producer-operated gathering lines
Production line hydrotest/preserving fluids utilizing produced water
Iron sulfide

* Modified after Navarro (1995).

Methane. The most common compounds and constituents associated with oil-field properties that may be considered hazardous include methane gas, crude oil, drilling mud, and refined petroleum products including volatile organic compounds. Methane gas is a colorless, odorless, tasteless paraffin compound that is less dense than air, formed as the by-product of organic decomposition. The concern surrounding methane is its flammability and explosive potential, particularly in man-made enclosed spaces such as poorly ventilated rooms, basements, and conduits. Since methane is lighter than air, it can migrate upward along natural or man-made conduits such as fractures in bedrock or along oil wells that have not been abandoned properly. When it reaches a confined space, the methane can be explosive when its concentration in air is in the range of 5 to 15 percent.

Methane in an oil field environment is typically biogenic (bacterial) or petrogenic (thermogenic) in origin. Biogenic gas typically is the result of decomposition of nonpetroleum organic deposits such as plants and landfill deposits. Petrogenic gas typically is a by-product of petroleum hydrocarbons. Background levels of methane are usually less than a few hundred parts per million (ppm). In situ values of 1000 to 20,000 ppm are considered to be potentially hazardous, and greater than 20,000 ppm is considered potentially dangerous. In 1985, an explosion and fire destroyed a department store and a number of adjacent structures in a portion of the abandoned Salt Lake Oil Field in the commercial Fairfax district of Los Angeles. More stringent regulations were subsequently developed to assess whether abandoned oil wells have been properly sealed, and to require mitigative measures as necessary.

Crude Oil. Although crude oil in itself is not considered a waste, some states such as California consider it a designated waste should it exceed certain maximum contaminant levels for arsenic, chloride, chromium, lead, or polychlorinated biphenols (PCBs) or flash point. Thus, its disposal off site is subject to regulation. Crude oil is not usually of concern with regard to air quality since it has a natural source, typically has a high boiling point (greater than 302°F), and maintains a very low vapor pressure. Soil containing crude oil has been left in place during many redevelopment projects throughout Los Angeles and Orange Counties in southern California, typically at depths of 5 to 10 ft below final grade; however, its presence can have a significant financial impact on developers and lenders during oil field property redevelopment or transfers.

Drilling Muds and Cuttings. During rotary drilling for oil and gas wells, two types of wastes are generated: used drilling fluids, commonly known as *muds,* and drill cuttings. Drilling muds are mixtures of water and other chemical additives used to lubricate the drill bit, remove cuttings from the well bore, maintain the integrity of the hole until casing and production equipment is installed or during well abandonment, and to prevent blowout.

During drilling, different additives are mixed with water to yield the desired properties for the mud. The consistency (density, viscosity, weight, gel strength, filtration, and salinity) and mineral content of drilling muds vary to accommodate the nature of the strata, oil, gas pressure, and other oil and gas field characteristics. Drilling muds can occasionally be of environmental concern because of the potential presence of heavy metals that may exceed certain regulatory standards.

Onshore, a pit or sump is typically excavated adjacent to the drill rig, which serves as a mixing area for the muds and as a settling pond. Since drill cuttings and muds may in some instances be considered a waste material, they must be handled in an appropriate manner. The waste muds and cuttings are thus either injected into the subsurface, reused, or disposed of. Offshore, such fluids cannot always be discharged into open waters, and are then either reinjected into the subsurface, reused, or transported to onshore facilities for disposal.

Drilling muds are characterized as water-based muds, oil-based muds, or synthetic-based muds. Water-based muds are the least expensive and most widely used drilling fluid. Water-based muds contain both organic and inorganic additives. Additives may include clays (bentonite or attapulgite), barite, dispersants (tannins, quebracho phosphates, lignites, and lignosulfonates), starch, sodium carboxymethyl cellulose, polymers (cypan or drispac), detergents, and defoamers. Less widely used are oil-based muds that may contain some of the same additives listed above.

Some of these constituents, including metals such as chromium, can be of environmental concern. Most of the wells drilled to depths less than 10,000 ft, and about 85 percent of deeper wells, utilize water-based muds. They are commonly used both onshore and offshore. Offshore, water-based muds and associated cuttings are typically discharged to the open waters, provided certain discharge limitations are not exceeded. If the limits are exceeded, offshore handling of water-based muds is an alternative that can be expensive, and presents logistical problems and environmental risks.

In more complex drilling situations, notably with drilling depths exceeding 10,000 ft, high angle, high temperature, or other special drilling circumstance, oil-based muds are commonly used. Oil-based muds contain a continuous liquid phase of oil. True oil-based muds contain 5 percent water or less by volume, and use crude oil as a major constituent. Oil-based muds are very similar in composition to crude oil, and along with the drill cuttings, cannot be directly discharged into open waters.

Synthetic-based muds were recently developed to replace oil-based muds. These synthetic-based muds combine the performance of oil-based muds with the easier and safer disposal and handling aspects of water-based muds, while minimizing pollution. Synthetic-based muds contain no polynuclear aromatic hydrocarbons, have lower toxicity and lower bioaccumulation potential, and biodegrade at a rate faster than oil-based muds. These newly developed muds cost more than oil-based muds, but also provide lower disposal costs since they can be directly discharged to open waters, an important factor in areas such as the Gulf of Mexico where drilling has increasingly moved to deeper waters. In comparison to water-based muds, synthetic-based muds provide higher performance (i.e., cleaner hole, less sloughing, lower drill cutting volumes), while in some instances performing better than oil-based muds, and can be recycled.

Regardless of the type of drilling mud used, typical contaminants of interest that require periodic monitoring for significant changes are pH, electrical conductivity, sodium adsorption ratio, cation exchange capacity, exchangeable sodium percentage and total metals. Other constituents of concern include oil and grease and total petroleum hydrocarbons. Drilling fluids usually have a pH that falls within the alkaline range (pH > 10). This high pH is a result from the addition of lye, soda ash, and other caustics, which allows for the dispersion of clay and increased effectiveness. Weathering and aging cause a decrease in the overall pH. Soil salinity is measured by determining the electrical conductivity. This is an important test for soils and waste because of the potential for high brine content that adversely affects plant growth and water quality. Soils exhibiting an electrical conductivity in excess of 8.0 mmhos/cm usually require some manner of management or remediation. Sodium adsorption ratios (SARs) are determined to assess potential sodium damage from a waste material. Used in conjunction with electrical conductivity, potential damage associated with sodium salts can be ascertained. An SAR < 12 can restrict such materials for land disposal. Acceptable metals loading in muds are evaluated as the cation exchange capacity (CEC). Measured in meq/100 g, CEC values are required to estimate the exchangeable sodium percentage (ESP). Excess sodium typically results in a general lack of structural stability among soil particles and impeded water infiltration. Combined excess salinity and sodic conditions can limit remediation efforts (i.e., remove excess salts from the root zone) because of inherent slow infiltration and percolation characteristics.

Total metals analysis provides a good indication for all metals except barium that is best analyzed under the protocol set forth by the Louisiana Department of Natural Resources. Total metals include arsenic, barium, cadmium, chromium, mercury, lead, selenium, and zinc. Although seldom a significant problem, elevated concentrations of certain metals in soil or waste materials are labile. The metals of most concern in drilling muds are barium, chromium, lead, and zinc.

The presence of petroleum hydrocarbons in drilling muds or waste are typically due to the introduction of crude oil from a producing formation and diesel or mineral oil that is added to drilling muds. Although diesel is likely to be the most common ailment, diesel-affected soil and waste materials can be easily remediated via a variety of options.

Chemical Additives. Secondary and tertiary methods of production generally require the use of injected fluids that may contain various production-enhancing chemicals such as surfactants and polymers. Production in marginally producing, generally older oil and gas fields becomes more attractive as the price of oil and gas moves upward. With a major increase in price of oil and gas, enhancements in field production are evaluated. One common enhancement in older oil and gas fields is the cleanup and stimulation within individual wells as part of a field workover program. This type of production enhancement program usually requires the use of chemicals, including a variety of acids.

Acidizing operations require the use of a variety of chemicals for pH adjustment and associated precipitation issues. The use of acids can create a number of production problems, including the release of fine particles that can plug a formation as well as the corrosion of steel drill pipe and casing. Highly corrosive produced waters require the use of corrosion-resistant tools and chemical inhibitors. These corrosion inhibitors, such as oil-wetting surfactants, slow down the reaction time of acid on the metal drilling and production pipe. Fluid loss control agents (silica flour and oil-soluble resins with natural gum) are added, to reduce leak-off in fracture acidizing operations. Diverting or bridging agents (graded salt, wax beads, sand) in fracture acidizing may be used as materials to prop up the newly created fractures. Particularly in dry gas wells, alcohol has been used as an additive to reduce the time required for well cleanup. Clay stabilizers are used to fix clays in situ, thereby minimizing migration of clays and subsequent plugged permeability. Iron sequestering agents (acetic, citric, and lactic acids) are used to inhibit the precipitation of iron after the acids are spent from an acidizing operation.

Hydraulic fracturing in the oil and gas fields uses nitrogen as a well stimulation. Nitrofied fracturing and acidizing uses a foam on fluid-sensitive wells for improved fluid-loss control and cleanup operations for better production. Five acid systems are frequently used with carbonate reservoirs, such as limestone and dolomite: mineral acids (hydrochloric and hydrofluoric/hydrochloric), organic acids (acetic and formic acids), powdered acids (sulfamic and chloroacetic acids), retarded acids (gelled acids, oil wetting surfactants, and emulsified acids), and mixed acids (combinations of acids). The strength of an acid will range from a few percent to less than 30 percent by weight in water.

Naturally Occurring Radioactive Material (NORM).   NORM is found at levels exceeding background at many oil and gas production and processing facilities. NORM originates in subsurface oil and gas formations, and is usually brought up to the surface with produced fluids and gases, including brine water, natural gas, and other oil field fluids. NORM forms as scales and precipitates on tubing and equipment, sludge, and sands with isotopes of radium, thorium, and uranium, as radon gas emitted from radium-contaminated materials and soils, and as deposits of lead Pb-210 on the interior of pipes from the transmission of natural gas and produced waters (Veil et al., 2000).

Isotopes of uranium and thorium, which originate in hydrocarbon-bearing formations, are parent isotopes of radium and radon. Occurring primarily as Ra-226 of the uranium U-238 decay series, and Ra-228 of the thorium Th-232 decay series, these isotopes have long half-lives. Therefore, the long-term potential for disposal is of concern. Oil wells that produce large quantities of produced water will also tend to accumulate the greatest amount of radium-bearing materials as a result of the solubility of radium and its chemical similarity to certain ions such as calcium, strontium, and barium. Gas wells precipitate radon daughters from natural gas streams and fluids. These wells tend to accumulate materials containing larger quantities of lead-210, polonium-210, and bismuth-210.

Other Hazardous Compounds. There are a variety of hazardous compounds associated with oil and gas facilities that are indirectly related to the produced hydrocarbons. These hazardous compounds are typically found in equipment maintenance and chemical storage areas in oil and gas fields: hydraulic fluids, painting wastes, used equipment lubrication oils, unused free fluids and acids, radioactive-tracer wastes, waste solvents, herbicides for vegetation control, and pesticides. In addition, PCBs, a dielectric fluid, are common in transformers built prior to 1979. Unless a transformer has a label stating "PCB-free," transformer oils are assumed to contain PCBs.

Lead, a durability agent, was added to paint and is commonplace in industrial paints and coatings. Lead may be present in the paint surfaces of rigs, tanks, and production equipment. Lead paint was phased out in the United States by December 1980. Unless tested, all metal surfaces older than December 1980 are presumed to contain lead. Metal products containing lead paint are still being imported into the United States on painted products as of today. Torch cutting on metals containing lead paints such as pipelines, tanks, and production equipment can release lead fumes, exposing workers to airborne lead. Dust from cutting lead paint in oil and gas fields is also an employee exposure risk.

Asbestos has been used for a variety of industrial purposes since the 1920s. In oil and gas fields, asbestos has been used in tar wrap for corrosion control of metal surfaces, such as those on tanks and pipelines. The fibrous nature of asbestos is similar to that of straw in bricks, adding strength to the wrap. Thermal insulation on tanks, pipes, or equipment containing asbestos may be present in oil and gas fields. In steam injection plants for the production of heavy oil, steam lines may have thermal wrapping containing asbestos. Unless tested, all suspected asbestos-containing materials dating from before 1980 are presumed to contain asbestos. Nonetheless, importation of asbestos or use of stored asbestos-containing materials may continue to the present.

8.3.3 Sources of Environmental Concern

In general, environmental concerns associated with oil and gas fields include accidental releases of waste fluids or produced petroleum; aesthetic impacts associated with physical facilities such as drilling rigs, storage tanks, and pipeline corridors; and potential conflicts with other land uses for the area. Primary sources for the release of hydrocarbon constituents that may result in hazardous conditions or generate materials considered hazardous include oil wells, sumps and pits, surface reservoirs and aboveground storage tanks, improperly abandoned wells, random spillage and spills, and leakage from storage units and pipelines (Fig. 9.12). Secondary sources include pumping stations, piping ratholes, transformers and capacitors, underground tanks, and well cellars (Fig. 8.8).

Cement and Annular Failures. Poorly cemented annular spaces in oil or gas wells may act as a conduit for production fluids, both brines and hydrocarbons, to leak into possible groundwater-bearing zones. Channeling is caused by the incomplete displacement of the drilling mud by the cement slurry, resulting in washed-out sections of the annular space. Secondary channeling is caused when annular voids are created after the cement slurry is in place. Shattering in perforation zones can create

(a)

(b)

FIGURE 8.8 Historic photo of leaks from (a) a pumping station and (b) a well cellar.

additional annular damage and possible leakage of contaminants or brines into groundwater-bearing zones (Fig. 8.9). Poor cement bonding between the interface of the casing and cement, or cement and the well bore wall, create leakage problems as well. Poor-quality cement may result if the wrong cement or additives are used, or if the cement is prepared improperly. The failure of the cement can cause void spaces

FIGURE 8.9 Photograph of perforated drilling pipes resulting from corrosion by brine fluids, Martha Oil Field, Kentucky. (*From Eger and Vargo, 1989.*)

and further cement failure, providing conduits into the subsurface of production fluids. Leakage can occur at the interface between the casing and cement and can cause conduits to form between the casing and the cement interface.

Aesthetic Concerns. Aesthetic concerns include those factors that affect our senses in an unfavorable manner. Visual impacts such as the sight of an oil rig located in what is considered a pristine wilderness or wetland area, or offshore rigs on the distant horizon, are offensive to some individuals. Visual evidence of spillage or leakage of petroleum or other compounds at an oil or gas field includes stained or discolored soil, dead vegetation, and petroleum sheen on water. Even produced waters with high salinity can kill vegetation.

In congested urban areas, oil-drilling activities have been camouflaged to have the appearance of a high-rise building (Fig. 8.10). Other concerns, such as dust, odorous fumes, noise, traffic, and the potential for fires, explosions, and spills, can also generate unfavorable aesthetic value, especially in urbanized areas.

Land Use Conflicts. Environmental concerns that affect oil fields are amplified in pristine, highly visible, sensitive areas such as wetland areas or on the North Slope of Alaska (Fig. 8.11) and many urban environments. As oil fields within urbanized areas reach the end of their productive lives, they are rapidly taken out of production and redeveloped. Nowhere is this more evident than in southern California. This highly aesthetic, densely populated area has a rich history of oil and gas exploration and exploitation dating back to the 1860s, the first year of commercial pro-

FIGURE 8.10   A modern drilling operation camouflaged as a high-rise building in southern California.

duction. As oil fields within the urbanized Los Angeles Basin area reach maturity and the end of their productive lives, the property associated with these fields faces high demand for more profitable land usage. In such areas, numerous high-volume refineries and tank farms also exist, which can contribute significantly to subsurface degradation and poor air quality. General disposal of waste materials during the active life of an oil field operation is certainly easier to manage in an urban setting with an abundance of service and support resources readily available. In nonurban environments, these issues remain; however, concern and emphasis on the potential impact of operations on natural habitats and sensitive environs such as wetlands and wildlife areas are usually greater (Fig. 8.12).

FIGURE 8.11   An exploration drilling rig set on a man-made gravel pad with modular housing, North Slope, Alaska.

FIGURE 8.12   Photograph showing an active oil field situated within the Bolsa Chica wetland area in southern California.

Military Conflicts.   Military conflicts provide additional large-scale environmental concerns regarding well drilling and oil and gas fields. This is no better exemplified than the situation that developed during the Gulf War in 1991. The largest recorded oil spill in history occurred during this event. Approximately 750,000 thousand tons of oil was spilled. This case history is further discussed later in this chapter.

8.4   WASTE HANDLING AND WASTE MINIMIZATION

8.4.1   Waste Handling and Minimization Onshore

Onshore waste handling and minimization at drilling sites is accomplished by a variety of means including the construction of reserve pits or facsimile (i.e., a multipit system), disposal onto the land surface with subsequent land farming, hauling off site for treatment and disposal, subsurface injection, and closed-loop systems.

Reserve Pits.   One of the most widely used methods of handling waste at onshore drilling locations is the use of reserve pits. Reserve pits are earthen depressions constructed to allow enough area where drilling fluids and cuttings can be stored and segregated for eventual disposal once the well is completed. Reserve pits come in a variety of shapes and sizes, but usually are square or rectangular. Solid separation is an important role, and thus the pit is usually large enough to provide enough retention time for adequate separation of solids (Navarro, 1995). The solids are then discarded, and the fluids are reused or discharged into the surrounding area, assuming all regulatory requirements are met. Reserve pits sometimes also serve as part of the drilling fluid circulation system, again allowing for solid separation and the reuse of clean drilling fluid. Today, a series of open-top steel tanks is commonly used for this purpose.

Land Farming.   Land farming involves the removal of drill cuttings and fluids from the reserve pits or holding tanks, spreading them on the land surface, then, once dried, tilling them into the soil. Impacted soils from oil exploration and production sites containing crude, fuels, lubricants or other hydrocarbon wastes can be land farmed, a form of bioremediation. The tilling and turning of the piles aerate and volatilize the lighter end hydrocarbon range. Addition of nutrients, soil amendments, microbes, and moisture have been used to enhance the land farming process. Since most cuttings and fluids are rich in nutrients, they can serve as a fertilizer, enhancing crop production (Deuel and Holliday, 1977). This method is relatively inexpensive, requiring only a bulldozer and dragline, and possibly a dump truck, depending on the distance from the reserve pits and the area to be land-farmed. Pilot testing is usually conducted prior to actual operation to assess potential problems and mix design parameters. From a regulatory perspective, permission from the landowner and lead regulatory agency must be obtained. In addition, an estimate of the amount of drill cuttings and fluids needs to be determined to allow for adequate space for the amount of material to be generated. This is important, since land farming of such materials below a certain depth in the well is not allowed, and the shallow portion of the well will account for about two-thirds of the materials to be land-farmed.

Confirmatory testing of representative soil samples is also performed after completion of land farming activities to assure that adequate mixing has been done, specific parameters are within regulatory limits, and the native soil was not contaminated.

Should crop yields become reduced in the future, such tests can be used to defend against unfounded claims.

Subsurface Injection.   Wastes are also commonly disposed of via injection wells (Fig. 8.13). The use of wells for the disposal of water and waste fluids has been known since the early 1930s, with only four injection wells reported prior to 1950. Since 1950, the injection of waste fluids into deep underground aquifers through the use of wells has been more prevalent. This reflects improved drilling technology and the inability to discharge waters to surface waters even after dilution and treatment. Fluids allowed for disposal into the subsurface through injection wells include coproduced water, wastewater, scrubber blowdown waters, drilling waters, and water softener regeneration brine water from steam generators and cogeneration facilities.

As of 1985, the petroleum industry accounted for about 25 percent of all injection wells, with over half of the fluids injected considered nonhazardous (Fig. 8.14). In

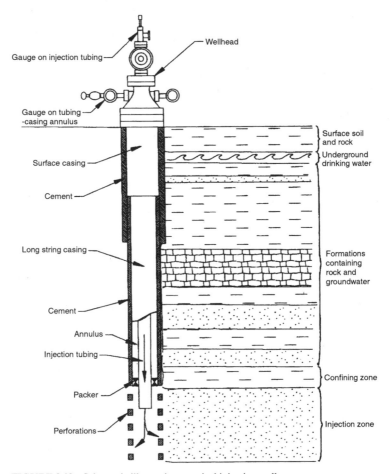

FIGURE 8.13   Schematic illustrating a typical injection well.

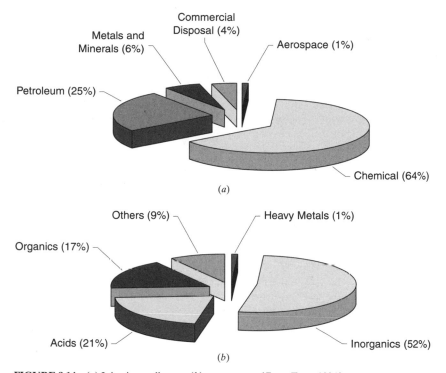

FIGURE 8.14   (*a*) Injection well users; (*b*) waste types. (*From Testa, 1994.*)

1986, an estimated 879,000 active oil and gas wells generated over 60 million barrels of oil field wastes, most of which was brine. The average daily production was on the order of 7.6 million barrels of oil, 40 billion cubic feet of natural gas, and 61 million barrels of produced water. These fluids were subsequently injected into 166,000 injection wells throughout the United States.

Injection wells must be designed in a manner to protect all geologic formations that are penetrated by the well and contain usable waters. Each injection well must thus be assessed with regard to geologic and hydrogeologic conditions such as stratigraphy, structure, permeability, and porosity. Changes in reservoir pressures and temperature with time, and residual oil, gas, and water saturation, are also important in understanding flow characteristics of injected fluids.

Closed-Loop Drilling Systems.   Closed-loop drilling systems are designed to minimize the amount of waste that is ultimately disposed of, and eliminate the discharging of materials into the environment. Such a system for drilling purposes is essentially a solids-control system. These systems consist of a series of screened shale shakers, desanders and desilters, and centrifuges. These mechanical units are used to remove as many of the solids from the drilling fluid as possible, ranging from relatively larger particles down to colloidal solids, allowing the drilling fluid to be recirculated into the drilling fluid system for reuse. Modern systems incorporate chemical injection of the fluid to enhance the removal of solids and assist in recovering much of the water used in the drilling fluid, which is then used as makeup water for the drilling fluid system.

8.4.2 Waste Handling and Minimization in Marine Environments

Producing oil and gas fields in marine environments present special challenges due to the proximity of sensitive ecosystems such as fisheries breeding grounds, coral reefs, wetlands, and salt marshes. Fortunately, operational practices and technology have improved such that there has not been a major oil spill from an offshore drilling platform for about 20 years. Oil spills to the open waters from discovery and recovery operations typically account for less than 5 percent of the total volume of all oil discharges to the marine environment. Natural seeps release much more petroleum to the marine environment than production facilities. In fact, seeps in the North Sea are estimated to contribute 4 times as much oil to the marine environment then all spills from discovery and recovery activities.

Offshore operations in the United States have only three alternatives for the disposal of waste materials and cuttings:

- Discharge from the drilling vessel or platform under existing governmental regulations
- Collect and transport waste materials to the shore for ultimate treatment and/or disposal
- Subsurface injection

Waste types typically generated in marine settings include drilling wastes, produced water, injection water, and various well treatment solutions and chemicals. Drilling wastes are composed of drilling fluids or muds and cuttings that are generated during the drilling of the oil or injection well. Historically, some oil was discharged in oil-based drilling muds, although this practice is not done as much these days with the development of non-oil-based muds. Produced water includes formation water that is brought up to the surface with the oil and gas. Produced water is routinely cleaned to about 30 parts per million before discharging, and accounts for the majority of oil contamination reported. Injection water is used for influencing the flow of subsurface fluids, disposal of waste fluids, and for secondary recovery. In addition, various well treatment fluids and chemicals are routinely added to the well during production and as part of the oil-water separation process and to control corrosion.

Storage Tanks. In lieu of reserve pits, drilling operations offshore commonly will use steel tanks designed for the storage of drilling fluids and cuttings. On inland water, a combination of storage and discharge is sometimes used, whereas offshore, direct discharge into open waters is commonly done.

Discharge. Because of environmental concerns, discharging of drilling cuttings and fluids is basically restricted to offshore rigs that operate in federal waters. In order to discharge, it is important to have knowledge of the area to be drilled, the type of drilling fluid and mud to be used, and the regulations governing the particular area. Discharged waste must meet very specific criteria set forth by the Mineral Management Service (MMS). Two tests are typically required: a sheen test and a Mysid shrimp mortality, or LC-50, toxicity test. With the sheen test, the drilling contractor must visually observe the discharge and, if sheen is observed, discontinue discharging until the problem is addressed; a report is also made to MMS. If the problem cannot be remedied, then the contractor must dispose of the waste into barges, or install a closed-loop system on the rig.

In addition, the contractor is required to obtain a sample of the drilling fluid and have it analyzed for LC-50 on a weekly basis. This test evaluates the mortality rate of the Mysid shrimp in a diluted solution of the drilling fluid for 96 hours. If greater than 50 percent of the shrimp die within the test period, then the fluids are considered too toxic for discharge directly into the open waters. Discharge of fluids that fail such tests can result in fines of $10,000 a day retroactive to the last day of a successful test.

Discharge can also be halted as a result of a stuck pipe. With new drilling technology, such as horizontal or directional drilling, top drive drilling motors, and aluminum drill pipe, the potential for a stuck pipe has increased. An oil-based spotting fluid, one of the most successful being Black Magic, is commonly used to free a stuck pipe. Time is of the essence, since there is an exponential correlation between the time the pipe is initially stuck and it becoming permanently stuck.

8.5 CHARACTERIZING ENVIRONMENTAL IMPACT IN RETIRED OIL AND GAS FIELDS

Environmental concerns associated with onshore oil and gas field properties are evaluated through a phased approach. Assessment of oil and gas field properties, certainly retired ones, typically consists of four phases: due diligence records review and site reconnaissance, subsurface assessment, detailed delineation and characterization, and remediation. A summary of the various phases and specific tasks performed in each phase is presented in Table 8.3.

8.5.1 Phase I—Preliminary Environmental Assessment

Phase I site assessment activities are initially performed to evaluate the potential for adverse environmental impact and to identify specific areas of concern. Activities conducted as part of Phase I are noninvasive and include a site reconnaissance; review of agency documents, records, and reports; review of historical aerial photographs, oil and gas field maps, and hydrogeologic setting generally obtained from the literature. Operator and owner interviews might be included to help to establish hazardous waste handling procedures and housekeeping efforts. Waste manifests and tracking forms might be reviewed to verify the location and management of hazardous materials at an oil or gas field.

8.5.2 Phase II—Preliminary Subsurface Assessment

The objective of the Phase II activities is to determine the actual presence of constituents or circumstances that may be considered hazardous or toxic, or pose a health and safety concern, and characterize the type and extent of contamination at each of the areas of potential environmental concern as identified during conduct of Phase I activities.

Subsurface assessments at such properties are conventionally performed to assess potential adverse impact to soil and groundwater and to identify the potential presence of vapors and of leaking wells that may have been improperly abandoned. Sampling of surficial soils, drilling of soil borings, and installing groundwater monitoring wells with retrieval of representative soil and groundwater samples address these objectives. Soil gas surveys are also routinely performed to evaluate the potential presence of volatile compounds in the vadose zone.

TABLE 8.3 Environmental Site Assessments*

Phase no.	Phase description	Task description
I	Due diligence	Historic records review Regulatory agency file review Historical aerial photograph review Oil field maps and records review Site reconnaissance Documentation of findings and recommendations

II	Preliminary subsurface assessment	Work plan preparation
		Drilling and sampling
		Groundwater monitoring well
		  installation and sampling
		Analytical program
		Oil/gas well abandonment
		Documentation of findings and
		  recommendations

III	Subsurface assessment	Workplan preparation
		Drilling and sampling
		Groundwater monitoring well
		  installation and sampling
		Analytical program
		Remedial strategy evaluation
		Pilot study to evaluate remedial
		  option effectiveness
		Documentation of findings and
		  recommendations

IV	Remedial strategy implementation	Work plan preparation
		Implementation of remedial
		  strategy
		Additional drilling and sampling
		Confirmatory soil and groundwater
		  sampling
		Documentation of findings
		Site closure

* Modified after Testa (1993).

Soil samples can be retrieved by trenching, by hand-augered borings for the re-trieval of shallow samples, or by drilling of soil borings for deeper samples. Trenches are usually dug with a backhoe, and are generally 2 ft wide, 10 to 15 ft long, and 10 to 12 ft deep. Hand-augered borings are used to collect shallow soil samples in areas not easily accessible to larger equipment, or when only a limited number of shallow samples are needed. Hand-augered borings are generally 2 to 3 in in diameter, and extend 5 to 10 ft in depth.

The geologist or engineer logging the boring or trench will note the location of stained or discolored soil and the presence of the various lithologic changes, as well as free product, if encountered. Soil samples may be collected for possible chemical analysis or lithologic or hydrogeologic characterization. Soil may also be collected to analyze on site. On-site analysis may include a screening tool, such as a photoioniza-tion detector (PID), a portable instrument that measures organic vapors. For a more detailed soil evaluation, soil gas surveys may be used as a screening tool to guide the subsequent drilling program, and the selection of soil boring and well locations. Occasionally used to assess the potential presence of hydrocarbon-impacted soil, sample locations are usually formulated in a grid pattern. At each location, a probe is inserted into the soil and vapor samples are retrieved. The samples are then analyzed in an on-site mobile laboratory by a gas chromatograph for such constituents as total petroleum hydrocarbon (TPH), volatile and semivolatile organic compounds, gasoline constituents such as benzene (and toluene, ethylbenzene, and xylenes), and methane gas concentrations.

Rotary Drilling Methods.   Rotary drilling methods have been used for decades in the environmental field for the retrieval of soil and groundwater samples to depths exceeding 150 ft below ground surface. A truck-mounted hollow-stem auger (HSA) drill rig is commonly used (Fig. 8.15*a*), fitted with a modified split-spoon sampler. The sampler is fitted with stainless steel or brass sleeves, and is attached to the cable or drill stem, which is lowered to the desired depth through the center pipe of the hollow-stem auger. For the collection of undisturbed samples, the sampler is advanced into the soil ahead of the auger bit to the desired depth using a 140-lb hammer. The sample is then withdrawn and the sleeves removed. After the soil samples are collected, the hollow-stem auger string is advanced by connecting to additional hollow-stem augers at the surface. The hollow-stem auger rig is most commonly used for the installation of groundwater monitoring wells, typically 2 in in diameter, or groundwater extraction wells, typically 4 to 8 in in diameter.

Direct Push Technology (DPT) Rigs.   Over the past decade, one of the most common methods of soil, groundwater, and soil vapor sampling is the use of direct push technology rigs (Fig. 8.15*b*). One person typically operates DPT rigs. These rigs, sometimes called *probe rigs,* are generally quicker and less costly than the more conventional hollow-stem auger rotary drill for collecting soil, vapor, and water samples for environmental projects. DPT equipment allows for fewer permanent monitoring wells, multiple-depth sampling programs, elimination or minimization of drilling-derived wastes, and minimal exposure of workers to potentially hazardous soil cuttings.

DPT sampling relies on dry impact methods to push or hammer boring and sampling tools into the subsurface for environmental assessments. This technology does not require hazardous chemicals, drilling fluids, or water during operation. A typical auger borehole to 60 ft would generate approximately 6 drums of soil cuttings. DPT equipment produces soil samples but generally does not produce significant drilling-derived wastes.

The most basic of all DPT equipment to collect undisturbed soil samples is the manually operated slide hammer. The hand-held slide hammers, typically weighing 12 to 30 lb, are dropped approximately 12 to 24 in onto steel extension rods. The soil sampler with retaining sample liner is connected to the leading edge of the extension rods. Some soil sampling systems have foot pedals attached to the rods that allow the operator to step down to push the DPT sampler into the ground. Sampling depth can be increased by using small hand-held augers to drill down to the target depth. In soft soil, maximum depth of manual DPT sampling is approximately 10 to 15 ft; in hard to moderately hard soil, depth of sampling is approximately 2 to 8 ft. The depth range can be increased greatly by using a narrow-diameter sampler. Specialized samplers have been developed for sand, mud, and boggy soil.

Benefits of the manual DPT sampling method are minimal setup time, low costs, and minimum disturbance of the site. The depth of sampling is the limiting factor of

(a)

(b)

FIGURE 8.15 Common environmental rigs: (a) conventional hollow-stem auger (HSA) rig and (b) direct push technology (DPT) drilling rig.

the manual DPT method, and the level of physical effort is large. The body of a manual soil sampler ranges from about ½ to 2 in in diameter and 6 in to 4 ft in length. For environmental sampling projects, clear plastic, stainless steel, or brass liners are commonly used with these DPT samplers. Hand-held or portable electric, hydraulic, or pneumatic rotohammers or jackhammers can be added to the manual DPT sampling system to extend the sampling depth. Reversing the direction of hammering on the slide hammer can provide enough force to extract the sampler and rods. As the depth of sampling increases, the side friction on the samplers and any sampling extension rods increases. For removal of the samplers at greater depth, manual-probe rod jacks supply approximately 2000 to 4500 lb of lift capacity needed for extraction. Hand-held DPT equipment can be used for sampling at an angle as well as for horizontal sampling.

Cone Penetration Testing (CPT).   Cone penetration testing rigs, a form of direct push technology, use the static weight of a vehicle to push the sampling rods into the ground. CPT rigs use a 20-ton truck and are capable of sampling to depths of 250 ft. CPT rigs, originally developed for use in the geotechnical field, typically push their sampling and testing probes from the center of the truck.

DPT Probe Soil Sampling.   Small, highly maneuverable DPT rigs were developed in the late 1980s. The probe rigs were placed on pickup trucks and vans. Probe rigs generally push the rods from the back of the truck. A percussion hammer has been added to these probe units to enhance the depth of sampling. These smaller probes have lowered the cost of DPT sampling projects to depths approaching 60 ft. Truck-mounted DPT probe rigs are typically hydraulically powered. The percussion/probing equipment pushes rods connected to small-diameter (0.8 to 3.0 in) samplers.

The DPT soil samples are commonly collected in 2- to 5-ft-long clear plastic (polyethylene or butyrate) liners contained within an outer sampler. The plastic liners are easily cut with a knife and are transparent for easy lithologic characterization. Brass, aluminum, stainless steel, or Teflon liners are also available, depending on the sampler. After removal from the sampler, the soil liner or core is immediately capped on both ends with Teflon tape, trimmed, and then capped with plastic caps. The samples are labeled and placed in individual transparent, hermetically sealed sampling bags. The samples are put in the appropriate refrigerated environment and shipped under chain-of-custody procedures to a state-certified laboratory.

Various DPT soil samplers have been designed and manufactured by numerous companies. The main sampler types used in DPT projects include split-spoon samplers, open-tube samplers, piston samplers, and dual-tube samplers. The split-spoon sampler consists of the sample barrel that can be split in two along the length of the sampler to expose soil liners. The split-spoon sampler without sample liners is useful for lithologic logging where chemical analysis is not required.

The open-tube sampler contains soil liners and has been designed for environmental sampling within the same borehole, providing that soil sloughing is minimal. Continuous coring with the open-tube sampler begins at the ground surface with the open-ended sampler. The open-tube sampler is reinserted back down the same borehole to obtain the next core. The open-tube sampler works well in stable soil such as medium- to fine-grained cohesive materials—silty clay soil or sediments, for example. The open-ended samplers are commonly ¾ to 2 in in diameter and 2 to 5 ft in length. The simplicity of the open-tube sampler allows for rapid coring.

Dual-tube sampling uses two sets of probe rods to collect continuous soil cores. One set of tubes is driven into the ground as an outer casing. A second, inner sampling rod is driven in the center of the outer casing to a depth below the outer tube and sufficient to fill the soil sampler. The inner sampling rod is then retracted and retrieved from the center of the outer casing, and the outer casing is driven one sampling interval. This sampling method is repeated to the total depth of the boring. The advantages of dual-tube sampling include continuous coring in both saturated and unsaturated zones and the virtual elimination of cross-contamination in sampling through perched water tables. The outer casing can be used as a tremie pipe when the boreholes are sealed, allowing bottom-up grouting.

For a discrete depth sample to be collected in unstable soil, a piston sampler is used. The piston sampler is equipped with a piston assembly that locks into the cutting shoe and prevents soil from entering the sampler as it is driven in the existing borehole. After the sampler has reached the zone of interest, the piston is unlocked from the surface and the piston retracts as the sampler is advanced into the soil.

DPT Probe Water Sampling. Various types of sealed samplers are available for DPT groundwater sampling. Many DPT probe water samplers use a retractable or expendable drive point. After driving to the zone of interest, the outer casing is raised from the borehole, exposing the underlying well screen. For a nondiscrete groundwater sample, the outer casing contains open slots. The open-slotted tool is driven from ground surface into the water table. Groundwater is collected by using an inner tube or smaller-diameter bailer inserted into the center of the open-slotted water sampler.

8.5.3 Phase III—Remedial Strategy Development

Phase III activities focus on specific delineation of the vertical and lateral extent of contaminant plumes, and generate data sufficient to develop a cost-effective and technically efficient remediation strategy.

A core component of Phase III is the development of the corrective action plan (CAP). The CAP evaluates practical remediation considerations including: space requirements, time constraints, regulatory acceptance, technical constraints (groundwater depth, soil type, contaminant levels, contaminant characteristics), client needs, potential liability, risk of uncertainty about existing data and assumptions, future site use, cost, funding sources, tax and accounting implications, and other site-specific issues.

Bench testing of the various proposed remedial technologies might be performed by using actual groundwater or soil collected from the site. In a bench test, various physical and chemical parameters, such as pH, temperature, concentration of target compounds, time, and concentration of treatment chemicals, would be evaluated. If successful, a pilot study in a small section of the remediation area might be selected to verify that the treatment technology will be successful. By approaching the remediation project in small increments, the chance for a large-scale remediation failure is reduced. Pilot studies may be performed to assess the feasibility and overall effectiveness of certain remedial strategies and technologies. Technologies considered for remediation of oil and gas field cleanups of petroleum hydrocarbons in the vadose zone (above the groundwater table) might include, but are not limited to, variations on bioremediation, chemical oxidation, excavation with treatment (soil washing, soil incineration, thermal desorption, soil recycling with asphalt, bioremediation), excavation without treatment (off-site disposal at a landfill), soil vapor extraction, fixation and surface capping. Aquifer treatment technologies might include, but are not limited to, passive systems such as funnel and gate, variations on bioremediation, chemical oxidation, pump and aboveground treatment, free product pumping, and skimming. A summary of selected remediation technologies is included in Table 8.4.

TABLE 8.4 Summary of Conventional Groundwater and Soil Remediation Methods*

Methods	Type	Advantages	Disadvantages
Conventional			
Excavation, transfer, and disposal to landfill	Ex situ soil	Effective at removal of shallow contaminants	Expensive, disruptive, relocates problem to another site (landfill)
Excavation and incineration, soil washing	Ex situ soil	Effective at removal of shallow contaminants	Expensive, disruptive
Bioremediation/ natural attenuation	In situ soil and groundwater	Works well when conditions are right	Costly in monitoring and time-consuming
Soil vapor extraction	In situ soil, groundwater with extraction pumping	Works well when conditions are right	Costly to operate
Pump and treat	Ex situ groundwater treatment	Works well for free product removal and providing hydraulic control	Generally ineffective at remediation, costly in investment and maintenance, and very time-consuming
Less conventional			
In situ jetting chemical delivery system	In situ injection using a lance to inject liquids for chemical oxidation, bioremediation, pH adjustment, and metals stabilization	Can be less costly than other methods and less time-consuming, with minimal disruption; can be combined with ozone injection for oxidation projects	Site-specific design required related to site soil/water chemistry; bench tests and pilot studies recommended to verify site specific conditions
Passive systems: funnel and gate		Low cost for maintenance; a truly passive groundwater treatment system	Very expensive for installation; works only with stable groundwater flow directions; treatment must be designed for site
Trench collectors with ozone treatment wall	The water passively moves into the interceptor trench located perpendicular to the flow direction	Ozone works well with hydrocarbons, and solvents and over time can oxidize volatile contaminants at oil and gas fields effectively	Works only with stable groundwater flow directions; ozone must be generated on site; high electrical usage of ozone generator; limited amount of ozone produced daily

* Modified after Testa and Winegardner (2000).

8.5.4 Phase IV—Remediation

Phase IV activities incorporate remediation of adversely impacted soil and groundwater and formulating a strategy for site closure. When dealing strictly with crude oil–impacted soil, groundwater quality is typically not a significant issue, and in many areas crude oil–impacted soil is not deemed a significant threat. For example, in several oil fields in southern California, crude oil–impacted soil has been allowed to remain in place, although covered by a minimum of 10 ft of clean soil in areas planned for residential development, provided groundwater issues are nonexistent. The presence of refined hydrocarbon–impacted soil and groundwater, however, requires a more sophisticated approach. Oil and gas fields may have vapors such as methane in the shallow soil that requires extraction or passive venting to the surface.

8.5.5 In Situ Remediation Delivery Systems

In situ remedial technologies often provide cost-effective and practical cleanup alternatives to the more conventional excavation programs for soil and the pump-and-treat approaches for groundwater contamination:

- Chemical Injection
- Enhanced bioremediation
- Soil vapor venting or extraction

The goal of in situ remediation is to reduce the mass, toxicity, mobility, volume, or concentration of contaminants in soil or groundwater by adding liquids to oxidize, bioremediate, neutralize, or precipitate contaminants in the subsurface without digging and handling of the soil or groundwater. In situ environmental remediation is optimized when geologic factors such as lithology, permeability, and porosity and contaminant, soil, and groundwater chemistry are fully evaluated and included in the design and implementation of a remediation program. Once the oil or gas lease has been characterized, remediation design can begin. Chemical injections for in situ remediation at oil and gas fields can be broken into two distinct aspects: delivery of the treatment chemicals and the chemistry of the treatment chemicals with the contaminants. Since oil and gas fields have numerous aboveground storage tanks, equipment pads, and pipelines, leakage into the subsurface hillsides, mud pits, and soil stockpiles. The flexibility and accuracy of this injection delivery system provides distinct advantages over both conventional in situ and ex situ remediation systems at oil and gas fields. Hot spots can be effectively treated by in situ methods. As a result, the DPT liquid injection and jetting technologies can provide appreciable savings in cost and time over traditional remediation technologies, such as excavation and off-site disposal. High-pressure jetting with tip pressures exceeding 5000 psi allow for hydraulic fracturing of soil as an enhancement to a variety of other in situ remediation methods.

In situ remediation uses chemical oxidizers to rapidly treat soils contaminated with toxic and persistent organic wastes. The two most common oxidizers used in soil and groundwater remediation are hydrogen peroxide and potassium permanganate to treat petroleum hydrocarbons (such as gasoline, diesel, motor oil, and jet fuel), volatile organic compounds, munitions, certain pesticides and wood preservatives. Fuel hydrocarbons and selected other organic compounds have been remediated by jetting using low concentrations of liquid oxidants, nutrients, and other amendments.

Oil and gas field drilling muds are stored in mud pits. Drilling muds have been known to contain toxic metals. Under the correct subsurface conditions, soluble metals, such as arsenic and cyanide, have been stabilized by sulfide compounds such as calcium polysulfide, precipitating metals into an insoluble sulfide. Calcium polysulfide reduces the toxic chromium VI into chromium III and then immobilizes the chromium molecule by binding it to the soil particle as chromium hydroxide. Alkalinity, pH, and organic content must be evaluated prior to any in situ metals stabilization project. Injection ports are grouted with bentonite or neat cement.

Jetting equipment with associated lances or DPT rigs have been used with high-pressure pumps to inject various chemicals on close spacing (2 to 10 ft) into the subsurface for in situ remediation. Chemical injections into monitoring wells or injection wells that are spaced far apart are frequently associated with nonreactive zones between injection ports. This subsurface phenomenon is especially problematic in deposits or lithologies having limited porosities.

Chemical Injections

Oxidation. In situ chemical oxidizers have the potential for rapidly treating soils contaminated with toxic and persistent organic wastes. In situ oxidation uses contact chemistry of the oxidizing agent to react with petroleum hydrocarbons, volatile organic compounds, munitions, certain pesticides, and wood preservatives. The most common oxidizers used in soil and groundwater remediation are hydrogen peroxide (and the hydroxyl radical), potassium permanganate, and ozone, which are nonselective oxidizers. Other oxidants are available, but are less commonly used because of cost, time, or potential toxic by-products (Table 8.5). Table 8.6 summarizes advantages and limitations of hydrogen peroxide and potassium permanganate.

TABLE 8.5 Comparative Oxidative Potentials

Species	Volts
Fluorine	3.0
Hydroxyl radical	2.8
Ozone	2.1
Hydrogen peroxide	1.8
Potassium permanganate	1.7
Hydrochlorous acid	1.5
Chlorine dioxide	1.5
Chlorine	1.4
Oxygen	1.2

Hydrogen peroxide, when in contact with a metal catalyst such as iron II, which is commonly known as Fenton's reagent, forms a more powerful oxidizer, the hydroxyl radical.

The hydroxyl radical ($OH^{\cdot}$) in the subsurface can be used to rapidly mineralize hydrocarbon, solvent, and other contaminants to water and carbon dioxide. This reaction is enhanced in the presence of iron. The metal catalyst can be usually provided by naturally occurring polyvinyl chloride (PVC) iron oxides within the soil or fill, or added separately as a solubilized iron salt, such as iron sulfate. Fenton's reagent has been well documented for over 100 years and has been in use in water treatment

TABLE 8.6 Summary of Advantages and Limitations or Concerns of Two Common Oxidizers

Hydrogen peroxide, H_2O_2		Potassium permanganate, $KMnO_4$	
Advantages	Limitations or concerns	Advantages	Limitations or concerns
Widely available— mix hydrogen peroxide liquid with water	Special handling and safety precautions, especially at higher concentrations	Stable and relatively safe to handle— mix white powder with water	Bright purple staining
Inexpensive	Low pH (2–4) optimal; requires acidic environments	Reactive under neutral pH (7)	Generally more expensive and less powerful oxidizer than hydrogen peroxide; concentrations used less than 7% because of saturation limit
Nontoxic by-products (CO_2, H_2O)	Nonselective oxidizer	Nontoxic by-products (CO_2, H_2O, MnO_2)	Nonselective oxidizer Increased MnO_2 can decrease permeability
Fenton's reagent produces one of the most powerful oxidizers available (hydroxyl radicals)	Short reaction period (seconds to minutes; up to hours)	Long reaction time (hours to days) compared with hydrogen peroxide	Hexavalent chromium can be produced under specific conditions— evaluate chemistry
Recommendations			
Evaluate chemistry; bench tests can provide important information prior to fieldwork			

plants for well over 50 years. The chemistry is well documented (Watts et al., 1991, 1992; Watts and Stanton, 1994) to destroy petroleum hydrocarbons and other organic compounds. Hydrogen peroxide arrives in the field as a liquid stored in PVC drums. When the chemical oxidant hydrogen peroxide (H_2O_2) is injected into the subsurface, it decomposes readily into reactive hydroxyl radicals ($OH^\bullet$), hydroxyl ions (OH^-), and water (H_2O). The oxidation of a contaminant by hydrogen peroxide involves complex reactions influenced by a number of variables, including pH, reaction time, subsurface temperature, available catalysts, and hydrogen peroxide concentration, which usually ranges from 10 to 25 percent. In subsurface environments having pH of 8.0 or greater, strong or weak acids can be used to lower pH and optimize the oxidation process, as determined by a laboratory bench test. Hydrogen peroxide works best in acidic environments. Because the hydrogen peroxide reaction time is seconds to minutes, close spacing of the injection ports is required.

The oxidation reaction is based on the principle of Fenton's reagent, where the iron and hydrogen peroxide initially react to form hydroxyl radicals and other by-products as shown in Eq. (8.1). The subsequent complex reactions of Fenton's process have been well documented by Barb (1950) and Spencer et al. (1996) and are not included here.

$$Fe^{2+} + H_2O_2 \rightarrow OH^{\bullet} + OH^- + Fe^{3+} \qquad (8.1)$$

Trichloroethene (TCE; C_2HCl_3) is a degreaser that might be encountered at oil and gas fields for use in equipment maintenance. The oxidation reaction of TCE is shown in Eq. (8.2).

$$4OH^{\bullet} + C_2HCl_3 \rightarrow 2CO_2^- + 3Cl^+ + 5H^+ \qquad (8.2)$$

Any hydrogen peroxide not used in the chemical oxidation process breaks down to water and oxygen rapidly. In addition to the reaction described in Eq. (8.2), there are also a large number of competing reactions involving the free radical scavengers, most importantly, carbonate and bicarbonate alkalinity, that will greatly affect the overall reaction scheme. Hydrogen peroxide at lower concentrations (1 to 5 percent) can also serve as an oxygen source for microbes in the subsurface to enhance bio-degradation of contaminants. Therefore, many in situ chemical oxidation projects are designed to move into a second, longer-term bioremediation phase due to all the newly available oxygen in the subsurface.

Although handling hydrogen peroxide and other oxidants requires significant safety training and planning, the oxidant is effective at remediation and relatively inexpensive. The reaction time for hydrogen peroxide in the subsurface is usually seconds to minutes, with occasional reactions being completed within minutes to hours. During injections, temperature of the reaction liquids is monitored to evaluate the success of the reactions. According to field observations, temperatures of the reaction liquids less than approximately 60°C indicate that the hydrogen peroxide is reacting properly. Excessive temperatures indicate that the exothermic reaction is consuming peroxide at a very rapid and uncontrolled rate. Forensic chemical analyses from various sites have shown that the hydrogen peroxide reaction tends to work first on the longer chain carbon sources, including total organic carbon (TOC), rootlets, and heavier-end hydrocarbons, prior to oxidizing the lighter-end hydrocarbons.

Trace chlorine from chlorinated compounds will likely combine with sodium or calcium ions to form salts or with hydrogen to form weak acids. Careful evaluation of the chemistry of the soil and water are required prior to the start of any injection process.

Potassium Permanganate.   Although a weaker oxidizer than hydrogen peroxide, potassium permanganate ($KMnO_4$) lasts longer and can react in an environment with much higher pH than hydrogen peroxide. For field use, potassium permanganate is shipped as a powder and is mixed with water, creating a deep purple liquid. The solubility of potassium permanganate is strongly influenced by temperature, and at 30°C, the solution has slightly over an 8 percent concentration of potassium permanganate. The pH range is critical in being able to determine whether the oxidation reaction will be fast or slow. The chemical formula for chemical oxidation of TCE by potassium permanganate is shown in Eq. (8.3):

$$2KMnO_4 + C_2HCl_3 \rightarrow 2MnO_2 + 2CO_2 + K^+ + 3Cl^- + H^+ + 2K^+ \qquad (8.3)$$

Of concern to some is the oxidation of trivalent chromium into the hexavalent variety. Oxidizers such as potassium permanganate, hydrogen peroxide, magnesium peroxide, and ozone will not create much hexavalent chromium, and, in the presence of other organic contaminants, hexavalent chromium would be reduced to trivalent chromium.

Ozone (O_3) is a powerful gas-phase oxidizer that can be used to treat hydrocarbons and chlorinated solvents, among other contaminants that might be found at an oil or gas field. It must be generated on site, and the gas cannot be stored; therefore all the ozone gas that is generated must be injected into the subsurface or destroyed by an ozone destruction unit on the ozone generator. For in-situ treatment, the ozone gas can be bubbled into closely spaced sparge points that release the bubbles into the aquifer for remediation. The smaller the bubbles, the more surface area and the faster they can travel through small pore spaces. Microbubbles can be produced by pushing the ozone gas through a diffusion material. The other option for in situ ozone treatment is to saturate water above ground and inject the treated water into the injection ports. Aboveground treatment of extracted groundwater using ozone generators has been documented (Testa and Winegardner, 2000).

For all types of in situ chemical oxidation methods, chemical compatibility of the injection equipment, personal protective equipment, and safety procedures become critical with the injection of potentially dangerous chemicals including oxidizers, acids, bases, and other chemicals.

Bioremediation. Bioremediation of soil and groundwater is a cleanup technology that uses the ability of micro-organisms to degrade hazardous organic compounds into nonhazardous compounds such as carbon dioxide, water, and biomass. Bioremediation can be utilized in an oxygen-rich (aerobic) environment or in an oxygen-deficient (anaerobic) environment. The micro-organisms responsible for degrading hazardous compounds are generally naturally occurring. The most common and widely accepted approach to bioremediation is the enhancement of environmental conditions to favor biodegradation. Aerobic biological degradation and natural attenuation of fuel hydrocarbons and selected other organic compounds have been well documented (Rice et al., 1995; Mace et al., 1997).

Aerobic bioremediation has been used successfully as a remediation method for well over a decade at sites ranging from Superfund sites to small gasoline stations. Aboveground methods using bioreactors to treat petroleum hydrocarbon–contaminated groundwater have been successfully used to optimize biodegradation parameters, including dissolved oxygen, pH, and temperature in combination with acclimated cultures of facultative bacteria. Aboveground bioreactors have been very effective for accelerating in situ biodegradation of dissolved-phase petroleum hydrocarbon compounds.

Aerobic biodegradation relies on indigenous microbes that utilize petroleum hydrocarbon compounds as their carbon and energy sources. Successful biodegradation projects largely depend on the presence of acclimated hydrocarbon-degrading bacteria along with sufficient dissolved oxygen, near-neutral pH, and adequate dissolved nutrients (ammonia nitrogen and orthophosphate) to metabolize the petroleum hydrocarbon compounds into biomass, carbon dioxide, and water. Monitoring these factors over time can verify subsurface conditions required for optimum biodegradation and reveal potential site deficiencies. Although oxygen is frequently the rate-limiting factor for in situ biodegradation, hydrocarbon biodegradation can still proceed slowly in low oxygen environments in the presence of nitrate or other alternative terminal electron acceptors utilized by facultative bacteria.

Enhanced passive biodegradation (i.e., "semipassive" biodegradation) of petroleum hydrocarbon–contaminated soil can be accomplished by using oxygenating agents and nutrients to stimulate in situ hydrocarbon-degrading bacteria in a saturated to moist environment. The oxygen can be supplied as a gas, liquid, or solid phase. Gas-phase oxygen is supplied by air sparging the aquifer with fine air bubbles by diffusion of molecular oxygen through membranes. A liquid form of oxygen is hydrogen peroxide, which can be poured into wells or trenches or injected into ports in the subsurface. Solid-phase oxygenation methods have been widely used, including magnesium peroxide or calcium peroxide. Site-specific liquid nutrient formulations can be injected under high pressure into separate boreholes within the bio-treatment zone to further enhance petroleum hydrocarbon compound degradation efficiency. Continuous monitoring and analysis of selected biological and chemical factors can then be used to optimize the enhanced passive bioremediation process. For gas-phase oxygen applications in the vadose zone, air is vented through the soil in a process called *bioventing*. Air flow rates for this process are much smaller than for the soil vapor extraction technique described in the next section. Aboveground and in-situ bioremediation has been used successfully to treat soil and groundwater at various oil and gas leases.

Soil Vapor Extraction.   The soil vapor extraction system (SVE) technique involves the extraction of volatile organic compounds (VOCs) from the subsurface by the creation of a vacuum in the vadose zone by an air blower connected to a vadose zone well (Fig. 8.16) or trench. Usually, the blower discharges through an air emissions device, such as a drum of granular activated carbon (GAC). A SVE can be used as an effective treatment of VOC contamination in the soil but is limited in the ground-water. A well-engineered SVE may include the following enhancement processes:

• Increase vacuum
• Circulation of treated hot air into the soil
• Reduced humidity in the air and moisture in the soil
• Increased air flow in the soil
• Desorption of VOC from clayey and silty soil

SVE removes volatiles from soil pores and enhances biodegradation of semivolatile and nonvolatile hydrocarbons.

8.5.6   Risk Analysis and Modeling

In addition to the various types of remediation that are available, an important aspect of remediating oil and gas fields is the acceptable level of risk associated with residual contamination. The allowable levels of residual contaminants might be established by the regulatory agency or by a consultant, who might use a computer model, such as the risked-based corrective action (RBCA) model to evaluate sub-surface risks to humans and the environment. Limitations of future use may be dictated by a regulatory agency as a condition of site closure of a former oil or gas lease. If the target chemicals are generally immobile and the health risk to humans and the environment is low, regulatory site closure might occur without the costly and time-consuming efforts of complete remediation.

8.6   ABANDONED OIL AND GAS WELLS

Of the approximately 3.4 million oil and gas wells drilled in the United States since 1859, more than 2.5 million have been plugged and abandoned. Approximately 17,000 wells are plugged and abandoned annually. Unplugged, or orphaned, wells with no existing owner or operator are largely a legacy of the past, when site re-

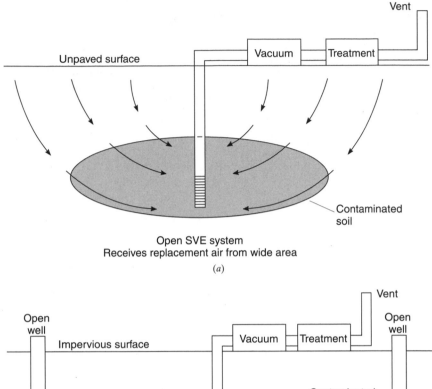

FIGURE 8.16 Conceptual soil vapor extraction system configuration for an (a) open and (b) closed system.

storation was not commonly deemed necessary. Today, site restoration following exploration and production activities is regulated, and integrates advanced technology, increased research and development, and generally improved cooperation and responsibility. However, some marginally productive wells are kept in operation to avoid the closure costs associated with environmental and oil and gas regulations.

Onshore, wells have been permanently plugged with cement to prevent any flow of subterranean fluids into the well bore, thereby protecting groundwater resources. Actual well abandonment commonly involves the redrilling of the well to the known production depth, and then injecting cement in the well bore to a level approximately 100 ft below ground surface. The well is then pinched closed, and additional cement is pumped down the well bore. Venting may be required to avoid methane accumulation. Other structures such as wastewater handling pits are closed, and storage tanks, well heads, processing equipment, and pumping jacks are removed.

Improperly abandoned oil and/or gas wells can be assessed through review of historic records, monitoring for certain gases such as methane, conducting a geophysical survey for the subsurface presence of metal, or excavation of an exploration pit. Such pits are typically 10 ft square by 10 ft deep, and centered on the suspected well location. Once the pit is excavated, and no visible signs of the well are evident (casing, cellar boards, etc.), a magnetometer is used to scan the pit for signs of metal. When the casing is located, the well is then redrilled, beginning inside the casing and extending to the total known production depth. Once the well bore has been reopened, the well is abandoned using the procedure described above. Once the well is properly abandoned, building permits can be obtained although sign-off from the lead regulatory agency may also be required.

Offshore, well bores are sealed below the seafloor, and platforms are fully or partially removed, or toppled in place as part of establishment of an artificial reef. Approximately 100 platforms are decommissioned each year.

8.7 PREVENTIVE MEASURES

8.7.1 Seismic Exploration

Seismic crews increasingly understand the baseline regional environments in which they work and the seasonal changes that impact those environments. This understanding allows them to conduct their surveys in a more efficient manner, and select tools and equipment more knowledgeably, thus avoiding downtime, fines, and adverse environmental impact (Perkins et al., 1999). Some survey crews now undergo training from biologists and environmental scientists who also work as crew members, maintain regulatory compliance, and interact with regulatory agencies. A systematic approach is developed that includes trafficability and sensitivity mapping, field inspection, and working with observers. Satellite imagery is used to produce color infrared images showing different types of living habitats, water zones, physical environmental features, and limits of operations. Environmentally sensitive areas such as bird rookeries, marshes, and nesting areas are identified, and seasonal changes observed and monitored. The environmental observer then works with survey crews and other field personnel to lower the environmental impact by a variety of means including minimizing extent of activities, flagging problem areas, and rerouting crews and equipment by existing roads and waterways.

8.7.2 Field Measures

Waste management efforts taking place throughout the industry during exploration and production include several practices that minimize potential environmental threat and legal liabilities (EPA, 1995). These include such practices as:

- Overdesign the size of reserve pits to avoid overflows
- Maintain a closed-loop mud system when practical, especially with oil-based muds
- Minimize materials stored and waste generation by using the smallest volumes possible
- Reduce the volumes of excess fluids entering reserve and production pits
- Design drilling pads to contain stormwater and rigwash fluids
- Recycle and reuse oil-based muds and high-density brines, and reclaim oily debris and tank bottoms, when practicable
- Avoid placement of nonexempt wastes in reserve and production pits
- Review material safety data sheets of materials used, and utilize less toxic alternatives when possible
- Train personnel and perform routine inspections of equipment, materials, and waste storage areas

8.8 CASE HISTORIES

8.8.1 Signal Hill, California

Oil fields within urbanized areas present unique issues especially when they reach the end of their productive life. In areas such as southern California, these properties are rapidly taken out of production and redeveloped. California has a rich history of oil and gas exploration and extraction dating back to the 1860s, the first years of commercial production. California comprises 56 counties, 30 of which are known to produce petroleum; 18 of these produce chiefly oil, while the remaining 12 produce mostly gas. Oil occurrence is generally found south of 37° north latitude, and within four basins: the Los Angeles Basin, Ventura Basin, San Joaquin Basin, and Santa Maria Basin (Fig. 8.17). Within the Los Angeles Basin, more than 10 giant oil fields (a giant being defined as one yielding an ultimate recovery in excess of 100 million barrels) have been identified. Within the City of Long Beach, about 12 square miles, or 3840 acres, are currently or have been petroleum-producing properties, whereas most of the city of Signal Hill falls into this category.

When the area known as Signal Hill was being considered for redevelopment, an assessment was made to evaluate what the impact of oil production activities over the years on overall soil and groundwater conditions, and whether the area could be safely redeveloped at reasonable cost. Concerns included the potential costs for remediation and potential liability associated with building above or adjacent to oil wells. The scope of work included a review of historic oil files and records, evaluation of subsurface geologic and hydrogeologic conditions, review of historic aerial photographs, and field reconnaissance of former and existing well and sump sites. Historic oil maps were reviewed, and about 57 former drilling mud sump sites were identified. An aerial view and a map of the location of all formerly abandoned and active well and sump sites are shown in Fig. 8.18a and b, respectively.

Wells abandoned prior to promulgation of the most current standards had to be reabandoned before building permits could be issued for any site with documented wells. This included oil wells within 100 ft, or wildcat wells within 500 ft, of a proposed construction or redevelopment site.

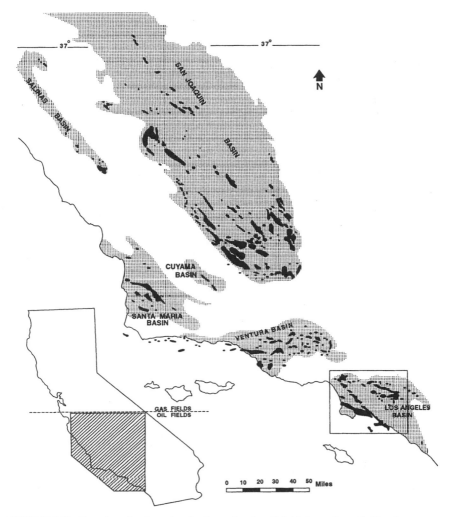

FIGURE 8.17   Location of sedimentary basins and major oil fields in southern California.

8.8.2   Old Salt Lake Oil Field, Los Angeles

Gas ventings from existing oil fields in an urban environment have been documented over the years in the densely populated Los Angeles area (Hamilton and
Meehan, 1992). The Old Salt Lake Oil Field, at one time the biggest producing field
in California, underlies the immediate area. Over 400 abandoned wells that formerly
tapped the field were situated throughout the surrounding area.

On March 24, 1985, an explosion partially collapsed the roof of a commercial structure, the Ross Dress-for-Less Store located on Third Street in the Wilshire-Fairfax district of Los Angeles (Fig. 8.19). The explosion, the result of accumulated methane, blew

(a)

FIGURE 8.18 (a) Aerial view and (b) map showing location of formerly abandoned and active wells and sump sites in Signal Hill, southern California.

out windows and reduced the store's interior to a heap of twisted metal, with 23 people requiring hospital medical treatment. Four blocks were closed off, and spouting gas flames continued through the night.

Following the event, a postexplosion exploration and gas-control drilling program was implemented and completed, and a special task force was convened by the Los Angeles City Council to evaluate what had happened and how future accidents could be prevented. A second event in the general area occurred on February 7, 1989, when explosive levels of methane gas were evident in a K-Mart store. Fortunately no explosion occurred, but this event revived interest and a second task force was initiated.

Two primary scenarios for the 1985 and 1989 gas ventings were postulated:

- Biogenic gas/rising groundwater scenario: Methane gas was generated from near-surface decay of organic matter in the shallow alluvial soil, and then displaced and pressurized by a rising water table. The presence of former oil field activities, sumps, and wells was viewed as a coincidence.

- Oil field gas/abandoned well conduit scenario: Isotopic analysis of gas samples suggested that the gas was of thermogenic origin, hence from an oil reservoir source.

The three-point program developed by the task force included venting by relief wells, use of an areawide monitoring survey team, and inspections by the city for code compliance. The task force conclusions discarded the potential for the abandoned wells and/or pressure injection operations to be a significant factor or cause, which raised further concern within the technical community.

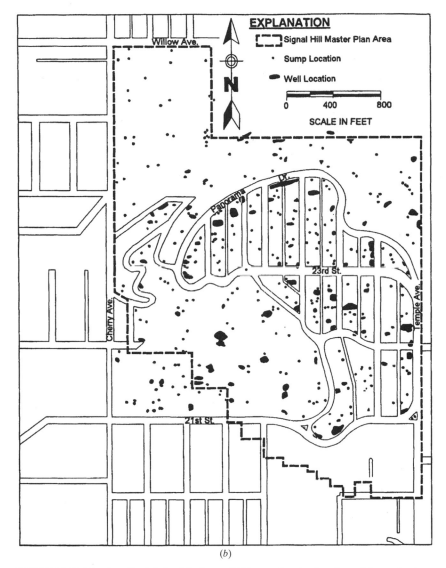

(b)

FIGURE 8.18 (*Continued*)   (*a*) Aerial view and (*b*) map showing location of formerly abandoned and active wells and sump sites in Signal Hill, southern California.

A later study suggested that increasing well head pressures at a well designated
as Gilmore no. 16, sanctioned by the California Division of Oil and Gas, was sufficient to fracture the underlying formation at a nearby fault. Referred to as the oil
field gas/intermittent fracture conduit scenario, this could result in the upward
migration of fluids and notably gas from the less confining area to the surface environment.

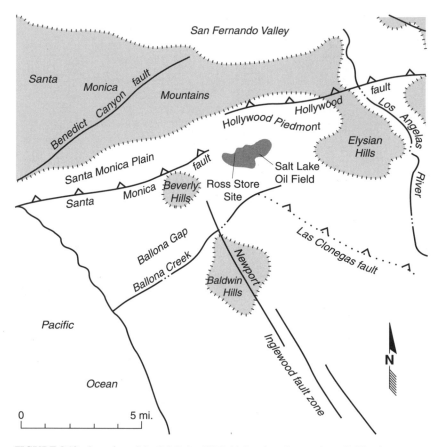

FIGURE 8.19   Location of the Salt Lake Oil Field, Los Angeles, southern California.

8.8.3   Martha Oil Field, Kentucky

The Martha Oil Field is located in Lawrence and Johnson Counties, southeastern Kentucky. The Martha Oil Field encompasses over 50 mi^2 (80.45 km^2). Oil production from the Weir Oil Sands (Lower Borden Formation, Osage Series and Mississippian System) began at the turn of the century. Secondary recovery, consisting of water flooding, commenced in 1955 in the eastern portion of the field. The Martha Main Field area occupies approximately 4500 acres (18.21 km^2), of which, until 1986, almost 3200 acres (12.95 km^2) were under the water flooding program. Within this field, the responsible party owned and operated 601 injection wells, 779 oil production wells, 26 industrial water wells, and 27 water-cooling wells, and owned 120 permanently abandoned wells within the Martha Main portion of the Martha Oil Field. Cumulative oil production through 1985 totaled over 22.5 million barrels.

Upwelling of brine and oil via breached well casings, uncemented well annuli, and improperly plugged and abandoned wells resulted in widespread contamination of the three underground sources of drinking water (USDWs: Alluvium, Breathitt, and Lee Formations). Causal effects include increased potentiometric head within the Weir Oil Sands by freshwater injection and cones of depression in the Lee Aquifer resulting from pumping of industrial water supply wells.

Investigations conducted by the U.S. Environmental Protection Agency (EPA) and Army Corps of Engineers in 1986 confirmed the prevalence of the contamination in all three USDW aquifers (Fig. 8.20). EPA determined that the responsible party was in violation of both the Safe Drinking Water Act and Underground Injection Control Regulations by (1) allowing the movement of fluids containing contaminants into the USDWs during operation and maintenance of their injection activities, (2) failure to properly plug and abandon injection wells that were temporarily abandoned for 2 years or more, and (3) injection at pressures that resulted in the movement of injected fluids and/or formation fluids into a USDW.

The responsible party subsequently proposed a program for the proper plugging and abandonment of all injection wells, oil production wells, industrial water supply wells, and most gas production wells within the Martha Main Field Area, in conjunction with development of a remediation and monitoring program. About 1433 wells and 45 injection wells were planned for abandonment.

8.8.4   The Gulf War of 1991, Kuwait

Society's first experience with environmental terrorism on a regional scale started on the January 25, 1991, in the Arabian Gulf on the eastern shores of Kuwait, adjacent to Saudi Arabia. Crude oil from pumping stations at Mina Al-Ahmadi was purposely dumped into the Gulf waters. In excess of two million barrels per day, crude oil was released as an act of war. Although of little if any practical value from a military perspective, ecologically the oil severely affected the ecosystem of the Gulf. Compared to the 260,000 barrels spilled by the *Exxon Valdez* in 1989, 4 million barrels of crude oil made it to the Gulf, making it the largest spill in history (Fig. 7.5).

The environmental impact was not restricted to the Gulf area alone. Between February 23 and 27, 1991, over 600 wells were set aflame, burning about 5 million barrels per day (Fig. 8.21a). An international effort was set into motion to remedy this situation. It was not until November 8, 1991, that the oil well fires were under control.

Although the oil wells received most of the media attention, these damaged wells also spilled large quantities of oil that spread across the desert, and accumulated in depressions. What remained was the so-called oil lakes (Fig. 8.22). Even though almost a decade has passed since they were initially formed, numerous oil lakes up to a kilometer in diameter still remain (Fig. 8.21b).

8.8.5   Santa Barbara Channel Blowout

The oil and gas industry has come a long way since the time when it was accepted practice to "blow" gas caps into the atmosphere. In the late nineteenth century and into the twentieth century, operators used this practice so that wells in a gas cap of an oil reservoir could be finally induced to produce a small amount of highly valued crude oil. Modern drilling rigs use blowout prevention equipment at the surface (ground surface or sea bottom) if a blowout in a borehole or a well occurs. The blowout equipment is designed to close the top of the borehole, control the release of fluids, allow movement in the drill pipe and permit pumping of liquids into the borehole or well, if the hole is cased. A blowout preventer stack is a series of large, high-pressure valves that can shut off the well at the surface when the formation

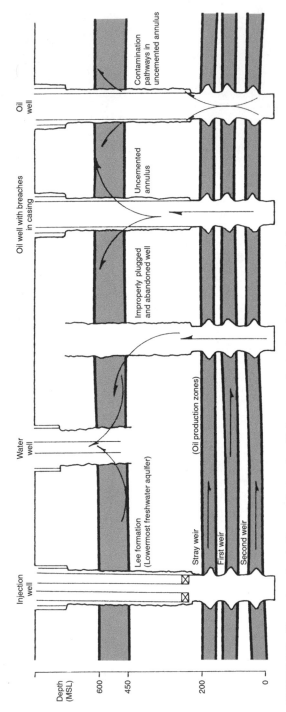

FIGURE 8.20 Illustration of probable conduits for oil and brine contamination into the Lee Aquifer during waterflooding activities in the Weir Sands, Martha Oil Field, Kentucky.

(a)

FIGURE 8.21   (a) Typical oil well set on fire during the Gulf War and (b) the resultant oil lake after the war.

pressures exceed pressure of the liquids in the borehole or well. Other blowout-prevention equipment includes a choke manifold, choke and kill lines, and various control panels.

Oil and gas released from a subsea blowout passes through three zones; the high-velocity jet zone at the wellhead, which is highly turbulent; the buoyant plume zone, where buoyancy takes over as the primary driving force; and the boil zone (shown with arrows on Fig. 8.23), which is at the air-water interface.

(*b*)

FIGURE 8.21 (*Continued*)   (*a*) Typical oil well set on fire during the Gulf War and (*b*) the resultant oil lake after the war.

Oil Spills.   On January 29, 1969, a Union Oil Company platform in Santa Barbara Channel, located 6 mi off the coast of Summerland, California, had a blowout. The well was drilled to about 3500 ft below the ocean floor. During the replacement of a worn drilling bit at the bottom of the drill string, the drilling mud used to maintain pressure in the borehole became low and the formation pressure exceeded the drilling mud pressure. A natural gas blowout occurred and an initial attempt to cap the borehole was successful. The formation pressure continued building in the subsurface, and the expanding mass of oil and gas created five breaks in an east-west fault on the ocean floor in the Santa Barbara Channel.

Impact.   For 11½ days, approximately 4800 bbl of crude oil bubbled to the surface and was spread into an 800 mi^2 slick by winds and swells. Thick tar was deposited on beaches over 35 mi of coastline from Rincon Point to Goleta. Wildlife was documented to have been impacted, most notably, seals, dolphins, and diving sea birds.

Cleanup.   The well was controlled by pumping special, heavy-weight drilling muds at a rate of 1500 bbl/h down to 3500 ft through the borehole. The borehole was capped with a cement plug. Residual amounts of oil and gas continued to escape in this area for months after the incident. Skimmers collected oil from the surface of the ocean. Airplanes dropped detergent on the floating oil slick, in hopes of dispersing it and breaking it up. Straw soaked up the tar on the shore and the impacted beach sands were raked up. Rocks were steam cleaned. Investigators determined that more steel casing inside the borehole could have minimized or prevented the incident. In addition, drilling mud weight was determined to be lower than recommended for the formations drilled.

FIGURE 8.22 Distribution of oil lakes after the Gulf War in Kuwait.

Summary.   The Santa Barbara Channel blowout created political challenges for the offshore oil industry in California. The public's concerns created an environment in California where subsequent federal and state oil and gas lease sales would be canceled or delayed significantly. The wildlife has since recovered in the area. However, the lasting impact of the Santa Barbara Channel spill is still felt in a state where an offshore moratorium for exploration drilling is still enforced. Ironically, the volume of the Santa Barbara spill was the equivalent of less than 6 days of normal, naturally occurring oil seeps in the Santa Barbara Channel area.

8.9 SUMMARY

Eventually all oil and gas wells have to be abandoned and the surfaces of oil and gas fields have to be reclaimed. These activities will for one reason or another require those involved to heighten their awareness and sensitivity to the environment in which they work. Oil or petroleum is society's primary extracted resource, and forms the foundation of modern civilization. There is little doubt that it will continue to be so as we enter into the future. As society continues its dependency on oil, exploration and production will encroach into more environmentally hostile territories (e.g., harsh working conditions, fragile environments, and locations with difficult access). Public confidence in the exploration and production of oil and gas in sensitive areas can be achieved only through continued and consistent prudent environmental management.

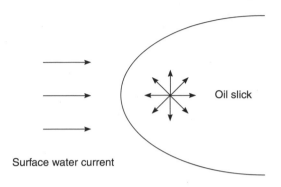

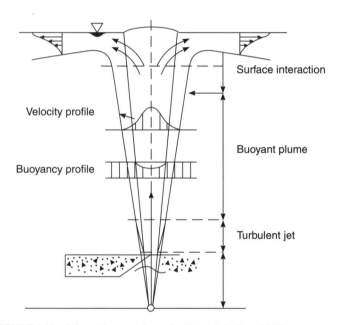

FIGURE 8.23   Subsea blowout schematic. (*From S. L. Ross, 1997.*)

REFERENCES

American Petroleum Institute, 1995, *Basic Petroleum Data Book,* vol. 15, no. 1, Washington, D.C.

Barb, W. G., Buxendale, J. H., Hargrave, K. R., 1950, "Reactions of ferrous and ferric ions with hydrogen peroxide. Part 1: The ferrous iron reaction," *Journal of the Chemical Society,* vol. 121, pp. 462.

Bowie, C. P., 1918, "Oil-Storage Tanks and Reservoirs with a Brief Discussion of Losses of Oil in Storage and Methods of Prevention," United States Bureau of Mines Bulletin No. 155, Petroleum Technology Report No. 41, 76 pp.

Deuel, L. E., and Holliday, G. H., 1997, *Soil Remediation for the Petroleum Extraction Industry,* 2d ed., PennWell, Tulsa, Okla., 242 pp.

Eger, C. K., and Vargo, J. S., 1989, "Prevention, Ground Water Contamination at the Martha Oil Field, Lawrence and Johnson Counties, Kentucky," in *Environmental Concerns in the Petroleum Industry,* S. M. Testa (ed.), Pacific Section of the American Association of Petroleum Geologists, Palm Springs, Calif., pp. 83–105.

Fakhoury, E., and Patton, K. R., 1992, "A Practical Approach for Environmental Site Assessment of Oilfield Properties Prior to Redevelopment," in *Association of Engineering Geologists 35th Annual Meeting Proceedings,* Los Angeles.

Hamilton, D. H., and Meehan, R. L., 1992, "Cause of the 1985 Ross Store Explosion and Other Gas Ventings, Fairfax District, Los Angeles," in *Engineering Geology Practice in Southern California,* B. W. Pipkin and R. J. Proctor, eds., Star Publishing, Belmont, Calif., pp. 145–157.

Howar, M., Hayes, D., and Gaskill, B., 1987, "Clearing and Abandonment of Deep Water Wells at the Tar Creek Superfund Site," in *Proceedings of the Hazardous Materials Control Research Institute 8th National Superfund '87 Conference,* Washington, D.C., pp. 439–443.

Jacobs, J., 2000, "Applications of Jetting Technology for In-situ Remediation," *Abstracts of the Association of Engineering Geologists and Groundwater Resources Association of California Annual Meeting,* September 24, San Jose, p. 92.

Jacobs, J., and von Wedel, R., 1997, "Enhanced In Situ Biodegradation Via High Pressure Injection of Oxygenating Agents and Nutrients," American Institute of Professional Geologists, 34th Annual Meeting, Houston, *Abstracts,* p. 27.

Jacobs, J., 1996, "Passive Oxygen Barrier for Groundwater," *Hydrovisions,* January/February, Groundwater Resources Association.

Jacobs, J., 1995, "Vertical and Horizontal Direct Push Technology and In-Situ Remediation Delivery Systems," *Abstracts,* 1995 Annual Meeting, Groundwater Resources Association, Sacramento, October 6.

Mace, R. E., Fisher, R. S., Welch, D. M., and Parra, R. A., 1997, "Extent of mass and duration of hydrocarbon plumes from leaking petroleum storage sites in Texas," Texas Bureau of Economic Geology, Geological Circular 97-1, 52 pp.

McFaddin, M. A., 1996, *Oil and Gas Field Waste Regulations Handbook,* PennWell Publishing, Tulsa, Okla. 390 pp.

Navarro, A., 1995, *Environmentally Safe Drilling Practices,* PennWell, Tulsa, Okla., 157 pp.

Orszulik, S. T. (ed.), 1997, *Environmental Technology in the Oil Industry,* Blackie Academic and Professional, New York, 400 pp.

Patin, S., 1999, *Environmental Impact of the Offshore Oil and Gas Industry,* EcoMonitor Publishing, East Northport, N.Y., 425 pp.

Perkins, A. F., Gagliano, M. H., and Moses, P., 1999, "Monitoring and Education Help Seismic Crew Protect Environment in Transition-Zone Survey," *Oil and Gas Journal,* February 22, pp. 33–37.

Rice, D., et al., 1995, "Recommendations to improve the cleanup process for California's leaking underground fuel tanks," submitted to the California State Water Resources Control Board and the Senate Bill 1764 Leaking Fuel Tank Advisory Committee, Lawrence Livermore National Laboratory Report, October.

S. L. Ross Environmental Research, Ltd., 1997, *Fate and Behavior of Deepwater Subsea Oil Well Blowouts in the Gulf of Mexico,* prepared for Minerals Management Service, October, 27 pp.

Sackett, R. L., and Bowman, I., 1905, "Disposal of Strawboard and Oil-Well Wastes," United States Geological Survey Water-Supply and Irrigation Paper, no. 113, 52 pp.

Spencer, C. J., Stanton, P. C., and Watts, R. J., 1996, "A Central Composite Rotatable Analysis for the Catalyzed Hydrogen Peroxide Remediation of Diesel-Contaminated Soils," *Journal of Air and Waste Management,* vol. 46, pp. 1067–1074.

Testa, S. M., 1993, *Geological Aspects of Hazardous Waste Management,* CRC Press, Boca Raton, 537 pp.

Testa, S. M., and Winegardner, D. L., 2000, *Restoration of Contaminated Aquifers,* 2d ed., CRC Press, Boca Raton, 446 pp.

U.S. EPA, 1995, "Crude Oil and Natural Gas Exploration and Production Wastes: Exemption from RCRA Subtitle C Regulation," EPA/530-K-95/003, 30 pp.

U.S. EPA, 1985, "Modeling Remedial Actions at Uncontrolled Hazardous Waste Sites," EPA/540-2-85/001, Washington, D.C.

U.S. EPA, 1988, "Corrective Action: Technologies and Applications," Office of Research and Development, Center for Environmental Research Information, April 19–20, Atlanta.

U.S. EPA, 1991a, Seminar on RCRA Corrective Action Stabilization Technologies, Technology Transfer, Center for Environmental Research Information, Washington, D.C.

U.S. EPA, 1991b, *Stabilization Technologies for RCRA Corrective Actions Handbook,* Office of Research and Development, EPA/625/6-91/026, Washington, D.C.

Veil, J. A., Daly, J. M., and Johnson, N., 2000, "DOE Helps the EPA Expedite Offshore Regulations for Synthetic-Based Mud," *USDOE's Oil & Gas Environmental Research Program,* winter 2000, vol. 5, no. 1, pp. 1–4.

Von Wedel, R., and Jacobs, J., 2000, "Overview of Aerobic Bioremediation and Future Development Potential," *Abstracts,* Association of Engineering Geologists and Groundwater Resources Association of California Annual Meeting, September 24, San Jose, p. 118.

Watts, R. J., Udell, M. D., Rauch, P. A. and Leung, S. W., 1990 "Treatment of Pentachlorophenol Contaminated Soils Using Fenton's Reagent." *Hazardous Waste Hazardous Materials,* Vol. 7, p. 335–345.

Watts, R. J., Smith, B. R., and Miller, G. C., 1991, "Treatment of Octachlorodibenzo-p-dioxin (OCDD) in Surface Soils Using Catalyzed Hydrogen Peroxide," *Chemosphere,* vol. 23, pp. 949–956.

Watts, R. J., et al., 1992, "Hydrogen Peroxide for Physicochemically Degrading Petroleum-Contaminated Soils," *Remediation,* vol. 2, pp. 413.

Watts, R. J., and Stanton, P. C., 1994, "Process Conditions for the Total Oxidation of Hydrocarbons in the Catalyzed Hydrogen Peroxide Treatment of Contaminated Soils," *WA-RD,* vol. 337.1, Washington State Department of Transportation, Olympia, Wash.

Williams, B., 1991, *U.S Petroleum Strategies in the Decade of the Environment,* PennWell Books, Tulsa, Okla., 336 pp.

World Oil Exploration Drilling and Production, 2001, Gulf Publishing, Houston, Tex.

Youngquist, W., 1997, *GeoDestinies—The Inevitable Control of Earth Resources Over Nations and Individuals,* National Book Company, Portland, Ore., 499 pp.

CHAPTER 9
OIL SPILLS AND LEAKS

Stephen M. Testa

Testa Environmental Corporation
Mokelumne Hill, California

James A. Jacobs

Fast-Tek Engineering
Redwood City, California

9.1 INTRODUCTION

The business of providing society's energy from petroleum employs more people than any other in the United States or the world. The sheer magnitude of this industry and the extensive infrastructure it requires to recover, process, and distribute petroleum products for our use makes it a daily influence on our lives. It is thus understandable that concerns about the environmental impact of petroleum in the environment have developed. From an environmental perspective, much attention has focused on the release of oil and petroleum products in the environment from accidental spills and other impacts such as air pollution. Responsible stewardship of our soil, water, and resources requires better understanding of the environmental consequences of petroleum production, processing, transporting, storing, and use.

The impact of petroleum in the environment can take many forms, and once an environmental impact has occurred, the significance of the impact can become very difficult to evaluate. Many of today's petroleum-related environmental problems are actually inherited from antiquated facilities or operational practices that are no longer in use (Fig. 9.1). However, considering the huge volume of oil and petroleum products that are moved, stored, and used every day, spills and leaks are inevitable. Overall, much progress has been made in understanding and mitigating the environmental impacts of oil and petroleum products. Technological advances, and new operational procedures, enable our use of petroleum to be increasingly safe and environmentally sound.

Despite our best efforts to reduce the number of spills, and minimize their overall impact to the environment, it is roughly estimated that 30 to 50 percent of oil spills are the result of human error and 30 to 40 percent are a result of equipment failure. Such spills result in high costs of cleanup, government fines, and litigation. Costs for cleanup have been estimated at $50 to $3000 per liter, depending on oil

FIGURE 9.1 Past leakage of crude oil from an aboveground storage tank farm at an antiquated refinery site. The tanks were set on gravel without bottoms within an unlined bermed area.

type and spill location (Environment Canada, 1978). Shoreline cleanup costs are by far the most expensive. In Canada, these costs average $200 per liter, whereas in the United States they average $1000 per liter.

Our ability to effectively address environmental issues is illustrated by examples of how undesirable releases of petroleum and petroleum-derived products to the environment are prevented and mitigated. The ability to effectively resolve environmental problems associated with our use of petroleum is continually being pursued.

This chapter provides a synopsis of the use of oil in society, sources of hydrocarbon contamination; subsurface behavior, including detection and occurrence; characterizing environmental impact; developing remediation and risk abatement strategies; and undertaking preventive measures. Case histories, both within the United States and international, are also presented.

9.1.1 Historic Use of Petroleum in Society

Petroleum has been known and used by civilizations for thousands of years. Several oil-producing regions throughout the world such as in the Middle East have been known for centuries. Marco Polo described in 1291 the use of petroleum as fuel for lamps. Early seafarers used asphalt from natural oil seeps to caulk and waterproof their sailing vessels. The Babylonians used asphalt as a cementing agent for bricks; the cradle of the Babylonian Moses was cemented by it, and the Walls of Jerico were held together by oil-based tar used as mortar. According to the Bible, Noah used two coats of tar to make his ark watertight. With the advent of the industrial revolution in the nineteenth century, a demand developed for an inexpensive fuel for lighting. It is somewhat ironic that the early use of petroleum, notably kerosene, was viewed as an effective solution to air pollution caused by the burning of coal (Fig. 9.2).

FIGURE 9.2 Remnants of the burning of coal during the nineteenth century is evident on this historic building in Glasgow, Scotland.

Upon introduction of the automobile, the demand for petroleum as an inexpensive and efficient fuel for transportation developed. Between 1900 and 1912, auto registration increased from 8000 to 900,000. Gasoline, originally a waste by-product in the production of kerosene that was commonly disposed of in pits and burned, filled this need. With the advent of World War I and the novel use of tanks, airplanes, and trucks for transport of troops and equipment, the need for oil dramatically increased. Oil today remains the energy medium of choice for transportation and a key ingredient in many other products important to our civilization.

9.1.2 Current Use of Petroleum in Society

Today, petroleum products and by-products permeate our society, ranging from fuels for transportation and heating to the raw material for literally thousands of products including plastics, paints, cosmetics, fabrics, and pharmaceuticals and medicine, to name just a few. If one considered how a barrel (42 gallons) of crude oil is processed, 1 barrel generates about 44.2 gallons of product, the majority (about 35 gallons or 83 percent) being gasoline and other fuels (Fig. 9.3). Oil is the key societal resource and formed the foundation of twentieth century civilization, with little doubt there that it will continue to be so as we enter into the new millennium.

The known recoverable world's oil resources have been increasingly abundant. This trend will increase because of technological advances that allow us to more effectively locate new subsurface deposits and more cost-efficiently recover petroleum from them. However, there are factors that also inhibit our ability to explore for and recover oil. Prospective areas for new discoveries may be environmentally sensitive and closed to exploration. Barriers to petroleum exploration and development also come from international political turmoil and unrest, institutional barriers, uncertain property rights and territorial disputes, and urbanization (Fig. 9.4). The location of new petroleum resources in remote regions without transportation facilities can also inhibit their development.

The ubiquitous use of petroleum in our society reflects its abundance, and the ease with which it can be produced, transported, and converted to other beneficial forms. Our society is dependent upon the availability of petroleum, with oil providing about 40 percent of the energy consumed in the United States, and 97 percent of our transportation fuels. Considering the current rate of consumption, it is estimated that about 200 billion barrels of recoverable oil exists in the United States, with a total world reserve on the order of 2 trillion barrels. This amount would sustain the United States for another half-century even if not another drop of oil was discovered or if no new technologies or techniques for the extraction of oil developed. This scenario of course is not very realistic, and it has been estimated that another 1.4 to 2.1 trillion barrels of recoverable oil remain to be produced worldwide. These estimated volumes should sustain the current rate of consumption for another 63 to 95 years, assuming current conventional petroleum consumption rates.

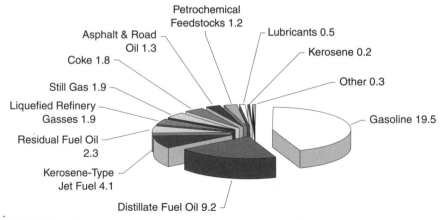

FIGURE 9.3 Pie diagram showing the variety of products in gallons produced from a barrel of crude oil.

FIGURE 9.4 Encroachment of urban sprawl in 1963 around refineries and tank farms in southern California. (*Courtesy of UCLA Department of Geography, Fairchild Collection.*)

9.1.3 The Environmental Challenge

Our continued dependence on petroleum for at least a few generations to come require that we understand and mitigate the environmental challenges that accompany its use. The environmental impact of our society's dependence on petroleum products can range from a localized issue such as the release of gasoline from an underground storage tank at a service station to regional and widespread impacts such as from offshore oil spills from tankers and production wells. A glaring example of release of petroleum to the environment occurred as a result of the Gulf War. The global population watched in horror at what Saddam Hussein perpetuated as 700 of Kuwait's 1500 oil wells released oil into the environment, 600 of which were set on fire. During this terrorist event, an estimated 11 million barrels of oil each day either burned or spilled. Over 60 million barrels collected in depressions, forming over 70 oil lakes on land (Fig. 9.5), and oil slicks at sea in the

FIGURE 9.5   The Al-Rawdhatayan Oil Field in northern Kuwait showing impact from crude oil on the desert environment at the close of the Gulf War in March 1991. (*From Testa, 1994.*)

Persian Gulf. This is an extreme example, but it highlights the principal environmental concerns associated with our use of petroleum, and its possible impacts on our environment.

9.2   DEFINING PETROLEUM

9.2.1   Hydrocarbon Chemistry

Hydrocarbons in general are simply compounds of hydrogen and carbon that are characterized according to their respective chemical composition and structure. Each carbon atom can essentially bond with four hydrogen atoms. Methane is the simplest hydrocarbon:

$$
\begin{array}{c}
\text{H} \\
| \\
\text{H}\!-\!\text{C}\!-\!\text{H} \\
| \\
\text{H}
\end{array}
$$

Methane (CH_4)

Each dash represents a chemical bond in which the carbon atom has four links and the hydrogen atom has one.

More complex forms of methane can be developed by adhering to the simple rule that a single bond exists between adjacent carbon atoms and that the rest of the bonds are saturated with hydrogen atoms. With the development of more complex forms, there is thus also increase in molecular size.

Hydrocarbons that contain the same number of carbons and hydrogen atoms, but have a different structure, therefore different properties, are known as *isomers*. As the number of carbon atoms in the molecule increases, the number of isomers rapidly increases. The simplest hydrocarbon having isomers is butane, as follows:

$$
\begin{array}{cc}
& CH_3 \\
& \backslash \\
CH_3 - CH_2 - CH_2 - CH_3 & CH - CH_3 \\
& / \\
& CH_3
\end{array}
$$

| Normal butane | Isobutane |

Hydrocarbon compounds can be divided into four major structural forms: (1) alkanes, (2) cycloalkanes, (3) alkenes, and (4) arenes. Petroleum geologists and engineers commonly refer to these structural groups as (1) paraffins, (2) naphthenes or cycloparaffins, (3) aromatics, and (4) olefins, respectively. The following discussions will focus on paraffins, naphthenes or cycloparaffins, and aromatics. Olefins are characterized by double bonds between two or more carbon atoms. Olefins are readily reduced or polymerized to alkanes early in their transformation in the subsurface, a process referred to as diagenesis, and are not found in crude oil and only in trace amounts in a few petroleum products. The more common chemical types and structural forms are illustrated in Fig. 9.6.

Paraffin-type hydrocarbons are referred to as *saturated* or *aliphatic* hydrocarbons. These hydrocarbons dominate gasoline fractions of crude oil and are the principal hydrocarbons in the oldest, most deeply buried reservoirs. Paraffins can form normal (straight) chains and branched-chain structures. Normal chains form a homologous series in which each member differs from the next member by a constant amount—that is, each hydrocarbon differs from the succeeding member by one carbon and two hydrogen atoms. The naming of normal paraffins is a simple progression using Greek prefixes to identify the total number of carbon atoms present. Branched-chain paraffins reflect different isomers (different compounds with the same molecular formula). Where only about 60 normal chain paraffins exist, theoretically, over a million branched-chain structures are possible with about 600 individual hydrocarbons identified.

Naphthenes or cycloparaffins, formed by joining the carbon atoms in a ring-type structure and are the most common molecular structures in petroleum. These hydrocarbons are also referred to as *saturated* hydrocarbons since all the available carbon atoms are saturated with hydrogen.

Aromatic hydrocarbons usually compose less than 15 percent of a total crude oil, although they often exceed 50 percent in heavier fractions of petroleum. The aromatic fraction of petroleum is the most important environmental group of hydrocarbon chemicals and contains at least one benzene ring comprising six carbon atoms in which the fourth bond of each carbon atom is shared throughout the ring. Schematically shown as a six-sided ring with an inner circle, the aromatics are unsaturated and thus can react to add hydrogen and other elements to the ring. Benzene is known as the *parent compound* of the aromatic series, and benzene, toluene, ethylbenzene, and the three isomers of xylene (ortho-, meta-, and para-), are significant (approximately 15 percent) constituents of gasoline.

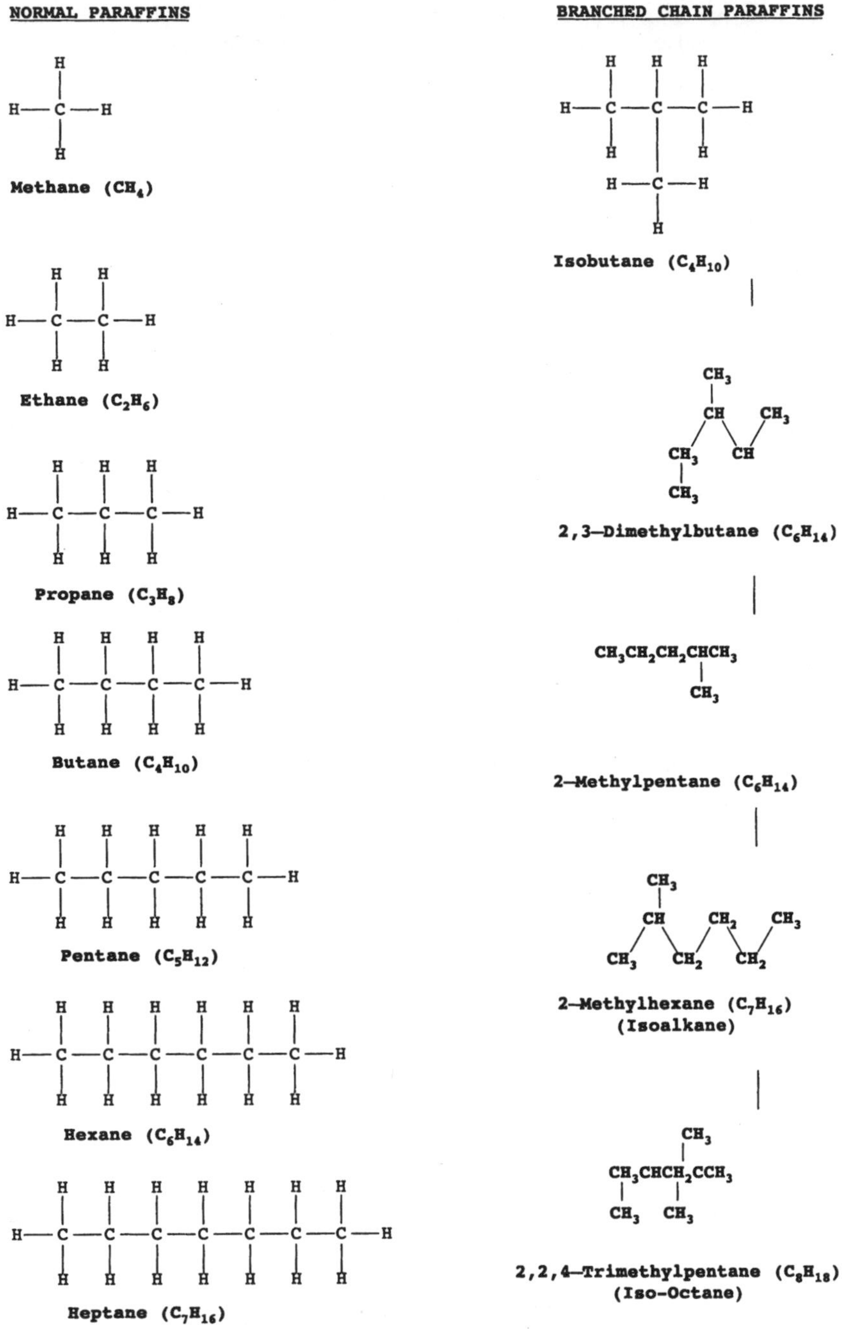

NORMAL PARAFFINS   **BRANCHED CHAIN PARAFFINS**

$$\begin{array}{c} H \\ | \\ H-C-H \\ | \\ H \end{array}$$

Methane (CH_4)

$$\begin{array}{ccc} H & H & H \\ | & | & | \\ H-C-&C-&C-H \\ | & | & | \\ H & H & H \end{array}$$
$$\begin{array}{c} H-C-H \\ | \\ H \end{array}$$

Isobutane (C_4H_{10})

$$\begin{array}{cc} H & H \\ | & | \\ H-C-&C-H \\ | & | \\ H & H \end{array}$$

Ethane (C_2H_6)

CH_3
|
CH CH_3
CH_3 CH
|
CH_3

2,3—Dimethylbutane (C_6H_{14})

$$\begin{array}{ccc} H & H & H \\ | & | & | \\ H-C-&C-&C-H \\ | & | & | \\ H & H & H \end{array}$$

Propane (C_3H_8)

$CH_3CH_2CH_2CHCH_3$
|
CH_3

2—Methylpentane (C_6H_{14})

$$\begin{array}{cccc} H & H & H & H \\ | & | & | & | \\ H-C-&C-&C-&C-H \\ | & | & | & | \\ H & H & H & H \end{array}$$

Butane (C_4H_{10})

CH_3
|
CH CH_2 CH_3
CH_3 CH_2 CH_2

2—Methylhexane (C_7H_{16})
(Isoalkane)

$$\begin{array}{ccccc} H & H & H & H & H \\ | & | & | & | & | \\ H-C-&C-&C-&C-&C-H \\ | & | & | & | & | \\ H & H & H & H & H \end{array}$$

Pentane (C_5H_{12})

$$\begin{array}{cccccc} H & H & H & H & H & H \\ | & | & | & | & | & | \\ H-C-&C-&C-&C-&C-&C-H \\ | & | & | & | & | & | \\ H & H & H & H & H & H \end{array}$$

Hexane (C_6H_{14})

CH_3
|
$CH_3CHCH_2CCH_3$
| |
CH_3 CH_3

2,2,4—Trimethylpentane (C_8H_{18})
(Iso-Octane)

$$\begin{array}{ccccccc} H & H & H & H & H & H & H \\ | & | & | & | & | & | & | \\ H-C-&C-&C-&C-&C-&C-&C-H \\ | & | & | & | & | & | & | \\ H & H & H & H & H & H & H \end{array}$$

Heptane (C_7H_{16})

FIGURE 9.6   Major structural forms for various hydrocarbon compounds.

NAPHTHENES (CYCLOPARAFFINS)

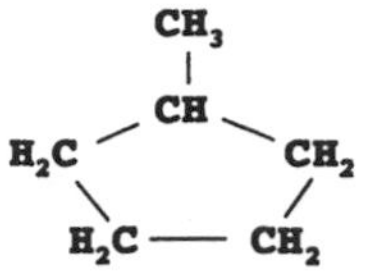

$$CH_3$$
$$|$$
$$CH$$
$$H_2C \qquad CH_2$$
$$H_2C \longrightarrow CH_2$$

Methylcyclopentane (C_6H_{12})

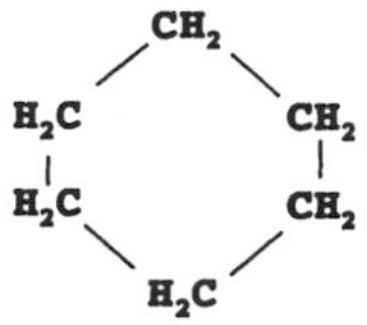

$$CH_2$$
$$H_2C \qquad CH_2$$
$$H_2C \qquad CH_2$$
$$H_2C$$

Cyclohexane (C_6H_{12})

$$CH_2 \qquad CH_2$$
$$H_2C \qquad CH \qquad CH_3$$
$$H_2C \qquad CH_2$$
$$H_2C$$

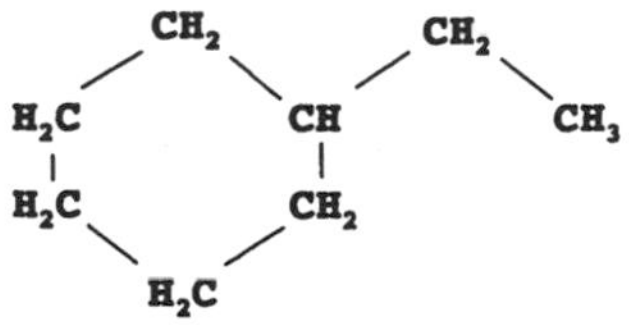

Ethylcyclohexane (C_8H_{16})

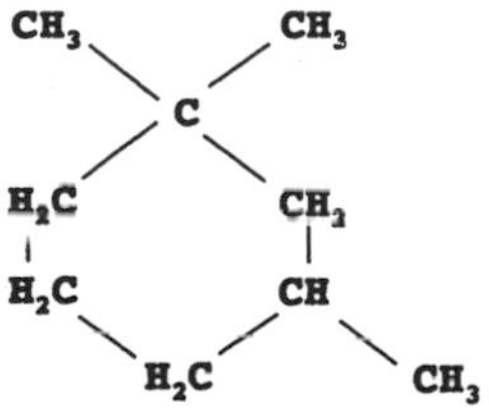

$$CH_3 \qquad CH_3$$
$$C$$
$$H_2C \qquad CH_2$$
$$H_2C \qquad CH$$
$$H_2C \qquad CH_3$$

1, 1,3—Trimethylcyclohexane (C_9H_{18})

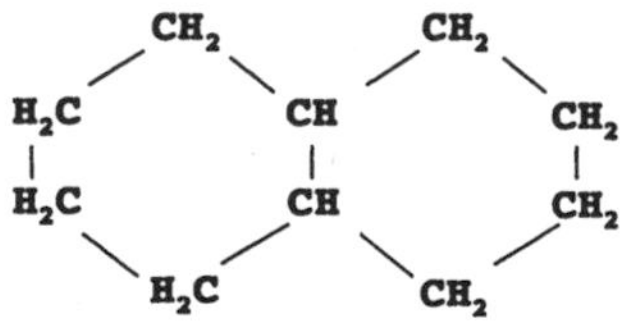

$$CH_2 \qquad CH_2$$
$$H_2C \qquad CH \qquad CH_2$$
$$H_2C \qquad CH \qquad CH_2$$
$$H_2C \qquad CH_2$$

Decalin ($C_{10}H_{18}$)

FIGURE 9.6 (*Continued*)   Major structural forms for various hydrocarbon compounds.

AROMATICS

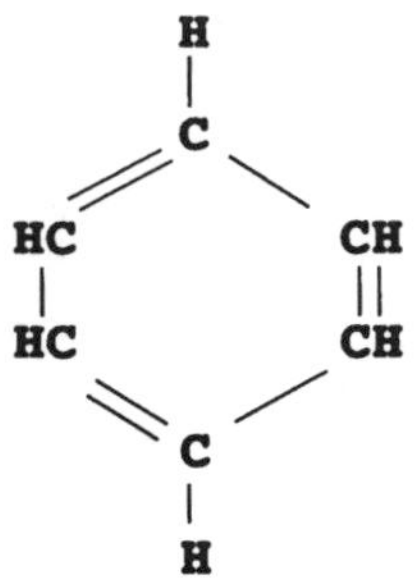

Benzene (C₆H₆)

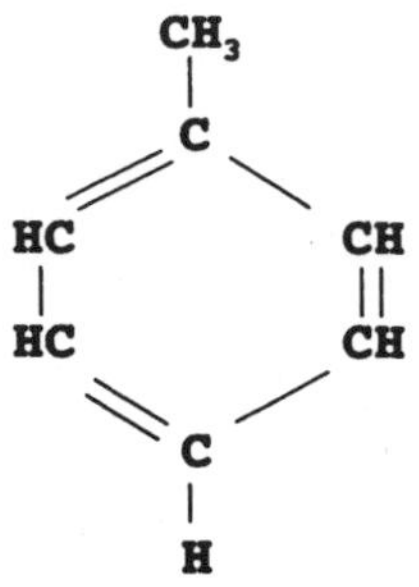

Toluene (C₇H₈)

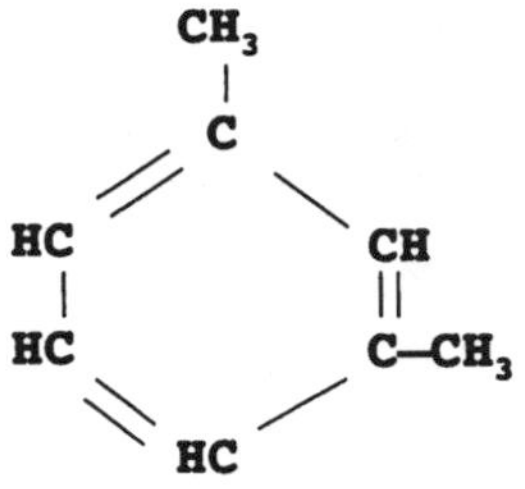

Metaxylene (C₈H₁₀)

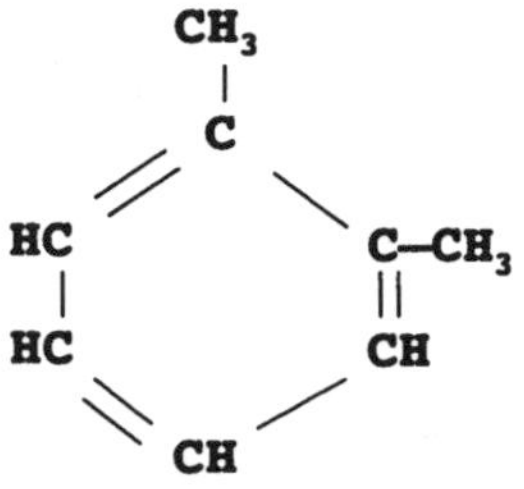

Orthoxylene (C₈H₁₀)

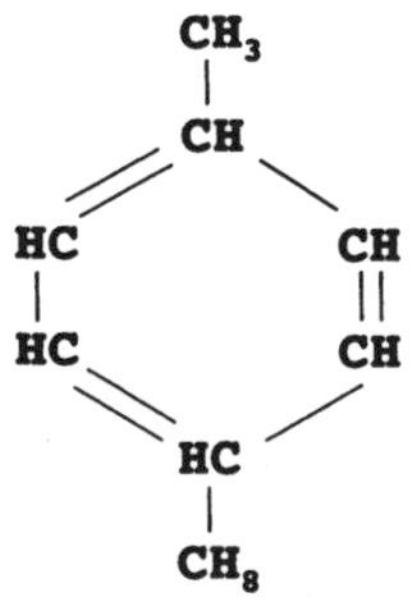

Paraxylene (C₈H₁₀)

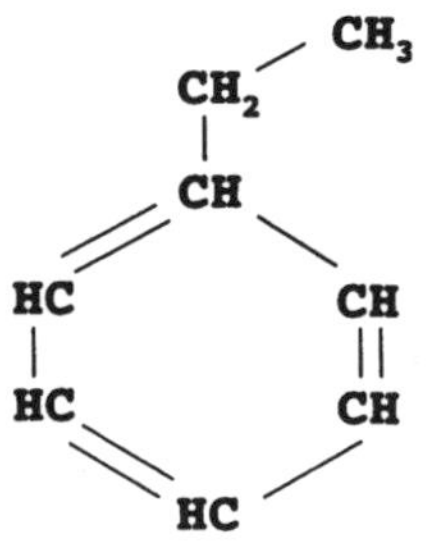

Ethylbenzene (C₈H₁₀)

FIGURE 9.6 (*Continued*)   Major structural forms for various hydrocarbon compounds.

9.2.2 Physical Characteristics of Petroleum

Petroleum is a naturally occurring mixture that usually exists in gaseous form (natural gas) or in liquid form (crude oil), but can also exist as a solid (waxes and asphalt). Primarily composed of hydrocarbons, which are compounds that contain only hydrogen and carbon, petroleum varies widely in chemical complexity and molecular weight. Pertinent physical properties for some of the more common petroleum products are presented in Table 9.1.

Petroleum is any mixture of natural gas, condensate, and crude oil. The term petroleum is derived from the Latin derivative *petra* for rock and *oleum* for oil. A petrochemical is a chemical compound or element recovered from petroleum or natural gas, or derived in whole or in part from petroleum or natural gas hydrocarbons, and intended for chemical markets. Petrochemicals and hydrocarbons are simply compounds of hydrogen and carbon that can be distinguished from one another based on chemical composition and structure.

The physical nature and chemical characteristics of petroleum are fundamental to understanding the impacts of its release to the environment. The physical characteristics of petroleum determine how it behaves in the subsurface as well as above ground where it can come in contact with soil, water, and life. The chemistry of petroleum in large part determines how it is dispersed in the environment and impacts life. Both the physical and chemical characteristics of petroleum are important foundations for the technology we use to mitigate unwanted environmental consequences of petroleum use.

9.2.3 Chemical Characteristics of Petroleum

Crude oil (commonly called just *crude*) is the initial oil extracted from the subsurface without any refinement into other liquid forms, or products. It is a naturally occurring heterogeneous liquid consisting almost entirely of hydrogen and carbon. The composition of crude oil can vary significantly, depending on its origin and age. Crude generally ranges from 83 to 87 percent carbon (by weight), 11 to 14 percent hydrogen, with lesser amounts of sulfur (0.1 to 5.5 percent), nitrogen (0.05 to 0.08 percent), and oxygen (0.1 to 4 percent). Trace constituents constitute less than 1 per-

TABLE 9.1 Pertinent Physical Properties of Common Petroleum Hydrocarbon Products

Oil type	No. 2 fuel	Light crude (S. La.)	Heavy crude (Bachaquero)	Bunker C
Specific gravity at 77°F	0.856	0.854	0.977	0.942
API at 77°F	33.8	34.2	13.3	18.9
Kinematic viscosity at 77°F, cSt	3.1	7.8	2600	2800
Pour point, °F	−10	10	15	65
Surface tension at 77°F, dyn/cm	37.1	34.2	38.6	39.9
Interfacial tension with synthetic seawater at 77°F, dyn/cm	36.0	24.9	37.8	46.2
Emulsification characteristics with synthetic seawater at 77°F	3 min	65 min	2 h	None after 2 weeks

cent of the total volume and include phosphorus and heavy metals such as vanadium and nickel.

Crude is classified on the basis of the relative content of three basic hydrocarbon structural types: paraffins (waxy crude), naphthenes, and aromatics (Fig. 9.7). About 85 percent of all crude oil can be classified as asphalt base, paraffin base, or mixed base. Sulfur, oxygen, and nitrogen contents are often relatively high in comparison with paraffin base crude, which contains little to no asphaltic materials. Mixed-base crude oil contains considerable amounts of both wax and asphalt. Chemically, crude oil is composed of some methane (normal straight chain paraffins), and isoparaffins (branched-chain paraffins) cycloparaffins or naphtenes (ring-structures), aromatics (benzene ring-structures), and asphaltics.

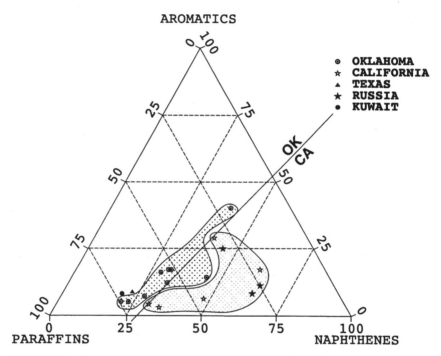

FIGURE 9.7 Ternary diagram showing representative crude oils and their respective composition. California crude shows a trend toward the naphthenes while Oklahoma crude shows a trend toward the paraffins. (*From Testa and Winegardner, 2000.*)

9.2.4 Hydrocarbon Constituents of Environmental Concern

Spilled hydrocarbons can consist of crude, refined petroleum products such as fuels (gasoline, diesel, and aviation and jet fuels), lubricating oil and fluids, and waste oil. These products are of environmental concern if accidentally released in the environment. From an environmental or regulatory perspective, the petroleum product, or a specific constituent, may drive the program developed to address the concern. When a specific constituent drives the remedial response, this can present a challenge, since each organic or nonorganic compound has specific physical, chemical, and biological properties.

The aromatic fraction of petroleum is the most important group of hydrocarbon chemicals. Benzene is known as the *parent compound* of the aromatic series with toluene, ethylbenzene, and xylenes. Benzene, being a carcinogen, if present, typically drives a remedial effort; however, other constituents, depending on the product released, may drive the remedial effort. These constituents may include certain volatile organic compounds or fuel additives such as methyl tertiary butyl ether (MTBE). MTBE is an octane booster and one of several synthetic fuel oxygenates used to meet regulatory oxygen mandates for reformulated fuels in areas not in compliance with federal standards for ozone pollution (Jacobs et al., 2001). Other constituents such as lead, cadmium, chromium, and sulfur are typical constituents of waste oil and can also be of environmental concern.

Oils are not all petroleum based. Such products as animal fats (e.g., butter) and vegetable oils (e.g., soybean oil) can also be of environmental concern. Inflamed butter can result in the generation of anhydrous ammonia and sulfuric acid gases, whereas vegetable oils spilled in water can reduce the dissolved oxygen level and adversely damage the local ecosystem.

9.3 SOURCES OF SPILLS

Huge volumes of oil are used each day. Worldwide, about 7 million tons of oil are used each day. The United States uses about 2,300,000 tons of oil and petroleum products each day. In Canada, about 230,000 tons of oil is used on a daily basis. With such huge volumes of oil and petroleum products, it is understandable that spills occur frequently and are inevitable. Spills are more frequent in the United States than Canada since more oil is imported by sea and more fuel is transported by barge. Some of the larger spills, excluding those that are a result of well blowouts, which are discussed in Chap. 8, are summarized in Table 9.2.

Statistics for spills are maintained by a number of agencies. In the United States, the Environmental Protection Agency and various state agencies maintain records of spills on land. In navigable waters, the Coast Guard maintains such records, whereas the Minerals Management Service maintains records of spills from offshore exploration and production activities. In Canada, provincial offices obtain data, while Environment Canada maintains a database of spills.

Statistics for spills can be misleading. For example, different estimates can result if the chemical or physical properties of the oil are uncertain. Offshore spills may involve uncertainty as to the exact volume contained in a particular vessel's compartment. In addition, amounts transferred to other vessels or burned may not get recorded.

Oil spills into oceans, excluding large accidental spills, primarily occur in navigable waters from near-shore operations (Fig. 9.8). Navigable waters under United States jurisdiction include bays, harbors, rivers, lakes, sounds, and oceans within 200 miles of the coastline. Sources for oil spills into the oceans, in descending order, include those introduced by rivers and oceans, tank operations, other transport units, coastal facilities, natural seeps, tanker accidents, and offshore production (Fig. 9.9). With a continued increase in oil tanker size and capacity over the years, the potential for accidents has also increased. Considering the occasional large spill, offshore spills from marine vessels (including tankers, barges, and freighters) account for about 62 percent of all spills, with the volume of product released to the environ-

TABLE 9.2   Summary of Major Oil Spills

Ship/incident	Country	Location	Year, Month/Day	Tons, 10^3
Gulf War	Kuwait	Sea Island	1991, January 26	750
Atlantic Empress/				
Aegean Captain	Off Tobago	Caribbean Sea	1979, July 19	280
Castillo de Bellveri	South Africa	Saldanha Bay	1983, August 6	260
Amoco Cadiz	France	Brittany	1978, March 16	230
Haven	Italy	Genoa	1991, April 11	140
Odyssey	Off Canada	North Atlantic	1988, November 10	130
Torrey Canyon	England	Land's End	1967, March 18	120
Sea Star	Oman	Gulf of Oman	1972, December 19	120
Texaco Denmark	Belgium	North Sea	1971, December 7	120
Storage Tanks	Kuwait	Shuaybah	1981, August 20	120
Pipeline rupture	Russia	Usinsk	1994, October 25	120
Urquiola	Spain	La Coruna	1976, May 12	110
Irene's Serenade	Greece	Pylos	1980, February 23	110
Julius Schindler	Portugal	Azores	1969, February 11	110
Pipeline rupture	Iran	Ahvazin	1978, May 25	110
Independenta	Turkey	Bosporus Strait	1979, November 15	103
ABT Summer	Off Angola	Atlantic Ocean	1991, May 28	100
Hawaiian Patriot	Off United States	West of Hawaii	1977, February 23	100
Storage tanks	Nigeria	Forcados	1979, July 6	90
Braer	United Kingdom	Shetland Islands	1993, January 5	85
Othello	Sweden	Vaxholm	1970, March 20	80
Jakob Maersk	Portugal	Oporto	1975, January 29	80
Aegean Sea	Spain	La Coruna	1992, December 3	80
Nova	Iran	Persian Gulf	1985, December 6	77
Sea Empress	United Kingdom	Milford Haven	1996, February 15	75
Kark 5	Morocco	Atlantic Ocean	1989, December 19	75
Katina P	South Africa	Indian Ocean	1992, April 17	72
Wafra	South Africa	Atlantic Ocean	1971, February 27	70
Fuel storage tank	Rhodesia	Salisbury	1978, December 11	60
Epic Colocotronis	United States	West of Puerto Rico	1975, May 13	60
Sinclair Petrolore	Brazil		1960, December 6	60
Fuel storage tank	Japan	Sendai	1978, June 12	58
Assimi	Oman	Ras al Had	1983, January 7	53
Andros Patria	Spain	Bay of Biscay	1978, December 31	52
Yuyo Maro 10	Japan	Tokyo	1974, November 9	51
Heimvard	Japan	Hokkaido	1965, May 22	51
Metula	Chile	Strait of Magellan	1974, August 9	50
Peracles GC	Qatar	Persian Gulf	1983, December 9	50
World Glory	South Africa	Indian Ocean	1968, June 13	48
Ennerdale	Seychelles	Indian Ocean	1970, June 1	45
British Ambassador	Japan	Iwo Jima	1975, January 13	43
Tadotsu	Indonesia	Strait of Malacca	1978, December 7	43
Mandoil	United States	Oregon	1968, February 29	43
Texaco Oklahoma		Northwest Atlantic	1971, March	38
Trader	Greece	Mediterranean Sea	1972, June 11	37
St. Peter	Colombia	Pacific Ocean	1976, February 6	37
Irene's Challenge		Pacific Ocean	1977, January 17	37
Exxon Valdez	United States	Valdez, Alaska	1989, March 24	36
Napier	Chile	Off west coast	1973, June 10	35
Storage tank	Japan	Mizushima refinery	1974, December 18	34

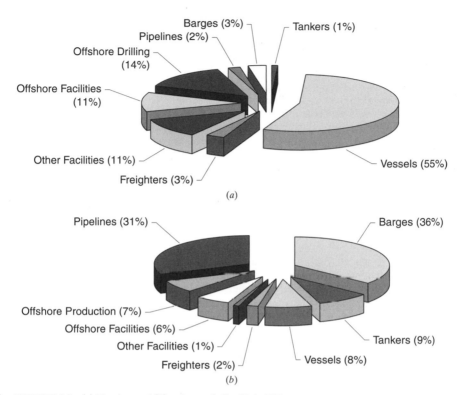

FIGURE 9.8 (*a*) Number and (*b*) volume of oil spills in U.S. waters.

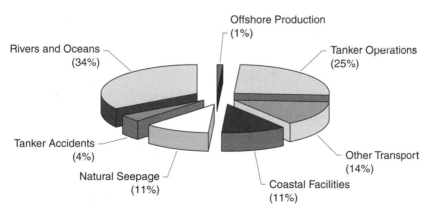

FIGURE 9.9 Primary sources for oil spills in oceans. (*From National Research Council, 1991.*)

ment accounting for about 55 percent of the total volume (Fig. 9.8). Causes of such spills in descending order include grounding, transfers, collisions and equipment failures, ramming, among others (Fig. 9.10).

Between 20 to 340 million gallons of oil are spilled onto the earth or into the ocean each year. The large range reflects episodic impact of a few very large spills that occur infrequently. Most of the largest spills are from tankers, barges, pipelines, and onshore facilities. For example, a 10.8 million gallon (257,143 barrel) tanker spill in 1989 accounted for over 75 percent of the total volume spilled for that year. In 1994, a 1.5 million gallon pipeline spill accounted for 37 percent of the annual total volume. One large spill can be a significant factor in determining the overall environmental impact from spills in any given year. The greatest number of spills occur within inland waters such as rivers, lakes, and bays during transportation by tanker, barge, pipeline, or rail.

The largest spill ever recorded was the estimated 240 million gallons (5,714,286 barrels) of crude oil that was released to the Persian Gulf, as a result of environmental terrorism during the Gulf War in 1991 (Table 9.2). Some of the greater magnitude spills go unnoticed because of their location, and others that are not so large become well known. Certainly oil spills that occur in coastal waters off western Europe and North America receive intense media attention and coverage. The grounding of the *Exxon Valdez* spilled 10.8 million gallons (257,143 barrels); however, this incident although occurring in an environmentally sensitive area, does not make the top 50 largest spills since 1972. The 16 largest spills since 1972 ranged from 23 to 140 million gallons (547,619 to 3,333,333 barrels, respectively).

The overall volume of spills in the United States has been significantly reduced. In the late 1980s, an average of 7.9 million gallons (188,095 barrels) of oil was spilled each year compared to 2.1 million gallons in the 1990s, representing a 74 percent decrease. Compared to the 272 billion gallons consumed by the United States, the volume spilled makes up about 0.0004 percent of that consumed. Another perspective is to look at the amount of product being inadvertently released. In the late 1980s, 63 percent of all spill incidents involved volumes of less than 10 gallons. In the 1990s, these small volume releases composed about 72 percent of all spills. Large spills account for about 69 percent of the total volume spilled over the past decade.

Most petroleum transport occurs on water. Although some of the larger spills are from tankers, tankers are not the primary source of oil pollution in the marine environment. Worldwide, about half of the oil spills in open waters is a result of runoff from land-based sources (Fig. 9.11a). Only about 24 percent of oil spilled in marine

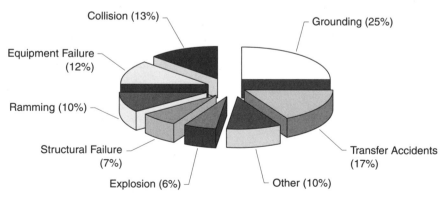

FIGURE 9.10 Primary causes of tanker spills in U.S. waters.

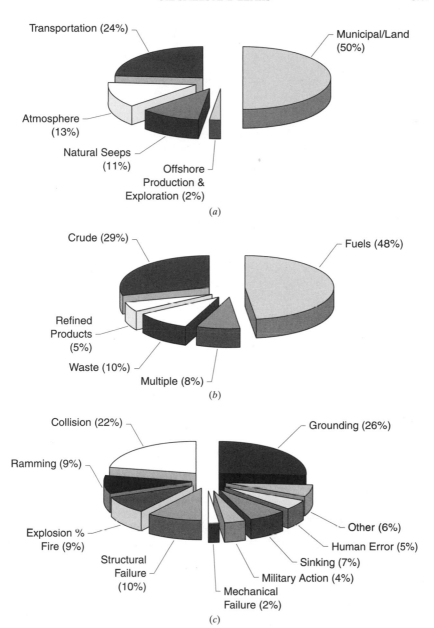

FIGURE 9.11 (*a*) Primary sources, (*b*) oil type, and (*c*) causes of oil spills in the world's oceans.

waters are derived from transportation, of which only about 10 percent are from tankers. Natural seeps account for about 11 percent, whereas offshore production and exploration activities and releases to the atmosphere account for 2 and 13 percent, respectively. Fuels (48 percent) and crude oil (29 percent) account for the majority of offshore spills (Fig. 9.11*b*). When one views the causes of vessel spills, grounding (26 percent) and collisions (22 percent) account for the majority of releases (Fig. 9.11*c*). Other causes include structural failure (10 percent), ramming (9 percent), explosions (9 percent), sinking (7 percent), military action (4 percent), mechanical failure (2 percent), and others (6 percent).

Oil can also be spilled during its transportation and distribution on land. Significant land-based spills commonly reflect unanticipated breached pipelines, bulk storage facilities, or the results of vehicle accidents (Fig. 9.12). Occasionally leakage is extensive. Often, however, leaks are smaller and more chronic in nature, reflecting slow releases over long periods of time (Fig. 9.13). Sometimes the constituent chemicals that make up a crude oil or refined product will travel at different velocities once they encounter groundwater. This is especially true of the gasoline additive MTBE, which travels much faster than other portions of gasoline, such as benzene, toluene, ethylbenzene, and xylenes (BTEX).

Inland spills can create immediate environmental impact as well as long-term potential consequences. However, with proper and prudent precautions, the number and size of spills can be and have been reduced substantially. For example, pipelines and storage tanks are routinely tested for tightness, and berms are typically constructed around enclosed aboveground storage tanks so that if a release occurs, the product is retained within the confines of the bermed area.

(*a*)

FIGURE 9.12 (*a*) Oil spill as a result of a breached valve in an unlined bermed containment area, (*b*) continuous discharge of crude oil directly onto the ground surface, and (*c*) one-time release of refined product from a breached pipeline.

(b)

(c)

FIGURE 9.12 (*Continued*)   (*a*) Oil spill as a result of a breached valve in an unlined bermed containment area, (*b*) continuous discharge of crude oil directly onto the ground surface, and (*c*) one-time release of refined product from a breached pipeline.

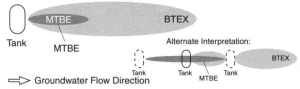

(a) Conceptual sketch of the map view of plume geometry for a point source release over three distinct release events

Tank   Last Event   Second Event   First Event

⟹ Groundwater Flow Direction

(b) Conceptual sketch of the map view of plume geometry for a point source release-steady state

Tank

⟹ Groundwater Flow Direction

Concentration of gasoline constituents

High   Medium   Low

(c) Map view of plume geometry for a release of gasoline with MTBE

BTEX and MTBE   MTBE

Tank
⟹ Groundwater Flow Direction

(d) Map view of plume geometry for a release of gasoline without MTBE followed by a later release of gasoline containing MTBE. Alternate interpretation: previously unidentified up-gradient MTBE or down-gradient gasoline plumes detected.

MTBE   BTEX

Tank   MTBE

Alternate Interpretation:

BTEX

Tank   Tank   Tank
⟹ Groundwater Flow Direction   MTBE

FIGURE 9.13   Dissolved contaminant gasoline with and without MTBE showing plume geometry under three distinct release events. (*From Jacobs et al., 2001.*)

9.3.1 Spills during Discovery and Production

Oil fields can cover tens to hundreds of square miles, and consist of hundreds of production wells. Oil field facilities may include, in addition to production wells, sumps for the storage of waste fluids (mostly water), injection wells for subsurface disposal of waste fluids, pumping facilities, storage tanks for recovered oil, and pipelines. With about 17 percent of the United States' oil production derived from offshore wells, marine settings for the exploration, discovery, and production of oil present special environmental challenges.

Producing oil and gas fields in marine environments present special challenges due to the proximity of sensitive ecosystems such as fisheries breeding grounds, coral reefs, wetlands, and salt marshes. Oil spills to the open waters from discovery and recovery operations typically account for less than 5 percent of the total volume of all oil discharges to the marine environment. However, there have been several large oil spills derived from offshore blowouts at production wells. The environmental impacts derived from the drilling for oil and gas are discussed in Chap. 8.

Natural seeps release much more petroleum to the marine environment than production facilities. In fact, natural seeps in the North Sea are estimated to contribute 4 times as much oil to the marine environment relative to all spills derived from discovery and recovery activities.

9.3.2 Spills during Transportation and Distribution

Petroleum and its products are distributed through a complex transportation system in order to deliver them where they are needed. Crude must be transported either by ocean tanker, tank truck, barge, tanker vessel, or pipeline to a refinery. Nearly half of the world's seaborne trade involves the transport of crude oil or associated products. Some ocean tankers, the largest ships in the world, can carry 3 million barrels of crude oil.

Much domestic crude and refined product is transported by pipeline. Pipelines, mostly buried beneath public rights-of-way, provide the most convenient way to move crude oil over land. Three-quarters of the domestic crude oil and one-half of the refined products are transported over 352,000 km of pipelines in the continental United States. Some pipelines can move product at rates of about 8 km/h. The trans-Alaska pipeline transports 1.5 million barrels per day, which equates to about 17 percent of the United States' daily production. Crude oil derived from the Overthrust Belt in Utah and Wyoming is transported by pipelines to refineries serving the Midwest and western markets. Other major pipelines connect Texas and the northeastern United States. Pipelines are commonly instrumented and computerized to monitor flow, and are tested for leaks by putting water or other liquids in the line under pressure. Special tools are also used inside pipelines to detect zones of weakness.

The distribution system is commonly divided into primary, secondary, and tertiary levels. Primary distribution systems include refining facilities, pipelines, tankers and barges, and large bulk storage terminals. Secondary distribution systems include primarily bulk plants (mainly wholesale storage facilities with less than 50,000 barrels capacity), that receive product via rail or truck. Also included in this category are service stations, truck stops, and retail oil dealers. There are an estimated 15,000 companies that own and/or operate bulk plant facilities, and about 17,000 service stations. The tertiary distribution system involves the ultimate consumer of the finished product. Consumer-controlled facilities such as personal vehicle fuel tanks, home heating oil tanks, generating plant fuel tanks, farm fuel storage tanks, and other small-capacity storage facilities make up the tertiary system. As of March 1998, there were an estimated 63 million barrels of petroleum products in tertiary-level storage facilities.

The primary environmental concern about the transportation and distribution aspects of the petroleum industry is spills. Accidental releases of hydrocarbon from pipelines occur primarily from breached pipelines, which account for about 79 percent of all accidents. Other origins include leaky valves, weld joints, bolted fittings, and pumps. External forces such as human error, third-party damage, and natural hazards are the major causes of accidents, although other factors include corrosion, failed pipe or weld joint, and operator error. Spills can be generally divided into three broad groups based on their point of origin: offshore spills originating either from boats or ships, inland water spills originating during transportation to or from storage facilities, and land-based spills originating from on-site facilities such as storage and transportation units.

Just about every aspect in the handling of petroleum presents the potential for spills. In the United States, the largest volume of spills involves transportation on barges (about 36 percent in total volume spilled). Transportation via barges, railroads, pipelines, tankers, and freighters accounts for about 67 percent of the total volume released (Fig. 9.14). Less than 10 percent involved offshore facilities and pipelines. The most typical products released into the environment are composed of relatively heavier industrial fuel oil (i.e., bunker oils and intermediate fuel oils) or residual oil (ranging from 45 to 95 percent in total volume), with the remainder being distillate fuel (22 percent) or crude oil (18 percent).

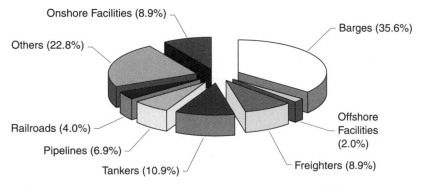

FIGURE 9.14   Primary sources of oil spills during transportation (for 1995).

9.3.3   Spills during Refining

In order to understand the environmental aspects associated with refining and processing of petroleum, it is important to understand what happens at a refinery. A refinery converts crude oil and other hydrocarbon feedstocks into useful products and raw materials for other industries. Refining involves the separation and blending, purifying, and quality improvement of desired petroleum products. In the early days of the petroleum industry, refineries were constructed close to producing fields. In recent years the development of extensive transportation systems enables refineries to be located just about anywhere. Refineries are large, complex facilities that use a variety of chemical and physical processes to make a diversity of end products (Fig. 9.15).

The typical refinery utilizes three primary processes in converting crude oil into these varied fuel products: separation, conversion, and treatment. The primary products of refineries are:

- Fuels such as gasoline and diesel fuels, aviation and marine fuels, and fuel oils
- Chemical feedstocks such as naphtha, gas oils, and gases
- Lubricating oils, greases, and waxes
- Asphalt
- Petroleum coke
- Sulfur

FIGURE 9.15   A typical refinery showing a mass of tanks and pipelines.

Transportation fuels are by far the primary products of refineries (Fig. 9.16). The variables that determine gasoline quality include blend, octane level, distillation range, vapor pressure ratings, and other considerations such as sulfur content or gum-forming tendencies. The petrochemical industry and other industries also utilize refinery products to manufacture plastics, polyester clothing, certain pharmaceuticals, and a myriad of other useful products.

The primary fuels we are all familiar with are gasoline, diesel (or no. 2 fuel oil), heavy fuel oils, and liquefied petroleum gas (propanes and butanes). Gasoline is composed of a mixture of volatile hydrocarbons suitable for use in internal combustion engines. The primary components are branched-chain paraffins, cycloparaffins, and

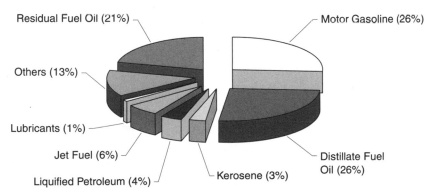

FIGURE 9.16   Refinery product production worldwide. (*From Department of Energy, 1997.*)

aromatics. Diesel fuel is composed primarily of unbranched paraffins with a flash point between 110 and 190°F. Fuel oils are chemical mixtures that can be distilled fractions of petroleum, residuum from refinery operations, crude petroleum, or a mixture of two or more of these, with a flash point greater than 100°F. These petroleum mixtures basically represent progressive fractions of a distillation column. These major petroleum hydrocarbon constituents can easily be represented by their increasing boiling points in a gas chromatograph separating column (Table 9.3).

TABLE 9.3 Common Petroleum Hydrocarbon Products Derived from the Refining of Crude Oil

Distillate	Boiling point, °F	Product
Gas	240	Fuel gases
		LPG
		Petrochemical feedstock
Light-heavy naphtha	335	Gasoline
		Petrochemical feedstock
		Solvents
		Jet fuel (naphtha type)
Kerosene	420	Jet fuel (kero type)
Light gas oil	500	Auto and tractor
		Diesel
		Home-heating oil
Heavy gas oil	600	Commercial oil
		Industrial oil
		Lubricants
Residuals (bottoms)	800	Bunker coal
		Asphalt
		Coke

Refineries in 1995 reported that about 1.3 billion pounds of toxic chemicals classified as refinery waste was produced. Of this total, 60 million pounds, or less than 4 percent, entered the environment as air emissions (Fig. 9.17a); 38 percent was burned, 36 percent was treated to reduce the volume of material considered toxic or to reduce toxicity prior to disposal, and 20 percent was recycled in some manner. When one views where the waste materials end up, about 80 percent is air emissions, with 8 percent released in water, 6 percent disposed of in underground injection wells, 4 percent released off site, and 1 percent released on the land (Fig. 9.17b).

Refineries and associated storage tank farms can have leaks that lead to hydrocarbon-impacted soil, dissolved hydrocarbon constituents in groundwater, or the presence of subsurface hydrocarbon product that accumulates as underground pools overlying the water table or as perched zones between the water table and land surface. These pools, which mainly reflect small leaks over long periods of time, can become large, in some cases being over a million barrels. Underground pools of this size can be a continuing source of soil and groundwater contamination for decades.

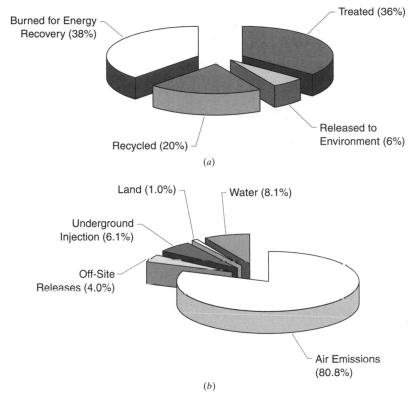

FIGURE 9.17   Pie diagrams (for 1995) showing how toxic materials generated at refineries are (*a*) managed and (*b*) released into the environment. (*From API, 1997a.*)

Over the past decades, progress has been made. Many refineries have taken the initiative to minimize and prevent releases by relocating underground structures such as pipelines and storage tanks to above ground, upgrading aboveground storage tanks and reservoirs, and increasing monitoring of storage units for leaks and emissions. Most active refineries, in the United States and elsewhere, have also implemented product recovery and soil and groundwater remediation programs, and have taken efforts to minimize emissions to the atmosphere. Such programs have, to a limited extent, included reuse and recycling programs.

9.3.4   Spills from Producer to User

The petroleum industry currently produces and markets more varieties of gasolines than ever. These fuels are designed to be more efficient in specific geographical areas and to reduce air emissions. The mode of delivery of fuels and other products to the end user varies, depending on the amount and type of product. End users of products can include large customers such as refineries, bulk liquid storage facilities, retail service stations, and, of course, the small-quantity end user that requires oil for heating or gasoline for driving.

Large-Volume End Users. Refiners and storage and transfer facilities purchase and handle various products for further refining and distribution, thus serving as clients to the petroleum industry as much as the individual who purchases gasoline for automobiles. At refineries and bulk liquid storage facilities, petroleum products are stored in aboveground storage tanks, reservoirs, and underground tanks. Many regulations and restrictions have been promulgated over the past decade to ensure safe storage at these facilities. For example, enclosed aboveground storage tanks, historically could be bottomless, allowing product to seep directly into the ground, or have steel bottoms that eventually would corrode, with similar results. Storage tanks are now required to have double bottoms, leak detection devices, and spill prevention systems. An impermeable liner or containment space may also be required beneath tanks. Now, monthly monitoring for compliance with corrosion and spill regulations is required.

Many of the old concrete-lined reservoirs built shortly after the end of the nineteenth century are now being taken out of commission, and at some refineries, all underground structures including pipelines are being relocated to above ground to minimize the likelihood that a leak would go undetected. An original rationale for putting pipelines and storage tanks below ground was to minimize safety hazards for fire and explosion, and take advantage of limited aboveground space.

Small-Volume End Users. The great number of small-volume end users are retail service stations (Fig. 9.18), repair shops, industrial operations, a multitude of other businesses, and all automobile drivers. In the United States, petroleum fuel products are chiefly sold through about 193,000 retail service stations, which in turn supply an estimated 173 million drivers. The environmental impacts associated with small-volume end users include leaks and automobile emissions.

FIGURE 9.18 Removal of an old corroded underground storage tank (UST) at a retail fuel service station has been a common scene over the past decade.

Leaks are the primary environmental concern associated with underground storage tanks at service stations. Such leaks can contaminate soil and groundwater, and have become the focus of recent regulatory concern. Millions of underground storage tanks (USTs) were installed in the 1950s and 1960s. Out of an estimated 2.5 to 3 million USTs throughout the United States, more than 400,000 have leaked directly or from associated piping or are currently leaking. Assuming the average estimated amount of contaminated soil as a result of such leaks to be on the order of 50 to 80 yd^3, the volume of contaminated soil solely attributed to USTs is on the order of 20 to 32 million yd^3. This is a conservatively low estimate, since it does not take into consideration the numerous cases where much larger volumes of soil have been impacted, or the thousands of unrecorded USTs and their impact. All states now have UST programs in place that ensures that USTs conform to current requirements and regulations, and that any unauthorized release of product into the subsurface is mitigated in a timely manner.

Another small-volume end-user environmental problem is used automobile oil. Used automobile oil is insoluble, persistent in the environment, and can contain chemicals and heavy metals that are considered toxic. It is estimated that the used oil from a single oil change can contaminate 1 million gallons of fresh water (a year's supply for 50 people) and make it unfit for public consumption. The problem is that about 60 percent of the nation's drivers change their own oil. This in turn generates over 200 million gallons (or over 4,761,900 barrels) of used oil annually. Many communities and agencies have initiated recycling programs. If all the used oil could be recycled, this effort would save about 1.3 million barrels of oil per day.

9.4   SPILL BEHAVIOR AND RISK ASSESSMENT

9.4.1   Spill Behavior

When an oil spill occurs, whether on land or on water, several physical, chemical, and biological transformations or changes occur. These processes commence immediately after a spill occurs. The effects of these processes as a whole on the oil, and its ability to migrate through the subsurface environment, is essentially what we refer to as *behavior*. The dominant processes will guide the cleanup strategy and allow one to assess the potential effects on the environment. Important processes that affect behavior of oil in the subsurface environment for land-based spills are summarized in Fig. 9.19.

Important controlling factors that affect spill behavior are dependent on the environment in which the spill occurred (i.e., land versus water). Weather conditions play a critical role in oil spills on water, whereas subsurface soil conditions play an important role with land-based oil spills. For example, in the case of an open water spill, removal of the floating product (or slick) is imperative before the slick impacts fragile coastal environments. With an underground land-based release (such as a breached pipeline or underground storage tank), escaping vapors could potentially present a safety hazard. These types of spills or releases can also result in the presence of free-phase hydrocarbon overlying the water table as pools, potentially impacting beneficial groundwater resources, or occurring in close proximity to water wells utilized for drinking water purposes. These types of scenarios adversely impact groundwater resources as well as soil.

Land-Based Spills.   When a spill on land occurs, or subsurface contamination is discovered, priorities and sequences of events vary with the type of spill or release, loca-

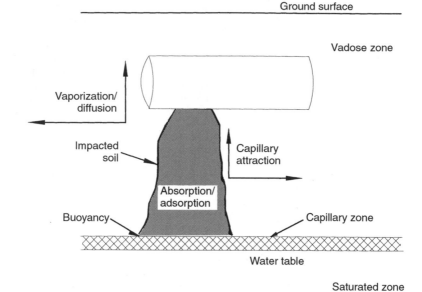

FIGURE 9.19 Subsurface processes affecting a release of refined product from an underground storage tank above the water table. (*From Testa and Winegardner, 2001.*)

tion, site-specific constraints (such as aboveground and underground structures), and site-specific limitations reflective of the subsurface media. Migration of petroleum hydrocarbons in the subsurface is dependent on several factors, including volume of release, time duration, area of infiltration, physical properties of the hydrocarbon, properties of the soil or geologic media, and subsurface flow dynamics.

Density or specific gravity is important in determining how oil or refined petroleum will behave in the subsurface. Nonaqueous-phase liquids (NAPLs), referred to in the federal regulations as *free product,* can occur in the subsurface in two forms (Fig. 9.20): light NAPLs (LNAPLs) are lighter than water and have a density less than 1, whereas dense NAPLs (DNAPLs) have densities greater than 1. Typical LNAPLs include most crude oil, used oil, and fuels (such as gasoline, diesel, and jet fuel), Stoddard solvents, and mineral oils. Densities for these substances range from about 0.6 to 1.0 g/mL.

DNAPLs are broadly classified on the basis of certain chemical properties such as density, viscosity, and solubility. Some of the more common DNAPLs are chlorinated solvents [e.g., trichloroethylene (TCE), tetrachloroethylene (PCE), and trichloroethane (TCA)], creosote, and coal tar [e.g., polycyclic aromatic hydrocarbons (PAHs) or polynuclear aromatic hydrocarbons (PNAs) such as anthracene, chrysene, fluorene, naphthalene, phenanthrene, and pyrene]. Although many chlorinated solvents are characterized by relatively high densities and low viscosities, creosote and coal tar compounds have relatively low densities and high viscosities. In comparison to water, chlorinated solvents are characterized by relatively high densities, low viscosities, and significantly high specific gravities. Creosote and coal tar compounds have viscosities 10 to 20 times greater than that of water, and specific gravities only slightly greater than water. Some of the more common LNAPLs and prevalent DNAPLs, and their respective physical properties, are presented in Table 9.4.

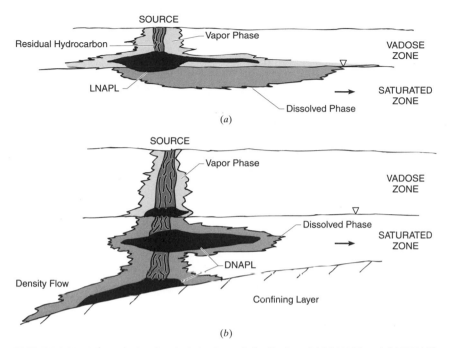

FIGURE 9.20   Schematic showing the behavior and distribution of (*a*) LNAPL and (*b*) DNAPL in the subsurface.

LNAPLs.   When a release occurs on land, regardless of the source, the movement of petroleum hydrocarbon through the subsurface can be divided into four general phases: seepage into and possibly through the vadose or unsaturated zone, lateral spreading in the zone immediately overlying the water table (or other significant permeability constants) with development of a "pancake" layer, stable accumulation within the capillary zone, and dissolved phase in groundwater. During seepage through the vadose zone, downward migration of the hydrocarbon can occur as bulk product zones of affected soil and/or as "fingers" (Fig. 9.21). Both of these conditions are important in assessing subsurface presence, the vertical and lateral extent of affected soil, and the potential to adversely affect groundwater quality. Facies architecture, stratigraphic controls imposed by the depositional environment (e.g., channeling), bedrock orientation, and fractures can all play a role in the extent of affected soil.

If the release is sufficiently large and/or enough time has elapsed, as in the case of a small but continuous release with time, the hydrocarbon will eventually approach saturated conditions above a perched or unconfined water table in the capillary fringe above the water table (Fig. 9.22). The capillary fringe rises above the water table to a height dependent upon grain size, the height increasing with decreasing grain size (Fig. 9.22*a*). Within this zone, the hydrocarbon begins to occupy pores not occupied by capillary or residual water. Additional accumulation of hydrocarbon eventually causes the development of a floating pool, with the lateral spreading that occurs being referred to as a *pancake layer*. Large accumulations can eventually depress the capillary fringe zone. When this occurs, the hydrocarbon is in

TABLE 9.4 Select Properties for Certain LNAPLs and DNAPLs*

Fluid	Density, g/mL	Dynamic viscosity, cP[†]	Kinematic viscosity, cS[‡]	Water solubility, mg/L	Henry's law constant, atm-m^3/mol	Vapor pressure, mm Hg
LNAPL						
Water	0.998	1.14				
Automotive gasoline	0.729	0.62				
Automotive diesel fuel	0.827	2.70				
Kerosene	0.839	2.30				
No. 6 jet fuel	0.844	—				
No. 2 fuel oil	0.866	—				
No. 4 fuel oil	0.904	47.20				
No. 5 fuel oil	0.923	215.00				
No. 6 fuel oil or bunker C	0.974	—				
Norman Wells crude	0.832	5.05				
Avalon crude	0.839	11.40				
Alberta crude	0.840	6.43				
Transmountain Blend crude	0.855	10.50				
Bow River Blend crude	0.893	33.70				
Prudhoe Bay crude	0.905	68.40				
Atkinson crude	0.911	57.30				
LaRosa crude	0.914	180.00				
DNAPL						
Halogenated semivolatiles						
1,4-dichlorobenzene	1.2475	1.2580	1.008	8.0×10^1	1.58×10^{-3}	6×10^{-1}
1,2-dichlorobenzene	1.3060	1.3020	0.997	1.0×10^2	1.88×10^{-3}	9.6×10^{-1}
Arochlor 1242	1.3850	—	—	4.5×10^{-1}	3.4×10^{-4}	4.06×10^{-4}
Arochlor 1260	1.4400	—	—	2.7×10^{-3}	3.4×10^{-4}	4.05×10^{-5}
Arochlor 1254	1.5380	—	—	1.2×10^{-2}	2.8×10^{-4}	7.71×10^{-5}
Chlordane	1.6	1.1040	0.69	5.6×10^{-2}	2.2×10^{-4}	1×10^{-5}
Dieldrin	1.7500	—	—	1.86×10^{-1}	9.7×10^{-6}	1.78×10^{-7}
2,3,4,6-tetrachlorophenol	1.8390	—	—	1.0×10^3	—	—
Pentachlorophenol	1.9780	—	—	1.4×10^1	2.8×10^{-6}	1.1×10^{-4}
Halogenated volatiles						

1,2-dichloropropane	1.1380	0.8400	0.72	2.7×10	3.0×10	3.93×10
1,1-dichloroethane	1.1750	0.3770	0.321	5.5×10³	5.45×10⁻⁴	1.82×10²
1,1-dichloroethylene	1.2140	0.3300	0.27	4.0×10²	1.49×10⁻³	5×10²
1,2-dichloroethane	1.2530	0.8400	0.67	8.69×10³	1.1×10⁻³	6.37×10
Trans-1,2-dichloroethylene	1.2570	0.4040	0.321	6.3×10³	5.32×10⁻³	2.65×10²
Cis-1,2-dichloroethylene	1.2480	0.4670	0.364	3.5×10³	7.5×10⁻³	2×10²
1,1,1-trichloroethane	1.3250	0.8580	0.647	9.5×10²	4.08×10⁻³	1×10²
Methylene chloride	1.3250	0.4300	0.324	1.32×10⁴	2.57×10⁻³	3.5×10²
1,1,2-trichloroethane	1.4436	0.1190	0.824	4.5×10³	1.17×10⁻³	1.88×10
Trichloroethylene	1.4620	0.5700	0.390	1.0×10³	8.92×10⁻³	5.87×10
Chloroform	1.4850	0.5630	0.379	8.22×10³	3.75×10⁻³	1.6×10²
Carbon tetrachloride	1.5947	0.9650	0.605	8.0×10²	2.0×10⁻²	9.13×10
1,1,2,2-tetrachloroethane	1.6	1.7700	1.10	2.9×10³	5.0×10⁻⁴	4.9
Tetrachloroethylene	0.8900	0.8900	0.54	1.5×10⁻²	2.27×10⁻²	1.4×10
Ethylene dibromide	1.6760	1.6760	0.79	3.4×10³	3.18×10⁻⁴	1.1×10
Nonhalogenated semivolatiles						
2-methyl naphthalene	1.0058	—	—	2.54×10	5.06×10⁻²	6.80×10⁻²
o-Cresol	1.0273	—	—	3.1×10⁴	4.7×10⁻⁵	2.45×10⁻¹
p-Cresol	1.0347	—	—	2.4×10⁴	3.5×10⁻⁵	1.08×10⁻¹
2,4-dimethylphenol	1.0360	—	—	6.2×10³	2.5×10⁻⁶	9.8×10⁻²
m-cresol	1.0380	21.0	20	2.35×10⁴	3.8×10⁻⁵	1.53×10⁻¹
Phenol	1.0576	—	3.87	8.4×10⁴	7.8×10⁻⁷	5.293×10⁻¹
Naphthalene	1.1620	—	—	3.1×10	1.27×10⁻³	2.336×10⁻¹
Benzo(a) anthracene	1.1740	—	—	1.4×10⁻²	4.5×10⁻⁶	1.16×10⁻⁹
Fluorene	1.2030	—	—	1.9	7.65×10⁻⁵	6.67×10⁻⁹
Acenaphthene	1.2250	—	—	3.88	1.2×10⁻³	2.31×10⁻²
Anthracene	1.2500	—	—	7.5×10⁻²	3.38×10⁻⁵	1.08×10⁻⁵
Dibenz (a,hh) anthracene	1.2520	—	—	2.5×10⁻³	7.33×10⁻⁸	1×10
Fluoranthene	1.2520	—	—	2.65×10⁻¹	6.5×10⁻⁶	4.8×10⁻⁵
Pyrene	1.2710	—	—	1.48×10⁻¹	1.2×10⁻⁵	6.67×10⁻⁶
Chrysene	1.2740	—	—	6.0×10⁻³	1.05×10⁻⁶	6.3×10⁻⁹
2,4-dinitrophenol	1.6800	—	—	6.0×10³	6.45×10⁻¹⁰	1.49×10⁻⁵
Miscellaneous						
Coal tar	1.028§	18.98§				
Creosote	1.05	1.08§				

* All measurements at 15°C unless indicated otherwise.
† cP = centipoise; water has a dynamic viscosity of 1 cP at 20°C.
‡ cS = centistokes.
§ 45°C (70°F).

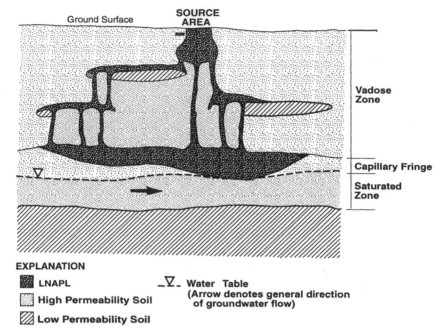

FIGURE 9.21 Schematic showing the conceptual distribution of LNAPL in the vadose zone. Note the development of both pancake and fingering distributions.

direct contact with the saturated zone, and the accumulation serves as a source of groundwater contamination.

DNAPLs. It is these contrasts that cause the subsurface behavior of both the immiscible phase and dissolved phase of DNAPLs to be different from that of LNAPLs. When released into the subsurface, DNAPLs behave much the same as LNAPLs within the vadose zone; however, once groundwater is encountered, LNAPLs will tend to form a pool or pancake layer, whereas DNAPLs will tend to continue to migrate vertically downward through the water column until a significant permeability contrast is encountered, provided enough of a volume of DNAPL was released with time. Immiscible chlorinated solvents are relatively mobile and strongly influenced by gravity, while the dissolved phases are also relatively mobile, reflecting sorption properties and significant solubilities of some constituents. The dissolved constituents can thus migrate large distances. Some dissolved compounds derived from DNAPLs such as creosote and coal tar are less influenced by gravity, and thus are less mobile. Lower solubilities and concomitant stronger sorption can also result in lower mobility.

DNAPLs' mobility is influenced by their respective density, viscosity, and interfacial tension with water. Mobility in the soil matrix is influenced by small-scale features, such as soil type, intrinsic permeability, mineralogy, pore size, pore geometry, and micropores, and large-scale features, such as heterogeneities and anisotropic conditions, structure, and stratigraphy (Fig. 9.23). Once released in the subsurface, DNAPLs migrate vertically downward through the vadose zone with some lateral spreading where significant permeability contrasts are encountered. Within the vadose zone,

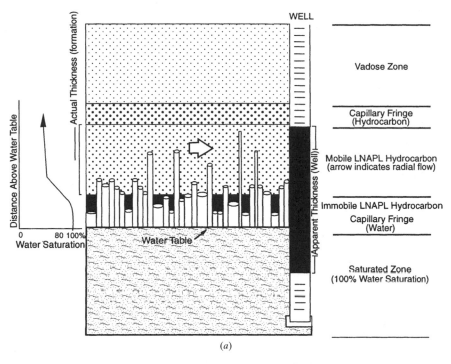

(a)

(b)

FIGURE 9.22 (a) Schematic illustrating actual LNAPL thickness in the subsurface in comparison to the exaggerated apparent LNAPL thickness as observed in a monitoring well; (b) sand box model showing similar phenomena.

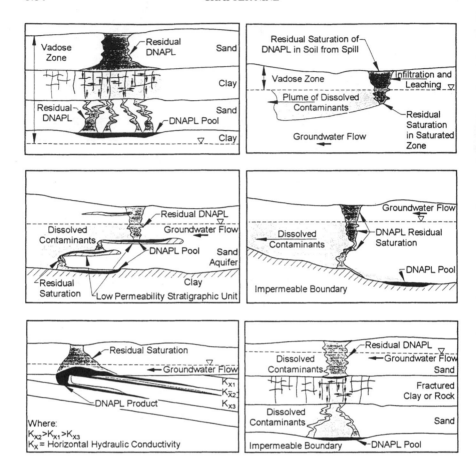

FIGURE 9.23   Conceptual distribution scenarios for the subsurface presence of DNAPL.

DNAPL residual hydrocarbon becomes trapped in pore space via surface tension, or as dissolved constituents in residual soil water, or as vapor. With significant releases, capillary or entry pressures are overcome, and the DNAPL will eventually reach the water table or saturated zone. At this point, DNAPLs continue to migrate downward by the influence of gravity occurring as dissolved constituents and within pore spaces. The pressure head required for penetration increases as grain size decreases. DNAPL are often gradient controlled, following the dip of the interface between the aquifer and top of the lower confining layer, regardless of groundwater flow directions, filling in topographic lows, or migrating along preferred pathways such as fractures, bedding planes, and zones of relatively higher permeability. DNAPLs will not necessarily pond at depth above some confining layer. In cases where significant vertical separation exists between the water table and the lower confining layer, the DNAPL may manifest itself in essentially thin lenses or layers within the saturated zone as evidenced at the U.S. Dept. of Energy Savannah River site, South Carolina.

NAPLs can be trapped or pooled in the subsurface by large-scale hydrogeologic features such as confining layers and bedrock, in addition to small-scale depositional features. For LNAPL, traps may consist of buried channels, gravel bars, eolian wedges, and lateral and vertical facies changes, among others. For DNAPLs, traps include scoured channel bases, volcanic collapse features, irregular topographic depressions on the upper surface of a confining layer, etc. Interface geometry and grain size are important factors in determining whether interface trapping of NAPLs is possible. In order to trap NAPLs, the height of the trapping feature (trap-closure height) must be greater than the capillary intrusion of water into the NAPL phase, and the trap boundary must be sufficiently fine-grained to prevent NAPL from entering its pores. For both LNAPLs and DNAPLs, the closer the fluid density is to that of water, the greater the closure height necessary to retain the NAPL in an interface trap (i.e., sand/shale interface). Likewise, NAPLs with densities that significantly differ from that of water require the least closure for trapping. Trap-closure heights can range from one to several tens of centimeters for coarse-grained material, and 1 to more than 5 m for fine-grained sands.

Where DNAPL enters fractured geologic media, it will preferentially enter the larger pore spaces, that is, the individual fractures. A small amount of DNAPL can migrate a significant distance if the surrounding rock is characterized by small pores having high displacement or entry pressure. Conversely, the DNAPL may penetrate into the rock matrix should a buildup of DNAPL pressure within the fracture network occur, thus reducing the extent of DNAPL migration within the fracture network.

Marine-Based Spills. When oil is spilled on water, the oil type and volume and the weather conditions play an important role during and after the spill. Those processes that dominate in the water environment include weathering (evaporation and emulsification), natural dispersion dissolution, photo-oxidation, sedimentation, and biodegradation. *Weathering* is the physical and chemical changes that occur when oil is spilled into the environment, and has the greatest effect of the fate of the oil. Weathering processes occur at varying rates, and are typically greater immediately after the spill occurs. The most important weathering process is evaporation, with oil type being a major factor on its effectiveness. Emulsification is the second most important weathering process, and has the greatest effect on the properties of the oil, causing liquid oil to become a viscous and heavy mass.

Spilled oil on the water surface initially forms a pool that slowly thickens near the source. The pool then tends to thin quickly while growing in lateral extent. Just how fast an oil slick migrates depends on its viscosity and overall temperature. A significant volume of the oil may evaporate. How fast it evaporates depends on its composition, temperature, and wind conditions. For surface-water spills, about 80 percent of the likely evaporation occurs within just a couple of days. The warmer and windier the conditions, the greater the evaporation that occurs, removing the more volatile components. A light fuel such as gasoline would evaporate completely above freezing temperatures; whereas only a small amount of a heavier product such as bunker fuel would evaporate. Bunker fuel is a mixture of bunker C and diesel fuel used primarily as propulsion fuel for ships; bunker C is a heavy residual fuel that remains after gasoline and diesel production at refineries. Although windy conditions can assist in the evaporation of up to 50 percent of the hydrocarbon in spilled crude oil in aqueous environments, they also cause the oil to subdivide and disperse, and contribute to the formation of emulsion by wave action.

Oil and various petroleum products also evaporate in different ways on land and water. On land, the spill is slowed by the formation of a skin or crust that develops on top on the surface. In water, however, the skin or crust forms when heavier hydrocarbon compounds (i.e., waxes and resins) separate and rise to the surface, minimizing evaporation underneath.

High winds and wave action can also cause an oil slick to drift at 3 to 3.5 percent of wind velocity, with the leading edge of the oil slick occurring where stronger wind conditions prevail. The residual oil consists of heavier hydrocarbons, waxes, sulfur-containing compounds and asphalts, and is denser, more viscous, and prone to sinking, especially since oil has an affinity for suspended particulates and sediment, which add to its density. Once the spill reaches the shoreline, it tends to coat whatever it comes in contact with.

Other processes of importance include emulsification, dispersion, dissolution, photo-oxidation, sedimentation, and biodegradation. Emulsification is the process where one liquid is dispersed into another in the form of droplets. These water-in-oil emulsions are sometimes referred to as "mousse" or "chocolate mousse." Natural dispersion occurs when small droplets (less than 0.020 mm) are injected into the water column. These droplets are relatively stable and can persist in the water for long periods of time. Certain oil types (light crude oils, diesel fuel, etc.) and favorable high-energy conditions such as high seas can cause a spill to disperse readily. The long-term fate of the dispersed droplets can vary. To some degree they degrade, but can also rise through the water column and form another oil surface, or precipitate to the bottom with time.

Soluble components of the oil, including lower-molecular-weight aromatics and some polar compounds, can dissolve within the water through a process known as *dissolution*. Like evaporation, dissolution occurs immediately following a spill, and its rate decreases quickly with time. The importance of dissolution is that the soluble compounds, especially those associated with light crude oil or diesel fuel, are typically toxic to aquatic life.

Photo-oxidation occurs when the oxygen and carbon molecules combine under the influence of the sun's rays, forming new compounds. Certain oils are susceptible to this process, forming resins that are partially soluble and thus can dissolve in water-forming emulsions.

Oil deposited on the bottom of the water column (i.e., seafloor) is called sedimentation. Case studies have shown that about 10 percent of a spill's volume may be deposited on the bottom of the water column. The presence of deposited oil may potentially adversely affect the biota and aquatic environment. Residual weathered oil may end up as tar balls or mats. Tar balls are agglomerates of thick pancake-shaped oil less than about 10 cm in diameter. Mats are larger in size, ranging from about 10 cm to 1 m in diameter. Tar balls and mats usually end up along shorelines, the result of spills or natural seeps.

Lastly, biodegradation relies on the presence of micro-organisms to degrade petroleum hydrocarbons. Hydrocarbons with lower molecular weights biodegrade more rapidly. The ability of hydrocarbons to degrade is also dependent on oxygen availability, in conjunction with the availability of certain macronutrients such as fixed nitrogen and phosphates, plus a number of micronutrients. Biodegradation is a very slow process, and thus cannot be relied upon as an important factor during the spill response and cleanup phase.

9.4.2   Remediation Constraints and Limitations for Land-Based Spills and Leaks

Several factors are considered in selecting the appropriate remediation strategy, which can be one technology or technique or, more likely, a combination of technologies and techniques. The most common constraints and requirements include such factors as:

- Hydrogeologic constraints
- Chemical composition constraints
- Physical constraints
- Time constraints
- Economic constraints
- Hydrogeologic constraints
- Space requirements
- Regulatory acceptance of remediation technology
- Regulatory treatment goals
- Client needs and expectations
- Potential liabilities and future site use
- Unforeseen and intangible costs
- Funding sources
- Tax and accounting implications
- Other unexpected considerations

The subsurface is a complex environment. The primary hydrogeologic constraint is the inability to recover fluids and hydrocarbons from low permeability environments. Chemically, higher-molecular-weight hydrocarbons are more difficult to recover than lighter-weight hydrocarbons, and take longer to evaporate and biodegrade. Physical constraints include the presence of above- and belowground structures that prevent or minimize access to key locations for remedial purposes. Hydrocarbon type, volume, and spill location will affect the time required for cleanup. Obviously the sooner a spill is responded to, the better; however, unless adequate spill preparation has been coordinated, response to spills is always a race against the clock in order to minimize their overall impact. All the financial resources available will not necessarily facilitate cleanup. At some point, hydrocarbons remain in the environment with diminishing returns for the dollars being spent. When this point is reached, further remedial activities do not advance the cleanup process, and some risk must be accepted.

9.4.3 Characterizing Risk for Land-Based Spills and Leaks

When a site or area has been impacted by petroleum hydrocarbons, and the need for corrective action has been determined, selection of the appropriate remedial strategy is vital to the success of the effort. With the recognition of the widespread and ubiquitous occurrence of petroleum hydrocarbon-affected soil and groundwater during the 1970s and 1980s, numerous approaches and innovative techniques, or combination of techniques, have been employed in the assessment and subsequent remediation of petroleum contamination. In order to be successful, a remediation strategy must be environmentally sound, cost-effective, and timely. To meet these three objectives, it is vital that the risks imposed by the presence of petroleum hydrocarbons in the environment, and the level of effort required to reduce such risks to reasonable levels, be clearly understood.

The mere presence of hydrocarbons in the environment does not in itself qualify for concern or present significant adverse impact and harm to the public health, safety and welfare, or to degradation of soil and groundwater resources. Nor does the mere presence warrant that remediation be performed. Before the potential risk can be adequately addressed, characterization of subsurface conditions in regard to the detection and occurrence of these constituents, assessment of fate and transport mechanisms and processes, and conceptualization of preferential migration pathways are required, followed by accountable and responsible development of an appropriate remedial strategy. The risks associated with the presence of petroleum hydrocarbons in the environment will vary. Factors include geologic and hydrogeologic conditions, the potential for adverse impact on groundwater resources considered to be of beneficial use, potential impact on existing ecosystems, existing and planned site usage, magnitude of the problem (i.e., local versus regional in extent), time limitations, and financial constraints.

9.5 ANALYTICAL APPROACHES TO DETERMINING ENVIRONMENTAL IMPACT

9.5.1 Characterizing Land-Based Spills and Leaks

Drilling and Well Installation. Subsurface geologic and hydrogeologic conditions, as well as the subsurface presence of hydrocarbons, can directly be determined for soil and rock by the drilling of soil and rock bores, and for groundwater by subsequent installation and construction of monitoring wells. Several techniques are available for the drilling and installation of wells, regardless of whether their eventual use will be for monitoring, gauging, delineation, injection, or recovery purposes. A summary of these techniques is provided in Table 9.5. A typical monitoring well construction detail is shown in Fig. 9.24.

Although well construction details for monitoring and recovery of NAPL are similar to those of conventional monitoring wells, several factors need to be emphasized.

1. Obviously, the well screen must overlap the mobile hydrocarbon interval and be of sufficient length to account for seasonal fluctuations or changes due to recovery or reinjection influences.

2. Filter pack design can also have a bearing on whether hydrocarbon presence is detected or confirmed. Filter packs must be designed to allow not only mobile hydrocarbon but also capillary hydrocarbon to migrate into the well. Otherwise, since hydrocarbon in the formation can exist at less than atmospheric pressure, a poorly designed filter pack can result in capillary hydrocarbon being unable to migrate into the well; in such cases, areal extent of subsurface hydrocarbon may be much broader than is being accounted for.

3. Well design and construction details among a network of wells, including the filter pack design and developing procedures, should remain consistent. With recovery wells, too coarse a filter pack can minimize the well's ability to attract capillary hydrocarbon to the well. For example, a typical hydrocarbon product with a density of 0.8 g/cm^3 and an interfacial tension with air of 30 dyn/cm will accumulate to a thickness of approximately 25 to 33 cm in a fine sand before exceeding atmospheric pressure. A part of this capillary hydrocarbon is recoverable with a properly designed, finer-grained filter pack.

When a well is installed through the liquid hydrocarbon/water interface, the thickness of the liquid hydrocarbon as measured in the well is significantly greater than that which actually exists in the formation. This can result in exaggerated estimates of

TABLE 9.5 Summary of Drilling Techniques for the Construction and Installation of Monitoring, Recovery, and Injection Wells

Drilling technique	Geologic material*	Depth limitations, ft	Well type[†]	Remarks
Geoprobe	U	50	M	Geologic and hydrogeologic characterization only; excellent sampling and analytical capabilities (soil, water, or vapor); 1-in-diameter well screen capability; accessibility excellent
Hand-augured	U	15	M, R	Accurate sampling; difficult in coarse sediments or loose sand; physically demanding; fluid levels easily detected; borehole variable because of friction of auger; depth limited; inexpensive
Driven	U	25	M	No sampling capability; quick and easy method to detect and monitor shallow fluid levels
Hollow-stem auger	U	180	M, R, I	Accurate sampling; continuous sampling available; diameter limitations; fluid levels easily detected; no drilling fluids required; smearing of borehole walls in fine-grained soils and sediments, causing sealing
Jet	U	200	M	Diameter limitations; fluid level (water and NAPL) difficult; sampling accuracy limited; produced fluids require handling (hazardous if NAPL is encountered)
Bucket auger	U	100	R, I	Sampling of borehole wall easy; can install large-diameter well; difficult to control caving
Cable tool	U	1000	M, R, I	Satisfactory sampling; fluid levels easily detected; drilling can be slow
Hydraulic rotary	U or C	2500+	M, R, I	Fast; retrieval of accurate samples requires special attention; knowledge of drilling fluids used to minimize plugging of certain formations is critical; good for recovery and injection well construction; continuous coring available; produced fluids require handling (hazardous if NAPL is encountered)
Reverse circulation	U or C	2000+	M, R, I	Formation relatively undisturbed compared with other methods; large-diameter boreholes can be drilled; no drilling mud usually required because of hydraulics associated with this method; good for recovery and injection well construction; produced fluids require handling (hazardous if NAPL is encountered)
Air rotary	U or C	2000+	M, R, I	Fast; cuttings removal rapid; poor sample quality; diameter limitations; formation not plugged with drilling fluids; dangerous with flammable fluids
Air percussion	U or C	2000+	M, R, I	Fast; cuttings removal rapid; good in consolidated formations

* U = unconsolidated; C = consolidated.
[†] M = monitoring well; R = recovery well; I = injection well.
Source: Testa and Winegardner (2000).

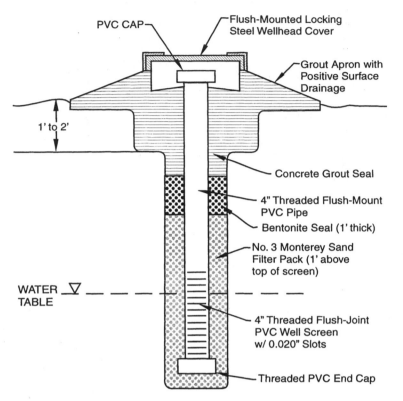

FIGURE 9.24   Schematic showing construction details of a conventional groundwater monitoring well.

the actual volume of hydrocarbon that is spilled or is in the subsurface, and a poorly designed, inefficient recovery and remediation system (Fig. 9.22).

Soil Vapor Monitoring.   Soil vapor monitoring is an important aspect of any subsurface investigation of a spill area where hydrocarbons are of concern. Monitoring of hydrocarbon vapors in soil provides a nondiscriminatory indication of the presence of a variety of organic compounds in the subsurface or ambient air, and can be used for the relative quantification of hydrocarbons. Soil vapor monitoring is commonly used for field screening of potential hydrocarbon presence in the field or during drilling or excavation. Instrumentation is hand held, lightweight, inexpensive, and easy to use.

Field results commonly need to be confirmed with further chemical testing at a certified laboratory. In addition, the detected presence of hydrocarbons does not necessarily confirm the presence of hydrocarbon-contaminated soil, since vapors tend to migrate within preferential pathways such as relatively coarse-grained soil layers and utility trenches. Thus, what is being measured is the presence of hydrocarbon vapors that have migrated from a nearby source area.

Geophysical Surveys.   Geophysical techniques have been conventionally used for the subsurface characterization of geologic and hydrogeologic conditions, detection of buried debris and utility structures, and information about where to best situate drill sites, monitoring wells and sampling locations. Over the past 2 decades, geophysical techniques have been developed to address environmental needs. Techniques now exist to detect and delineate contaminant plumes, buried drums, and in some cases free-phase hydrocarbons.

Nonobstructive techniques are referred to as surface geophysical techniques. This is important where disturbance may pose a health and safety concern. In this case, surveys are performed above ground such that no subsurface penetration of equipment or instrumentation is required. Surface geophysical techniques include use of ground-penetrating radar, electromagnetic induction, electric resistivity logs, seismic reflection and refraction, gravity measurements, and magnetometry.

Downhole geophysical techniques have been developed for use where boreholes or monitoring wells exist. They are usually for stratigraphic correlation and evaluation of geologic and hydrogeologic conditions in a borehole and well. This can be important where contaminant plumes exist or are suspected. The more conventional downhole geophysical techniques include use of acoustic logs, caliper logs, electric logs, fluid logs, and nuclear logs.

Two downhole techniques, dielectric well logging and use of a optoelectronic sensor, can be used to detect free-phase hydrocarbon thickness in a well under certain conditions (Keech, 1988; Wagner et al., 1989). Dielectric well logging allows detection of (1) interfaces between dissimilar fluids such as water, hydrocarbons, and air, (2) relative fluid dielectric constants, and (3) well casing reflectance. This technique is not applicable for relatively small subsurface oil thickness, but may prove promising for large thickness. The use of optoelectric sensors allows the detection of rising oil droplets that enter through the well screen openings and rise to the water surface in the well. Both of these techniques are experimental and have been used to determine actual free-phase hydrocarbon thickness in the subsurface.

Forensic Chemistry. Forensic chemistry for environmental purposes involves the geochemical characterization, or "fingerprinting," of leaked crude oil or refined products. Forensic analysis over the past decade has increasingly been used to characterize and distinguish different hydrocarbon substances that have been spilled or leaked into the environment, relative timing of the release, and source areas. The array of supportive chemical tools, from simplest to relatively more sophisticated, include determination of API gravity, development of distillation curves, and trace metal analysis. Gas chromatographic analysis is used for isotope fingerprinting.

9.5.2 Characterizing Marine Spills

Marine spills are characterized with the help of several tools and instruments for detection, tracking, measurement of oil properties, and determining potential impact on the environment. With the use of marine vessels, aircraft, and satellites, technology is available today to detect and map oil spills on water, at night, in wetlands, along shorelines, and on ice and snow. Thickness, degree of weathering, and source evaluation can also be determined under certain conditions.

Chemical Testing. Like techniques available for hydrocarbon-impacted soil and groundwater, chemical field screening and laboratory techniques are commonly used in the identification and fingerprinting of oil and petroleum products. The most common analysis performed in the field is determination of total oil in a sample. This is accomplished by several means for analyzing the sample for total petroleum hydrocarbons (TPH). TPH is typically measured in parts per million or kilograms per milligram. Gas chromatography is conventionally used in the laboratory.

Visual Surveillance.   Aircraft are commonly used for the surveillance of offshore oil spills. Part of the reason for this is that thin oil layers or sheens are very difficult to detect visually from an oblique angle, notably when vision is limited or interfered with by the presence of fog, mist, high seas, etc.

Detection.   In relatively small areas, such as those around shorelines and harbors, fluorescence and oil sorbent techniques have been used. Fluorescence is commonly used to detect oil on water and, once detected, a radio signal can be transmitted to an oil spill response agency. An ultraviolet light is used to detect the presence of oil. When oil is detected, the oil that is present fluoresces, that is, absorbs the ultraviolet light and re-emits it as visible light. Another technique is to use oil sorbent. Any change in the physical properties of the sorbent triggers a device. Limitations include false readings from very small amounts of oil, or lack of sensitivity from large quantities of oil. In addition, unless these types of detectors are located at a place where a leak or spill may occur, the entry point or source/direction of the oil remains difficult to determine, especially under poor weather conditions.

In large areas such as open waters, buoys have been used for detection. Some buoys have been developed to move along the surface of the water with the moving oil slick. The buoy transmits a signal as to its position, corresponding to the position of the oil slick, directly to receivers that are located on a ship, aircraft, or satellite. Some buoys are equipped to receive Global Positioning System (GPS) data, and this information is transmitted with the signal. The location of the spill can then be located and monitored by a remote receiver. Difficulties occur if the buoy does not respond to wind and surface current conditions in the same way as an oil slick. This level of precision can be difficult to achieve; although some buoys are available that can effectively locate a range of crude oil and bunker C.

Remote Sensing.   Remote sensing is the collection of information about an object by a recording device that is not in physical contact with the object. Remote sensing is usually restricted to methods that record reflected or radiated electromagnetic energy in lieu of penetrative methods. Remote sensing usually involves such devices as cameras, infrared detectors, microwave frequency receivers, and radar systems. A variety of sensors for the purpose of detection and mapping of oil spills are available. When poor weather conditions, evening and other conditions, and other phenomena make detection and mapping of an oil spill difficult, then remote sensing techniques can prove a powerful tool for locating large spills. Remote sensing instrumentation is usually carried aboard aircraft or a satellite. A variety of sensor types have been developed for environmental purposes with only a few applicable to oil spills on water, and fewer to those on land.

Remote sensing sensors applicable to oil spills include visual and ultraviolet sensors, infrared sensors, laser fluorosensors, passive microwave devices, thickness sensors, radar, and satellites. Visual spectrum devices include video cameras, which are limited to the same constraints as visual surveillance, though useful for documentation or providing a frame of reference for another sensor type. Sensors operating within the ultraviolet spectrum can be useful under certain conditions for mapping thin oil sheens.

Infrared sensors are relatively inexpensive and can pick up thick oil slicks, greater than about 100 mm, by differences in temperatures. In the daytime, the oil absorbs infrared radiation from the sun, and thus would appear as hot on a cold water surface. At night, the oil would appear colder than the surrounding water surface. Infrared sensors can provide information on thickness, although other substances also can show a similar infrared response, including biogenic oil, debris, weeds, and oceanic and riverine fronts. An advantage is that the sensors can simply be placed on a ship's mast and used to support cleanup efforts by, for example, positioning the ship for effective oil recovery. Combined with ultraviolet images that show low thickness, a relative thickness map can be developed.

Laser fluorosensors are one of the most powerful remote sensing tools available. They use a laser in the ultraviolet spectrum to detect oil at sea or on land, and in some cases determine whether the target substance is a light, heavy, or lubricating oil. They operate on the principle that oils containing aromatic compounds have a tendency to absorb ultraviolet light and give off visible light in response. Also reliable on snow or ice, their limitations are that they are expensive and large in size and weight.

Passive microwave sensor devices are used to detect natural background microwave radiation. Oil slicks on water absorb some of this radiation in proportion to their thickness; thus, with calibration, relative thickness can be determined. Although this sensor is applicable through fog and in darkness, it has poor spatial resolution and high cost.

Actual oil thickness can be determined by using sonic generators that send sound waves through oil. Although reliable oil thickness can be determined, this approach is experimental, and the sensor is large and heavy.

Radar can be used to detect oil on water, provided the water is calm with small waves on the order of centimeters in length. Good for large search areas, evenings, or in poor visibility, radar is limited to wind speeds of about 2 to 6 m/s. At winds below 2 m/s, not enough small waves exist to differentiate between oil and water. At wind speeds above 6 m/s, waves can propagate through the oil and not be picked up by the radar. Other constraints include the inability to use radar near shorelines or in areas where wind shadows may look like oil, and the size and cost of radar.

Satellites can provide images within the visible spectrum, for determining areal extent, provided the oil spill is very large and sea conditions allow enough contrast between the oil and water.

9.6 LAND-BASED REMEDIATION, TREATMENT, AND REUSE/RECYCLING TECHNOLOGIES

9.6.1 Soil Remediation, Treatment, and Reuse and Recycling

Over the past decade, numerous technologies or combinations of technologies have been used to protect and remediate petroleum-contaminated soil (Table 9.6). The remediation approach can vary significantly depending on site-specific factors (i.e, type and volume, depth, obstructions, etc.). Shallow soil contamination, that less than about 10 ft below ground surface, typically involves more conventional approaches to the remediation of petroleum hydrocarbon-contaminated soil including vapor extraction, bioremediation, washing, stabilization, and natural attenuation. When hydrocarbon-affected soil extends to deeper depths, the cost of physical removal for subsequent treatment is cost-prohibitive; thus, the risk of using nonconventional technologies as well as conventional ones becomes more acceptable by reducing the overall concentration in place. Such in situ technologies include vapor extraction,

TABLE 9.6 Summary of Conventional Soil Remediation Strategies to Minimize, Contain, or Remediate Petroleum Hydrocarbons- and Organics-Impacted Soil

Remedial strategies	Process	Practical constraints	Remarks
In situ strategies			
Soil vapor extraction (SVE)	Evaluate hydrogeologic conditions Assess extent of impacted soil Conduct pilot study Design and install system Monitor effectiveness Confirm effectiveness	Fine-grained soils and low-volatility hydrocarbons limit effectiveness	Not technically viable for clayey soils Requires disposal of air medium
Air sparging	Evaluate hydrogeologic conditions Assess extent of impacted soil Conduct pilot study Design and install system Monitor effectiveness Confirm effectiveness	Fine-grained soils and low volatile hydrocarbons limit effectiveness	Not technically viable for clayey soils Requires disposal of air filtration medium
Steam injection and stripping	Evaluate hydrogeologic conditions Assess extent of impacted soil Conduct pilot study Design and install system Monitor effectiveness Confirm effectiveness	Fine-grained soils and permeability contrasts limit ability to inject steam and recover fluids from subsurface	Overall effectiveness difficult to assure pending pilot study Relatively high operation and maintenance costs
Soil washing/ extraction (in-place leaching)	Evaluate hydrogeologic conditions Assess extent of impacted soil Excavate and crush Mix with wash fluids Treat wash water Construct infiltration and recovery system Irrigate washing fluids Retrieve and treat fluids	Limited to granular soils and moderate to high solubility hydrocarbons	High costs; limited applicability Often used in biotreatment practices Permit approval difficult
Bioventing	Evaluate hydrogeologic conditions Assess extent of impacted soil Conduct pilot study Design and install venting system Monitor effectiveness Confirm effectiveness	Fine-grained soils and low-volatility hydrocarbons limit effectiveness	Not viable for clayey soil Requires disposal of air filtration medium

TABLE 9.6 Summary of Conventional Soil Remediation Strategies to Minimize, Contain, or Remediate Petroleum Hydrocarbons- and Organics-Impacted Soil (*Continued*)

Remedial strategies	Process	Practical constraints	Remarks
		In situ strategies	
Bioremediation (chemical degradation)	Evaluate hydrogeologic conditions Assess extent of impacted soil Conduct pilot study Design and install pumping and injection system Monitor effectiveness Confirm effectiveness	Fine-grained soils or permeability contrasts limit effectiveness Requires ongoing operation and maintenance	Overall effectiveness difficult to assure Extensive on-site monitoring required
Natural attenuation	Evaluate hydrogeologic conditions Assess extent of impacted soil Document source elimination Confirm plume stability Conduct feasibility study Perform periodic monitoring	Requires regulatory approval	Not usually viable for sensitive and/or beneficial-use groundwater
		Ex situ strategies	
Excavation/ disposal as waste	Evaluate hydrogeologic conditions Assess extent of impacted soil Excavate and transport Maintain documentation	Cradle-to-grave liability as generator	High cost; easy to overexcavate
Excavation/ aeration and disposal	Evaluate hydrogeologic conditions Assess extent of impacted soil Excavate, spread, and turn Import and compact clean fill Transport and dispose of aerated soil Maintain documentation Confirm effectiveness	Cradle-to-grave liability as generator Emissions considerations Chemical testing can be extensive	Permitting can be difficult

TABLE 9.6 Summary of Conventional Soil Remediation Strategies to Minimize, Contain, or Remediate Petroleum Hydrocarbons- and Organics-Impacted Soil (*Continued*)

Remedial strategies	Process	Practical constraints	Remarks
		Ex situ strategies	
Excavation/land farming and replacement	Evaluate hydrogeological conditions Assess extent of impacted soil Excavate, aerate, and add nutrients and water Replace and compact Maintain documentation Confirm effectiveness	Emisions considerations	Permitting can be difficult
Mechanically enhanced Volatilization	Evaluate hydrogeologic conditions Assess extent of impacted soil Excavate, crush, aerate, and replace Maintain documentation Confirm effectiveness	Requires dust control Requires vapor treatment	High costs but suitable under certain circumstances
Soil washing (above ground leaching)	Evaluate hydrogeologic conditions Assess extent of impacted soil Excavate and place over collector bed Flush with wash fluid Replace and treat fluids Confirm effectiveness	Requires total fluid collection Requires temperature and odor control Requires sufficient open area	May be used in association with biotreatment Permitting not very difficult
		Others	
Reuse/recycle (asphalt incorporation)	Evaluate hydrogeologic conditions Assess extent of impacted soil Conduct pilot test Develop mix design Determine end use Excavate and produce product Confirm effectiveness Construct end use (road base, pavement, etc.) Maintain documentation	Soil must pass flash test	Requires comprehensive understanding of reuse/recycling regulations On-site end use easier to deal with

TABLE 9.6 Summary of Conventional Soil Remediation Strategies to Minimize, Contain, or Remediate Petroleum Hydrocarbons- and Organics-Impacted Soil (*Continued*)

Remedial strategies	Process	Practical constraints	Remarks
		Others	
No action	Evaluate hydrogeo-logic conditions Assess extent of im-pacted soil Demonstrate low or minimal risk	Requires regulatory acceptance	Site specific

Source: Testa and Winegardner (2000).

thermal treatment, solidification and stabilization, and natural attenuation. Since it is rare that the contamination can be reduced to zero levels unless physically removed, some hydrocarbon will remain in place.

The primary concerns regarding hydrocarbon-impacted soil are potential health risks associated with direct contact with the affected material, the potential for explosion or fire hazard (Fig. 9.25) from hydrocarbon vapors, and potential adverse impact on overall groundwater resources. Petroleum hydrocarbon–impacted soil can be handled in a variety of ways. The principal means of handling hydrocarbon-impacted soil include:

FIGURE 9.25 Photograph showing the release of volatile hydrocarbon vapors derived from a LNAPL pool overlying a shallow water table.

- Excavation and off-site disposal
- Excavation and treatment
- In situ treatment
- Reuse and recycling or resource recovery

Excavation and Off-Site Disposal. Excavation and off-site disposal, commonly referred to as "dig-and-haul," is the physical removal of impacted soils by conventional excavation techniques for subsequent disposal or treatment. This option is typically restricted to shallow soils (less than approximately 10 ft below ground surface) and can be expensive when indirect costs such as transportation and treatment and/or disposal are considered. Backhoes are generally used to depths of about 15 to 20 ft below ground surface. Greater depths for soil removal typically require larger earthmoving equipment, such as excavators. Excavating near surface or subsurface obstacles such as tanks, buildings, or roads frequently requires the use of wood or metal shoring, which greatly increases the price and duration of the remediation project. Excavation and off-site disposal does not remove the long-term financial and regulatory liability of the generator of the material, since the soil eventually ends up in a landfill and the generator shares in the long-term safe containment of the material.

Excavation and Treatment. Excavation and treatment of impacted soils reduces the hazardous nature on site, and can be accomplished either above ground or in situ (in place). On-site treatment of excavated material may include such strategies as land treatment or bioremediation, vapor extraction, thermal treatment, washing, and chemical extraction, or some form of solidification/stabilization. Land spreading involves aeration by spreading of the soil in thin layers, applying moisture or nutrients if necessary, and allowing a combination of aeration and degradation of hydrocarbons to occur. Bioremediation uses micro-organisms that feed on hydrocarbons as an energy source in their metabolic process. Petroleum is thus reduced to biomass, water, and carbon dioxide. Vapor extraction technology such as soil vapor extraction works best for volatile organic compounds such as gasoline, and is accomplished through the induction or injection of air. Volatile hydrocarbons occurring in solution or as residual saturation are transferred to the injected air or steam, and withdrawn by extraction wells. Semivolatiles as well as volatiles can be removed by using steam or heated air. Thermal treatment utilizes heat to volatilize the hydrocarbons (typically at 300 to 700°F) so that they can be extracted in vapor form or applied directly for soil incineration (at much higher temperatures). Solidification/stabilization technologies use processes that encapsulate the contaminated soil in a solid of higher structural integrity and durability so that it is more compatible for reuse, storage, or disposal, and so that contaminants do not leach out.

In Situ Treatment. In situ treatment includes many of the same technologies described for aboveground treatment. Treatment depends on a variety of factors. The way a spilled liquid moves in the soil and later in the groundwater is determined by both the hydraulic conductivity of the soil and the physical and chemical characteristics of the contaminant. The partitioning of petroleum chemicals between the aqueous, vapor, and the sorbed phases is a characteristic of the individual constituents of the spilled hydrocarbon (Fig. 9.26).

The Henry's law value is the air-water partition coefficient, which is a measure of the tendency of a chemical to volatilize when moving through the soil. As the Henry's law value increases for a specific chemical, it is more likely to remain in the soil pore spaces in the vapor phase. Since the lighter-end and highly volatile compounds such as methane, butane, hexane, and benzene are easily found in soil pore spaces, these com-

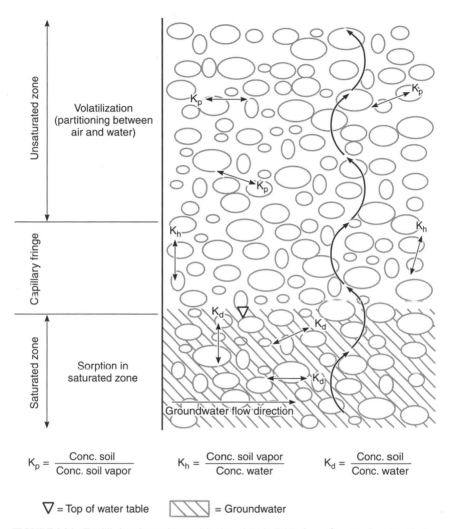

Unsaturated zone

Volatilization
(partitioning between
air and water)

Capillary fringe

Saturated zone

Sorption in
saturated zone

Groundwater flow direction

$$K_p = \frac{\text{Conc. soil}}{\text{Conc. soil vapor}} \qquad K_h = \frac{\text{Conc. soil vapor}}{\text{Conc. water}} \qquad K_d = \frac{\text{Conc. soil}}{\text{Conc. water}}$$

∇ = Top of water table   = Groundwater

FIGURE 9.26 Equilibrium forces for partitioning of the individual constituents of a spilled hydrocarbon between aqueous, vapor, and adsorbed phases.

pounds are generally more amenable to soil vapor surveys for assessment and soil vapor extraction for remediation than are heavier, less volatile chemicals, such as heavy crude oil or diesel fuel. These chemical characteristics show the propensity of a compound to be found in specific subsurface locations and consequently must guide the decisions on assessment and in situ remediation strategies of spilled hydrocarbons.

In situ remediation methods include venting technologies such as vapor extraction (Fig. 9.27), bioremediation (Fig. 9.28), and isolation/containment. Soil flushing (with water or solvent that is recovered for treatment) is sometimes applied where an aquitard below the contaminated soil plume prevents infiltration into a aquifer. Any in situ approach that involves the use of wells to inject air, steam, heat, or nutrients also requires that the soil be of moderate to high permeability in order to be

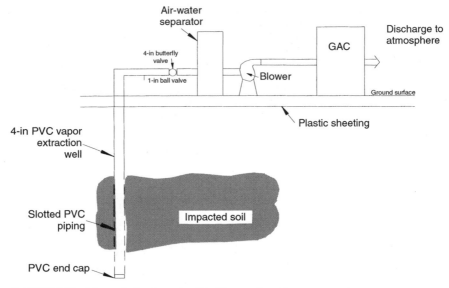

FIGURE 9.27 Schematic showing a simplified layout of a soil vapor extraction system.

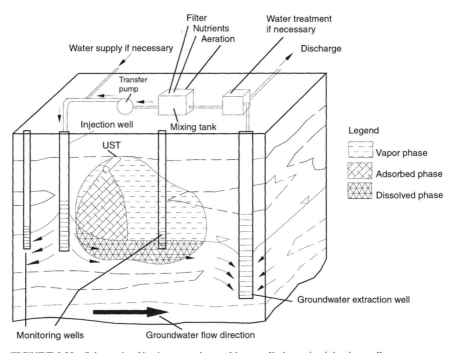

FIGURE 9.28 Schematic of in situ groundwater bioremediation using injection wells.

effective. Soils such as clay that have high porosity, but low permeability, are generally not conducive to these technologies. In such cases, isolating and containing the contaminated material through the construction of subsurface barriers and surface caps can be effective. Solidification/stabilization is practiced at depths down to 120 ft by injecting agents such as portland cement through hollow stem augers approximately 12 ft in diameter. Successive, adjacent soil columns are thus remediated.

Bioremediation uses subsurface microbes to degrade hydrocarbons. To enhance the naturally occurring aerobic degradation of hydrocarbons, oxygen in the form of air, hydrogen peroxide, or magnesium peroxide can be injected into the ground. Nutrients can also be added to enhance degradation activity. Bioremediation enhancements and chemical oxidants can be introduced into the subsurface with or without high pressure by using injection ports, trenches, filter galleries, or wells. In situ injection treatment works best in gravels and sands, although silty sands have been successfully remediated.

Hydrogen peroxide (H_2O_2) is a liquid and is a commonly used oxidizer for aboveground spills as well as in situ remediation applications. Hydrogen peroxide reacts within seconds to minutes and oxidizes hydrocarbons on contact. The end products of the reaction are oxygen, carbon dioxide, and water. Related to hydrogen peroxide remediation, Fenton's reagent was developed in the late 1890s when it was noted that iron added to hydrogen peroxide produced enhanced exothermic reactions. Fenton's reagent is known to oxidize and, in some cases, completely mineralize a variety of organic substrates. The reaction involving Fenton's reagent produces hydroxyl radicals (OH•), which are unstable and, in turn, react with available organic substrates, such as spilled hydrocarbons. Iron is naturally occurring in soil and groundwater or can be added. Factors favoring the success of in situ chemistry using Fenton's reagent include generally low pH range, available iron content, low percentage of total organic carbon, and low alkalinity.

Potassium permanganate ($KMnO_4$) is a solid oxidizer that is hydrated prior to injection for use in remediation of hydrocarbons. This chemical, which is widely known for a distinctive bright purple staining, lasts much longer in the environment and tends to be safer to handle than high concentrations of hydrogen peroxide. Hexavalent chromium can be produced by potassium permanganate under specific conditions if the subsurface chemistry is not fully understood. Both hydrogen peroxide and potassium permanganate are injected in a liquid form into the soil and groundwater (Table 9.7).

TABLE 9.7 Comparative Oxidation Potentials

Species	Volts	Commonly used for chemical oxidation of hydrocarbon spills?
Fluorine	3.0	No
Hydroxyl radical	2.8	Yes; associated with Fenton's reagent
Ozone	2.1	Yes; generated on site as a gas
Hydrogen peroxide	1.8	Yes; commonly used liquid
Potassium permanganate	1.7	Yes; long-lasting, purple stain
Hydrochlorous acid	1.5	No
Chlorine dioxide	1.5	No
Chorine	1.4	No
Oxygen	1.2	No

Another oxidizer used to treat hydrocarbons in situ is a gas, ozone. Ozone (O_3) is a powerful and unstable oxidizer that must be generated on site. Reaction time is rapid, generally seconds to minutes. The gas is typically bubbled into the subsurface at the injection ports or bubbled into a stream of water that is injected into the subsurface. All three oxidizers are nonselective and will oxidize hydrocarbons and solvents, as well as rootlets, specific metals, and carbonaceous materials. Abundant free oxygen is one of the by-products of the oxidation reactions and, consequently, aerobic bioremediation treatment generally follows an in situ oxidation treatment phase.

Since each site has a unique set of hydrogeologic, chemical, and biological factors, the key to success for any in situ remediation project is having enough detailed subsurface data to develop a sound remediation strategy. A bench test is suggested prior to a pilot-scale or full-scale remediation program. To adequately evaluate the success of an in situ treatment program, several soil and groundwater samples collected after the treatment process is recommended.

Reuse and Recycling or Resource Recovery.   The concept of resource recovery views contaminated soil and other materials as a resource rather than a waste. Resource recovery is preferred over remediation options and incorporates recycling, which is the use, reuse, or reclamation of all or part of a waste. Reclamation involves technologies that result in the removal of the contaminant from its matrix for the purpose of collection and reuse. Use and reuse technologies involve the incorporation, not removal or modification, of the contaminant and/or its matrix as an ingredient to produce a commercially viable product. The main difference between remediation and recycling is in the value of the end product. Remediation and treatment strategies are rarely 100 percent effective. Thus, a residue results that typically requires further treatment and/or disposal, or justification for no further action, whereas, recycling produces an end product that is marketable and can be sold commercially.

Asphalt incorporation involves the incorporation of petroleum hydrocarbon–impacted soil as an ingredient in the production of asphaltic products (Fig. 9.29) such as road bases and roadways, berms, and liners (Fig. 9.30). Two conventional processes

FIGURE 9.29   Core of cold-mix asphalt (CMA) retrieved from a roadway. The asphalt was made by incorporation of petroleum-contaminated soil as part of the mix-design.

(a)

(b)

FIGURE 9.30 Photographs of (a) an operations road and (b) reconstruction of a containment berm at an active refinery.

exist for the reuse and recycling of such soil: cold-mix asphalt (CMA) and hot-mix asphalt (HMA). The more widely used method is CMA (Fig. 9.31), where contaminated soil is mixed with aggregate and a water-based emulsion.

Petroleum-contaminated soil has also been used in the production of cement and bricks. In the production of cement, petroleum-contaminated soil is processed along with the raw materials in a kiln. Subsequent heating at temperatures up to 2700°F essentially breaks down the petroleum in the soil and enables the residue to be incorporated into the clinker. In the manufacture of bricks, the impacted soil essentially is used to replace one or more of the needed raw materials such as shale or firing clay.

9.6.2 Remediating Petroleum-Impacted Groundwater

As with the remediation of petroleum-contaminated soil, numerous technologies have been used to remediate contaminated groundwater. The approach to the remediation of petroleum-contaminated groundwater can vary significantly, depending on site-specific factors such as the type and volume of hydrocarbons and the subsurface geology. Such technologies can be divided into two groups, depending on whether free-phase or dissolved hydrocarbon is present.

- Removal of free-phase hydrocarbons
  - Linear interception systems
  - Well point systems
  - Skimming systems
  - Pumping systems

FIGURE 9.31 Photograph of a mobile pug mill being used to incorporate petroleum- and metal-contaminated soil into cold-mix asphalt (CMA) at a metal recycling yard. The material was eventually utilized on site for pavement.

- Removal of dissolved-phase hydrocarbons
  - Pump-and-treat
  - In situ air sparging
  - In situ bioremediation
  - Natural attenuation

Removal of Nonaqueous-Phase Liquids.   When a release into the subsurface is large enough to cause an accumulation of free-phase hydrocarbon, the liquid hydrocarbon (whether it's gasoline, diesel, or some other type of fuel) eventually forms an underground pool overlying the water table. Some of the more conventional approaches to removal of NAPL pools include (Table 9.8):

- Linear interception systems
- Well point systems
- Skimming systems
- Pumping systems

Linear interception systems involve the construction of a ditch or trench that intercepts the underground flow of the liquid hydrocarbon. These systems can incorporate a pneumatic or electrical pump to enhance flow of liquids to the trench. Well point systems incorporate surface pumps attached to certain well heads, with the amount of flow to the well and the horizontal extent of influence predetermined (Fig. 9.32). Linear interception systems and well point systems are commonly used under shallow water table conditions.

Skimming systems involve the placement of an extraction pump downhole in the well at a depth slightly above the water–liquid hydrocarbon interface. The rate of pumping is slow, with low volumes typically recovered, since no inducement to enhance the rate of groundwater flow is implemented.

For large volumes or under deeper water table conditions, recovery of liquid phase hydrocarbon can be conventionally accomplished with either one-pump or two-pump systems (Fig. 9.33). One-pump systems are fully automatic and incorporate one downhole pump that retrieves both liquid hydrocarbon and water. Because of mixing, these liquids require separation above ground, and the liquid hydrocarbon is recycled, while the water is sometimes reinjected or disposed of following treatment for dissolved hydrocarbon constituents, if necessary.

In situations where very large volumes of liquid hydrocarbon exist, as at some refinery sites, a two-pump system may be used. Two-pump systems use a downhole upper pump that retrieves solely liquid hydrocarbon and a lower pump dedicated to retrieving solely water. The lower pump generates a cone of depression in the groundwater, thus intercepting the lateral extent of the liquid hydrocarbon, while the hydrocarbon pump is set slightly above the liquid hydrocarbon–water interface. The liquid hydrocarbon and water are produced separately and do not need to undergo separation prior to being reused or treated.

Aquifer Restoration.   Strategies for remediation of groundwater containing dissolved hydrocarbons can be divided into three basic approaches (Table 9.9):

- Physical
- Biological
- Chemical

TABLE 9.8 Summary of Conventional NAPL Hydrocarbon Recovery Alternatives

LNAPL recovery approach	System type	Hydrogeologic conditions* (permeability)	Depth to water, ft	LNAPL presence	DNAPL presence	Dissolved hydrocarbon presence	Restrictions
Linear interception	Passive	Low to high	0–10	Yes	No	No	Loosely consolidated formations; underground structures
	Active	Low to high	0–10	Yes	No	No	
One-pump systems	Submersible turbine	Moderate to high	Unlimited	Yes	Yes	Yes	
	Mechanical	Moderate to high	Unlimited	Yes	Yes	Yes	
	Positive displacement	Moderate to high	Unlimited	Yes	Yes	Yes	
Two-pump systems	Submersible turbine	Moderate to high	Unlimited	Yes	No	Yes	
Skimming units	Floating	Irrelevant	Unlimited	Yes	No	No	
	Suspended	Irrelevant	Unlimited	Yes	No	No	
Other systems	Timed bailers	Irrelevant	0–100	Yes	Yes	No	
	Rope skimmers	Irrelevant	0–15	Yes	No	No	Surface water use only
	Belt skimmers	Irrelevant	0–15	Yes	No	No	
	Vacuum-assisted	Low to moderate	0–30	Yes	Yes	Yes	
	Vapor extraction	Moderate to high	0–100	No	Yes	No	
	Air sparging	Moderate to high	0–100	No	Yes	No	Product volatility
	Biodegradation	Moderate to high	0–50	Yes	Yes	Yes	Availability of oxygen and nutrients

* Low = $<10^{-5}$ cm/s; moderate = 10^{-5}–10^{-3} cm/s; high = $>10^{-3}$ cm/s.
Source: Testa and Winegardner (2000).

9.56

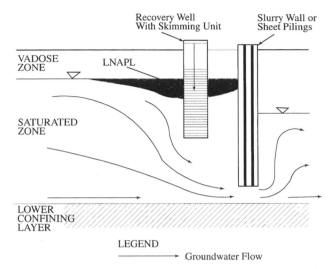

FIGURE 9.32   Schematic showing typical barrier-type system.

Physical pumping strategies involve the pumping of hydrocarbon-impacted ground-water in conjunction with some form of aboveground treatment similar to the one-pump systems described for the recovery of liquid hydrocarbon, except that there is no liquid hydrocarbon. One or several pumping schemes can be applied, and the impacted groundwater extracted and treated above ground in a manner dependent upon the geology and the type, volume, and concentration of the hydrocarbon constituents.

Some of the more conventional modifications of groundwater pump-and-treat remediation and aquifer restoration strategies for petroleum hydrocarbons include vapor-enhanced pumping, and bioenhanced degradation. The ability to withdraw hydrocarbon-impacted groundwater from low-permeability soils can be accomplished to some degree with vacuum enhancement. Bioenhanced biodegradation can be done in situ as well as air sparging. Air sparging involves the introduction of air into an aquifer stripping the volatile organic chemicals from the groundwater. The volatiles are removed from the vadose zone by vapor extraction. Bioenhanced degradation involves the introduction of nutrients and air or peroxide to enhance the natural biologic processes (Table 9.10). Aboveground treatment can include a variety of strategies including chemical precipitation, evaporation, reverse osmosis, and ion exchange to remove metals, and carbon absorption, air stripping, ultraviolet oxidation and use of bioreactors to remove organics. Common dissolved hydrocarbon constituents treated by these methods are the aromatic fuel hydrocarbons such as benzene, toluene, ethylbenzene, and xylenes. Other constituents of concern are fuel additives such as lead and methyl tertiary-butyl ether (MTBE is an additive used to boost gasoline octane rating and to help reduce pollutant emissions from automobiles).

All hydrocarbons degrade in the environment with time. Natural attenuation consists of unenhanced physical, chemical, and biological processes that act to limit the migration and reduce the concentration of contaminants in the subsurface. Nat-

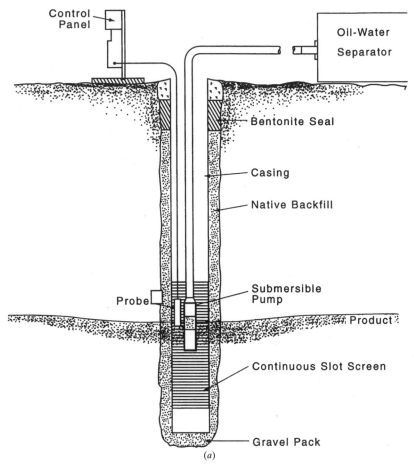

FIGURE 9.33 Schematic showing conventional (*a*) one-pump and (*b*) two-pump LNAPL recovery systems.

ural attenuation is very important because it is often technically infeasible to clean a hydrocarbon-impacted site to regulatory cleanup levels for a variety of conditions including the presence of low-permeability soils, and the inability to remove all the hydrocarbons from the individual soil particles. Natural attenuation is an approach that has gained much acceptance over the past few years. The most important process is in situ aerobic bioremediation, which destroys a large percentage of the hydrocarbons when bacteria oxidize hydrocarbons to obtain energy. The bacteria are indigenous micro-organisms that slowly degrade the hydrocarbons with time under the right conditions (Fig. 9.34). This process can also be enhanced by increasing the hydraulic gradient (thus dissolved oxygen) or nutrients in the subsurface (Fig. 9.35). When permitted as a remediation strategy, extensive monitoring is required.

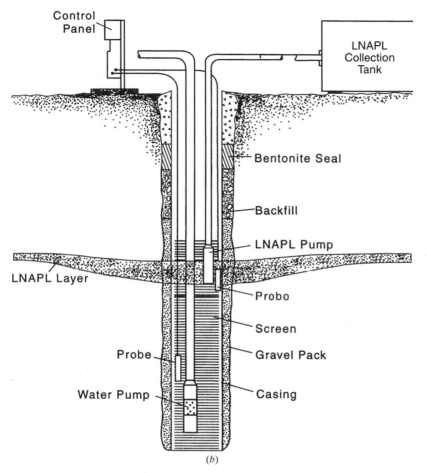

FIGURE 9.33 (*Continued*) Schematic showing conventional (*a*) one-pump and (*b*) two-pump LNAPL recovery systems.

9.6.3 Site Closure

All hydrocarbons degrade in the natural environment, albeit slowly. Since some residual hydrocarbon remains in the environment despite our best efforts, the means in which site closure (no further action) is reached is through a combination of processes commonly referred to as *natural attenuation*.

Aerobic bioremediation is the most important process in regard to petroleum hydrocarbons because this process is capable of destroying a large percentage of hydrocarbon contaminant mass under certain conditions. Destruction occurs as a result of bacteria oxidizing reduced materials such as hydrocarbons to obtain energy. Their metabolism removes electrons from the hydrocarbon donor via a number of enzyme-catalyzed steps along respiratory or electron transport chains to the final

TABLE 9.9 Summary of Conventional Aquifer Restoration Strategies for Dissolved Phases

Strategy	Description	Remarks
Physical strategies		
Pump-and-treat	Pumps water via wells for above-ground treatment and subsequent discharge	Cleanup to promulgated standards difficult to accomplish once asymptotic conditions are reached Generates large volumes of water relative to contaminant mass
Pulsed or variable pump-and-treat	Varied pumping rate allowing contaminants to dissolve, desorb, and/or diffuse from stagnant areas	Same as pump-and-treat May increase cleanup time
Air sparging	Injects air below the water table and captures it above the water table to extract volatile contaminants and promote biodegradation	Can be inefficient in low-permeability zones and complex geologic settings Typically limited to depths less than 30 ft Multicomponent mixtures can adversely affect extractability
Steam-enhanced extraction	Injects steam above and/or below water table to promote volatilization of contaminants	Can be inefficient in low-permeability zones or complex geologic settings
In situ thermal	Injects heat above the water table via Joule heating, radio-frequency heating, or means to promote volatilization of contaminants	Difficulty in attaining uniform heat distribution
Natural attenuation	Allows contaminants to biodegrade naturally without human intervention other than monitoring	Requires demonstration that biodegradation is occurring Periodic monitoring required Source elimination required Plume stable or reducing Nonbeneficial water or naturally poor water quality preferred
Physical containment	Physically contains contaminant plume with use of cutoff walls, caps, liners, etc.	Site specific Limited in depth
In situ reactive barriers	Treat contaminated water as it passes through a physical barrier containing reactive chemicals, organisms, or activated carbon	Site specific Slow process
Biological strategies		
In situ bioremediation	Pumps nutrients through subsurface to promote growth of micro-organisms that biodegrade contaminants	Can be inefficient in low-permeability zones and complex geologic settings NAPLs can impede progress and growth of micro-organisms

TABLE 9.9 Summary of Conventional Aquifer Restoration Strategies for Dissolved Phases (*Continued*)

Strategy	Description	Remarks
	Biological strategies	
		Compound specific when chlorinated solvents are present Accumulation of intermediate compounds considered hazardous may result with chlorinated solvents
	Chemical strategies	
Soil flushing	Flushes surfactants or cosolvents below water table to promote recovery of contaminants with low water solubility	Can be inefficient in low-permeability zones or complex geologic settings Slow reaction rates Adverse chemical reactions a concern
In situ chemical treatment	Injects chemicals to transform contaminants in place	Inefficient in low-permeability zones and complex geologic settings Slow reaction rates Adverse chemical reactions a concern

Source: Testa and Winegardner (2000).

TABLE 9.10 Bacterial Classification Based on Energy Source

Source	Source process	Description
Direct energy source	Phototrophic Chemotrophic Lithotrophic Organotropic	Uses sunlight or other light source Chemical reactions provide energy Inorganic chemical reactions Organic chemical reactions
Carbon source	Heterotrophic	Uses organic compounds (e.g., hydrocarbons) as carbon source
	Autotrophic	Uses inorganic compounds (e.g., carbon dioxide) as carbon source
Terminal electron acceptor (TEA)	Aerobic Anaerobic Facultative	Oxygen is TEA A compound other than oxygen is TEA Can function in the presence or absence of oxygen

Reaction	Typical redox potential (Eh volt)	Electron acceptor	End product
Oxygen reduction	$+0.80 \rightarrow +0.20$	O_2	H_2O, CO_2
Nitrate reduction/nitrate respiration	$+0.40 \rightarrow -0.20$	NO_3	NO_2, N_2
Iron reduction	$+0.05 \rightarrow -0.60$	Fe_2O_3	FE, O
Sulfate reduction	$-0.10 \rightarrow -0.75$	SO_4	HS
Methogenesis	$-0.20 \rightarrow -0.80$	CO_2	CH_4

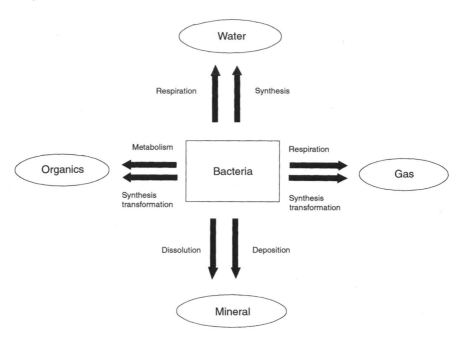

FIGURE 9.34   Schematic illustrating microbial interactions in the saturated zone.

electron receptor, typically oxygen. The metabolized hydrocarbon ends up as new cell mass, with the by-products being carbon dioxide, water, and the growth of additional micro-organisms. Other significant natural attenuation processes include volatilization, dispersion, and adsorption. Volatilization, for example, can significantly reduce hydrocarbon contaminant mass in soil. Light-end hydrocarbons are typically degraded in the vadose zone or slowly released to the atmosphere. Dispersion is the primary mechanism in groundwater for transporting soluble contaminants away from their source areas where little degradation occurs, to where they can be readily degraded, typically at the fringes of a plume where oxygen levels in groundwater are not depleted. Alternatively, adsorption can limit migration. Organic-rich soils such as peat may be effective adsorbers of petroleum, implying containment, that can be an effective way to manage low-risk sites. These secondary processes are not viewed as significantly important because they do not result in contaminant destruction, but rather in mass transfer. However, these secondary processes should still be considered as playing an important role in natural attenuation.

Although natural attenuation has been around for some time, it was not until the mid-1990s that agencies responsible for establishing cleanup levels for soils and groundwater at impacted sites began moving away from a position of cleaning up to regulatory levels or pristine conditions. At least at some sites, agencies started to move toward a position of taking no further action despite less than pristine conditions prevailing, and where cleanup to regulatory levels (i.e., background levels, maximum contaminant levels, or more stringent levels based on cancer risks, etc.) was not economically or technically feasible.

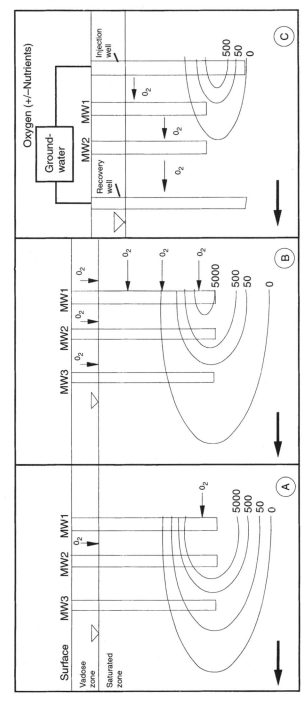

FIGURE 9.35 Schematic showing various natural attenuation scenarios for a dissolved-petroleum-contaminant plume. Scenario A illustrates a stable plume under high-permeability conditions, scenario B shows a shrinking plume under low-permeability conditions, and scenario C shows reducing conditions via enhanced biodegradation.

The difference between the use of "natural attenuation" and "no further action" as a remedial strategy is that natural attentuation needs to be demonstrated. This implies that additional site characterization and development of a groundwater monitoring phase for an acceptable period of time may be necessary. By *demonstration,* it is meant that there is evidence that contaminant reduction has been achieved, and that intrinsic bioremediation is likely responsible for that reduction. *Natural attenuation,* if properly demonstrated, increases the overall protection of the environment by either containment or destruction of contaminants. *No further action,* on the other hand, implies no additional investigation is required regardless of whether the contaminants of concern are degrading or migrating. Natural attentuation also serves as (1) an interim measure until future technologies are developed, (2) a managerial tool for reducing site risks, and (3) a bridge from active engineering (i.e., pump-and-treat, vapor extraction, etc.) to no further action.

No further action, however, may be preferable to natural attenuation in certain instances. Very low risk situations may be better served since it eliminates the need of continued monitoring and further documentation. Sites with low levels of contaminants or nondiscernible plumes may be better candidates for no further action. Furthermore, very minor releases of hydrocarbons to the subsurface may not be sufficient to support bioremediation. Alternatively, sites with elevated levels of contaminants in nonportable aquifers may be better addressed through conduct of a risk assessment.

Evaluation of Parameters. There exist several methods or lines of evidence to demonstrate whether natural attenuation is occurring. Because sites can vary dramatically in their complexity and level of effort required, the level of documentation that can be reasonably obtained will vary. What is important is that, because many impacted sites will be difficult to completely clean up to the satisfaction of all parties, the use of natural attenuation will likely be considered at some stage of the project or remedial action. It is thus prudent to generate evidence and documentation to assess the suitability of natural attenuation as part of the site subsurface characterization process.

The fundamental parameters for evaluation of natural attenuation can be divided into four general groups (Table 9.11):

* Hydrogeologic factors
* Chemical characteristics
* Biological characteristics
* Circumstantial factors

Hydrogeologic Factors. Hydrogeologic factors for consideration include aquifer type, hydrogeologic gradient, permeability, recharge capability, depth to groundwater, moisture content/field capacity, dissolved oxygen, depth to contamination, extent of contamination, and plume stability.

Aquifer Type. Aquifers or water-bearing zones characterized by relatively high quantities of naturally occurring carbon can favorably influence biodegradation of highly chlorinated solvents, whereas high quantities of anthropogenic carbon will influence biodegradation of highly chlorinated solvents, but at a faster rate. Low concentrations of carbon with high concentrations of dissolved oxygen may influence the biodegradation of vinyl chloride, but not favorably influence biodegradation of highly chlorinated solvents that require anaerobic conditions.

TABLE 9.11 Primary Parameters for Evaluation of Natural Attenuation as a Remedial Strategy

Parameter group	Parameter for evaluation
Hydrogeologic	Gradient
	Permeability
	Recharge
	Moisture content/field capacity
	Depth to impacted (contaminated) area
	Groundwater depth
	Dissolved oxygen
	Soil gas
	Extent of contamination/plume stability
Chemical	Hydrocarbon type
	Chromatographic evidence
	Hydrocarbon concentration
	Soil pH
	Nitrogen and phosphorous
Biological	Microscopic examination
	Plate counts
	Total heterotrophs
	Petroleum degraders
	Total organic carbon
Circumstantial	Time required for cleanup
	Age of release

Source: Testa and Winegardner (2000).

Groundwater Gradients.   Groundwater gradients that are favorable for natural attenuation are consistent seasonally, with moderate steepness such that a steady flow of electron acceptors is supplied to the plume, without being too steep to cause migration of a plume beyond the ability of microbes to contain it.

Permeability.   The rate of microbial ability to metabolize hydrocarbons is limited primarily by the availability of electron acceptors and nutrient supply. In general, uniform soil zones of moderate to high permeability are more favorable for natural attenuation because of their ability to transmit fluids. Deposits that tend to channelize groundwater flow may be undesirable. Hydraulic conductivities $>10^{-9}$ are considered acceptable.

Recharge.   Strong recharge of meteoric water to subsurface water-bearing zones provides an annual source of oxygen-enriched water, and, in fertilized areas, also provides nutrients for microbial growth. In addition, for releases to the vadose zone, the downward infiltration of wetting fronts displaces oxygen-depleted and CO_2-rich soil gas with fresh, oxygenated gas. However, excessive ponding of water or heavy-precipitation types of climates can cause water sealing to occur. This stops diffusion of oxygen, and may cause microbial activity to become curtailed.

Moisture Content/Field Capacity.   Within the vadose zone, moisture content is important, since microbial growth is limited by excessively wet or dry soil. Moisture content, expressed as a percentage of the field (or holding) capacity, indicates the ratio of moisture to air in the soil. The recommended range for optimal growth is between 40 and 70 percent.

Depth to Contamination. In general, the shallower the release in the vadose zone, the more rapid the diffusion of soil gas, and the greater the indigenous microbial density.

Dissolved Oxygen. For groundwater plumes, the presence of dissolved oxygen (DO) is critical for maintaining aerobic conditions. A DO concentration of at least 1 to 2 milligrams per liter (mg/L) is considered a minimum value to sustain a microbial population. Anaerobic conditions, if present, may cause growth of bacteria capable of degrading hydrocarbons using alternate electron acceptors such as iron (Fe^{3+}) and nitrate (NO_3^+). DO should be a standard field parameter and measured as part of any groundwater program where remediation may potentially be required. DO measurements do need to be measures in a closed cell, and should not be measures on groundwater samples retrieved with bailers. Submersible low-flow or bladder-type pumps are recommended for this purpose.

Soil Gas. The minimum O_2 concentration that can support aerobic metabolism in unsaturated soil is approximately 1 percent. O_2 diffuses into soil because of pressure gradients, and CO_2 moves out of soil because of diffusivity gradients. Excess water restricts the movement of O_2 into and through the soil. A minimum air-filled pore volume of 10 percent is considered adequate for aeration. Soil gas surveys using a mobile, small-truck-mounted geoprobe unit are a valuable tool to demonstrate a zone of enhanced microbial metabolism in the subsurface.

Extent of Contamination/Plume Stability. Defining the extent of subsurface contamination, both in the vadose zone and in groundwater, is a fundamental objective of any investigation, and lays the foundation for the natural attenuation alternative. An adequate groundwater monitoring well network is necessary to determine the lateral and vertical extent of hydrocarbon-impacted soil and groundwater. If the release is relatively recent, it becomes important to demonstrate that natural attenuation is limiting plume migration. Older releases are anticipated to have generally stabilized. For groundwater, asymptotic concentration limits should be achieved. Regardless, it is critical to demonstrate plume stability through the use of soil borings, soil gas data, and monitoring wells, among other techniques.

Chemical Characteristics. Chemical characteristics of importance to natural attenuation processes include petroleum hydrocarbon or organic compound type, concentration, pH, and nitrogen and phosphorus content.

Hydrocarbon Type. The light hydrocarbons generally degrade more readily than heavier hydrocarbons. A significant percentage of hydrocarbons found in light to medium distillates (i.e., gasoline, jet fuels, diesel, etc.) are all amenable to biodegradation. The monoaromatic hydrocarbons (i.e., BTEX) are the most soluble and degrade the easiest, followed by the straight-chained alkanes. Compounds that are resistant to degradation include the heavier isoprenoids (branched-chained alkanes) such as pristane and phytane hydrocarbons, and asphaltenes. For diesel-range hydrocarbons, a useful indication of degradation is the C_{17}/pritane ratio. Biodegradability of gasoline additives such as MTBE is considered poor and characterized by much lower natural attenuation potential relative to dissolved BTEX. This is evidenced by MTBE's persistence in groundwater and the extensive size of MTBE-dissolved plumes reported in the literature.

Chlorinated solvents behave differently relative to fuel components in the subsurface, and may require both aerobic and anaerobic conditions at varying times during the natural attenuation process for complete biodegradation to carbon dioxide and chloride ions to occur. Hydrocarbon compounds associated with highly chlorinated (or oxidized) solvents such as TCE and PCE have a limited number of hydrogen atoms (or electrons) and thus do not serve as electron doners but rather electron acceptors. Biodegradation of the highly chlorinated solvents can be accomplished through reductive chlorination, as electron donors or through cometabolism. With reductive chlorination, dissolved oxygen is the preferred electron acceptor from a thermodynamic perspective. Thus, oxygen, if present, will undergo reduction reactions prior to the less thermodynamically favorable chlorinated compounds. Highly chlorinated solvents thus biodegrade more favorably under anaerobic conditions, and not typically proceed to biodegrade under aerobic conditions. Under anaerobic conditions, biodegradation is electron-donor and carbon-source limited. Other sources for electron donors such as naturally occurring organic carbon or anthropogenic carbon must be present or anaerobic biodegradation will be limited.

Vinyl chloride is the least-oxidized chlorinated aliphatic hydrocarbon, and may serve as an electron donor. A vinyl chloride molecule consists of more hydrogen atoms than chloride atoms (3 to 1); thus, reductive dechlorination is not favorable to biodegradation. However, under aerobic conditions, vinyl chloride can serve as an electron donor with oxygen as an electron acceptor.

Cometabolism refers to the degradation of the chlorinated solvent as a by-product of the degradation of other substrates by micro-organisms, and does not benefit the micro-organism. As the degree of dechlorination decreases, the cometabolism rates increase. Thus, less oxidized or chlorinated solvents such as chlorinated ethenes (excluding PCE) biodegrade more favorably under aerobic conditions.

Chromatographic Evidence. Sample chromatograms often contain evidence that natural attenuation has occurred. If chromatograms of freshly released petroleum can be obtained, then a comparison can be made between the chromatograms of fresh and weathered samples. This comparison can be done relatively quickly after the chemical release.

Hydrocarbon Concentration. The concentrations of hydrocarbons in both soil and groundwater are important to consider. High concentrations of BTEX (>10 mg/kg) and other volatile hydrocarbons have solvent-type properties that are toxic to membranes. High total petroleum hydrocarbon (TPH) concentrations require more oxygen and nutrients for degradation. TPH concentrations over 10,000 mg/kg can also be detrimental by inhibiting water and air flow by obstruction of soil pores, and may be biologically unavailable. For these reasons, active remediation of release source areas by excavation, LNAPL recovery, or vapor extraction is always recommended when practical. Conversely, low concentrations of hydrocarbons may not be enough to stimulate microbial growth.

Soil pH. Soil pH should be in the range of 6 to 8, to maintain cell turgidity and promote enzymatic reactions. Soil buffers, such as carbonate minerals, can be valuable in neutralizing acidic groundwater as a result of high CO_2 concentrations because of microbiological activity.

Nitrogen and Phosphorus. Soil with carbon/nitrogen/phosphorus ratios of 100/10/1 is recommended for optimal bacterial growth. However, suboptimal ratios are not thought to be an impediment to intrinsic bioremediation, as oxygen is typically the factor limiting microbial growth in the subsurface. Low concentrations of these nutrients are typically recycled and made available to new microbial colonies through growth, death, and decay of older colonies.

Biological Characteristics. Tests for biological characteristics and characterization techniques include microscopic and chromatographic examinations, plate counts, total heterotrophs, petroleum-degrading bacteria, total hydrogen–degrading bacteria, and total organic carbon.

Microscopic Examination.   Microscopic examination involves a slurry derived from fresh soil that is examined under high magnification for bacterial type, cell health, protozoan, and approximate number of bacteria.

Plate Counts.   Plate counts involve indirectly determining the number of organisms in soil (direct counting is difficult and tedious). Plate-count techniques are selective and designed to detect micro-organisms with particular growth forms. Soil suspensions undergo serial dilution, and each dilution is placed in a single substrate (agar). After incubation, single colonies are counted, and each colony is associated with a single viable microbial unit capable of propagation, otherwise termed a *colony-forming unit* (CFU). Plate counts tend to underestimate the actual number of bacteria by typically an order of magnitude. The most common tests are for total heterotrophs and petroleum degraders.

Total Heterotrophs.   This test is for all bacteria capable of using organic carbon as an energy source. The number obtained from this test is compared to background samples from similar soil types of similar depths, and compared to the number of hydrocarbon-degrading bacteria within the same sample.

Petroleum Degraders.   This test cultures bacteria that are able to grow by using petroleum as the sole carbon source. A popular variation of this test is called the *sheen screen method,* which relies on the ability of micro-organisms to emulsify oil. It is important that the petroleum type used as the carbon source resembles the petroleum encountered at the site. If the concentration of the CFUs is a significant percentage of the total heterotrophs, and significantly elevated above background levels, then the indigenous microbial population has adapted to the release of hydrocarbons. CFUs greater than 1×10^{-5} are considered capable of supporting significant biodegradation.

Total Organic Carbon.   Formations with a significant total organic carbon content greater than the petroleum hydrocarbon content, such as peat-rich soil, can compete with the hydrocarbon-degrading microbes for oxygen and nutrients. A representative sample should thus be analyzed for total organic carbon should natural organic matter be observed in soil samples or borings.

Circumstantial Factors.   Circumstantial factors include the time required for cleanup and age of release. Time required for natural attenuation to effectively reduce contaminant levels is difficult to predict, although certain models are available. Time should not, however, be a controlling factor. Sites in industrial areas and airfields, for example, are not likely to change land use in the foreseeable future and thus have the advantage of time. If it can be demonstrated that contaminant levels are declining, then absolute time for cleanup should not be critical, provided the site is being monitored. However, natural attenuation may not be appropriate for sensitive sites that must be remediated in accordance with best available technology to cleanup standards within compressed time frames.

The relative age of the release can play an important role in the use of natural attenuation. Microbial populations need to adapt to the presence of petroleum hydrocarbons before they can metabolize the hydrocarbons and reproduce in concentrations above background levels. With older releases, if natural attenuation is occurring, associated groundwater plumes will hopefully have stabilized and begun to shrink, and maximum seasonal water table fluctuations have occurred and distributed LNAPL over a greater area. This maximum distribution results in less obstruction to groundwater flow through the zone of contamination, dissolves out the more toxic compounds, and brings electron acceptors (oxygen) to the microbes. It is thus much easier to gather evidence for natural attenuation in older releases than in more recent ones. Natural attenuation is more difficult to document in recent releases or when the release has not stabilized.

9.7 MARINE REMEDIATION STRATEGIES

9.7.1 Offshore Strategies

The effectiveness of any mitigation action includes: the quantity spilled, type and composition of the oil, proximity of the spill to land and other sensitive areas, how much time will transpire before action can be taken, and the weather. When oil spills on water, attention is initially focused on the immediate containment of the spill, recovery of as much oil as possible, and preventing and minimizing adverse effects and damage to natural resources. The major steps in this process are:

- Stopping the spill
- Containing the spill
- Primary cleanup
- Secondary cleanup

Stopping the spill involves stabilizing the ship, breached pipeline, or oil well. Stopping spills on water begins with the repair of the leak that may involve repair of the leak, removal of as much oil from the vessel as possible, or the placement of booms around the vessel. Lightering, which is the removal of oil or water from damaged or other compartments, may be performed, thus minimizing the potential for the ship to topple or capsize. Lightering is also routinely performed on many supertankers when they enter shallow water channels and ports.

Spill containment may involve the placement of booms, skimming of surface oil, and/or bombing of a vessel to burn remaining oil (Burger, 1997; Shaheen, 1992; Sittig, 1978). Containment of an oil slick can be accomplished by the use of booms, which are long rolls of absorbent material strung end-to-end, that float on the water surface. Primary cleanup may involve the placement of additional booms and skimming, and possible burning of an oil slick or use of dispersants to break up the oil before it reaches sensitive receptors such as beaches and fisheries. The biggest hurdle is to get the booms to the vessel or to sensitive areas for protection, before the oil has had an opportunity to spread. Sometimes surface skimmer units can be used, especially when the oil is heavy in nature. Another means of removing the oil is by burning, although this method can be technically challenging, and is dependent on the freshness of the oil, weather conditions, mixing of oil with water, location, availability of special equipment, and potential downwind environmental impacts.

9.7.2 Shoreline Strategies

Primary cleanup on shorelines uses several methods, including physical removal either by hand or with heavy equipment, vacuuming, washing with hot water and high pressure, dissolving with solvents and chemical cleaners, and natural attenuation. Physical removal has the least effect on intertidal organisms, whereas vacuuming, which performs well on large rocks and few small rocks or sand, can be very time-consuming and is disruptive to organisms. The use of washing techniques also damages plants and animals, and may result in more oil in these zones.

Secondary cleanup may incorporate the use of power sprayers or mechanical means to clean up shoreline beaches, or use of bioremediation with nutrients to speed up natural attenuation of oil. Dispersants are sometimes used when the oil must be dispersed to protect resources. Dispersants are chemicals that break up the oil into smaller droplets, and spread the oil throughout the water column. The use of dispersants does not clean up the oil or change the oil, rendering it harmless, but merely helps control a spill. They work best in about 30 ft of water, and are not effective by themselves in handling larger spills. Dispersants, although a valuable tool in some cases, can have a negative effect on marine organisms by increasing and prolonging the exposure of oil to organisms. Regardless of the strategy used, any cleanup decision involves issues about ecosystems and public policy. Decisions made must balance the potential damage associated with intertidal zones relative to other organisms and animals higher up on the food chain, local economies in regard to fisheries and tourism, etc. Solvents such as kerosene, mixed with some detergents, are not considered environmentally sound in most cases.

As previously mentioned, all hydrocarbons degrade; thus, released oil, if left in the environment, will eventually naturally biodegrade and attenuate. This process can prove to be a viable remedial alternative under certain circumstances and conditions. Micro-organisms or nutrients can be introduced to enhance the breakdown of the spilled oil. The amount of oil that is biodegradable can range greatly from about 11 to 90 percent. Some dependency on natural processes, albeit slow, such as dispersion, emulsification, evaporation, photo-oxidation, and turbulence, is required for that portion of the oil considered nonretrievable.

9.8 SPILL PREVENTION

Considering the large quantities of crude oil and petroleum products that are transported, stored, or used everyday, oil spills are bound to happen. It is impossible to know when and where an oil spill is going to occur; however, certain prudent steps can be taken to avoid and minimize the potential adverse effects from such an event. Under the best of conditions, the response is quick and well organized, and the only way to accomplish this is to plan in advance. A contingency plan serves this purpose and addresses all conceivable scenarios such that, when a spill occurs, contacts, resources, and strategies can be put into place immediately (Fig. 9.36).

The contingency plan consists of four essential elements:

- Hazard identification
- Vulnerability analysis
- Risk assessment
- Response actions

Hazard identification involves the compiling of certain information such as the types of oil used, stored, and/or transported in a given area, and the locations where oil is stored in large quantities. The mode of transportation, extreme and seasonal weather conditions throughout the year, and the location of response equipment and trained personnel are data that would also be deemed important in assessing the potential hazards.

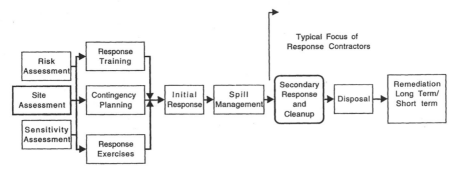

FIGURE 9.36 Flow diagram illustrating the various components of spill response for a marine oil spill.

Vulnerability analysis provides information about resources and communities that could be adversely impacted, and allows for reasonable decision making, in the event of an oil spill. Information pertinent to a vulnerability analysis may include lists of public safety officials in a particular community, and lists of vulnerable facilities such as hospitals, nursing homes, schools, prisons, and recreational areas. A list of special events or environments that may be adversely susceptible to oil or water contamination may also be of interest (i.e., wetlands, waterway, etc.).

Risk assessment allows the contingency planner to compare the hazard and the vulnerability in a particular location to the risk posed to a community. Risk assessments can be used to determine how best to control a spill, prevent certain elements in a community from exposure, and mitigate the damage done as a result of a spill.

Response actions are developed to address the risks identified as part of the risk assessment. Important actions required to minimize the hazards to the community and environment may include notifying all private companies and government agencies that are responsible for the cleanup effort, mobilizing trained personnel and equipment to the spill site, and assuring the safety of response personnel. Response actions would also encompass defining the oil spill size, location, and content; stopping the oil from leaving its container; containing the spill to a limited area; removing the oil; and disposing of the oil once removed from the water or land.

9.9 CASE HISTORIES

9.9.1 Regional LNAPL Release, Los Angeles Coastal Plain, California

California has a rich history of oil and gas exploration and development dating back to 1876, the first year of commercial production. During the early 1920s, numerous refineries and associated aboveground bulk liquid storage tanks, farms, and terminals were constructed in close proximity to both petroleum production areas and the nearby shipping facilities of the Los Angeles harbor. The majority of these facilities continued to expand their operations and areal extent through the late 1940s, while many others were initially isolated or located in moderate to heavily industrialized areas. However, with the encroachment of urban development, many are now in close proximity to densely populated light-manufacturing, commercial, and residential-zoned areas.

Regulatory Framework. Several refineries, tank farms, and other petroleum-handling facilities nationwide are included on the EPA 1996 National Priorities List, or are regulated under RCRA. Such facilities present several potential subsurface environmental concerns, reflecting approximately 80 years of continued operation. These concerns include:

- Hydrocarbon-affected soil that could potentially be characterized as a hazardous material
- Accumulation of hydrocarbon vapors posing a potential fire or explosion hazard
- Dissolved hydrocarbons in groundwater that may adversely affect beneficial-use water-bearing zones or drinking water wells
- LNAPL hydrocarbon pools occurring as perched zones and generally overlying the water table that serve as a continued source of both soil and groundwater contamination

Sixteen major refineries and 33 aboveground bulk-liquid tank farms are situated on the Los Angeles coastal plain. The locations of these refineries and tank farms, and associated major pipeline corridors, are shown in Fig. 9.37. In early 1985, oil droplets

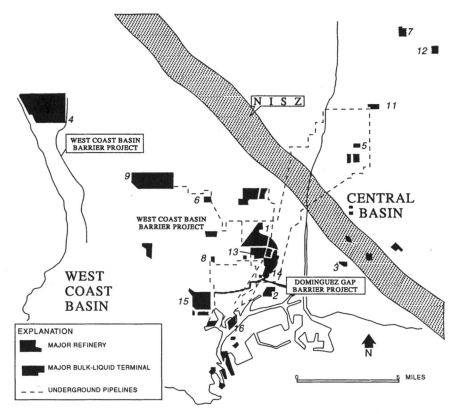

FIGURE 9.37 Location of major refineries, tank farms, and pipeline corridors in the Los Angeles County coastal plain. (*From Testa, 1992.*)

were evident on beach sands near high-priced beachfront real estate located just west of the Chevron refinery in El Segundo. The immiscible hydrocarbons evidently leaked from the nearby refinery, migrated downward to the shallow water table, and then moved downward with the regional groundwater flow toward the ocean and beachfront properties. Elevated concentrations of hydrocarbon vapors presenting a potential explosion hazard were also detected at the bottom of a construction pit under excavation near the refinery and several adjacent homes. The environmental regulatory community recognized a significant pollution problem. At that time, an estimated (although exaggerated) 6,000,000 barrels of LNAPL hydrocarbon product were thought to exist beneath most of the 1000-acre, 80-year-old refinery.

The highly visible presence of petroleum and potential hazard to public health, safety, and welfare subsequently prompted a minimum of 16 oil refineries and tank farms to be designated as health hazards by the California Department of Health Services. This designation reflected both potential and documented occurrences of leaked hydrocarbon product derived from such facilities during 70 years of operations, which migrated through subsurface soils and accumulated as LNAPL hydrocarbon pools overlying the water table. Several of these refineries were listed as hazardous waste sites, with assessment and remediation required under RCRA. The majority of the facilities also fell under the jurisdiction of the Los Angeles Region California Regional Water Quality Control Board (CRWQCB) Order 85-17, which was adopted in February 1985, reflecting the potential regional adverse impact on overall groundwater quality.

Fifteen refineries were included under CRWQCB Order 85-17. By December 1985, this order was expanded to include aboveground bulk-liquid storage facilities. Order 85-17 was the first regulatory mandate nationwide to address large-scale regional subsurface environmental impacts by the petroleum-refining industry. This order required, in part, the assessment of the subsurface presence of hydrocarbons and other associated pollutants that could affect subsurface soils and groundwater beneath and adjacent to such facilities. Specifically, the following items required addressing:

- Characterization of subsurface geologic and hydrogeologic conditions
- Delineation of LNAPL pools including volume and chemical characterization
- Implementation of LNAPL recovery
- Overall restoration of the aquifer contaminated with dissolved hydrocarbons and associated contaminants
- Eventual remediation of soils contaminated with residual hydrocarbons

At the time this order was issued, approximately six refineries had already commenced LNAPL hydrocarbon recovery programs, although such programs were limited in scope. In these cases, the large volumes of LNAPL present and initial ease of hydrocarbon recovery proved economically favorable, since the recovered LNAPL product could be easily recycled and sold.

Hydrogeologic Setting.   The Los Angeles coastal plain extends approximately 50 miles in a northwest-southeast direction, and roughly 15 to 20 miles in width between the Pacific Ocean and the base of the Puente Hills and Santa Ana Mountains. Encompassing approximately 775 square miles, the coastal plain is characterized by low relief and gentle surface gradients seaward. The low relief is interrupted by the Newport-Inglewood structural zone (NISZ), characterized by a northwesterly trending line of gentle topography extending roughly 40 miles in length. The NISZ is the major structural feature in the area and consists of en-echelon faults, anticlines, and domes, and is marked by a series of low hills and coastal mesas, which are broken by six topographic gaps or low areas.

The Los Angeles coastal plain is divided into four principal groundwater basins. The Santa Monica Basin and West Coast Basin are located southwest of the NISZ; the Central Basin and Hollywood Basin are situated northeast of the NISZ. The NISZ separates the West Coast Basin from the Central Basin. Hydrostratigraphic units underlying these portions of the West Coast and Central Basins where major refineries and bulk-liquid terminals occur are shown in Fig. 9.38. Salient hydrostratigraphic units beneath the area encompassing sites 1, 13, and 14 (Fig. 9.38) in the West Coast Basin are, in descending stratigraphic position, the Gaspur, Gage ("200-foot sand"), Lynwood ("400-foot sand"), and Silverado aquifers. The Silverado aquifer, which occurs within the Lower Pleistocene San Pedro Formation, is of very good quality and is the primary source of beneficial use groundwater from the West Coast and Central Basins. Beds of relatively low permeability soils that act as aquicludes or confining layers separate these aquifers in some, but not all, places. In particular, angular unconformities developed along the flanks of folds within the NISZ form contact zones between younger aquifers and one or more older saturated zones.

Pertinent to the occurrence of LNAPL is the Gage aquifer, which is encountered under water-table conditions. In this portion of the coastal plain, the hydrogeologic regime is influenced by numerous factors, all of which play a role in remediation strategy. These factors include regulated pumpage from the Silverado aquifer as well as continued artificial recharge into the Gaspur, Gage, and Lynwood aquifers through injection wells associated with the Dominguez Gap Barrier Project and seawater intrusion. The NISZ also has a significant, multifaceted impact on groundwater occurrence, quality, and usage.

The area encompassed by site 4 (Fig. 9.38) is situated near the western margin of the West Coast Basin. Sand dune deposits of Holocene age, and marine and continental deposits of Pleistocene age, underlie this portion of the coastal plain. In descending stratigraphic order, the key hydrostratigraphic units include the Old Dune Sand aquifer, the Manhattan Beach aquiclude, the Gage aquifer, the El Segundo aquiclude, and the Silverado aquifer (Fig. 9.38). Pertinent to the occurrence of LNAPL hydrocarbon is the Old Dune Sand aquifer, which is encountered under water table conditions. This portion of the West Coast Basin is also influenced by artificial recharge as part of the West Coast Barrier project.

Site 3, situated in the southern portion of the West Coast Basin, is underlain by, in descending stratigraphic position, the Semiperched, Gage, Lynwood, and Silverado aquifers (Fig. 9.38). Most pertinent to LNAPL hydrocarbon occurrence is the Semiperched aquifer, which is encountered under water-table conditions.

Sites 5, 7, 11, and 12 are situated in the Central Basin northeast of the NISZ (Fig. 9.38). This area is generally underlain by shallow perched and semiperched aquifers, and nine distinct regional aquifers. In descending stratigraphic position, these aquifers are the Semiperched, Gaspur, Exposition-Artesia, Gardena, Gage, Hollydale, Jefferson, Lynwood, and Silverado (Fig. 9.38). The Exposition-Artesia and deeper aquifers are the primary water-bearing zones tapped for municipal and industrial uses. The shallow aquifers, primarily the Semiperched but also to a large extent the Gaspur, in general are polluted with brines and industrial wastes rendering them unsuitable for domestic use throughout the region. LNAPL in the Central Basin occurs chiefly within shallow, perched zones of limited lateral extent and within the Semiperched aquifer, which is encountered under water-table conditions. The deeper aquifers are generally of good water quality, although some saltwater intrusion and localized occurrences of contaminated groundwater has been reported within the Exposition-Artesia aquifer.

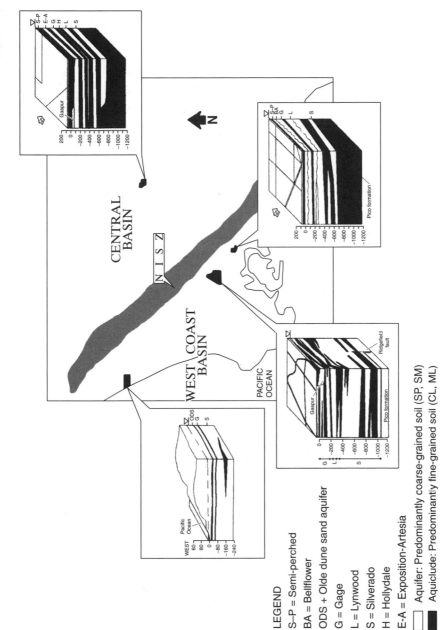

FIGURE 9.38 General subsurface hydrogeologic setting beneath major petroleum-handling facilities in Los Angeles County. (*From Testa, 1992.*)

LEGEND

S–P = Semi-perched

BA = Bellflower

ODS + Olde dune sand aquifer

G = Gage

L = Lynwood

S = Silverado

H = Hollydale

E-A = Exposition-Artesia

☐ Aquifer: Predominantly coarse-grained soil (SP, SM)

■ Aquiclude: Predominantly fine-grained soil (CL, ML)

LNAPL Occurrence.   The ubiquitous occurrence and areal extent of major LNAPL
hydrocarbon pools beneath these facilities situated on the Los Angeles coastal plain
are shown in Fig. 9.39. For purposes of this discussion, a pool is defined as an aerially
continuous accumulation of LNAPL. Two or more pools that have distinct differences in their respective physical and chemical properties are referred to as *coalesced
pools.* Individual accumulations of relatively uniform product are referred to as *subpools,* since it is inferred that they have coalesced to form areally continuous occur-

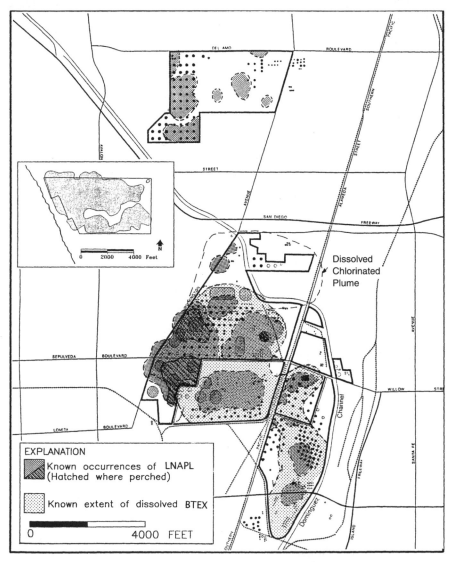

FIGURE 9.39   General spatial distribution of LNAPL pools and dissolved BTEX and chlorinated
plumes beneath portions of the West Coast Basin, Los Angeles coastal plain. (*From Testa, 1992.*)

rences. The occurrence of several pools and subpools at a particular site reflects releases from multiple sources at various times. The combined areal extent of these LNAPL pools is on the order of 1500 acres. The estimated cumulative minimum volume is on the order of approximately 1.5 million barrels; an estimated cumulative maximum volume is on the order of 7.5 million barrels. The discrepancy in LNAPL hydrocarbon volume reflects varying methodologies involved in the estimates.

The largest known subsurface accumulation of LNAPL hydrocarbon in the coastal plain area occurs beneath an active refinery (refer to site 4, Fig. 9.38). Groundwater beneath this site occurs under unconfined water-table conditions within the Old Dune Sand aquifer. The LNAPL hydrocarbon pool is elongated toward the northwest and encompasses approximately 690 acres in areal extent. Apparent LNAPL thickness in this pool ranges up to approximately 12 ft. The LNAPL pool is composed of a variety of refined petroleum product types, including light oil, diesel fuel, gasoline, jet fuel, and reformate.

One of the more complex LNAPL-impacted areas is in the vicinity of sites 1, 13, and 14 (Fig. 9.38). Several major refineries, terminals, and pipeline corridors are located in this area. Both groundwater and LNAPL beneath this area occur within the Gage aquifer under both perched and water-table conditions, at depths of 40 to 50 ft and 30 to 60 ft below ground surface, respectively. Some facilities are underlain by several pools, which in turn represent two or more coalescent pools of diverse product types. The release of bulk-liquid hydrocarbon product in this area has occurred over time via several pathways, including surface spillage, breached pipelines, corroded storage tank bottoms, and failed reservoir bottoms. Many of these occurrences have resulted from very slow releases over long periods of time.

Source identification of dissolved hydrocarbons in groundwater and refined LNAPL is difficult because of several factors. These factors are not limited simply to varying microbiological, chemical, and physical processes. The complex historical industrial development of the area, the close proximity of several crude and petroleum-handling facilities (including clusterings of refineries, bulk storage tank farms, and underground pipelines) in industrialized areas, and the numerous underground storage tanks associated with gasoline service stations in less industrialized areas also make source identification difficult to ascertain. Several methodologies have been used to identify not only the crude and/or refined product type, but also the brand, grade, and in some instances, the source crude. These methods have included routine determination of API gravity, development of distillation curves, and analysis for trace metals (notably, organic lead and sulfur). More sophisticated methods have included gas chromatography, statistical comparisons of the distribution of paraffinic or n-alkane compounds of specific molecular weight, and determination of isotopic ratios of carbon and hydrogen ($^{13}C/^{12}C$ and $^{2}H/^{1}H$ or D/H, respectively) for lighter gasoline-range fractions, and $^{15}N/^{14}N$ and $^{34}S/^{32}S$ ratios for the heavier petroleum fractions.

LNAPL Hydrocarbon Recovery. Most major petroleum-handling facilities have implemented aquifer restoration programs, with some facilities further ahead in their respective programs than others. Effective recovery of LNAPL hydrocarbon can be accomplished by several means. Where perched or water-table conditions are deep (greater than 30 ft below ground surface), the approaches to LNAPL hydrocarbon recovery have conventionally included both single- and two-submersible-pump systems. Under shallow perched water-table conditions, as commonly experienced at port facilities, or shallow perched conditions, passive skimmer-type systems have also been utilized. In low-yielding formations with minimal LNAPL hydrocarbon presence, simply bailing the LNAPL periodically has been performed. More conventional one- and two-pump submersible systems have been utilized within the water-table aquifer.

Overall efficiency and effectiveness of the several LNAPL hydrocarbon recovery programs in operation in the Los Angeles coastal plain are limited by numerous factors. The most important is the inability to handle and treat coproduced waters which, depending upon the size of the facility and the scale of the recovery program, can potentially exceed 1000 gal/min. This reflects antiquated wastewater treatment systems (or lack thereof) at the facilities, in conjunction with existing systems functioning at capacity, cutbacks in the volume of hydrocarbon-affected groundwater that the Los Angeles County Sanitation District is willing to accept, and increased regulatory pressure against injection without treatment.

The overall efficiency and effectiveness of LNAPL hydrocarbon recovery programs have also been impacted by other factors. These factors include limitations associated with LNAPL recovery from low-yielding formations, inability to gain access to optimal recovery and off-site locations, coproduced water handling constraints, and economic constraints.

Regional Long-Term Remediation Strategy. The large extent of petroleum-affected groundwater, complexities of the regional hydrogeological setting, and the numerous factors involved in developing short- and long-term remediation strategies have resulted in an avoidance of a site-specific approach to formulating regional objectives in the Los Angeles area groundwater basins, and prioritizing these objectives. The primary objective is the protection of the Silverado aquifer and other regional beneficial-use aquifers.

Regulatory emphasis has recently focused on the following issues:

* Completely delineate the areal extent of LNAPL, and the horizontal and vertical extent of its respective dissolved constituents.
* Enhance efficiency and effectiveness of existing LNAPL recovery programs.
* Develop analytical and numerical groundwater models to (1) predict the fate and transport of LNAPL and its dissolved constituents, (2) provide more reliable LNAPL volume determinations, and (3) enhance design for optimal groundwater cleanup strategies.
* Identify the extent and sources of DNAPLs and other organic compounds (e.g., chlorinated solvents and pesticides).
* Develop a source elimination program to detect leakage from aboveground tanks and underground piping in the early stages of LNAPL release.
* Review current technologies and develop a soil cleanup strategy consistent with the depth and quantity of contaminants present.

In response to the issues outlined above, horizontal delineation of LNAPL is near completion, with off-site delineation required at a few localities. The horizontal and notably the vertical delineation of dissolved hydrocarbon constituents have yet to be addressed for most of the affected sites. Despite the large volumes of LNAPL released over time, the lateral extent of dissolved hydrocarbon constituents, notably benzene, toluene, ethylbenzene, and total xylenes (BTEX) has been limited to distances of about 200 ft or less hydraulically downgradient of most of the larger LNAPL pools at those sites where plume delineation has been addressed. The vertical extent of these constituents has not been fully ascertained. Delineation programs to address the vertical extent of dissolved hydrocarbon constituents in groundwater is currently being implemented at some sites. Enhancing the efficiency and effectiveness of existing LNAPL recovery programs is more difficult in nature, notably for the larger LNAPL recovery programs. The primary factor limiting most recovery programs is the inability of the facilities to handle the generated coproduced water.

Certainly if protection of the Silverado aquifer is the primary objective, and the water-table aquifers (i.e., Gage, Old Dune Sand, and Semiperched aquifers) are essentially a write-off, then the intervening aquifers, although not used for water supply, may serve the purpose of "guardian" aquifers.

Two of the larger LNAPL hydrocarbon occurrences, sites 1 and 4 (Fig. 9.38), formerly reinjected coproduced groundwater into generally the same hydrostratigraphic zone from which it was withdrawn; site 1 reinjected without treatment into the Gage aquifer, whereas site 4 reinjected into the Old Dune Sand aquifer. Because of the presence of dissolved hydrocarbons, notably benzene, in the coproduced water that is typically returned to the aquifer during LNAPL recovery operations, immediate application of EPA's toxicity characteristic rule may result in classification of the reinjected water as disposal of a hazardous waste. This in turn would terminate use of underground injection control (UIC) Class V wells (which many of these operations currently use). These wells would then be automatically reclassified as UIC Class IV wells, which would be prohibited under RCRA. An extension was given by EPA to allow oil companies time to comply with this rule so that LNAPL recovery efforts involving the return of benzene-affected groundwater via reinjection wells will not be readily interrupted. Although this relieves certain facilities from immediately complying with this ruling, many of these LNAPL hydrocarbon recovery programs will continue for tens of years, thus, despite the extension, the effects of future regulatory requirements and restrictions are uncertain.

Only rudimentary modeling has been performed to date to predict the fate and transport of dissolved hydrocarbons in groundwater over time. In addition, much uncertainty exists concerning the actual LNAPL volume present. Other organic compounds, notably chlorinated hydrocarbons, are of significant concern because of their high mobility in groundwater systems. These constituents are not typically derived from petroleum-handling facilities, and are inferred to be derived from old landfill cells and chemical-related industry-type sources nearby. In any case, assessment of the occurrence and lateral and vertical distribution of chlorinated hydrocarbons in the area has been limited.

Petroleum-affected soil and groundwater primarily reflect historic releases, although present contributions also exist at some sites. Source elimination programs are thus currently being developed on a site-specific basis. The objectives of this program are to assure that a majority of releases are historic in nature and to identify current releases. This program in general will include:

• Periodic piping testing and replacement, as appropriate.

• Eventual conversion of underground piping to aboveground piping.

• Periodic inspection of surface and/or aboveground storage tanks. Tanks will be periodically taken out of service for inspection and repaired as needed. Double bottoms are also being installed, as appropriate.

• Testing, removal, and replacement, if appropriate, of underground storage tanks.

Cleanup strategies for hydrocarbon-affected soil will most likely be the last issue to be mandated from a regulatory perspective and will certainly be the most difficult technically to address. This difficulty reflects the large deep-seated volumes of residual hydrocarbon presence, and the current lack of efficient, cost-effective methodologies for in situ remediation of residual hydrocarbons in low-permeability, fine-grained soils.

Regional hydrogeologic setting also plays an important role in formulating a long-term remediation strategy. Within the NISZ, structures such as folds and faults are critical with respect to the effectiveness of the zone to act as a barrier to the inland movement of salt water. A nearly continuous set of faults is aligned along the general crest of the NISZ, notably within the central reach from the Dominguez Gap to the Santa Ana Gap. The position, character, and continuity of these faults are fundamental to the discussion of groundwater occurrence, regime, quality, and usage. In addition, delineation and definition of aquifer interrelationships with a high degree of confidence is essential. The multifaceted impact of the NISZ is another aspect of the level of understanding required prior to addressing certain regional groundwater issues. A primary issue is which aquifers are potentially capable of being of beneficial use versus those that have undergone historic degradation. Those faults that do act as barriers with respect to groundwater flow, may, in fact, be one of several factors used in assigning a part of one aquifer to beneficial use status as opposed to another. A second issue, based on beneficial use status, is the level of aquifer rehabilitation and restoration to be required in association with the numerous LNAPL recovery and aquifer remediation programs being conducted within the Los Angeles coastal plain.

Overall understanding of this regional hydrogeologic setting, including proper delineation of the relationship between the various aquifers, is essential to implementing both short- and long-term regional aquifer restoration and rehabilitation programs. The Silverado aquifer is recognized presently as the primary source for municipal supplies in the area. The NISZ, where many of the oil fields beneath the coastal plain are situated, has a significant multifaceted impact on groundwater occurrence, quality, usage, and remediation strategy. Present groundwater management programs incorporate regulated pumpage that is limited by court order, but which tends to lower the water levels within the Silverado aquifer (and Lynwood aquifer). Future land use and site-specific concerns such as wastewater handling will all play a role in the approach to remediation, thus, remediation strategies need to be established on a case-by-case basis as well as the regional situation.

9.9.2 *Exxon Valdez* Spill

The *Exxon Valdez,* a 2-year old oil tanker with a capacity of 1.46 million barrels (62 million gallons) of oil, was the newest and largest of Exxon's 19-ship fleet. On the evening of March 23, 1989, with 1.26 million barrels of oil (54 million gallons) in cargo, the *Exxon Valdez* departed for Long Beach, California, from Valdez, Alaska. As the *Exxon Valdez* approached Prince William Sound at about 12 miles per hour, an effort was made to avoid chunks of ice from the nearby Columbia Glacier. The captain tried to turn the ship into an empty inbound shipping channel when it struck underwater rocks in Prince William Sound, which tore huge holes in 8 of the vessel's 11 giant cargo holds, releasing more than 11 million gallons of oil within 5 hours. After 7 hours, the resulting oil slick was 1000 ft wide and 4 miles long. To complicate matters, although about 80 percent of the oil remained on board, the ship was resting in an unstable position and in danger of capsizing. Removing the remaining oil from the vessel, and controlling and initiating cleanup of the spill, was of immediate concern.

Initial Response—The First Day. The spill occurred in coastal waters; thus, the United States Coast Guard's Office of Spill Control (OSC) maintained authority over all cleanup-related activities. OSC closed the Port of Valdez to all traffic, and along with representatives from the Alaska Department of Environmental Conservation, conducted a reconnaissance of the spill site to assess damages. By noon, the National Response Team, composed of representatives from 14 different federal agencies based in Washington, D.C., was activated. The Alyeska Pipeline Service Company was first to assume responsibility for the cleanup. Alyeska is owned by several oil companies, including Exxon, and operates the trans-Alaska pipeline and shipping terminal at Valdez, and maintained responsibility for implementing plans for responding to oil spills in the area. Alyeska immediately initiated emergency communications centers in Valdez and Anchorage.

Goals established by Exxon and the OSC included taking steps to prevent additional spillage of oil from the vessel into the ecologically sensitive Prince William Sound, and protect four fish hatcheries that were being threatened. Other concerns centered on the health and safety of emergency response personnel who would be working around highly flammable and toxic fumes. Initial problems centered around insufficient equipment being available to contain an 11 million gallon spill (i.e., booms and other mechanical equipment); the remoteness of the spill site, which resulted in significant logistical problems associated with the transportation of equipment and personnel over great distances; and lodging. Planes had to fly into Anchorage, a 123-mile drive from Valdez. A temporary tower was set up by the Federal Aviation Administration in Anchorage, to manage the increase in flights to the area. To exacerbate the situation, the barge usually used by the Alyeska's response team was undergoing repairs and thus, was not available for about 12 hours, including 2 hours just to reach the spill site.

The Response Team—The Second Day.   Exxon assumed responsibility for the cleanup and associated costs on the second day. Exxon immediately set up a communication network; four weather stations around Prince William Sound, which were critical for planning cleanup efforts; and a refueling station in Seward. Exxon directed another ship, the *Exxon Baton Rouge,* to remove the remaining oil from the vessel. By the fourth day, Exxon coordinated more than 240 tons of additional equipment including booms, skimmers, and dispersants. As for personnel, hundreds of people were brought in, including over 1000 Coast Guard personnel, nine Coast Guard cutters, and eight aircraft. Other agencies included the National Oceanic and Atmospheric Administration, U.S. Fish and Wildlife Service, and the Environmental Protection Agency. The Hubbs Marine Institute of San Diego, California, was also brought in to set up a facility to clean oil from otters, and the International Bird Research Center of Berkeley, California, for the rehabilitation of waterfowl.

Cleanup Methods Employed.   Three methods were employed in the effort to control and clean up the spill: in situ burning, chemical dispersants, and mechanical cleanup. A trial burn was performed in the early phases. A fire-resistant boom was placed on tow lines, with the ends of the boom each attached to a ship. The ships towed the booms away a safe distance from the main oil slick, and the oil was ignited. Although the fire did not endanger the vessel or main slick, unfavorable weather prevented additional burning.

Dispersants were also sprayed on the main slick by helicopters, and mechanical cleanup commenced with booms and skimmers. The dispersants were controversial, since limited wave action existed to mix the dispersant with the oil in the water. Limited supplies and application equipment also complicated the use of dispersants. In any case, it was deemed ineffective by the Coast Guard. Bad weather continued to plague the cleanup efforts. Thick oil and heavy kelp tended to clog equipment, and repairs to skimmers became time-consuming. Transferring oil from temporary oil storage vessels into more permanent containers was also difficult.

Ecological Impact. Sensitive areas were identified and categorized according to the degree of damage. The areas were ranked and priority for cleanup was established. High-priority sites included seal pupping and fish hatcheries areas, where special cleaning techniques were used. Wildlife efforts were slow, however, because of inadequate resources in the early stages. Some studies completed since estimate that between 100,000 and 300,000 birds were killed, along with an estimated 2650 sea otters. Exposure to the spill adversely affected the natural resources of the area in several ways including mortality, sublethal effects, and degradation of habitat. Mortality included death caused immediately or within a certain period of time after exposure to oil, cleanup activities, reduction in critical food sources, or other related causes. Sublethal effects included injuries that affected the health and physical condition of organisms, their eggs or larvae, but did not result in the death of the juvenile or adult organisms. Degradation of habitat included any alteration of or contamination of flora, fauna, and the physical components of the habitat.

Aftermath. On June 12, 1992, more than 3 years after the spill, the Coast Guard announced an end to the cleanup efforts. Pools of oil still remained in certain areas, but the harm caused to the ecosystem from the presence of the residual oil was deemed too small to justify the cost of further cleanup efforts. The ecological disaster created by this spill demonstrated quite clearly that, in combating oil spills on a large scale in surface water, speed of action, preparedness, and establishment of a chain of command is vital. Without these in place, chaos results.

9.9.3 Ashland Oil Spill

On January 2, 1988, a four million-gallon oil storage tank, owned by the Ashland Oil Company, was in the process of being filled to capacity when it suddenly spilt apart and virtually collapsed at an oil storage facility adjacent to the Monongahela River, in Floreffe, Pennsylvania. Diesel oil flowed over the containment dikes, across a parking lot on the adjacent property, and into an uncapped storm drain that emptied directly into the river. Within minutes, the oil slick migrated several miles down river, washing over two dam locks while dispersing throughout the width and depth of the Monongahela River, and on into the Ohio River.

Although less than one-half the size of the *Exxon Valdez* spill, the Ashland Oil Spill was the largest inland oil spill in United States history. More importantly, the spill temporarily contaminated the drinking water sources for an estimated 1 million people, and affected populations in Pennsylvania, West Virginia, and Ohio.

Quick notification from Ashland to the local response entities and the National Response Center was fundamental to the establishment of a command post on the evening of the spill. Local authorities implemented the initial on-site response during the night, with EPA taking control of the spill response and cleanup efforts the following day. An Incident-Specific Regional Response Team (RRT) formally commenced activities 2 days after the spill. The RRT was composed of numerous environmental and health-related agencies from both the federal level and the impacted states. Contractors employed by Ashland performed the actual work using booms, vacuum trucks, and other equipment to retrieve the spilled oil. About 20 percent of the oil that spilled and flowed into the river was recovered. EPA monitored the overall cleanup efforts, whereas state personnel developed a river monitoring system to track the spill, and a sampling and chemical testing program to assess water quality and protect water supplies. EPA also followed up with a compliance inspection program and a spill prevention control and countermeasure (SPCC) plan inspection of the facility where the spill occurred.

The river ecosystem was significantly affected, with thousands of birds and waterfowl killed. Although many birds were saved, mortality estimates for waterfowl, including ducks, loons, cormorants, and Canada geese, ranged from 2000 to 4000. Overall cost for cleanup of the spill was on the order of $11 million to the Ashland Oil Company, which was later charged with violation of the Clean Water Act. Total fines were on the order of $2.25 million.

9.9.4 Gulf War, 1991

Society's first experience with environmental terrorism on a regional scale started on January 25, 1991, in the Arabian Gulf on the eastern shores of Kuwait, adjacent to Saudi Arabia. Crude oil from pumping stations at Mina Al-Ahmadi was purposely dumped into the Gulf waters. In excess of 2 million barrels per day of crude oil was released as an act of war. Although of little if any practical value from a military perspective, ecologically the oil severely affected the ecosystem of the Gulf. Compared to the 260,000 barrels spilled by the *Exxon Valdez* in 1989, 5.7 million barrels of crude oil made it to the Gulf, making it the largest spill in history.

The oil slick thinned rapidly as it migrated southward with prevailing winds and currents on the order of 12 miles per day. After 4 days, the oil slick was 50 miles long, 12 miles wide, and 3 mm (approximately 0.1 in) thick. The oil was rich in light components that readily evaporated to the atmosphere. However, the heavier components became what was described as "chocolate mousse." Some of it settled to the bottom while some reached distant shores.

One of the main concerns was the potential for contamination within the intake zone (10 to 16 ft below the water surface) for the desalination plants that existed along the Arabian Gulf. Another concern was the potential impact on water cooling supplies for the electric generating plants.

Booms were utilized to contain the oil slick and protect these sensitive facilities and the shoreline. The booms extended 3 ft below the water surface. Intake waters were continuously monitored while ships were used to pick up the oil contained by the booms. Efforts were also made to protect wildlife and shorelines.

In comparison with the *Exxon Valdez* oil spill, which occurred in the Prince William Sound with an average water depth of 330 ft and with rejuvenation nearly every month, the Arabian Gulf setting is much different. The Arabian Gulf is enclosed with only one narrow outlet about 34 miles across at the Strait of Hormuz. With a water depth around 110 ft, the Gulf rejuvenates itself about once every 4 years.

9.10 SUMMARY

At least for the next few generations, the petroleum industry will continue to provide the primary energy needs for society. The industry has increasingly improved the means by which petroleum is explored, recovered, transported, processed, marketed, stored, and used. In addition, through a combination of efforts by the regulatory community and the petroleum industry, many of the inherited problem sites have been assessed and with programs set forth to mitigate and manage these areas. Furthermore, the operational practices that were once commonplace but harmful to the environment have been significantly improved through technological advances.

The technology and capability exists to protect the environment and produce the needed oil for a healthier economy and quality of life. As consumers of this vital resource, the primary environmental challenges we face in the future are:

- Practicing sound stewardship of this nonrenewable natural resource
- Preventing the uncontrolled and accidental release of petroleum and its derivatives in the environment
- Improving our understanding of the overall impact of petroleum, including health risks of petroleum and its derivatives in the environment
- Developing better strategies and technologies for the recycling, restoration, and remediation of petroleum-impacted soil, water, and air

REFERENCES

API, 1996, "Reinventing Energy—Making the Right Choices," American Petroleum Institute, Washington, D.C.

API, 1997a, *Petroleum Industry Environmental Performance,* 5th Annual Report, American Petroleum Institute, Washington, D.C.

API, 1997b, "How Much We Pay for Gasoline—1996 Annual Review," American Petroleum Institute, Washington, D.C.

Baker, K. H., and Herson, D. S., 1994, *Bioremediation,* McGraw-Hill, New York, 375 pp.

Burger, J., 1997, *Oil Spills,* Rutgers University Press, New Brunswick, N.J., 261 pp.

Department of Energy, 1997, *International Energy Annual Report,* DOE, Washington, D.C.

Devel, L. E., Jr., 1994, *Soil Remediation for the Petroleum Extraction Industry,* PennWell Books, Tulsa, Okla.

Durgin, P. B., and Young, T. M. (eds.), 1993, *Leak Detection for Underground Storage Tanks,* American Society for Testing and Materials, STP 1161, Philadelphia.

Environment Canada, 1978, "The Basics of Oil Spill Cleanup."

Fenster, D. E., 1990, *Hazardous Waste Laws, Regulations, and Taxes for the U.S. Petroleum Refining Industry,* PennWell Books, Tulsa, Okla., 215 pp.

Gregston, T. G., 1993, *An Introduction to Federal Environmental Regulations for the Petroleum Industry,* University of Texas, Petroleum Extension Service, Division of Continuing Education, Austin, Texas.

Jacobs, J., Guertin, J., and Herron, C. (eds.), 2001, *MTBE: Effects on Soil and Groundwater Resources,* Lewis Publishers, Boca Raton, Fla., 245 pp.

Keech, D. H., 1988, "Hydrocarbon Thickness on Groundwater by Dielectric Well Logging," in *Proc. of the NWWA of AGWSE and API Conf. on Petroleum Hydrocarbons and Organic Chemicals in Ground Water: Prevention, Detection and Restoration,* vol. I, Nov., pp. 225–289.

National Research Council, 1991, *Tank Spills: Prevention from Design,* National Academy Press, Washington, D.C.

Nielson, D. M. (ed.), 1991, *Practical Handbook of Ground-Water Monitoring,* Lewis Publishers, Chelsea, Mich., 717 pp.

Nyer, E. K., et al., 1996, *In Situ Treatment Technology,* Lewis Publishers, Boca Raton, Fla., 329 pp.

Orszulik, S. T., 1997, *Environmental Technology in the Oil Industry,* Blackie Academic and Professional, New York.

Patin, S., 1999, *Environmental Impact of the Offshore Oil and Gas Industry,* EcoMonitor Publishing, East Northport, N.Y., 425 pp.

Shaheen, E. I., 1992, *Technology of Environmental Pollution Control,* 2d ed., PennWell Books, Tulsa, Okla., 557 pp.

Sittig, M., 1978, *Petroleum Transportation and Production—Oil Spill and Pollution Control,* Noyes Data Corp., Park Ridge, N.J., 360 pp.

Testa, S. M., 1992, "Groundwater Remediation at Petroleum-Handling Facilities, Los Angeles Coastal Plain," in *Engineering Geology Practice in Southern California,* R. W. Pipkin and R. J. Proctor (eds.), Association of Engineering Geologists, College Station, Tex., pp. 67–79.

Testa, S. M., 1994, *Geological Aspects of Hazardous Waste Management,* Lewis Publishers, Boca Raton, Fla., 537 pp.

Testa, S. M., 1997, *The Reuse and Recycling of Contaminated Soil,* Lewis Publishers, Boca Raton, Fla., 268 pp.

Testa, S. M., and Winegardner, D. L., 2000, *Restoration of Contaminated Aquifers—Petroleum Hydrocarbons and Organic Compounds,* 2d ed., Lewis Publishers, Boca Raton, Fla., 446 pp.

United States Department of Energy, 1997, *Oil and Gas Research and Development Program, Securing the United States Energy, Environment and Economic Future,* Office of Natural Gas and Petroleum Technology, Washington, D.C.

United States Department of Transportation, 1999, Code of Federal Regulations (CFR) Title 49D, Part 195, Transportation of Hazardous Liquids by Pipeline, Accident Report Database 1982–1997.

United States Environmental Protection Agency, 1999, "Understanding Oil Spills and Oil Spill Response—Understanding Oil Spills in Freshwater Environments," EPA 540-K-99-007, 48 pp.

Wagner, R. B., Hampton, D. R., and Howell, J. A., 1989, "A New Tool to Determine the Actual Thickness of Free Product in a Shallow Aquifer," in *Proc. of the NWWA of AGWSE and API Conf. on Petroleum Hydrocarbons and Organic Chemicals in Groundwater: Prevention, Detection and Restoration,* Nov., pp. 45–59.

Williams, B., 1991, *U.S. Petroleum Strategies in the Decade of the Environment,* Pennwell Publishing, Tulsa, Okla., 336 pp.

Youngquist, W., 1997, *GeoDestinies—The Inevitable Control of Earth Resources over Nations and Individuals,* National Book Company, Portland, Ore.

CHAPTER 10

GROUNDWATER REMEDIATION AT FORMER MANUFACTURED-GAS PLANT SITES

David S. Lipson

Blasland, Bouck & Lee, Inc.
Golden, Colorado

10.1 INTRODUCTION

Restoring contaminated groundwater to pristine or even background conditions at former manufactured-gas plant (MGP) sites is impossible within short time frames when residual tars are present at or below the water table. This is due to the high percentage of pitch in MGP tars, which cannot be removed from below the water table except in some cases when it is within about 25 ft of the ground. The physicochemical properties of pitch in groundwater render it extremely immobile and resistant to weathering. In many instances, groundwater restoration within most regulatory time frames (i.e., 30 years) is simply unobtainable because of a lack of cost-effective remedial technologies that are able to remove residual MGP tars from saturated media below depths of about 25 ft. Because of this, trends in groundwater remediation at former MGP sites have focused on risk management, or plume management, rather than plume removal, by implementing multicomponent remedial strategies. Remedy components typically consist of a combination of engineered source removal, containment, and long-term monitoring activities.

This chapter examines groundwater contamination and remediation at former MGP sites. It discusses the nature, fate, and transport of dense, nonaqueous-phase liquids (DNAPLs) in groundwater, and provides physicochemical data on MGP tar samples. The chapter includes a summary of groundwater remedial activities at former MGP sites that are on the United States Environmental Protection Agency's (U.S. EPA's) Superfund list.

10.2 BACKGROUND

10.2.1 History of Manufactured-Gas Plants

The history of MGPs in the United States is discussed in detail by Harkins et al. (1987) and Hayes et al. (1996). The following discussion is drawn from information derived mainly from these sources.

MGPs produced combustible gases from coal and oil throughout the United States from about 1850 through 1950. The gases were used by consumers and industry as a source of energy for lighting, heating, and cooking. By-products of gas production in whole or in part were often recycled as feedstock for other processes, fuels for heating production ovens (e.g., retorts), or sold on the market as raw material for other processes. Economics generally dictated the extent to which MGP by-products were recycled, sold, or disposed of as wastes.

It is estimated that 11,000 gas-producing facilities were present in the United States in 1921. By the early 1940s, natural gas became more cost-effective compared with MGP gases, and MGP production rapidly declined. At this point, MGPs were either closed, sold, demolished, or refitted for other uses. As a result of a lack of strict environmental regulations at the time, much MGP process residual, by-products, and waste remained at the former MGPs.

With the authorization of CERCLA and RCRA legislations in the 1970s, and state-initiated regulatory programs, many former MGP sites were identified as contaminated sites and environmental issues are being addressed. A recent U.S. EPA publication indicates that currently there may be 3000 to 5000 former MGP sites across the country (U.S. EPA, 2000).

10.2.2 MGP Processes

Combustible gases were produced by pyrolysis cracking, a process in which hydrocarbon feedstock was heated, thereby producing combustible gases that were collected for industrial and household use. This was accomplished by heating coal or oil in a closed vessel at temperatures ranging from approximately 500 to 1600°F and collecting the offgases for distribution. Although many processes were employed, three main processes for gas production were in widespread use:

- Coal carbonization, which produced coal gas
- Carburetion processes, which produced carbureted water gas (CWG)
- Oil carbonization, which produced oil gas

Coal carbonization was an MGP process in which coal gas was produced by heating bituminous coal in a closed vessel and collecting the resultant gases. The closed vessel (e.g., retort) was heated with a variety of fuels including coal, process gas, coal tar, and, on rare occasions, coal gas. Coal gas was reported to be a mixture of carbon dioxide, carbon monoxide, oxygen, methane, ethane, hydrogen, nitrogen, and illuminants with a heating value of approximately 600 British thermal units (Btu) per cubic foot. By-products of coal carbonization generally included coke, coal tar, and process waters.

Carburetion was an MPG process in which CWG was produced by passing steam through a bed of solid carbon (e.g., anthracite coal, bituminous coal, coke from bituminous coal). Steam reacts with carbon and creates water gas, which is also referred to as *blue gas*. Liquid hydrocarbons would then be thermally cracked into the water gas, creating CWG. CWG was reported to be a mixture of primarily carbon monoxide and hydrogen with a heating value ranging from approximately 500 to 600 Btu/ft^3. As a result of the growth of the petroleum industry near the turn of the century, petroleum fuels provided a cheap alternative to coal as a feedstock and CWG became the predominant manufactured gas in the United States as many MGPs switched to carburetion processes during this time. By-products of the carburetion process generally included tars, uncracked portions of liquid hydrocarbons, and process waters.

Oil carbonization was an MGP process in which oil gas was produced by passing steam and atomized oil through a heated, brick-lined oven. Through this process, the oil and steam were gasified, producing oil gas. Crude oil was the original feedstock in these processes, but crude oil usage was gradually replaced by refinery by-products after approximately 1919 because refinery by-products were cheaper. Because of the lower heating value of refinery by-products, greater quantities were required to produce the same amount of product gas compared with crude oil. Refinery by-products also generated larger quantities of MGP by-products compared with crude oil and coal. Oil gas was reported to generally be a mixture of carbon dioxide, carbon monoxide, oxygen, hydrogen, methane, ethane, nitrogen, and illuminants, with a heating value ranging from approximately 600 to 700 Btu/ft^3. The primary by-products of oil gas production were lampblack, tar, and light oils.

Many different types of manufacturing equipment were used at MGPs to process feedstock, product gases, and by-products. Minor variations in the processes and the manufacturing equipment led to different component mixtures in the product gases. But it is notable that the basic gasification process (i.e., pyrolysis cracking) was used at most MGPs. This is important because former MGP site residuals have similar characteristics regarding their physical and chemical properties, location in the subsurface, and mode of environmental transport. These similarities allow former MGP sites to be lumped together as a category of hazardous waste site that is usually distinct from other types of hazardous waste sites. Knowing that former MGP sites have similar waste characteristics allows practitioners to approach the challenge of groundwater remediation at these sites with a methodical and efficient approach.

10.2.3 MGP Wastes

MGP by-products typically consisted of solid-, liquid-, and gaseous-phase materials, and their composition varied with the feedstock, processes, and oils used in gas production. Economic conditions dictated whether MGP by-products would be recycled, sold, or disposed of as wastes. Wastes identified at former MGP sites during environmental investigations can be classified as follows:

- Petroleum oils
- Tars
- Coke
- Spent oxides
- Ash and clinkers
- Emulsions
- Lampblack

Petroleum oils were liquid by-products of CWG and oil gas production processes and generally consisted of aromatic and volatile organic compounds (VOCs), poly-aromatic hydrocarbons (PAHs), heavier-molecular-weight hydrocarbons, and inorganic compounds such as metals, sulfur compounds, and nitrogen compounds. Petroleum oils were condensed with water from tar during tar separation processes.

They typically were lighter than water and were recovered from condensate mixtures by mechanical skimming. Once recovered, petroleum oil by-products could be mixed with other light oils, mixed with carburetor feedstock, or disposed of as liquid wastes.

Tars were liquid by-products produced during coal carbonization, carburetion, and oil carbonization processes. They primarily consisted of mixtures of hydrocarbon compounds formed during pyrolysis cracking processes, with minor fractions of ammonia, cyanide, phenolic compounds, and hydrogen sulfide. Whenever feasible, residual tars were recycled as a fuel or carbon source, but they could also be sold on the market. Residual tars were also disposed of as liquid wastes as dictated by economics. Most MGP residual tars were referred to as either coal-tar or oil-tar, depending on their origin and processing. Tars are one of the most prevalent (and recalcitrant) wastes found in groundwater at former MPG sites, and their presence below the water table has a profound impact on their remediability. Because of this, MGP residual tars are more thoroughly characterized in Secs. 10.3 and 10.4.

Coke was a solid by-product during coal carbonization, formed as a coal residual after most of the volatile material had burned off. Coke was reused as a fuel material, sold on the market, or disposed of as a solid waste.

Spent oxides were solid by-products generated during product gas refinement. Iron oxide was the primary material used to "scrub" product gases, and was sometimes used in conjunction with arsenic and wood chips. These materials were used to remove impurities such as ammonia, cyanide, and hydrogen sulfide from product gases prior to distribution. Oxide materials were continually reused until their capacities were spent. Spent oxides had little value as process materials, and therefore were usually considered wastes and disposed of accordingly.

Ash and clinkers were solid by-products composed of inorganic and other uncombustible by-products of MGP processes. No references were found indicating that ash and clinkers were recycled, and therefore it can be assumed that they were disposed of as solid wastes.

Emulsions were liquid by-products of the CWG and oil gas processes, but not during coal carbonization processes. They consisted of various mixtures of condensate water, tar, and petroleum oils that could not be readily separated. Consequently, they had little resale or recycling value and were disposed of as liquid wastes.

Lampblack was the term applied to petroleum coke, as opposed to coal coke. Therefore, lampblack is mainly associated with oil gas production processes. Carburetion methods did not generally produce lampblack although they used petroleum oils. Lampblack could be compressed into briquettes and either recycled as a fuel or carbon source, sold on the market, or disposed of as a liquid or solid waste.

10.3 PHYSICOCHEMICAL PROPERTIES OF MGP TARS

Residual tars are of particular interest at former MGP sites because they present the greatest hurdle to groundwater restoration when present at or below the water table. Most MGP residual tars are DNAPLs because they are denser then water by definition. The term *DNAPL* is used to distinguish between liquid-phase tars observed in environmental samples and dissolved-phase chemicals that dissolve or leach out of the tars into groundwater. This distinction is important from a fate, transport, and remediation perspective because the physical processes governing DNAPL fate and transport are different from those governing dissolved chemical fate and transport in groundwater. As discussed in Secs. 10.4 and 10.5, different remedial strategies are needed to manage DNAPL plumes and dissolved-chemical plumes.

This section discusses the physicochemical properties of MGP residual tars that can influence the fate and transport of both the DNAPL tars and their dissolved chemical plumes in groundwater, and examines the effect of these properties on the remediability of groundwater at former MGP sites. For purposes of this discussion, MGP residual tar and DNAPL are used interchangeably. It is important to remember that the specific physical and chemical properties of MGP tars will vary from site to site, depending on the feedstock of the MGP processes, the MGP processes themselves, and the postdisposal history of the tars (i.e., weathering). However, there is enough similarity between MGP residual tars that the general properties of tars can be reasonably characterized for remedial planning and plume management purposes.

10.3.1 Physical Properties

The key physical properties governing the fate, transport, and remediability of MGP residual tars in groundwater include:

- Density
- Viscosity
- Interfacial tension with water
- Wettability

Density is a fluid property defined as the mass of DNAPL per unit volume at a specified temperature. The density of MGP residual tars is reported to range from 1.01 to 1.42 milligrams per cubic centimeter (mg/cm^3) (Harkins et al., 1987; Cohen and Mercer, 1993; Electric Power Research Institute, 1996; and Pankow and Cherry, 1996). DNAPL density is temperature dependent, and since many MGP residual tars have densities close to 1 mg/cm^3, it is possible that some tars could be lighter than water (referred to as *LNAPLs*) during warmer seasons and denser than water (referred to as *DNAPLs*) during colder seasons. Furthermore, the density of MGP residual tars can change over time as the more soluble chemicals in the tar preferentially dissolve into groundwater, leaving a DNAPL mixture enriched in less soluble chemicals. The change in MGP residual tar density due to preferential dissolution by groundwater may render it either more or less dense with time, depending on the density of the individual chemicals. Knowing the density of MGP residual tars at a site is important because it can be used to evaluate the potential for a given tar to sink below the water table. Tar density should be measured at field temperatures during site investigation activities because of the variable composition of tars from site to site. Tar density can also be used to evaluate capillary pressure gradients that affect DNAPL tar migration. Capillary pressure gradients influencing DNAPL migration are particularly important in evaluating remedial technologies.

Viscosity is a fluid property defined as the ratio of stress to strain within a DNAPL, and can be considered the sheer resistance of a DNAPL to flow. Viscosities for MGP residual tars are reported to range from 10 to 650 centistokes (cSt) (Harkins et al., 1987; Cohen and Mercer, 1993; Electric Power Research Institute, 1996; and Pankow and Cherry, 1996). Like density, the viscosity of MGP residual tar is temperature dependent and it can also change with time because of preferential dissolution of the more soluble tar chemicals into groundwater. The change in tar viscosity due to preferential dissolution by groundwater renders the tar more viscous with time because the less soluble chemicals are also more viscous. In fact, many of the less-soluble chemicals present in MGP residual tars are solids when in their pure phase. Thus, tar viscosity in groundwater will tend to increase. Tar viscosity measurements should be made at field temperatures during former MGP site characterization activities. Understanding the viscosity of MGP residual tars at a site is needed to evaluate tar mobility, migration, and remediability in groundwater.

Interfacial tension with water (IFT) is a fluid property resulting from the attractive forces between two or more fluids in contact with each other, such as tar and groundwater, or tar and air. Understanding the IFT of tars and groundwater at a site is needed to evaluate tar mobility, migration, and remediability in groundwater. IFT for DNAPLs can range from 5 to 40 dynes per centimeter (dyn/cm) (Waterloo Educational Services, 1997). IFT of a coal tar sample collected by the author from a former MGP site in New York state was measured at 25 dyn/cm.

Wettability is a fluid property describing the affinity of a liquid to preferentially coat a solid surface in the presence of another liquid, such as tar coating soil grains or bedrock fracture walls in groundwater. Understanding the wettability of MGP residual tars in soils and bedrock at a site is needed to evaluate tar mobility, migration, and remediability in groundwater. The wettability of a DNAPL tar also depends on the physicochemical properties of the soil or bedrock. MGP residual tars can be wetting or nonwetting in a particular geologic medium, and therefore wettability of tars at former MGP sites should be evaluated during site characterization activities.

10.3.2 Chemical Properties

Composition.   MGP residual tars are mixtures of thousands of chemicals consisting primarily of hydrocarbons with lesser quantities of other organic and inorganic compounds. It is impossible to determine the precise chemical composition of MGP residual tars at the time they were produced because the composition would have depended on the composition of the feedstock and the specific MGP processes employed at the time. Additionally, analytical methods in use during the early part of the twentieth century were not precise by today's standards. Several references were found that provide chemical compositions of MGP residual tars based on recent analyses (Tables 10.1 and 10.2). The chemical compositions of the tar samples listed

TABLE 10.1   Composition of MGP Residual Tars (Percent)

| | Tar source | | | | | Range |
Tar composition	1	2	3	4	5	(Sources 1 to 4)
Volatiles/aromatics	1.3	0.72	0.55	5	25*	0.6–5
Acid extractables	2.0	1.15	1.94	2.5	2.5	1.2–2.5
Base/neutrals	24.6	21.5	16.0	30.6	8.5	16–31
N, S, O—heterocyclics	1.1	1.3	0.6			0.6–1.3
Pitch	62	59.8	63.5	62	60	60–62
Total	91	84	83	100	96	83–100

* May include base/neutrals.
Sources:
1. Coal tar reported in Cohen and Mercer (1993).
2. British coal tar reported in Cohen and Mercer (1993).
3. U.S. coal tar reported in Cohen and Mercer (1993).
4. Coal tar reported in Hayes et al. (1996).
5. Coal tar at carbonization temperature of 1000°C. Harkins et al. (1987).

TABLE 10.2 Properties of Coal Tar Chemicals Used in Evaluating Fate, Transport, and Remediability in Groundwater

Chemical	Molecular weight,* g/mol	Log K_{oc},* mL/g	Effective solubility,[†] mg/L	Retardation factor in groundwater[‡]
		VOCs		
Benzene	78	1.81	10–50	1–4
Toluenc	92	1.98	5–20	1–5
Ethylbenzene	106	2.40	0.1–1	2–12
Xylenes (ave.)	106	2.29	4–16	2–10
Styrene	104	2.96	0.1–3	5–41
		PAHs		
Napthalene	128	2.94	25–110	5–40
2-Methylnapthalene	142	3.93	1–10	38–380
Acenphthene	154	3.59	0.1–2	18–180
Acenapthylene	152	3.75	1–4	25–250
Benz(a)anthracene	228	5.30	10^{-4}–10^{-3}	880–8,800
Benzo(a)pyrene	252	5.95	10^{-4}–10^{-3}	3,900–39,000
Chrysene	228	5.12	0.002–0.02	580–5,800
Pyrene	202	4.80	0.1–1.5	270–2,700
Fluorene	166	3.45	0.1–1.5	13–130
Phenanthrene	178	4.27	0.2–2	84–840
1,6-Dimethylnapthalene	156	3.88	0.2–2	35–250
Fluoranthene	202	4.62	0.01–0.1	180–1,800
Benzo(j)fluoranthene	252	4.34	10^{-5}–10^{-4}	98–980
Indeno(1,2,3-cd)pyrene	276	6.20	10^{-6}–10^{-5}	7,000–70,000
Benzo(g,h,i)perylene	276	5.61	10^{-4}–10^{-3}	1,700–17,000
Pyridine	79	0.70	1,000–5,000	1–1.2
Carbazole	167	2.80	1–15	3–35
Dibenz(a,h)anthracene	278	6.31	10^{-4}–10^{-3}	9,000–90,000
Anthracene	178	4.20	0.03–0.3	70–700
		Acid extractables		
Cresol (avg)	108	1.31	300–3,000	1–2
Xylenol (avg)	122	1.26	30–300	1–2
Phenol	94	1.48	1,300–13,000	1–3

* K_{oc} is the organic carbon partitioning coefficient. Data from Syracuse Research Corp. (2000).
    [†] Estimated with a modified form of Raoult's law using the average chemical composition of tar samples provided in Cohen and Mercer (1993), Priddle and MacQuarrie (1994), Hayes et al. (1996), Electric Power Research Institute (1996), and a sample collected by the author from a site in New York. Range based on average molecular weight of unknown tar chemicals, assumed to be 400 to 1600 g/mol. Subcooled liquid solubility used where applicable.
    [‡] Estimated with methods provided in Freeze and Cherry (1979) assuming soil bulk density of 2.65 g/cm^3 and porosity of 0.3. Range based on amount of organic carbon in soil, assumed to be 500 to 5000 mg/kg.

in Tables 10.1 and 10.2 are useful for remedial planning and plume management purposes because they provide an "as is" snapshot of recent tar composition. But it must be remembered that it has been many decades since tars at former MGP sites were disposed.

A distinguishing characteristic of MGP residual tars is the high percentage of pitch, which apparently made up more than half of MGP residual tars by weight (Table 10.1). Pitch is a generic term that describes a mixture of heavier-molecular-weight chemical compounds that are difficult to identify and quantify because of their extremely low solubilities in water and other standard analytical solvents. The heavier-molecular-weight chemical compounds in pitch typically exist as solids at standard conditions, and this is why pitch was sometimes used in making road and roofing tars, and used for waterproofing wooden structures. Table 10.2 lists individual chemical compounds found in MGP residual tars that are currently of interest in environmental investigations and remedial efforts. Some of these chemical compounds can be toxic above certain concentrations, and others are known or suspected carcinogens.

Effective Solubility and Retardation Factor.   Table 10.2 provides chemical properties for tar chemicals that can be useful in evaluating the fate, transport, and remediability of these compounds in groundwater. VOCs, PAHs (including some that have nitrogen substituted in their ring structures, e.g., pyridine and carbazole), and acid-extractable compounds including cresol, xylenol, and phenol are included. The molecular weights and organic carbon partition coefficient (K_{oc}) values were obtained from a good physical chemistry database (Syracuse Research Corp., 2000). Effective solubility and retardation factor values were estimated by standard methods. From this information, it is obvious that the heavier-molecular-weight chemicals found in MGP residual tars (e.g., PAHs) are generally insoluble at concentrations of environmental interest, and they are also the least mobile in groundwater.

Effective Solubility.   Effective solubility of a tar chemical is defined as the dissolved concentration of the chemical when the tar is in equilibrium with groundwater. It is estimated by using a modified form of Raoult's law (Pankow and Cherry, 1996):

$$C_{eff} = X_m C \tag{10.1}$$

where C_{eff} is the effective solubility of the tar chemical, X_m is the mole fraction of the chemical in the tar, and C is the pure-phase solubility of the chemical. Several authors have indicated that Eq. (10.1) provides a reasonable estimate for predicting the effective solubility of organic compounds (Cohen and Mercer, 1991; Priddle and MacQuarrie, 1994; Pankow and Cherry, 1996). Most of the PAHs in MGP residual tars exist as solids in their pure phase at standard conditions, but exist as liquids when they are components of tar mixtures. In these cases, the subcooled liquid solubility (C_{sub}) should be used in place of C. Pankow and Cherry (1996) provide a method for estimating C_{sub} of an organic chemical based on its melting point. C_{sub} can be significantly higher than C, and therefore it is not surprising to encounter groundwater samples with PAH concentrations at or above their pure-phase solubilities. Using Eq. (10.1) to estimate C_{eff} of tar chemicals requires the average molecular weight of the unknown chemicals in the mixture.

The effective solubility values presented in Table 10.2 are for coal tar and were estimated by using Eq. (10.1). The mole fractions of the chemicals were based on the average mass fractions of seven coal tar samples reported in the literature and collected at former MGP sites. Subcooled liquid solubility values were estimated where appropriate. The average molecular weight of the unknowns was taken to range from 400 to 1600 grams/mole (g/mol). This represents a wide range in the average molecular weight of the unknowns; however, as noted above, the mass fraction of unknowns (pitch) often exceeds 50 percent.

Equation (10.1) is a reasonable screening tool for predicting effective solubility of tar chemicals and evaluating the fate, transport, and remediability of MGP residual tar chemicals in groundwater. As discussed in Secs 10.4 and 10.5, effective solubility values can be used at former MGP sites to:

- Define the limits of DNAPL zones below the water table
- Estimate excavation costs during feasibility studies
- Estimate water treatment costs during feasibility studies
- Estimate the length of time required for DNAPL dissolution in groundwater
- Estimate initial concentrations needed in fate-and-transport analyses and risk evaluations

It is notable that tar chemical composition changes with time in groundwater because of preferential dissolution of the more soluble chemicals from the tar. This is important because, in groundwater, tars become more viscous and less mobile with time, and they also become enriched in PAHs and other heavier-molecular-weight chemicals, and depleted in VOCs. Thus, another potential use for effective solubility calculations is as chemical boundary conditions, or starting points, for predicting the compositional changes (dissolution history) of tar mixtures in groundwater.

Retardation Factor.   Migration of dissolved tar organic chemicals in groundwater can be retarded by several orders of magnitude relative to the average linear groundwater velocity mainly as a result of hydrophobic sorption. Hydrophobic sorption in groundwater is a process where chemicals become sorbed to solid organic carbon within the soil. There is an inverse correlation between molecular weight and sorption: Higher-molecular-weight chemicals are more sorptive than lower-molecular-weight chemicals. Retardation of organic chemicals in groundwater can be described by a retardation factor, which is a number greater than 1 that indicates the ratio of the average linear groundwater velocity to a chemical plume's migration velocity.

Retardation factors are typically evaluated by using the organic carbon partition coefficient (K_{oc}), which is a physicochemical property of chemicals catalogued in standard reference books and databases (Syracuse Research Corp., 2000). K_{oc} can be combined with the fraction of organic carbon in soil to estimate the distribution coefficient (K_d) for a chemical, which indicates the mass of chemical per volume of groundwater that can be sorbed, or stored, in the solid organic carbon portion of a groundwater zone. This is important because hydrophobic sorption can retard chemical migration in groundwater by many orders of magnitude relative to the average linear groundwater velocity.

The retardation factors presented in Table 10.2 were estimated by using methods provided in Freeze and Cherry (1979), assuming a soil bulk density of 2.65 g/cm^3 and a porosity of 0.3. The amount of organic carbon in the soil was taken to range from 500 to 5000 milligrams per kilogram (mg/kg). This represents a reasonable range for most saturated soils. It is evident from the information in Table 10.2 that many PAHs are essentially immobile in most soils because of their high K_{oc} values and corresponding retardation factors. Given a typical average linear groundwater velocity of say, 500 feet per year in a normal saturated soil, most PAHs would not migrate more than a few feet in a year. Furthermore, this information shows that most PAHs are essentially immobile even in porous media with low organic contents.

10.4   FATE AND TRANSPORT OF MGP RESIDUAL TARS IN GROUNDWATER

10.4.1   Tar Migration

Tar by definition is a DNAPL, and therefore both terms are used in this discussion. Tar mobility in groundwater depends on its saturation. Saturation is defined as the ratio of DNAPL volume to pore (or fracture) volume, and ranges from 0 to 1, with 0 being no DNAPL present and 1 being the situation where the pores (or fractures) are completely filled. Above a certain residual saturation the DNAPL is mobile and can migrate as a result of a combination of gravity, capillary, and hydrodynamic forces. In this context, the term *residual* is used to indicate the particular DNAPL saturation at which the DNAPL is no longer able to migrate under ambient conditions. When DNAPL is at or below residual saturation it is not mobile. Thus, it is useful to think of subsurface DNAPL in terms of its mobile and immobile components. Residual-phase DNAPL exists as discrete, separate ganglia in soil pores that are not connected to other ganglia. Pankow and Cherry (1996) provide a thorough quantitative analysis of the physical processes governing DNAPL migration in porous media.

DNAPL migration is also governed by the quantity, timing, and nature of the source, and by the physicochemical properties of the DNAPL and the medium in which it resides. Following a release, a quantity of mobile tar can invade the pore spaces between soil grains and migrate downward through unsaturated soils to the water table, and if present in sufficient quantity it can migrate below the water table through saturated soils and even enter bedrock fractures. Tar migration patterns are controlled by the geologic structure, and they are generally limited to the higher-permeability pathways within the media. Tar saturation decreases with distance from the source until it is at residual saturation and no longer mobile. For a given quantity of source material, there will be an equilibrium distribution of tar in the subsurface whereby it occupies a finite volume and is no longer mobile. The time required for a given quantity of tar to reach its maximum extent in groundwater is governed primarily by its viscosity. Because of its relatively high viscosity, tar migration can be expected to be approximately 10 to 100 times slower than water migration.

Previously immobile tar in groundwater can be remobilized in some instances by increased hydraulic gradients across the tar ganglia. Immobile tar can also be remobilized by reducing its interfacial tension (IFT), such as by flushing with surfactants. Lowering tar viscosity by adding heat will likely not remobilize previously immobile tar. Additionally, tars have been found to be wetting in some geologic settings. In these cases the tar preferentially wets the soil grains or bedrock walls.

10.4.2   Extent of DNAPL Zone

This section outlines a method for delineating the extent of DNAPL zones in groundwater at former MGP sites. It is useful to know the horizontal and vertical extent of DNAPL below the water table for remedial planning purposes when groundwater remediation strategies are being evaluated. This is because different remedial strategies are needed to manage DNAPLs as opposed to dissolved chemical plumes. For example, sheet-pile cutoff walls can be used at sites where DNAPL may be mobile in groundwater, but may not be useful for containment of dissolved chemical plumes. It is also useful to know the extent of the DNAPL zone for cost-estimating purposes during feasibility studies, for example, estimating the magnitude of proposed soil excavations and evaluating groundwater treatment costs. Additionally, delineating the extent of DNAPL zones is typically required when waivers are requested for groundwater MCLs, as discussed in Sec. 10.5.2 under "Soil Excavation and Capping."

DNAPL zones at former MGP sites can be established on the basis of subsurface areas where DNAPL is confirmed or very likely to be present. DNAPL zone delineation is best accomplished by using the following criteria:

- Direct visual observations of DNAPL
- DNAPL chemicals in groundwater samples at concentrations greater than 10 percent of their effective solubilities
- DNAPL chemicals in soil samples at concentrations that, when partitioned into the water phase, are greater than their effective solubilities
- Known DNAPL entry points
- Anomalous dissolved chemical plume configurations

With these criteria, a DNAPL zone boundary can be identified by plotting and contouring relevant observations on site maps and geologic cross sections. This process can also help identify important data gaps.

As a safety factor, the DNAPL zone should be expanded to include subsurface areas where DNAPL may be present but conclusive evidence is lacking. This can be based on DNAPL chemicals in groundwater samples at concentrations greater than 1 percent of their effective solubilities; DNAPL chemicals in soil samples at concentrations that, when partitioned into the water phase, are greater than 10 percent of their effective solubilities; and anomalous plume configurations. The 1 percent solubility guideline is not definitive evidence of the presence of DNAPLs because of physicochemical heterogeneities. But it is useful for remedial planning purposes because it serves as an objective, quantitative tool for focusing active remedial activities on those areas posing the greatest risk.

10.4.3   Dissolved Plume Migration

DNAPLs present in unsaturated soils and below the water table will eventually dissolve in groundwater until they are gone. The rate of DNAPL dissolution is governed by DNAPL physical and chemical properties and hydrogeologic characteristics. DNAPLs can serve as continuing sources of dissolved chemical plumes for decades and centuries (Pankow and Cherry, 1996).

When DNAPLs or DNAPL leachate infiltrates downward through the unsaturated zone to the water table, a plume of dissolved chemicals can develop in the groundwater and migrate in a direction consistent with the groundwater flow direction. With time, dissolved chemical plumes can increase in size both parallel and perpendicular to the groundwater flow direction. Plume spreading of this nature is governed by a combination of factors including the physical properties of the waste materials, hydrogeological characteristics, and the nature and shape of the source area.

The size and spreading of a plume of dissolved chemicals in groundwater can be influenced by a variety of naturally occurring physical phenomena including hydrologic features (e.g., streams), geologic features (e.g., clay layers), transport properties (e.g., dispersion), and chemical properties (e.g., effective solubility). Plumes typically are limited in size because of combinations of these physical properties. When a plume has reached its maximum size, whereby its areal extent is no longer increasing with time, the plume can be said to be in a state of dynamic equilibrium. *Dynamic equilibrium* of chemical plumes in groundwater is an important concept because it implies two conditions:

- The areal extent of the plume does not change with time. In this condition the total volume of saturated soils containing chemicals of interest is no longer increasing with time.
- The chemical mass within the plume does not change with time. This situation occurs when the rate of chemical mass loading into the plume at the source is equal to the rate of chemical mass lost from the plume (usually at the plume fringe) because of chemical reactions (e.g., degradation, precipitation), volatilization into soil-air in unsaturated zone soils, dispersion, and groundwater discharge to surface water.

Understanding the stability of chemical plumes in groundwater at former MGP sites is important from a remedial planning perspective because, in extreme cases, increasing plumes can migrate to sensitive receptors. In these cases, measures such as institutional controls and hydraulic containment might be implemented to eliminate potential exposure pathways. However, given the many decades since MGPs were last operational, it is likely that most dissolved chemical plumes associated with MGP residues are currently stable and in a state of dynamic equilibrium. The exception to this would be former MGP sites at which recent activities may have mobilized previously immobile DNAPL, or otherwise altered factors influencing plume stability. Plume stability is further discussed in Sec. 10.5.2 under "Hydraulic Containment by Pump-and-Treat."

Groundwater plumes at former MGP sites typically consist of mixtures of soluble chemicals derived from residual tars and other waste types (Sec. 10.2.3). It is prudent to delineate and monitor individual plumes of chemicals that pose the greatest risk to human health and the environment via groundwater migration. The typical conceptual model of plume migration at former MGP sites involves multiple, overlapping plumes of soluble tar chemicals including VOCs such as benzene and related compounds, lighter-molecular-weight PAHs such as naphthalene and related compounds, and acid extractables such as phenol and cresol. Other chemicals and metals such as cyanide, ammonia, arsenic, and chromium can also be present in groundwater plumes at former MGP sites. For each soluble chemical of interest, plume size, shape, and spreading depend on source and constituent characteristics as well as hydrogeologic controls, and it should be a priority at former MGP sites to delineate and monitor chemical plumes in groundwater.

Plume migration in groundwater at former MGP sites is readily evaluated with analytical calculations and numerical modeling simulations. Fate and transport modeling of plumes can be performed for risk evaluation, remedial planning, cost estimating, and remedial monitoring purposes. Effective solubility estimates of tar chemicals (Table 10.2) should be considered in assigning initial concentration in plume modeling efforts. Plume modeling can be useful in designing appropriate monitoring strategies, identifying critical data gaps, hypothesis testing, and other remedial planning and design activities.

10.4.4 Matrix Diffusion in Bedrock

At some sites, DNAPL tars and dissolved chemicals have migrated vertically downward through saturated soils and invaded fractured bedrock. At some point when the DNAPL tar has reached residual saturation in fractured bedrock, tar migration will be arrested and the tar will essentially become nonmobile. Dissolved chemical plumes can continue in migrate in bedrock groundwater long after the DNAPL tars are below their residual saturation (nonmobile).

Plume migration in fractured bedrock can be retarded below the average linear groundwater velocity by physicochemical processes in both the fractures and the matrix (Lipson et al., 2001). These processes include hydrophobic sorption onto fracture walls and within the rock matrix, degradation in the aqueous and sorbed phases, and matrix diffusion. Diffusion of chemicals from the fractures into the bedrock matrix dominates plume retardation in fractured rock, and can retard the migration of dissolved chemicals associated with former MGP sites by several orders of magnitude relative to the groundwater velocity in the fractures. For example, modeling results suggest that naphthalene migration in fractured bedrock can be retarded by up to 100,000 times slower than the average linear groundwater velocity in some sandstones. Matrix diffusion must therefore be considered in evaluating groundwater migration of chemical plumes in fractured bedrock settings.

Matrix diffusion of groundwater chemicals in fractured bedrock represents a serious limitation to the effectiveness of currently available remedial technologies. No references were found where groundwater contamination at former MGP sites has been remediated to MCLs when the dissolved chemical plumes were located in bedrock groundwater. This is due in part to the process of reverse diffusion, whereby dissolved chemicals in the bedrock matrix diffuse back out into fracture groundwater and produce long-lasting, asymptotic concentrations above MCLs, otherwise known as the "tailing" effect.

10.5   GROUNDWATER REMEDIATION

From the foregoing discussion it is clear that groundwater restoration to MCLs within regulatory time frames at many former MGP sites is unlikely, particularly at those sites where residual tars are present below the water table and deeper than about 25 ft below the ground. This is due in part to differential dissolution of residual tars in groundwater, in which the more mobile tar chemicals are leached out of the tars sooner than the less mobile and immobile tar chemicals. This process can result in the formation of increasingly immobile and resilient tars that behave more as solids than as fluids when present in groundwater for any appreciable amount of time. This is why groundwater restoration at many former MGP sites is impracticable from an engineering perspective. The fact is that there do not exist any technologies that can remove residual tars from below the water table, with the exception of soil excavation technologies, given reasonable budgets.

When DNAPL tars are present at or below the water table and greater than approximately 25 ft below the ground, their complete removal is essentially impossible because of their characteristically high viscosity and the fact that they can preferentially wet soil grains and bedrock fracture walls in the presence of groundwater. In fact, in situ remedial technologies applied at former MGP sites have invariably failed to remove all of the MGP residual tars from below the water table when these materials migrated deeper than about 25 ft below the ground. We did not locate a single literature reference describing a former MGP site where groundwater contaminants associated with tars have been remediated to MCLs.

10.5.1 Remediation Trends at Former MGP Sites on the Superfund List

To gain some perspective on the outlook for groundwater restoration at former MGP sites, U.S. EPA Superfund Records of Decision (RODs) were queried in order to examine remediation trends. Eleven former MGP sites were identified on the Superfund list spanning the period from 1990 through 1999 (Table 10.3). RODs, explanations of significant differences, and ROD amendments focused on groundwater remediation at these sites have been issued at a rate of about one per year, with the exception of 1998, in which two RODs were finalized: the Calhoun Park and Pine Street Canal sites. Total remediation costs for the sites averaged approximately \$10 million and ranged from approximately \$4 million to \$27 million.

Groundwater remedies at the sites generally consisted of combinations of four to seven remedy components (Table 10.3). From this information, it is obvious that there is not one technology that can restore groundwater at former MGP sites. At the Superfund level, groundwater remedies are geared toward risk management rather than plume removal in order to protect human health and the environment. Therefore, Superfund RODs for former MGP sites represent plume management strategies that rely on numerous remedy components.

At the former MGP sites on the Superfund list, soil excavation, groundwater pump-and-treatment systems, and institutional controls were the most prevalent remedy components, being used more than 75 percent of the time. Long-term monitoring, capping, MCL waivers, NAPL pumping, and monitored natural attenuation (MNA) were sometimes used, between roughly 25 and 75 percent of the time. In situ bioremediation, barrier walls, and phytoremediation were seldom used, less than 25 percent of the time (Table 10.3).

The four highest-cost former MGP Superfund sites had soil excavation, pump-and-treat, and capping in common as remedy components. The three lowest-cost sites had excavation, institutional controls, and long-term monitoring in common. Other than this, it is difficult to draw specific conclusions regarding costs versus benefits for each of the groundwater remediation components from ROD information.

10.5.2 Groundwater Remediation Strategies at Superfund Sites

Hayes et al. (1996) provide fairly comprehensive descriptions of in situ site restoration technologies that may be used to aid in managing groundwater contamination at former MGP sites. The following sections provide brief descriptions of the various groundwater remediation components used at former MGP sites that are on the Superfund list, and discuss their role in groundwater remediation. Phytoremediation is discussed in greater detail because it is a relatively new, innovative groundwater remedial strategy that is currently underutilized but may exhibit more widespread use in years to come.

Soil Excavation and Capping. Soil excavation was a groundwater remedy component at all of the former MGP sites on the Superfund list. Soil capping was a component at five sites. While technically not considered groundwater remediation, soil excavation and capping are considered groundwater remedy components because contaminated soils above the water table can serve as a long-term source of groundwater contamination. Soil excavation is therefore used as a *source-removal* strategy, and soil capping is a *source-stabilization* strategy. Source-removal and -stabilization strategies are important because the longevity of groundwater contamination at former MGP sites depends directly on the presence and nature of the source.

TABLE 10.3 Summary of Groundwater Remedy Components at Former MGP Sites (Superfund)

	Fairfield Coal Gasification, Iowa (1990)	Peoples Natural Gas Co., Iowa (1991)	Central Illinois Public Service (1992)	Utah Power and Light (1993)	Dover Gas and Light, Delaware (1994)	Niagara Mohawk Power, New York (1995)	Ohio River Park Site, Pennsylvania (1996)	Brodhead Creek Site, Pennsylvania (1997)	Pine Street Canal, Vermont (1998)	Calhoun Park Area, South Carolina (1998)	Waukegan Coke, Illinois (1999)	Percent of sites using component
Excavation	X	X	X	X	X	X	X	X	X	X	X	100
Institutional controls	X	X	X	X		X	X	X	X		X	82
Pump-and-treat	X	X	X	X	X	X		X		X	X	82
Long-term monitoring		X	X	X		X	X	X	X			64
Capping				X	X	X	X		X		X	55
ARAR waivers	X	X			X			X				3
NAPL pumping					X	X		X		X		36
Natural attenuation				X			X				X	36
In situ bioremediation	X	X				X		X				18
Subsurface barrier wall										X		18
Phytoremediation										X		9
Total cost, $1000	5,815	8,000	9,346	**10,538**	19,200	15,300	3,258	4,120	4,379	7,743	**26,500**	

Note: Information from RODinfo™ (The Olewine Company, 2000). Dates shown are last date of ROD, ESD, or ROD amendment. **Bold** values are above average.

Appropriately designed soil excavations can remove groundwater contamination sources at sites where DNAPL migration is fairly shallow (i.e., less than 25 ft below the ground). Soil capping serves to minimize precipitation infiltration at former MGP sites, thereby reducing the quantity of leachate potentially reaching the water table. At sites where DNAPL has migrated below 25 ft, about the best thing soil excavation and capping can do is reduce the strength of groundwater contamination sources. But these strategies cannot eliminate sources entirely when the DNAPL is about 25 ft or more below the ground.

Excavating soils containing residual tars and other wastes has improved groundwater quality at many sites. For example, groundwater quality was shown to improve markedly 3 years after soil excavation at a former MGP site in New York (Electric Power Research Institute, 1996). However, soil excavations have generally failed to restore groundwater to MCLs or background concentrations at former MGP sites where wastes have migrated below the water table and beyond the reach of conventional soil excavation equipment. This is why delineation of DNAPL zones is so important at former MGP sites. If DNAPL is found or inferred to be present below 25 ft, then the probability of groundwater restoration is extremely low, and alternative remedial strategies must be employed.

As groundwater remedy components, it is reasonable to continue performing soil excavations and soil capping at former MGP sites. This is because removing and/or capping contaminated soils minimizes the risk of humans directly contacting surface wastes, reduces the mass of chemicals in the source area, and reduces the total mass and groundwater concentrations of tars and tar chemicals. In conjunction with other groundwater remedy components such as institutional controls and long-term monitoring, soil excavation and capping can be a reasonable approach for managing groundwater contamination plumes at former MGP sites.

Institutional Controls. Institutional controls were used as groundwater remedy components at nine of the former MGP sites on the Superfund list. Institutional controls are a *plume management strategy* intended to protect human health and the environment by eliminating exposure pathways to contaminated groundwater by preventing direct contact. They typically consist of deed restrictions that prevent groundwater extraction and consumption at contaminated properties. Property deeds can be modified or amended in order to restrict the use of water wells at the property.

Hydraulic Containment by Pump-and-Treat. Groundwater pump-and-treat systems were remedy components at nine of the former MGP sites on the Superfund list, despite the fact that this technology fails to restore groundwater to MCLs. Pump-and-treat technology theoretically can remove groundwater contamination, and therefore can serve as a plume removal strategy. However, this strategy has invariably failed to restore groundwater at former MGP and many other types of contaminated sites (e.g., retail gasoline stations, chlorinated solvent sites) because of physicochemical characteristics of the contaminants as well as hydrogeologic characteristics.

Probably the best use of pump-and-treat systems is as a *hydraulic containment strategy* to hydraulically isolate DNAPL source areas and prevent dissolved chemical plumes from expanding beyond their known limits. Hydraulic containment using pump-and-treat systems can be a necessary plume management strategy in extreme cases when plumes are growing or when there are potentially sensitive receptors nearby. For example, a plume migrating toward a potable supply well could be effectively captured by a strategically placed groundwater pump-and-treat system, and the potable supply well would be suitably protected. However, stable plumes that are in a state of dynamic equilibrium may not pose an imminent threat to human health and the environment, particularly if there are no receptors nearby, and it can be argued that hydraulic containment is not needed at sites with stable plumes. This is why plume delineation and monitoring are critical at former MGP sites.

Pump-and-treat technology can be extremely costly because of the often very expensive operation and maintenance costs, especially if the pump-and-treat systems are to be operated for several decades. There are alternative technologies that can be employed to accomplish hydraulic containment of DNAPL zones and dissolved chemical plumes, such as phytoremediation.

It can be argued that hydraulic containment strategies are not necessary at former MGP sites if it can be demonstrated that the DNAPL zone and dissolved chemical plumes are stable or shrinking as a result of hydrogeologic controls. Following a release to groundwater, DNAPL spreading will be governed by the nature and quantity of the source, the physical and chemical properties of the DNAPL, and hydrogeologic characteristics. Eventually DNAPL spreading below the water table will cease and there will be a stable DNAPL zone in which the DNAPL is essentially immobile. Naturally occurring stratigraphic "traps," such as depressions in clay layers and low-permeability layers, can also aid in immobilizing DNAPL at former MGP sites. Therefore, in cases where the DNAPL is immobile, hydraulic containment is not needed.

Long-Term Monitoring and Monitored Natural Attenuation. Long-term monitoring (LTM) was a groundwater remedy component at seven of the former MGP sites on the Superfund list, and monitored natural attenuation (MNA) was a remedy component at four sites. LTM and MNA of MGP residuals in groundwater represent *plume management strategies* that are intended to minimize risk by tracking plume stability and providing an early warning system should formerly stable plumes begin to migrate downgradient. Both are recognized as viable groundwater remedies or remedy components by the US EPA and many state regulatory agencies.

MNA relies on a combination of naturally occurring physical, chemical, and biological processes present in groundwater that cause chemical concentrations to decrease with time and groundwater plumes to stabilize or shrink. LTM and MNA have gained wide acceptance as the state of the science has improved, as shown by nearly 64 percent of the former MGP Superfund sites that have LTM and MNA built into their RODs. LTM and MNA are expected to play a major role in remediating many (if not most) groundwater plumes as a "polishing" strategy to follow up active remedial technologies such as pump-and-treat.

The generally accepted approach to evaluating MNA of chemicals in groundwater at former MGP sites involves evaluating three lines of evidence:

- Plume stability
- Geochemical conditions
- Mirobiological studies

Plume stability is considered to be the primary line of evidence because stable or shrinking plumes provide an empirical demonstration that natural attenuation is working and will continue to be protective of human health and the environment in the future. Conversely, growing plumes typically must be addressed with a combination of remedial technologies regardless of secondary and tertiary lines of evidence. Geochemical conditions within a plume are considered to be a secondary line of evidence because they can be used to identify some (but not all) of the specific geochemical and microbiological processes that lead to plume stability or shrinkage. Appropriate geochemical conditions, as a singular line of evidence, are not sufficient to demonstrate the efficacy of MNA. Similarly, the microbiology within a residual plume is considered to be a tertiary line of evidence because it can be used to identify some (but not all) of the specific microbiologic processes that lead to plume stability or shrinkage. An appropriate microbiology as a singular line of evidence is insufficient to demonstrate the efficacy of natural attenuation. Therefore, multiple lines of evidence are typically needed to demonstrate that MNA is working and to identify specific geochemical and microbiologic processes that contribute to plume stability or shrinkage.

MNA can also be used to manage dissolved chemical plumes not associated specifically with residual tars, but may be the result of other MGP wastes such as metals (arsenic, lead, chromium), sulfides, ammonia, and cyanide. Each of these chemicals can be immobilized or destroyed in groundwater, prior to off-site migration, by naturally occurring geochemical and microbiological processes. For example, Meehan et al. (1999) found that cyanide in groundwater can be biodegraded in situ under both aerobic and anaerobic conditions.

MCL Waivers. MCL waivers were groundwater remedy components at four of the former MGP sites on the Superfund list. They are sometimes termed *ARAR waivers.* MCL waivers represent *plume management strategies* that are used to waive the MCL regulations for certain chemicals within an explicitly defined boundary. The federal environmental regulatory framework (primarily CERCLA and RCRA) allows the U.S. EPA to grant MCL waivers for groundwater restoration at sites where it can be demonstrated that groundwater restoration is technically impracticable (TI) from an engineering perspective (U.S. EPA, 1993). Many states also have a regulatory framework that allows MCL waivers to be granted at contaminated properties.

MCL waivers are not a do-nothing approach because they typically are granted only for a specific chemical or group of chemicals, and only within a specifically defined volume of site soils, sometimes termed the *TI zone.* Groundwater restoration must still occur outside of the TI zone. U.S. EPA guidance indicates that the TI zone must be hydraulically contained by pump-and-treat or some other containment strategies if MCL waivers are granted (U.S. EPA, 1993).

TI waivers are typically granted in DNAPL zones (defined in Sec. 10.4.2) and other zones where groundwater restoration is demonstrated to be technically impracticable (e.g., matrix diffusion zones). However, regulatory guidance indicates TI zones should be hydraulically contained by an appropriate technology. However, it can be argued that hydraulic containment of TI zones may not be necessary at sites where the DNAPL and dissolved chemical plumes are stable and not migrating. MCL waivers have been demonstrated to be cost-effective in many cases (Saroff et al., 1997).

NAPL pumping. NAPL pumping was a groundwater remedy component at four of the former MGP sites on the Superfund list. NAPL pumping is a *source removal strategy* that relies on physically removing NAPL from wells by conventional pumping technologies. In some cases, residual tars are present below the water table in sufficient quantities that some portion of the NAPL can be removed by conventional pumping technologies. In fact, most regulatory policies require tar removal from site wells to the extent practicable, even though these remedial efforts are insufficient to restore groundwater to MCLs. Even under ideal conditions, NAPL pumping can remove only approximately one-half to two-thirds of the mobile NAPL (Pankow and Cherry, 1996). This is insufficient to restore groundwater to

MCLs at former MGP sites. Pump-and-treat technologies have invariably failed to remove sufficient quantities of NAPL. Even when all of the mobile tar has been removed from a well, there is usually enough immobile tar left at residual saturation in groundwater to cause MCL exceedences for a significant length of time. It is the immobile tar present below the water table at or below residual saturation that is nearly impossible to remove.

In Situ Bioremediation.   In situ bioremediation was a groundwater remedy component at two of the former MGP sites on the Superfund list. It is a *plume management strategy* that attempts to contain dissolved chemical plumes and minimize their spreading in groundwater by creating in situ biological treatment zones that destroy or immobilize dissolved chemicals in groundwater. This is accomplished by delivering nutrients into the treatment zone below the water table and maintaining optimal environmental conditions such as groundwater pH and temperature. Since biodegradation generally takes place in the aqueous phase, in situ bioremediation is not effective for addressing MGP residual tars enriched with insoluble organic compounds. In situ bioremediation will therefore not likely restore groundwater to MCLs within short time frames at sites where MGP residual tars are present below the water table.

Subsurface Barrier Walls.   Subsurface barrier walls were groundwater remedy components at two of the former MGP sites on the Superfund list. Subsurface barrier walls are a *plume containment strategy* generally used to isolate and hydraulically contain DNAPL zones as well as dissolved chemical plumes. Subsurface barrier walls are commonly constructed at sites adjacent to surface water features in an attempt to prevent DNAPL migration, and are typically used in conjunction with pump-and-treat systems in order to maintain an inward hydraulic gradient behind them. Subsurface barrier walls can be constructed as sheet-pile walls or slurry walls and typically are limited to depths of approximately 60 ft below the ground. It is notable that the performance of subsurface barrier walls is often based on aesthetics rather than groundwater MCLs, and the presence of a sheen on the wrong side of the wall can render the wall a failure even if groundwater MCLs are being achieved.

Phytoremediation.   Phytoremediation was a groundwater remedy component at only one of the former MGP sites on the Superfund list. It is a promising new groundwater remedial strategy that is currently being evaluated at many hazardous waste sites throughout the United States. The main advantage of implementing phytoremediation strategies at former MGP sites would be to reduce or eliminate the need for pump-and-treat systems. In many cases trees are capable of removing a sufficient amount of groundwater necessary to hydraulically control dissolved chemical plumes. In addition, phytoremediation has the following advantages:

• Wastes are reduced

• Costs are reduced

• Roots can penetrate and contact a larger volume of soil than a typical pumping well

• Root growth enhances biodegradation in the root zone

• Root growth also enhances sorption and retardation of organic solute plumes

• Leaching of source materials from the unsaturated zone can be reduced

• Trees generally improve aesthetics

Phytoremediation systems currently being designed and installed consist of dense stands of water-loving trees cultivated with practices that promote deep root growth. The goal is to enable trees to use groundwater as their primary moisture source.

Hydraulic control of dissolved chemical plumes in groundwater can be accomplished by planting trees in dense multiple rows at the leading edge of plumes, with the rows set perpendicular to the direction of groundwater flow. Many tree species can transpire water and thereby "pump" groundwater at significant rates, depending on hydrophysical characteristics of the site. Groundwater removal by trees can create capture zones below the water table and prevent, minimize, or reverse plume expansion. Dissolved MGP residuals that enter tree-induced capture zones are drawn into the rhizosphere of the trees, where their fate may include biodegradation, plant uptake followed by metabolic transformation, or volatilization or immobilization in the root zone.

Transpirational water use by trees has two important effects on the local water budget:

- Recharge reduction
- Groundwater removal

Recharge reduction is the loss of groundwater recharge due to evapotranspiration. In many temperate regions, a net positive recharge condition exists. As long as the net recharge rate is positive, the net movement of water across the water table is downward. Maximizing tree transpiration so that the rate of evapotranspiration is greater than the rate of recharge can be achieved by planting dense stands of phreatophytes, thereby resulting in a net negative recharge and concomitant removal of groundwater from below the water table. If the rate of evapotranspiration is high enough, the net water flux across the water table can be upward, out of saturated soils. The U.S. EPA (2000a) indicated that a stand of poplar trees at one site reportedly transpired 6 feet of water per year. In many regions this transpiration rate would produce a net negative recharge and result in groundwater extraction. Using phytoremediation strategies for hydraulic control may be more feasible in arid regions where the pre-existing recharge rate is limited and abundant sunlight, high temperatures, and low humidity enhance water use by trees.

Groundwater removal occurs when high evapotranspiration rates create a net negative recharge, thereby causing removal of water from the capillary fringe and upward wicking from the saturated zone. Deep-rooted phreatophytes can survive with their roots below the water table and also may be able to extract some water directly from the saturated zone (Freeze and Cherry, 1979). In some cases, deep-rooted phreatophytes can remove significant quantities of groundwater. Water removal rates by mature deep-rooted phreatophytes in North America during the growing season range from approximately 5 to 400 gallons per day (gal/day) *per tree* (Ferro et al., 2001). A 1-acre tree stand with trees planted every 6 ft would contain approximately 1200 trees. Thus, even at the low-end estimate of 5 gal/day per tree, such a tree stand would remove water at an average rate of approximately 5 gal/min during the growing season.

Trees are likely to produce a demonstrable area of hydraulic containment, also known as a *groundwater capture zone*. Water-table drawdown values of approximately 3 to 6 ft reportedly have been achieved by deep-rooted phreatophytes (Ferro et al., 2001). At a site in New York State a stand of poplars and willows produces feet of water-table drawdown in the glacial till every summer, completely dewatering the overburden and creating a localized "cone of depression" in an upgradient area of the site. Phreatophytic consumption likely influences groundwater flow at many sites and causes apparent anomalies in the flow field. While the literature contains few or no well-documented field demonstrations showing mapped water-table depressions caused by trees, it is reasonable to conclude that a dense plantation of deep-rooted trees would cause a localized water-table depression.

Transpirational water use by trees can be analogous to water removal by pumping wells for purposes of estimating costs, remedial planning, or remedial design. Thus trees may be visualized, on an individual or group basis, as low-flow pumping wells that intersect the top of the saturated zone. Clearly the geometry of root systems differs from the geometry of a pumping well. An individual tree has a very large number of individual roots with a variety of lengths and diameters, all constituting a relatively flat dendritic system that extracts water from the unsaturated zone and water table. Therefore, care must be taken during design activities to limit the potential for hydraulic gaps to occur within the root system. A root system must extract all of the groundwater within a plume within the footprint of the plantation to hydraulically control the plume.

The area of water-table depression resulting from transpirational groundwater removal is analogous to the cone of depression produced around a drilled, partially penetrating pumping well, but there are some significant differences. The area of depression produced by trees is much more diffuse than the area of depression produced by a pumping well. Water removal by trees occurs over a comparatively larger area, producing a broad, flat depression at the water table. In contrast, the cone of depression produced by a pumping well is narrower and deeper because a drilled pumping well has a relatively small diameter compared with a tree plantation. Also, a well can produce a deeper cone of depression because the water level in the well can be lowered significantly below the water table. In either case, a practical radius of influence exists beyond which the depression caused by the trees or the well is negligible, and groundwater flow is essentially unaffected by groundwater extraction.

Designing hydraulic control systems with pumping wells is fairly straightforward because the flow-system geometry is relatively simple and the mathematics of well hydraulics and classical capture zone analysis have been used for decades. In contrast, there are few published examples of hydraulic containment design using trees (e.g., Schneider et al., 2000; Al-Yousfi et al., 2000; Halford, 1998). We are not aware of any field applications where hydraulic containment by trees has been clearly demonstrated.

Conventional groundwater containment systems rely on one or more pumping wells to extract groundwater and intercept plumes. Some plumes currently being contained by pumping wells could also be contained to some extent with trees. Trees could completely replace pumping wells in some situations, or reduce the extraction rate of existing pumping wells in other situations. In either case, using trees to help manage groundwater contaminant plumes can reduce the overall cost of remediation.

Groundwater pump-and-treat systems typically require a significant upfront capital expense and continued operation and maintenance (O&M) costs, which include monitoring. Because they commonly are installed at sites with long-lasting sources containing nonaqueous-phase liquids, pump-and-treat systems are commonly designed to operate for many years. Cost analyses are typically performed on a 30-year net present value basis.

Assuming either phytoremediation or pump-and-treat could control the same plume, the economics of hydraulic containment may be simplified to assessing the cost per unit volume of groundwater extracted. At a former solvent recycling facility in New England, a sophisticated groundwater extraction and treatment system was installed at a capital cost of approximately US$5 million. The annual O&M cost is approximately US$0.5 million. Assuming a 5 percent discount rate for 30 years (present worth factor of 15.37), the net present value of O&M is approximately US$7.7 million, for a total present value cost of US$12.7 million. This system extracts approximately 78 L/min continuously. Thus, the pump-and-treat unit cost for 30 years would be approximately US$0.0104/L. In contrast, an equally effective phytoremediation remedy covering 2.5 acres may cost US$250,000 in capital, with a conservative (high) assumed O&M cost of US$30,000/year. The total net present value for 30 years is approximately US$710,000. This plantation, after a 5-year maturation period, could extract approximately 100 L/min during the growing season (assumed to be 6 months), which theoretically would contain the plume during the growing season. The plantation would also be thick enough in the direction parallel to groundwater flow to account for 6 six-month dormant season. The phytoremediation unit cost for 30 years could be approximately US$0.0012/L, which is roughly an order of magnitude less than the unit cost for pump-and-treat. Thus, the unit volume cost for phytoremediation pumping and treatment is less than the typical cost to discharge chemically impacted groundwater to public sewers for treatment by municipal wastewater systems, which typically ranges from approximately US$0.003/L to US$0.015/L.

Nyer and Gatliff (1996) presented a similar cost comparison for pump-and-treat versus phytoremediation with a 4000-m^2 site and 6-m-deep aquifer. On the basis of the estimated capital cost and an assumed 5-year operating period, the phytoremediation option was projected to cost 62 percent less than pump-and-treat, or a cost savings of US$410,000. The 30-year net present value savings for the phytoremediation remedy would be approximately US$730,000.

The economics of phytoremediation are obviously site specific. In some cases phytoremediation will not prove economical—for example, if pump-and-treat is required to supplement water removal by trees to maintain hydraulic control even during the growing season. For the most part, phytoremediation will be economically beneficial where trees can hydraulically control the plume without assistance during the growing season, and where sufficient room exists to extend the plantation parallel to the groundwater flow direction to account for the dormant season. Phytoremediation will also be economical when applied inside a completely encompassing barrier wall containment system, with or without a cap, in which case a relatively low extraction rate may be sufficient to maintain an inward hydraulic gradient. Phytoremediation in these cases would have the added benefit of creating an upward hydraulic gradient, which promotes hydraulic control along the bottom of the containment area and also assists in containing DNAPLs.

REFERENCES

Al-Yousfi, A.B., R.J. Chapin, T. A. King, and S.I. Shah. 2000. "Phytoremediation—the Natural Pump-and-Treat Hydraulic Barrier System." *Practical Periodical of Hazardous, Toxic, and Radioactive Waste Management,* vol. 4, no. 2, pp. 73–77.

Cohen, R. M., and J. W. Mercer. 1993. *DNAPL Site Evaluation.* Boca Raton, Fla.: CRC Press, Inc.

Electric Power Research Institute. 1996. "Characterization and Monitoring before and after Source Removal at a Former Manufactured Gas Plant (MGP) Disposal Site." EPRI TR-105921.

Electric Power Research Institute. 1998. "Chemical and Physical Characteristics of Tar Samples from Selected Manufactured Gas Plant (MGP) Sites." EPRI TR-102184.

Ferro, A., M. Gefell, D. S. Lipson, R. Kjelgren, N. Zollinger, and S. Jackson. 2001. "Maintaining Hydraulic Control Using Deep Rooted Tree Systems," in *Advances in Biochemical Engineering/Biotechnology*, special volume: *Phytoremediation*, G. T. Tsao and D. Tsao (eds.). Heidelberg: Springer-Verlag. In press.

Freeze, R. A., and J. A. Cherry. 1979. *Groundwater*. Englewood Cliffs, N.J.: Prentice Hall.

Gas Research Institute (GRI). 1987. "Management of Manufactured Gas Plant Sites." GRI-87/0260.

Halford, K. J. 1998. *Assessment of the Potential Affects of Phytoremediation on Ground-Water Flow Around Area C at the Orlando Naval Training Center, Florida*. U.S. Geological Survey Water Resources Investigation Report 98:4110. Tallahassee, Fla.

Harkins, S. M., R. S. Truesdale, R. Hill, P. Hoffman, and S. Winters. 1987. "U. S. Production of Manufactured Gases: Assessment of Past Disposal Practices." Research Triangle Institute, North Carolina. EPA contract no. 68-01-6826 D.O. 35. October.

Hayes, T. D., D. G. Linz, D. V. Nakles, and A. P. Leuschner (eds.). 1996. *Management of Manufactured Gas Plant Sites*, vol. 1. Amherst, Mass.: Amherst Scientific Publishers.

Lipson, D. S., B. H. Kueper, M. J. Gefell, and B. R. Thompson. 2001. "Effect of Fracture and Matrix Characteristics on Plume Retardation in Fractured Bedrock," *Proceedings of the Fractured Rock 2001 Conference*. Toronto, March 26–28.

Meehan, S.M.E., T. R. Weaver, and C. R. Lawrence. 1999. "The Biodegradation of Cyanide in Groundwater at Gasworks Sites, Australia: Implications for Site Management," *Environmental Management and Health*, vol. 10, no. 1, pp. 64–71.

Nyer, E.K. and E.G. Gatliff. 1996. "Phytoremediation." *Ground Water Monitoring and Remediation*, vol 16, no. 1, pp. 58–62.

Olewine Company. 2000. "RODinfo Compilation of EPA Records of Decision."

Pankow, J. F., and J. A. Cherry. 1996. *Dense Chlorinated Solvents and Other DNAPLs in Groundwater*. Portland, Ore.: Waterloo Press.

Priddle, M. W., and K. T. B. MacQuarrie. 1994. "Dissolution of Creosote in Groundwater: An Experimental and Modeling Investigation." *Journal of Contaminant Hydrology*, vol. 15, pp. 27–56.

Saroff, S. T., M. J. Gefell, and D. S. Lipson. 1997. "Hazardous Waste Site Remediation Using Technical Impracticability Guidance." *Proceedings of the POWER-GEN International 1997 Conference*. December 11.

Schneider, W. H., J. G. Wrobel, S. R. Hirsh, H. R. Compton, and D. Haroski. 2000. "The Influence of an Integrated Remedial System on Groundwater Hydrology. Bioremediation and Phytoremediation of Chlorinated and Recalcitrant Compounds." *Proceedings from the Second International Conference on Remediation of Chlorinated and Recalcitrant Compounds*. Monterey, California. May 22–25. Battelle Press, Columbus, Ohio. C2-4:477.

Syracuse Research Corp. (SRC). 2000. "Environmental Fate Database." Available on the World Wide Web at www.syrres.com.

USDA. 1980. *The Biologic and Economic Assessment of Pentachlorophenol, Inorganic Arsenicals, Creosote*, vol. 1: *Wood Preservatives*. U.S. Department of Agriculture Technical Bulletin 1658-I.

US EPA. 1993. "Guidance for Evaluating the Technical Impracticability of Ground-Water Restoration." Office of Solid Waste and Emergency Response. Directive 9234.2-25. September.

US EPA. 2000a. "Introduction to Phytoremediation." Office of Research and Development. EPA/600/R-99/107. February.

US EPA. 2000b. "A Resource for MGP Site Characterization and Remediation, Expedited Site Characterization and Source Remediation at Former Manufactured Gas Plant Sites. Office of Solid Waste and Emergency Response." EPA 542-R-00-005. July.

Waterloo Educational Services (WES). 1997. "DNAPLs in Fractured Geologic Media: Behavior, Monitoring, and Remediation." Course notes from the short course presented by the University Consortium Solvents-in-Groundwater Research Program.

CHAPTER 11

NATURAL RESOURCE DAMAGES FROM THE GROUNDWATER PERSPECTIVE

Tyler E. Gass

Blasland, Bouck & Lee, Inc.
Golden, Colorado

Richard W. Dunford

Triangle Economic Research,
A BBL Company
Durham, North Carolina

11.1 INTRODUCTION

To the public and the media, the Comprehensive Environmental Response, Compensation, and Liability Act (CERCLA), or as it has been termed by the media, Superfund, is viewed as the regulating hammer that enables the federal government, and in many cases state agencies, the power to clean up the legacy of contamination and environmental impairment that has been caused by releases of hazardous substances, whether through accepted waste and material handling practices of past decades, or by accident, or by those who knowingly and willfully permitted such releases to the environment.

Most people view CERCLA as the tool to remedy these legacies of contamination either by having responsible parties perform or pay for the cleanup, or in cases where the responsible party no longer exists or is unknown, by using federal funds to achieve the clean up. However, there is another component of CERCLA, as well as a lesser-known regulatory statute, the Oil Pollution Act (OPA), that allows the federal government and state governments to seek reasonable compensation for injury to, destruction of, or loss of natural resources, including the reasonable costs of a damage assessment [CERLCA §§101(b); 107(a)(4)(c); OPA §§1001(5); 102(b)(2)].

Compensation of this kind is referred to as *natural resource damages* (NRDs), and relates to those resources held in trust for the public. Perhaps the NRD case of greatest notariety is that of the *Exxon Valdez* release (which led to the enactment of OPA). In this instance, compensation for permanent and interim impacts to the environment and ecosystem was over $900 million. Many natural resource damage assessments (NRDAs) are not as costly, and compensatory damages may range in only the tens of thousands of dollars.

Another artifact of media attention is that many people view NRDs to be associated primarily or only with damages to ecosystems. However, NRDs can be associated with injury to air, surface water, and groundwater. This chapter focuses on NRDs associated with impacts to groundwater, and how assessments are performed to achieve a value for the loss of a groundwater resource. However, to frame the picture of the groundwater NRDs, some background is provided on the relevant aspects of the regulatory statutes to assist the reader in understanding the assessment and compensation associated with injuries to groundwater resources.

NRDs associated with injury to groundwater resources are generally more closely associated with the Department of Interior (DOI) rules rather than the rules and guidance of the National Oceanographic and Atmospheric Administration (NOAA), which has more applicability related to OPA.

11.2 BACKGROUND

Natural resource damages (NRDs) are addressed in Sec. 311 of the Clean Water Act and Secs. 107 and 1006 of the OPA. These regulatory statutes provide authority for assessment and restoration of natural resource damages held in the public trust for resources that have been injured by hazard substances or discharges of oil, and to collect monetary damages from responsible parties for residual losses incurred by the public as a result of injuries to natural resources (such as groundwater). Damages to "private" property are excluded since there are other mechanisms available for private property owners to claim and be compensated for damages.

Both CERCLA and OPA define *natural resources* broadly to include fish, wildlife, biota, surface water, groundwater, drinking water supplies, and other such resources. To date, most regulatory agencies have focused their NRD efforts on cost recovery of injuries to wetlands, sediments, and aquatic ecosystems. However, a growing number of states, as well as the U.S. Environmental Protection Agency (EPA), are turning their attention to injury to groundwater and drinking water supplies. Although the definition of *natural resources* is broad, it contains the important limitation that the natural resources to which CERCLA applies must be *public* resources. If the resources are not *owned* by a government, they must at least be under some substantial form of government regulation, management, or control. Private individuals cannot recover damages for personal injury, property damage, or economic loss related to natural resource injuries under CERCLA; only natural resource trustees may do so (see discussion below). CERCLA's citizen-suit provision does, however, allow private parties to sue to enforce only CERCLA requirements or to compel federal officials to perform nondiscretionary duties under the law. This provision can be, and has been, used for "natural resource trustees" to fulfill their natural resource assessment (NRDA) obligations (Sharples et al., 1993).

11.2.1   Natural Resource Trustees

CERCLA states that "the President . . . shall act on behalf of the public as a Trustee of such natural resources to recover for such damages" [CERCLA Sec. 107(§)(1)]. The National Contingency Plan (NCP) states that federal officials will act as public trustees for natural resources, and that the governor of each state shall designate state officials to act as trustees for state trust resources.

Trustees have been given responsibility for restoring injured or damaged natural resources. The two major areas of trustee responsibility under CERCLA and OPA are: (1) assessment of injury to natural resources and (2) restoration of natural resources injured or services lost due to a release or a discharge. To meet these responsibilities, both statutes provide several mechanisms to permit the trustees to fulfill their obligations. The trustees can:

- Sue in court to obtain compensation from potentially responsible parties (PRPs) for NRDs and the costs of assessment and restoration planning.
- Conduct assessments or restorations in accordance with certain standards specified by the federal government and file a claim for reimbursement from the federal government.
- Participate in negotiations with PRPs to obtain PRP-financed or PRP-conducted assessments and restorations of the NRD.

11.2.2   NRD Assessments

One of the primary responsibilities of trustees under both CERCLA and OPA is to assess the extent of an injury to a natural resource and determine appropriate ways of restoring and compensating for that injury. A natural resource damage assessment (NRDA) is the process of collecting, compiling, and analyzing information to make these determinations. Trustees have the option of using the methodologies prescribed by the Department of the Interior (DOI), 43 CRF Part 11, or the Department of Commerce's National Oceanic and Atmospheric Administration (NOAA), 15 CFR Part 990. The DOI regulations are to assess NRD under CERCLA, while the NOAA methodologies are applicable for NRDAs under OPA. NRDAs have been performed or are underway in a variety of locations, many of which involve one or more Superfund sites.

The overall intent of the assessment regulations is to determine appropriate restoration and compensation for injuries to natural resources. If a federal or state trustee goes into federal court and sues a potentially responsible party (PRP) for NRDs under CERCLA, an assessment done in accordance with the DOI regulations is given the force and effect of a "rebuttable presumption" [CERCLA §107(f)(2)(C)]. A federal, state, or tribal trustee who sues a PRP for NRD under OPA, as assessment done in accordance with the NOAA regulations, is given a rebuttable presumption [OPA §1006(e)(2)]. This means that the burden of persuasion in court shifts to the PRP. It will be the task of the PRP to disprove the trustee's assessment.

11.2.3   NRD Restorations

Under CERCLA, monies recovered from an NRD claim are to be used only for restoration or replacement of the injured natural resource, or for acquisition of an equivalent resource (hereinafter called *restoration* unless otherwise noted) [CER-CLA §107(f)(1)]. Under OPA, recovered sums are to be used only to reimburse or pay costs incurred by the trustee with respect to the natural resources [OPA §1006(f)]; these include costs incurred while conducting NRDAs and developing and implementing plans for "the restoration, rehabilitation, replacement, or acquisition of the equivalent, of the natural resources" [OPA §1006(C)]. Any amount in excess of these costs must be deposited in the Oil Spill Liability Fund [OPA §1006(f)].

Restoration actions are principally designed to return injured resources to baseline conditions, but may also compensate the public for interim loss of injured resources from the onset of injury until baseline conditions are re-established. Restoration activities have been successfully completed at several sites.

Natural resource trustees are required to develop and implement plans for the restoration of natural resources. A trustee's plans form the basis of calculating NRD for court actions or claims against the OPA trust fund [OPA §§1006(c), (d)(1)-(2), 1012(a)(2)].

11.3 FACTORS TO BE CONSIDERED IN THE VALUATION OF NATURAL RESOURCE DAMAGES

Before a valuation of damages can be performed, the trustees must determine that all of the following criteria have been met:

* A release of a hazardous substance has occurred.
* Natural resources for which federal, state or native American trustees are responsible for have been or are likely to have been adversely affected by the release.
* Sufficient data is available or that data can be collected at a reasonable level of effort to perform an assessment of damages.
* Response actions, if any, will not completely remedy the injury to the natural resource.

In addition, DOI rules set out several types of injuries that are excluded from the NRDA process including, but not necessarily limited to, the following:

* Injuries that occurred wholly before the enactment of CERCLA.
* Injuries resulting from the application of agricultural chemicals registered in the Federal Insecticide, Fungicide, and Rodenticide Act (FIFRA).
* Injuries resulting from the release of recycled oil, if the oil is not mixed with another hazardous substance.

Finally, the trustees must be able to document the injury from the release, including the aerial extent and the impacts that have occurred. 43 CFR 11.62 provides detailed guidance on how injury determinations are to be conducted for surface water, groundwater, air, and geologic and biologic resources. DOI regulations provide specific definitions of injury for different resource groups.

11.4   APPLICATION OF THE NATURAL RESOURCE DAMAGE ASSESSMENT PROCESS TO GROUNDWATER CONTAMINATION

The NRDA regulations promulgated by the U.S. Department of the Interior (DOI) for hazardous-substance releases have three phases: injury determination, quantification of effects, and damage determination [Code of Federal Regulations (CFR), Sec. 43, 11.13]. In the first phase, scientists measure changes in the physical or chemical quality of viability of natural resources as a result of hazardous-substance releases (CRF, Sec. 43, 11.14). For groundwater resources, the following information is collected in the first phase to measure the extent of the injury attributable to the hazardous-substance release:

- Regulatory standards for the maximum concentration of relevant hazardous substances in groundwater
- Concentration of released contaminants in the groundwater
- Areal extent of the contaminated groundwater over time
- Cubic meters of contaminated groundwater over time
- Direction and rate of flow of groundwater contaminants over time
- Connection of groundwater to other natural resources and any resulting injuries to those resources

The quantification-of-effects NRDA phase requires identifying the natural resource services adversely affected by the injuries from the hazardous-substance release and measuring the concomitant reduction in these services. Natural resource services, according to the DOI regulations, are the physical and biological functions of natural resources (CFR, Sec. 43, 11.14). In general, there are two types of natural resource services: human-use and ecological. Human-use services are services that resources provide directly to people (such as the provision of drinking water to households). Ecological services are services that natural resources provide to other natural resources (such as support of living organisms in wetlands through groundwater discharges).

After identifying all of the affected services, the analyst must choose the units of measurement for these services. For most human-use services, this is relatively straightforward. For example, the drinking water services provided by groundwater can be measured in terms of the total cubic meters of extracted water that meets drinking standards. The water for irrigation services can be measured in terms of the total cubic meters of extracted water that meets agricultural standards. It is more challenging to determine an appropriate measure of ecological services because they often depend on complicated physical and/or biological relationships. To the extent possible, analysts must choose a measure that captures quality as well as quantity. For example, suppose that the water table is close to the surface at a site, such that some plants have their roots in the groundwater. A hazardous-substance release will probably not change the total cubic meters of water available for the plants, but the contaminated water may reduce the density of the plants, which in turn reduces the amount of food and nesting material for animals that depend on these plants. Clearly, the number of cubic meters of contaminated groundwater will not properly measure the reduction in ecological services as a result of groundwater injuries. A better measure may be changes in the stem density of the plants that have their roots in the groundwater.

The reduction in services attributable to natural resource injuries is quantified by comparing the services provided following the injury to the services that would have been provided in the absence of the injury (i.e., baseline services). The with-injury services are the services actually provided after the injury, which are potentially observable. However, baseline services cannot be observed; they must be estimated in some way. One approach for estimating baseline services is to use the services provided in an uncontaminated reference area to predict the baseline services in the injured area. The other approach for estimating baseline services is to predict such services using historical information on the services in the injured area prior to the injury.

There are three components of natural resource damages in the DOI process: restoration costs, compensable values, and assessment costs (CRF, Sec. 43, 11.80). The restoration costs are the costs of actions that return natural resource services to their baseline (i.e., without-injury) levels sooner than natural recovery. In general, restoration costs include the cost of restoring, rehabilitating, replacing, and/or acquiring the equivalent of the injured natural resources and/or services (CRF, Sec. 43, 11.80). The following example illustrates the meaning of "replacing and/or acquiring the equivalent" in a real-world situation. Suppose that an aquifer no longer provides water for household uses because of contamination. An existing desalination plant that pumps water from the ocean could increase its water output to replace the water no longer provided by the contaminated aquifer. The restoration costs for this option are the costs of obtaining additional water from the desalination plant, which includes operation and maintenance costs.

The second component of natural resource damages, compensable value, is the amount of money that compensates the public for the reduction of natural resource services prior to their return to baseline levels. Finally, damage assessment costs are the costs associated with the work required to implement the damage assessment process.

It is important to understand the economic relationship between restoration costs and compensable values. Restoration actions that return natural resource services to baseline sooner than natural recovery usually will decrease compensable values, because the public will experience a smaller loss of natural resource services over time. However, if restoration costs are far larger than the value of the services being restored, then the natural resource services should be allowed to recover naturally, in which case restoration costs will be minimal, but compensable values may be high. (Restoration costs depend on existing technology. If a low-cost restoration technology is available, then restoration actions may be implemented even if compensable values are low in the absence of restoration actions.) Thus, restoration costs and compensable value tend to move in opposite directions. Although not required by the DOI regulations, economic efficiency requires the selection of restoration actions that minimize the sum of restoration costs and compensable values.

As noted above, natural resource damages are limited to losses incurred by the public. Consequently, many groundwater damages are not part of natural resource damages. In particular, losses incurred by private entities are excluded from natural resource damages, because these losses are potentially recoverable under common law. For example, suppose a winery uses groundwater in its production process. If a hazardous substance contaminates the groundwater, then the winery can seek compensation under common law for the losses it incurs from the contamination. Those losses are excluded from natural resource damages.

The appropriate technique for estimating compensable value depends on the types of services affected by the natural resource injuries. The following two sections describe the principal valuation techniques used for human-use and ecological services provided by groundwater, regardless of whether the losses are included in natural resource damages. (It should be noted that many valuation studies do not clearly distinguish between human-use and ecological services. However, very different techniques are typically used to value these two types of services provided by groundwater.)

11.5   VALUING REDUCTIONS IN HUMAN-USE SERVICES

Human-use services of groundwater mainly involve the extraction of water for various uses, such as household (for drinking, bathing, cooking, and cleaning), agricultural (mainly irrigating crops and watering livestock), and commercial/industrial (e.g., cooling water, power production, food processing, and manufacturing processes). (Groundwater also can indirectly support other human-use services. For example, groundwater discharges in a stream can indirectly support fishing and other recreational activities in that stream.) The focus of this chapter is on valuing the direct human-use services of groundwater. The main approaches for valuing human-use services, which are discussed below, are averting-behavior cost, production cost, demand functions, and contingent valuation.

11.5.1   Averting-Behavior Costs

This method is based on observed behaviors by individuals who are trying to avoid or reduce damages associated with a groundwater injury. In general, individuals may adopt one or more of three types of averting behavior:

- Purchases of durable goods, such as a point-of-use water treatment system
- Change of personal routines to avoid contact with contaminated water, such as boiling water
- Purchases of nondurable goods, such as bottled water

For all three types of averting behavior, information is required on the costs that individuals incur to mitigate the groundwater injury.

To illustrate this approach, consider the following example. Suppose groundwater is contaminated such that it has an unsightly color but poses no health risks. Some individuals may begin purchasing bottled water because they dislike drinking the discolored tap water. Suppose that, on average, the households in the area increase their purchases of bottled water by $10/week. [There also may be more subtle costs associated with averting behavior, such as the cost of individuals' time to implement averting behaviors. Strasma (1996) describes the cost of individuals' time in this context.] One can conclude that clear water is worth at least $10/household/week because people are willing to pay that much to obtain clear water. One must say at least because "such expenditures do not measure all of the costs related to pollution that affect household utility" [National Research Council (NRC), 1997, p. 79]. Furthermore, interpreting these costs as values requires the following assumptions: (1) the averting behavior should have no other benefits besides lessening the impact of contamination, (2) individuals should not enjoy engaging in the averting behavior, (3) the contamination should not cause a large loss of income such that it would influence purchasing behavior, and (4) averting behavior does not lower costs (NRC 1997, p. 79). Thus, the example would be more than a lower bound if individuals like the bottled water more than the uncontaminated tap water or if individuals save on their water bill by switching to bottled water.

11.5.2 Production-Cost Techniques

Agricultural, industrial, and commercial enterprises produce goods that are sold in markets, which distinguishes them from households and makes additional valuation techniques possible. These techniques rely on the fact that water is an input to production and therefore influences profits. In general, the reduction in profits experienced by commercial enterprises is the appropriate measure of damages. As explained below, the change in production cost is often a good proxy for the reduction in profits.

To illustrate the production-cost approach, Fig. 11.1 shows the market price MP of widgets faced by a firm under perfect competition, which means the firm cannot affect the price of widgets. The notation MC_0 represents the firm's marginal cost of producing widgets, which one assumes involves the use of some groundwater. Suppose the presence of contaminants in the groundwater causes the firm to acquire water from a more costly, alternative source. This shifts the marginal cost of producing widgets to MC_1, which reduces the quantity of widgets sold by the firm from Q_0 to Q_1. The shaded area in Fig. 11.1 is the loss resulting from the increased cost of water. Specifically, the shaded area represents the firm's loss of producer surplus (i.e., profits). More technically, producer surplus is the difference between the market price of a unit of a good or service and the cost of producing that good or service.

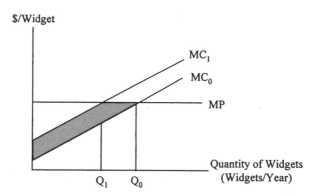

FIGURE 11.1 Reduction in producer surplus from groundwater contamination.

The reduction in units of output (i.e., $Q_0 - Q_1$) multiplied by the market price of the output MP provides a rough approximation of the lost producer surplus. In general, this approximation will overestimate the value of water for agriculture and other productive enterprises, because it ignores the potential substitution of other inputs for water in the production process. For example, entrepreneurial ability and management skill may partially offset some of the increased water costs. For agriculture in particular, another potential problem is that government programs and price supports often inflate the price of farm outputs, which in turn leads to an over-estimate of the value of water (NRC, 1997).

In some instances groundwater contamination may lead to costly responses by commercial enterprises that do not increase the marginal cost of production. For example, if an irrigation well becomes contaminated, the farmer may develop a new well nearby and abandon the contaminated well. If the cost of withdrawing water from both wells is the same, then the farmer's marginal cost of production will not increase. In this case the cost of developing the new well equals the farmer's loss of profits. A similar result occurs whenever any commercial or industrial enterprise incurs capital costs, but no additional operation or maintenance costs, as a result of groundwater contamination. Some analysts classify the capital costs in this situation as averting-behavior costs (NRC, 1997).

11.5.3 Demand Functions

Demand-function techniques are applicable to municipal water services, which include residential, governmental, and small commercial uses. This type of approach estimates a demand function for water, using information gathered over time and across geographic regions about both the price of water and associated quantity purchased. When the groundwater used by a municipality is contaminated, the water authority may have to undertake actions to treat or replace the water. Usually, these increased costs are passed on to ratepayers in the form of higher prices, which results in a loss experienced by water consumers. The measure of the loss is not the increase in the price of the water, but rather the reduction in consumer surplus, which reflects the value of goods and services in excess of their price. More technically, consumer surplus is the difference between the maximum amount that an individual is willing to pay for a unit of a good or service and the amount they actually pay for that unit.

Figure 11.2 shows the example just presented. The demand for water by consumers is represented by the line labeled D. The notation WR_0 represents the water rate in the absence of contamination, which results in Q_0 units of water being used by consumers. The groundwater contamination increases water rates to WR_1, lowering the amount of water used by consumers to Q_1. The shaded area in Fig. 11.2 represents the loss of consumer surplus associated with the water rate increase. In this

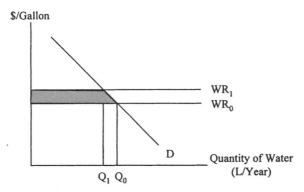

FIGURE 11.2 Reduction in consumer surplus from groundwater contamination.

example, the shaded area equals the difference in water rates ($WR_1 - WR_0$) multiplied by the average of Q_0 and Q_1. (A curved demand curve would result in a slightly different estimate of lost consumer surplus.)

Other scenarios are equally possible. In one case, the water authority identifies an alternative supply capable of providing water at lower costs, after the initial capital investment is absorbed. For example, in Fresno, California, some municipal water supply wells had become impacted by agricultural chemicals. However, some of the wells that were impacted were over 40 years old, had numerous service problems, and therefore, they were already costly to maintain and scheduled to be abandoned. The wells that replaced the wells to be abandoned were already scheduled to be installed, and were designed to provide water more efficiently and at lower cost. In this case, there may be no impact on the cost to the consumer. The second alternative involves actions taken by the responsible parties (RPs) to treat the water from the effected wells. In the second scenario, the RPs pay the cost of installing groundwater treatment units at the well head, and pay the operation and maintenance costs of operating the treatment unit. Therefore, cost impacts to the water authority and the consumer are averted, and the groundwater resource remains usable.

11.5.4   Contingent-Valuation Method

The contingent-valuation approach differs from those discussed above in that it depends on individuals' responses to hypothetical survey questions instead of observed behavior or market prices. The use of contingent valuation for valuing natural resource services has generated substantial controversy (Hausman, 1993; Diamond and Hausman, 1994; Hanemann, 1994; Portney, 1994). Although the more heated controversy involves using contingent valuation to estimate nonuse values, there remain significant concerns and cautions about using it to estimate use values, which are discussed here.

With contingent valuation, researchers elicit a household's willingness to pay for groundwater services through a hypothetical question. The survey first provides a scenario of groundwater contamination. This scenario may take an ex ante perspective by including a probability of contamination or an ex post perspective by presenting the effects of groundwater contamination. Once the scenario is presented, respondents are asked what they would be willing to pay to achieve or avoid the given scenario. (The form of the valuation question also is the subject of some debate. Some economists favor asking the respondent for the maximum amount that they would agree to pay, whereas other economists advocate asking the respondent if they would agree to pay a specified amount, which is varied for different respondents.) This approach, even when applied to human-use services, can lead to value estimates that are substantially biased. Some of the most important sources of bias include questions that induce respondents to either overstate or understate their willingness to pay for the services, survey designs that provide value cues, or poor specification of the valuation scenario, which leads to meaningless responses. [See NRC (1997) for more discussion of the issues associated with the contingent valuation method.]

11.6   VALUING REDUCTIONS IN ECOLOGICAL SERVICES

Ecological services are services that one resource provides to another resource. Because these services are not associated with any existing economic markets or human behaviors, they are extremely difficult to value in dollars. The only existing technique that can directly provide a dollar value to these services, the contingent-valuation method, has proved to be unreliable to date for estimating these nonuse values, as discussed below. However, there are two in-kind approaches for determining the appropriate compensation for reductions in ecological services: conjoint analysis and habitat equivalency analysis (HEA). These methods are recognized as viable alternatives for compensation determination under the National Oceanic and Atmospheric Administration (NOAA) damage-assessment regulations (Scaling, 1997). Although the DOI regulations currently focus on the valuation of ecological services, future revisions in those regulations may endorse these in-kind compensation approaches.

11.6.1 Conjoint Analysis

One of the key issues associated with the in-kind compensation approach is determining what increase in services the public considers equivalent to the lost services. Conjoint analysis, which has been used for more than 20 years in marketing and transportation applications [e.g., Cattin and Wittink (1982), Green and Srinivasan (1990), and Louviere 1994], is a promising technique for making this determination. Conjoint analysis can be used to explore the equivalence of different ecological services. The term *habitat services* will be used to describe the ecological services that humans value. This is consistent with the commonsense notion that humans value the overall condition of a habitat, not necessarily all of the individual ecological services within the habitat (such as the ecological services provided by microorganisms). Specifically, respondents in the conjoint analysis are asked to evaluate several pairs of habitat scenarios, each with different attributes. For example, these attributes might include the size of the wildlife population and number of different species of wildlife supported by the habitat and specific time periods involved.

Figure 11.3 shows an example of a conjoint question using the elicitation format referred to as a *graded-pair comparison*. (Other elicitation formats used in conjoint analysis include ranking, rating, and dichotomous choice.) Suppose that scenario A describes the services provided by a potential restoration project involving 20 km of riverine habitat that supports 20 different wildlife species for 2 years. Suppose that scenario B is another potential restoration project, which would provide 10 ha of salt

Please indicate your rating for the two service scenarios shown below								

Scenario A
•20 km of river habitat
•Habitat only available for 2 years
•Habitat supports 20 different
   wildlife species
					Scenario B
					•10 ha of salt marsh
					•Habitat available for 50 years
					•Habitat supports 15 different
					   wildlife species			

1	2	3	4	5	6	7	8	9
A is much better		A is somewhat better				B is somewhat better		B is much better

FIGURE 11.3 Example of graded-pair comparison conjoint questions.

marsh habitat capable of supporting 15 different species for 50 years. If this pair of scenarios receives a score of 5 from a respondent then, in the opinion of that respondent, the value of the ecological services provided in scenario B are equivalent to those provided in scenario A. Alternatively, if the respondent gives an 8 to this comparison, then the respondent prefers scenario B over A.

Statistical analysis of the resulting ratings of various pairs of scenarios yields estimates of respondents' tradeoffs among different habitat attributes. These estimates can then be used to calculate preference levels for the baseline services of the injured habitat and anticipated services of the restored habitat, or the services associated with the purchase/creation of new habitat. Comparison of these estimates will reveal the extent to which any loss in welfare associated with habitat injuries would be offset by various restoration alternatives.

There are several potential advantages of using conjoint analysis rather than contingent valuation (CV) for ecological services compensation. First, the conjoint-analysis approach circumvents the difficult and perhaps impossible task of determining a reliable dollar value for ecological services. Instead, the affected public indicates the specified combination of alternative services that would compensate it. Second, the restoration monies can be spent on something that the public has indicated that it wants, according to its responses to the conjoint survey. Third, the conjoint-analysis approach describes tradeoffs in a way that is consistent with economic theory. In particular, a conjoint survey elicits a willingness-to-pay (WTP) response. Economic theory indicates that the willingness-to-accept (WTA) criterion is the appropriate measure of value in a damage context because it recognizes the public's implicit property rights for natural resources (Freeman, 1993). The DOI also acknowledges that the WTA criterion is the appropriate measurement of damages (*Federal Register,* 1986b). However, DOI advocates use WTP because in practice it is impractical or infeasible to apply the WTA criteria to CV studies. The opportunity to use WTA as a measure of value makes the conjoint-analysis approach particularly attractive. Fourth, the conjoint-analysis approach may identify several possible restoration alternatives that will make the public whole (i.e., fully compensate the public). Then, the lowest-cost alternative that makes the public whole can be implemented, which is consistent with economic efficiency. In contrast, the CV approach focuses exclusively on one possible restoration alternative.

Despite some important advantages, conjoint analysis is not necessarily a cure-all for problems with the CV approach. See Mathews et al. (1995) for further discussion of the advantages, limitations, and potential disadvantages of the conjoint-analysis approach. However, conjoint analysis is a promising new technique.

11.6.2 Habitat Equivalency Analysis

The HEA approach for assessing natural resource damages is based on the premise that the public can be compensated for past losses of ecological services through the provision of additional ecological services of the same type and quality in the future (Unsworth and Bishop, 1994). As in the conjoint-analysis approach, the main advantage of HEA is that neither lost nor "replacement" ecological services need to be valued in dollars. However, unlike the conjoint-analysis approach, HEA

- Requires many restrictive assumptions, such as the unobserved value of lost and replacement services being equal and constant in real terms over time.

- Is appropriate only in cases where lost and replacement services are "of comparable quality and value," which can be difficult to determine.

- Should only be used when a majority of damages are lost ecological services, as opposed to human-use services.

Thus, compared to the conjoint-analysis approach, HEA is appropriate for a more restricted set of circumstances. However, because the injury determination phase of a damage assessment may provide the data required for the HEA approach, the cost of using this method will likely be low. Consequently, it should be considered when its assumptions are met [See Jones (2000) for additional information on HEA. See Unsworth and Bishop (1994) and Habitat (1995) for empirical applications of the HEA approach.]

To illustrate this approach, consider the following example. Suppose a shallow aquifer located in the sandy shoreline area provides sufficient water to support dune willows. Furthermore, suppose that a hazardous substance contaminates the groundwater, adversely affecting 9 ha of dune willows in the area for 20 years. The HEA approach allows analysts to estimate the hectares of enhanced or new dune willows required to compensate for the 20 years of lost ecological services on the 9 ha of dune willows. Specifically, the following projects might provide appropriate compensation based on the HEA analysis:

- *Project 1:* Plant 3 ha of dune willows adjacent to the injured site such that the stem density matches that found in an uninjured habitat. Design the project such that the 3 ha will sustain dune willows for 100 years. Allow the injured site to recover naturally to its full potential, which will take 20 years.
- *Project 2:* Within 5 years, restore the quality of the groundwater at the injured site to better-than-baseline conditions, which would be possible if historic contaminants were present, and replant dune willows at a 30 percent greater density than was possible given the historic contaminants.

Under both proposed projects, the injured ecological services are restored; however, the first project involves a longer natural recovery, whereas the second project accelerates recovery. By assumption, the public is fully compensated for lost ecological services in both cases. Under the first project, the required amount of compensation in the form of increased ecological services is greater because the resource remains injured for a longer period of time, namely 20 years. Under the second project, the public endures a smaller loss because recovery is 4 times faster; hence, less compensation is required in the form of increased ecological services.

11.7 NONUSE VALUES CONTROVERSY

Nonuse (or passive-use) values are values not linked to direct uses of natural resources, such as the value individuals derive from knowing that a natural resource exists (*Federal Register,* 1995). The measurement of nonuse values is one of the most controversial topics facing environmental economists today (Carson et al., 1993; Desvouges et al., 1993; Randall, 1993). Most economists agree that people may value the existence of unique wetlands, lakes, plants, animals, and other natural resources, although they may not actually use them. However, although use values can be readily estimated by observing people's actual behavior, nonuse values have no associated behavior. The main empirical challenges are determining which resources have nonuse values and estimating reliable values for those resources.

Both the DOI and NOAA regulations accept the existence of nonuse values and allow them to be included in damages, provided that these values can be measured "reliably." Trustees have included forgone nonuse values as a portion of compensable value in several damage assessments. Nonuse damages for groundwater injuries have a prominent role in at least one of those damage assessments—the Clark Fort River case in Montana (Schulze et al., 1995). Because potentially large monetary settlements are at stake, determining whether nonuse values can be reliably measured is very important. Much of the discussion surrounding nonuse values has focused on the CV method, which is used to measure these losses. This is the only currently available technique for measuring these losses in dollars. As discussed above, a CV survey describes a hypothetical market for an environmental commodity and elicits a (WTP) response from respondents for that commodity.

Although economists have undertaken many CV surveys, these past studies are of limited value for assessing the reliability of CV for measuring NRDA-related nonuse losses, because most previous CV studies have estimated-use values, not the more complex nonuse values. Furthermore, most of the studies that specifically address nonuse values involve circumstances that are very different from most damage assessments. In particular, these studies evaluate nonuse values associated with large changes in unique natural resources, such as the Grand Canyon. In contrast, NRDAs usually involve small, temporary changes in more common resources. Therefore, the extensive CV literature does not offer much assurance that this method can measure lost nonuse values reliably in damage-assessment contexts.

For guidance on measuring nonuse values in damage-assessment contexts, NOAA formed a blue-ribbon panel in 1992 headed by two Nobel laureates (Kenneth Arrow and Robert Solow) to evaluate the reliability of CV for measuring nonuse damages in NRDAs and make suggestions on how to improve the CV methodology. In its 1993 report (*Federal Register,* 1993), the NOAA panel highlighted several limitations and problems with the use of CV. According to the NOAA panel, natural resource injuries should meet the following criteria in order for nonuse damages to be potentially significant and reliably estimated:

- Affect a relatively unique resource
- Impact the resource in a sizable and noticeable fashion
- Diminish familiar well-defined services that individuals can understand
- Be permanent or long lasting

Because NRDAs often fail to meet most, if not all, of these criteria, the NOAA panel questions whether people can understand the situations presented in nonuse CV surveys. Misunderstandings may arise from the difficulty in placing a value on small, temporary changes in natural resource services and will lead to unreliable value estimates. For example, a damage assessment is more likely to involve a 1 percent reduction in a bird population over a 5-year period as opposed to the permanent extinction of an entire species. Clearly, people will have more difficulty valuing the small temporary change compared to the extinction scenario. However, these changes are precisely the type of resource service that needs to be valued in NRDA. [See Desvousges et al. (1992) for a nonuse experiment in which respondents did not express different values for migratory waterfowl deaths spanning 2 orders of magnitude in absolute terms, although just a small percentage of the total population.]

Groundwater exemplifies the type of resource that poses significant challenges to the analyst seeking a nonuse value. Unlike surface resources, such as the Grand Canyon, groundwater is completely invisible, making it more difficult for people to conceptualize. Thus, people have no direct experience with groundwater unless they have a private well. Further studies have shown that people have many misconceptions about groundwater. Mitchell and Carson (1989) found that survey respondents thought that groundwater moved much faster than it actually does. This would lead respondents to believe that contamination would spread very quickly. Respondents were also skeptical that the contamination could be contained and the aquifer would not be used in the future. Some people even believe that groundwater is an underground pool of water that sits in a cave, when instead it is simply saturated soil. This indicates that individuals are not very familiar with the services provided by groundwater, which violates one of the NOAA panel's criteria.

Although a particular aquifer may be a unique source of drinking water for a specific area, groundwater resources outside that area are plentiful. Because nonuse values are values associated with resources that survey respondents do not use, groundwater resources are not unique from a nonuse-value perspective. This violates another criterion specified by the NOAA panel.

A group of researchers at the University of Colorado, Boulder (McClelland et al., 1992) released the results of a nationwide study conducted for the U.S. Environmental Protection Agency (U.S. EPA) on nonuse values for groundwater. After careful review of the study and comments from the public, the U.S. EPA's Science Advisory Board (SAB) rejected the study's results. Among the reasons for this decision, the SAB indicated that "people answering the study's survey instrument could have interpreted the services provided by cleaning up contaminated ground water a number of different, conflicting ways. There is no way to know which of the meanings these respondents adopted in answering the valuation questions" (SAB 1993, p. 2).

The above problems and limitations associated with the estimation of nonuse values indicate that CV does not produce reliable estimates of the nonuse damages resulting from injuries to groundwater. Furthermore, if contaminated groundwater is cleaned up and its services return to baseline levels, then any nonuse damages from the contamination will no longer exist. Finally, if restoration projects are implemented, then the additional or enhanced groundwater services provided by these projects may result in gains in nonuse values over time, offsetting some or all losses of nonuse values resulting from groundwater injuries.

11.8 GROUNDWATER DAMAGE ASSESSMENT ISSUES

Many difficult issues can arise in estimating groundwater damages at Superfund and other sites where hazardous substances have contaminated groundwater. This section addresses three important issues.

11.8.1 Integrating Remedial Actions and Restoration Actions

The goal of remedial actions under the Comprehensive Environmental Response, Compensation, and Liability Act (CERCLA) and Resource Conservation and Recovery Act (RCRA) is to remove some or all specific hazardous substances from natural resources such as groundwater. Natural resource damages cover damages resulting from residual injuries, if any, following CERCLA or RCRA remedial actions, as well as damages prior to and during such actions (*Federal Register,* 1986a).

In effect, natural resource damages compensate the public for damages not mitigated by CERCLA or RCRA remedial actions. Thus, any restoration actions as part of natural resource damages are to compensate the public for losses residual to remedial actions.

Because natural resource damages compensate the public for losses not mitigated by CERCLA or RCRA remedial actions, there is an inverse relationship between the two types of actions. Specifically, removing hazardous substances from the environment more quickly and/or thoroughly via remedial actions tends to reduce the need for restoration actions and/or the magnitude of compensable values, other things being equal. Thus, higher remediation costs tend to yield lower natural resource damages, and vice versa. From an economic perspective, society is best served by minimizing the combined cost of remedial actions and natural resource damages. This implies that the selection and timing of remedial actions should take into account the associated natural resource damages.

The U.S. EPA leads most CERCLA and RCRA remediation actions. However, the U.S. EPA is not a trustee in natural resource damage actions. Furthermore, many natural resource damage actions are not initiated until after remedial actions have been selected at CERCLA and RCRA sites. Thus, it may be difficult to integrate natural resource damage considerations into the remediation process. Nevertheless, remediation and restoration actions should be integrated whenever possible.

11.8.2 Amount of Contaminated Groundwater

While groundwater damages should be based on the value of forgone groundwater services and/or the cost of restoring the forgone services, some trustees base groundwater damages on the amount of contaminated groundwater. For example, New Jersey has promulgated a groundwater damage formula based on the amount of contaminated groundwater, and California focused on the amount of contaminated groundwater in a recent case. This leads to the issue of how to properly measure the amount of contaminated groundwater.

Some trustees (e.g., New Jersey and California) base the amount of contaminated groundwater on the annual recharge volume of the contaminated aquifer (i.e., the amount of uncontaminated water going into the contaminated aquifer), under the assumption that this is the amount of water that will be contaminated each year. This measure of the amount of contaminated groundwater for the purposes of assessing natural resource damages is incorrect for several reasons. First, contaminated groundwater provides many services just as well as uncontaminated groundwater, such as preventing saltwater intrusion and subsidence. Second, only a small fraction of the annual recharge volume provides ecological or human-use services. Third, annual recharge volume will not reflect increases or decreases in the size of the contaminated aquifer over time.

If a contaminated aquifer is in hydrologic equilibrium (i.e., the amount of recharge each year equals the amount of discharge), then one measure of the total amount of contaminated groundwater is the volume of the contaminated aquifer. For example, if an aquifer always contains 10,000 m^3 of contaminated water, then the amount of contaminated groundwater is simply 10,000 m^3. This is equivalent to viewing the aquifer as a stock resource.

An alternative approach is to view the contaminated aquifer as a flow resource (i.e., as providing flow of services over time). If the annual flow of services was dedicated entirely to human uses, then the safe yield of injured water from the aquifer would be one measure of the amount of contaminated groundwater. In contrast, if all the annual flow of services was focused on supporting other natural resources, then the annual discharge of injured water from the aquifer into the surface waters or wetlands would be a measure of the amount of contaminated groundwater. Of course, some mix of ecological and human uses also would be possible. The main point is that annual withdrawals and/or discharges of injured water are the appropriate measures of the amount of contaminated groundwater under a flow perspective.

11.8.3 Damages to Unused Groundwater

In many instances, a hazardous-substance release will contaminate an aquifer that is not currently providing any human-use services. What are the human-use damages in this situation? The answer to this question depends on the reason that the aquifer is not providing any human-use services.

If an aquifer is not providing any human-use services because its water is not potable, then contaminating that water with hazardous substances may not result in any human-use damages. In theory, the only situation in which contaminating non-potable water will result in human-use damages is when the future value of water would have increased enough to justify treating and using the nonpotable water. If the treatment also would remove the hazardous substances, then the hazardous substances would not increase treatment costs. In this case, there are no incremental human-use damages from the presence of the hazardous substances. Alternatively, if the treatment would not remove the hazardous substances, then the present value of the incremental cost of removing the hazardous substances when the water is withdrawn in the future would be the appropriate measure of human-use damages.

Figure 11.4 shows a supply curve for groundwater that depicts the marginal cost of annually withdrawing water from four different aquifers. For simplicity, one assumes a constant marginal cost of extraction for each aquifer in Fig. 11.4, which is appropriate if annual withdrawals do not exceed annual recharge rates. If annual withdrawals exceed annual recharge rates, then each aquifer would have an upward-sloping marginal cost of extraction.

In Fig. 11.4, the marginal cost of withdrawing water from aquifer A is P_A, from aquifer B it is P_B, and so on. The maximum annual water withdrawal is Q_A, from aquifer A, $Q_B - Q_A$ from aquifer B, $Q_C - Q_B$ from aquifer C, and so on. Figure 11.4

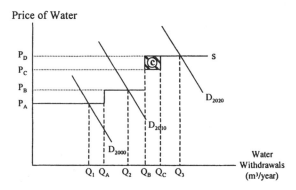

FIGURE 11.4 Incremental cost of contaminating unused aquifer.

also shows three demand curves for water—for the years 2000, 2010, and 2020. In 2000, aquifer A provides all water demanded at a price of P_A. Suppose that a hazardous-substance release contaminates aquifer C and the contamination cannot be removed by any known treatment process. In this situation, the contamination effectively eliminates the water in that aquifer (i.e., $Q_C - Q_B$) for potential use in the future. As shown in Fig. 11.4, aquifer C would have been used to supply water sometime after 2010, but aquifer D would have to be used instead because aquifer C is contaminated. Water from aquifer D is more costly than water from aquifer C, resulting in the loss of shaded area C each year. In other words, the shaded area represents the additional cost of getting water from aquifer D instead of aquifer C. The present value of this annual stream of incremental costs starting at some future date is the damages from contaminating the currently unused aquifer C.

In summary, a theoretically correct estimate of the human-use damages from contaminating an unused aquifer requires a detailed analysis of the cost of water from various sources (including surface water sources) and demand for water over time. From a more practical perspective, estimating human-use damages from contaminating unused aquifers may not be necessary. First, if the cost of withdrawing water from a contaminated aquifer is similar to the cost of withdrawing water from the next cheapest source, then the incremental cost of elimination of the contaminated aquifer will be very small. Second, if a contaminated aquifer would not be used for at least 50 years, then the present value of the incremental cost associated with the contamination of that aquifer will be very small. Furthermore, the uncertainties associated with estimates so far into the future may suggest that such estimates are not reliable enough for a damage assessment. Finally, the DOI regulations (CRF, Sec. 43, 11.84) limit natural resource damages to losses associated with "committed uses" of natural resources (i.e., current or planned uses for which there is a documented commitment). Thus, the DOI regulations exclude speculative future uses of natural resources, arguably eliminating damages to currently unused aquifers that might not be used until far into the future.

11.9   SUMMARY AND CONCLUSIONS

Natural resource damages focus on the damages incurred by the public from injuries to natural resources prior to and/or after the completion of remediation actions. Estimating damages arising from groundwater contamination poses several challenges. First, several reliable economic techniques can be used to value forgone human-use services resulting from groundwater contamination, but valuing forgone ecological services is problematic. Consequently, an in-kind compensation technique, such as HEA, is typically used for determining the appropriate amount of compensation for forgone ecological services. However, HEA has several limitations, as noted above. The reliability of other techniques, such as conjoint analysis, for estimating the appropriate compensation for forgone ecological services is still unproven.

Any potential nonuse values for groundwater should be very small from a conceptual perspective because groundwater is ubiquitous. Thus, groundwater contamination should not produce significant nonuse damages. Empirically, the characteristics of groundwater pose several difficulties in using contingent valuation to estimate nonuse values. A prominent attempt by the U.S. EPA to estimate nonuse values for groundwater was rejected by its SAB. Thus, it is very unlikely that a reliable estimate of nonuse damages could be developed for groundwater contamination.

The value of forgone services and/or the cost of restoring forgone services is the proper measure of groundwater damages from an economic perspective. However, some trustees base damages on the amount of contaminated groundwater, ignoring the type and magnitude of forgone services. This leads to uncertainty regarding how to properly measure the amount of contaminated groundwater.

The final challenge in estimating groundwater damages is associated with the contamination of groundwater not currently providing any human-use services. As discussed above, the appropriate measure of damages in this situation depends on the reason that the groundwater is not providing any human-use services.

REFERENCES

Carson, R. T., Meade, N. F., and Smith, V. K. (1993). "Contingent valuation and passive-use values: Introducing the issues." *Choices,* 2nd quarter, 4–8.

Cattin, P., and Wittink, D. (1982). "Commercial use of conjoint analysis: A survey." *J. Marketing,* **46**(3), 44–53.

Desvousges, W. H., Gable, A. R., Dunford, R. W., and Hudson, S. P. (1993). "Contingent valuation: The wrong tool to measure passive-use losses." *Choices,* 2nd quarter, 9–11.

Desvousges, W. H., Johnson, F. R., Dunford, R. W., Boyle, K. J., Hudson, S. P., and Wilson, K. N. (1992). "Measuring nonuse damages using contingent valuation: An experimental evaluation of accuracy." Monograph 92-1 for Exxon Corp., Research Triangle Institute, Research Triangle Park, N.C.

Diamond, P. A., and Hausman, J. A. (1994). "Contingent valuation: Is some number better than no number?" *J. Economic Perspectives,* **8**(4), 45–64.

Federal Register. (1986a). 51 (no. 148, August 1), 27692.

Federal Register. (1986b). 51 (no. 148, August 1), 27721.

Federal Register. (1993). 58 (no. 10, January 15), 4601.

Federal Register. (1995). 60 (no. 149, August 3), 39811.

Freeman, A. M., III (1993). The measurement of environmental and resource values, *Resources for the Future,* Washington, D.C.

Green, P. E., and Srinivasan, V. (1990). "Conjoint analysis in marketing: New developments with implications for research and practice." *J. Marketing,* **54**(4), 3–19.

Habitat equivalency analysis: An overview. (1995). National Oceanic and Atmospheric Administration, Washington, D.C.

Hanemann, W. M. (1994). "Valuing the environment through contingent valuation." *J. Economic Perspectives,* **8**(4), 19–43.

Hausman, J. A., ed. (1993). *Contingent valuation: A critical assessment,* Elsevier Science, Amsterdam.

Jones, C. A. (2000). "Economic valuation of resource injuries in natural resource liability suits." *J. Water Resource Planning and Management,* ASCE, **126**(6), 358–365.

Louviere, J. J. (1994). "Conjoint analysis." *Advances in marketing research,* R. Bagozzi, ed., Blackwell Publishers, Cambridge, Mass.

McClelland, G. H., et al. (1992). Methods for measuring non-use values: A contingent valuation study of groundwater cleanup, University of Colorado, Boulder, Colo.

Mathews, K. E., Johnson, F. R., Dunford, R. W., and Desvousges, W. H. (1995). "The potential role of conjoint analysis in natural resource damage assessments." TER General Working Paper No. G-9503, Triangle Economic Research, Durham, N.C.

Mitchell, R. C., and Carson, R. T. (1989). "Existence values for groundwater protection." Final Rep. to the Office of Policy Analysis, Cooperative Agreement CR814041-01, U.S. Environmental Protection Agency, Washington, D.C.

National Research Council (NRC). (1997). *Valuing ground water: Economic concepts and approaches,* National Academy Press, Washington, D.C.

Portney, P. R. (1994). "The contingent valuation debate: Why economists should care." *J. Economic Perspectives,* **8**(4), 3–17.

Randall, A. (1993). "Passive-use values and contingent valuation—Valid for damage assessment." *Choices,* 2nd quarter, 12–15.

Scaling compensatory restoration actions. (1997). National Oceanic and Atmospheric Administration, Washington, D.C.

Schulze, W. D., Rowe, R. D., Breffle, W. S., Boyce, R. R., and McClelland, G. H. (1995). *Contingent Valuation of natural resource damages due to injuries to the Upper Clark Fork River Basin,* RCG/Hagler Bailly, Boulder, Colo.

Science Advisory Board (SAB). (1993). "Contingent valuation methodology (CVI)." SAB Report, U.S. EPA, Washington, D.C.

Sharples, F. E., Dunford, R. W., Bascietto, J. J., and Suter, II, G. W. (1993). "Integrating Natural Resource Damage Assessment and Environmental Restoration at Federal Facilities." *Federal Facilities Environmental Journal,* Autumn, 1993.

Strasma, J. (1996). "Estimating losses and indemnization for victims of pollution of well-water supplies: Some objective components of subjective losses." *Proc., Annu. Meetings of the National Association of Forensic Economists,* Kansas City, Mo.

Unsworth, R. E., and Bishop, R. C. (1994). "Assessing natural resource damages using Environmental Annuities." *Ecological Economics,* Amsterdam, The Netherlands, **11**(1), 35–41.

CHAPTER 12
POLLUTION PREVENTION/WASTE MINIMIZATION

Kim Fowler

Battelle, Pacific Northwest Division
Richland, Washington

Marve Hyman

Bechtel National, Inc.
Richland, Washington

12.1 INTRODUCTION

The purpose of pollution prevention is to find opportunities that reduce or eliminate waste or conserve resources while avoiding costs and not sacrificing quality. This chapter provides an overview of what pollution prevention is, defines a pollution prevention hierarchy and common techniques within that hierarchy, and provides methodologies for incorporating pollution prevention into design, manufacturing/construction, operations, and deconstruction/environmental restoration.

The U.S. Pollution Prevention Act of 1990 defined *pollution prevention* as source reduction. *Source reduction* refers to any practice that "reduces the amount of any hazardous substance, pollutant, or contaminant entering any waste stream or otherwise released into the environment (including fugitive emissions) prior to recycling, treatment, or disposal; and reduces the hazards to public health and the environment associated with the release of such substances, pollutants, or contaminants" (U.S.C. Title 42). Simply put, the term pollution prevention means to reduce or eliminate the creation of waste or pollutants. In practice, pollution prevention activities tend to be broader than just source reduction and include reuse, recycling, purchase of environmentally preferable products, treatment, and environmentally safe disposal if no other options exist. For the purposes of this chapter, the terms waste minimization and waste reduction will be assumed to be the same as pollution prevention.

The field of pollution prevention is essentially all the techniques that will eliminate or reduce waste volume and/or toxicity. Pollution prevention specialists often refer to a pollution prevention hierarchy as a means to prioritize the ideas put forth to reduce waste. The typical order of the hierarchy is source reduction (or *reduce*), reuse, recycle, treatment, then disposal. Additionally when items need to be purchased, the consideration of the material/product's source should be considered—that is, whether it contains recycled content, whether an environmentally preferable alternative is available, whether the item is biodegradable or recyclable, etc. This technique is often referred to as *closing the loop,* as it provides markets for the recyclables and other environmentally preferable products. Generally speaking, techniques at the top of the hierarchy are preferred to those at the bottom. However, pollution prevention opportunities are typically evaluated for cost impact in addition to environmental impact, with the lowest cost being selected. Source reduction is typically the least expensive option. The hierarchy (Fig. 12.1) provides some structure to the thought process of identifying opportunities.

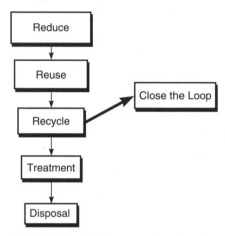

FIGURE 12.1   Pollution prevention hierarchy.

12.1.1 Techniques

Source reduction, or *reduce,* is the primary goal of pollution prevention activities. Identifying opportunities that reduce or eliminate the potential of waste before it is generated are ideal because they typically mean fewer resources were used, less waste was generated, and as a result, more costs were avoided. Pollution prevention techniques may be limited by the quality and/or cost of the technology involved. Continuous improvement of the technologies and reduced costs over time mean that pollution prevention opportunities should be reconsidered periodically. For example, microscale laboratory equipment has increased in availability and quality while costs have typically decreased since the early 1990s.

Source Reduction Techniques.   Source reduction is the effort to reduce the quantity and/or the toxicity of a material or product used.

Redesign. Modify the procedures of the process to reduce the number of steps and to incorporate other source reduction opportunities.

Reduce the scale. Perform the process using smaller-scale techniques or technology.

Substitute materials. Replace a hazardous substance with a nonhazardous substance.

Full utilization of materials. Identify the appropriate use of materials for the process considering its expected life length and use. Design the process and purchase materials appropriate for the life of the process or product. For example, where appropriate, eliminate use of disposable or single-use items, use more durable goods, and purchase long-shelf-life items.

In-depth characterization. Identify materials requiring remediation and target only those that require action.

Go electronic. Replace a process that historically was performed with chemicals with one that can be performed on a computer.

Segregate materials. Keep hazardous and nonhazardous materials separate, from delivery to storage to use to disposal.

Minimize unnecessary packaging. Purchase items with the minimum possible packaging and prepare products with the minimum possible packaging.

Practice good housekeeping. Keep the space clean, redistribute or dispose of materials when they are no longer needed, maintain equipment so that it is working at its optimal performance level.

Manage inventories. Order only those chemicals needed in the quantity needed; that is, do not order in bulk just because the chemical purchasing cost appears to be less expensive.

Design for flexibility. Consider what the "appropriate" level of adaptability/flexibility is for the expected activities, and design to that level. This technique may increase the useful life of the process, product, or facility and/or reduce the waste associated with modifications.

Evaluate level of technology. Consider the level of resource use in comparison to the outcome/results for a low or no-technology option versus the state-of-the-art technology. Determine needs prior to selecting appropriate level of technology.

Repair or upgrade. Evaluate whether modifications or upgrades to the existing equipment or facility would be more cost-effective than full replacement.

Modernize equipment. Review the equipment catalogs to identify whether or not up-to-date versions of equipment use less material, energy, water, or generate less waste.

Supply chain management. Develop guidance for the supply chain that encourages the incorporation of environmental considerations into product design.

Resource conservation. Select and use processes or technologies that minimize the overall consumption of resources (e.g., energy, water, materials, etc.).

Source Reduction Example: Redesign of Analytical Method. At the Pacific Northwest National Laboratory, the previous method for studying plasma protein binding used a radiological counting method, which produced low-level radioactive waste. Researchers redesigned the method by developing a gas chromatograph technique, which eliminated the need for radiological materials. This modification to the process also allowed for reduction in scale of the research and is replacing various biological tissue binding assays. (Cannon, 1999)

Reuse Techniques. Reuse is using an item again without changing its form.

Exchange materials. Communicate with other organizations about the availability and/or need for specific chemicals and materials so that excess materials can be used instead of being disposed of as waste.

Reusable materials. Use equipment that can be cleaned and reused versus disposable materials wherever possible.

Adapt. Adapt or modify materials or facilities so that they can be reused (e.g., remodel an old building to meet current needs rather than demolishing it and building a new one).

Design for reuse. Design the process, product, or facility so that it is adaptable to multiple uses after its initial use.

Reuse Example: Material Reuse during Environmental Restoration. At the U.S. Department of Energy's Argonne National Laboratory (ANL), several environmental restoration projects resulted in the reuse of materials, in addition to recycling and treatment efforts. During the decommissioning of a building at ANL, three plutonium gloveboxes were identified for reuse by other divisions at ANL. During the decommissioning of a small research reactor at ANL, two facilities were converted to other uses instead of being demolished. One building is being reused by ANL's waste management organization for low-level mixed waste storage, and the second building is being reused for road salt storage (U.S. DOE, 1999a).

Recycle Techniques. Recycling is modifying materials/products after primary use into new, useful products.

Recycle the recyclables. Where materials cannot be reused, coordinate with local recyclers or vendors for a recycling program that accepts a variety of materials.

Segregate recyclables. Separating each of the recyclable products increases the value of the materials for the recycling vendor. Recyclable materials are a commodity; thus, without a cost-effective market, recycling will not occur.

Design for recyclability. Design the product so that it can be easily recycled. For example, use only one type of plastic or metal in a product, make it easy to disassemble and segregate the recyclables, make it easy to return the product to a recycler, etc.

Disassembly/deconstruction. Take apart a facility, equipment, etc. to salvage the useful parts for reuse.

Recycle Example: Recycling of Fluorescent Lamps. An estimated 500 million fluorescent bulbs are produced and consumed annually in the United States. Fluorescent bulbs contain small quantities of elemental mercury, which is considered a hazardous substance. The first step fluorescent lamp manufacturers took to reduce this hazard was to minimize the quantity of mercury contained in the lamps. The next step was to develop recycling techniques to not only recover the mercury but also other valuable bulb contents like cadmium and yttrium. Some of the recycling techniques being used or under development include:

- A crush and separation process involves placing the bulbs in a crusher and then separating by material. The materials then go through a thermal process that vaporizes the mercury so that it can be captured. Next the materials go through a process that changes them from raw material into the desired products.
- A crush and water process involves crushing the bulbs under water and capturing the escaping mercury gas. The remaining parts of the bulbs are stripped by using an acid and then precipitated into a filter cake that is further processed so that the raw material can be made into the design products.

- A robotic, dry recycling process involves placing bulbs in a vacuum autoclave. A robotic gripper grabs and heats the end caps of the bulbs, which releases the glue so that the end caps can be removed for reuse. A brush cleans out the inside of the bulb, releasing the mercury vapor, cadmium, yttrium, and phosphor powder that were inside the bulb to ventilation tubes. The chemicals then go through a series of filters so that they are separated for reuse (Shahinpoor and Lantz, 1996).

Close-the-Loop Techniques. Closing the loop involves creating markets for environmentally preferable products. Items made with recycled content or environmentally preferable materials are worth manufacturing only if there is a market for the product. The product price typically decreases as the volume of items purchased increases, which may contribute to further growing the market.

Purchase items that are recyclable. Purchase items that can be recycled into existing markets, especially if one-time use of the material is required.

Purchase items made from renewable materials. Purchase items made from renewable sources (e.g., solar energy, wind energy, forest products, agricultural products, etc.) rather than nonrenewables (e.g., oil, coal, precious metals, etc.).

Purchase items that contain recycled content. Items that contain recycled content should be purchased, where possible, so that a continued market for recycling exists.

Purchase items made in an environmentally preferable manner. Consider how a product is made and select products that have a smaller environmental footprint (e.g., minimal use of hazardous chemicals, water, energy, and other natural resource use). According to the U.S. Executive Order 13010, environmentally preferable means "products or services that have a lesser or reduced effect on human health and the environment when compared with competing products or services that serve the same purpose. This comparison may consider raw materials acquisition, production, manufacturing, packaging, distribution, reuse, operation, maintenance, or disposal of the product or service" (E.O. 13101).

Return packaging materials to suppliers. Require suppliers to take back and reuse undamaged packaging.

"Waste equals food." Identify the redeeming qualities of the waste product(s) so that it can be used for another purpose (Hawkin, 1993).

Purchasing controls. Implement purchasing controls that require the purchase of environmentally preferable materials where quality and cost are comparable to other products.

Closed-loop systems. Institute process systems that allow for minimal or no inputs or outputs once started (e.g., closed-loop cooling systems).

Close-the-Loop Example: Environmentally Preferable Purchasing with Biobased Products. Replacement of petroleum-based hydraulic fluids with soybean-based alternatives was field tested by the University of Iowa and selected vehicle manufacturers. The test results indicated that the soybean oil performs equal to or better than the petroleum-based product, including excellent lubrication and wear protection properties. The soybean alternative requires no equipment modification or special handling and is the environmentally preferable product. As a result of the success of this field test, the U.S. Department of Energy's Sandia National Laboratory converted all of its fleet vehicles over to the soybean-based hydraulic fluid.

Treatment Techniques.   Treatment is the act of modifying an existing waste to render it less toxic, to reduce the total waste volume, and/or other techniques that modify the form of the waste after it has been generated.

> *Neutralize.* Adjust the pH of a liquid waste stream so that it is no longer considered hazardous and can be disposed into the sewer.
>
> *Evaporate.* Remove the water from the waste stream prior to disposal. This technique reduces volume, but may increase the toxicity of the now concentrated waste.
>
> *Compact.* Reduce the volume of the wastes by compressing it into a smaller size prior to disposal.
>
> *Crush.* Reduce the volume by physically breaking down the waste.
>
> *Recover waste.* Turn end-of-the-pipe waste into a marketable product.
>
> *Bioremediation.* Use biological methods to reduce the volume or toxicity of a waste or contaminated area.
>
> *In situ treatment.* Use a treatment technique at the location of the waste stream or contaminated area. This technique reduces the wastes associated with moving wastes, such as packaging and transportation.

Treatment Example.   A good example of evaporation to reduce waste volume is the use of solar drying basins for industrial waste sludges that do not contain volatile hazardous compounds. In many locales, the yearly average evaporation rate from basins and ponds is significantly higher than the precipitation rate. Industrial sludges are often high in water content. By evaporating most of the water, a semisolid cake is formed that has a relatively small volume. The cake is removed mechanically for disposal after some months of operation have ensued. If sludge generation is continuous, multiple basins or compartmented basins are employed—one basin or compartment receives newly generated sludge while others undergo evaporation or sludge removal.

Examples of in situ treatment are techniques used for remediation of soil that has been contaminated with organic compounds:

- Venting the soil with induced airflow for extraction of volatile organics via air stripping
- Venting the soil with reduced airflow rates (*bioventing*) or injecting peroxide solution to enhance biodegradation
- Soil heating with radio-frequency waves, electric blankets, or direct resistance heating from implanted electrodes to remove semivolatile and/or volatile organics

When such techniques are practiced in situ, the contaminated soil does not have to be pre-excavated for treatment, and the treated soil does not have to be backfilled or disposed of. Also, excavation equipment does not have to be decontaminated.

12.1.2 Benefits

The benefits of implementing pollution prevention opportunities have been well documented by industry, government, and environmental groups. When pollution prevention opportunities are implemented, the benefits may include:

- *Cost avoidance,* including reduced material purchase costs, reduced chemical management and waste disposal costs, reduced regulatory requirements and compliance costs, and increased data quality and work efficiency.

- *Safety improvements and risk reduction,* including reduced staff exposure to chemicals and reduced potential for accidents and spills. These benefits may also lead to reduced liability.

- *Process efficiency,* including more efficient use of raw materials and better design of processes and projects.

- *Environmental stewardship,* including reduced potential for releases to the environment, reduced raw material and energy consumption, reduced waste generation, and reduced ecological degradation.

- *Good public relations,* including reduced inventories and releases reported, increased positive activities to share with the community, and improved relations with regulators.

- *Enhanced operations,* including improved quality and quantity of product by utilizing new technologies or processes.

- *Increased regulatory compliance,* including potentially reducing the number of permits needed, number of chemicals that need to be reported, and the number of regulations that apply.

12.1.3 When Does Pollution Prevention Apply?

Pollution prevention can be applied at any time. That is, when a project/activity is in its initial stages of thought or when a waste site is in the process of being cleaned up, pollution prevention opportunities can be considered and are likely to offer alternatives to the current thought process or plan. The life cycle of a product, process, and facility are typically defined by these stages:

- Material selection/resource extraction
- Design
- Manufacturing/construction
- Distribution (transportation and packaging)
- Use
- Disposition (recycling, treatment, disposal, etc.)

The earlier environmental considerations are incorporated into a product, process, or project, the greater potential for impact. That is, design the product so that it uses "environmentally preferable" resources, has a long use life or is reusable, and ultimate disposition is as a feedstock for another product, and the result will be reduction of the overall environmental impact of the product. If the product is already in the manufacturing process, the options may be limited to optimizing the manufacturing process and modifying resource use because major changes to the product design would be very costly. Interface, Inc. has continued to optimize its existing products, but with the goal of becoming a restorative enterprise, giving more to the earth than it is taking, it has had to redesign its products and processes from scratch (Anderson, 1998).

12.2   POLLUTION PREVENTION/WASTE MINIMIZATION IN DESIGN

Pollution prevention can be incorporated into all phases of a project or facility's life—from material selection, design, manufacturing/construction, operation, and final decontamination/decommissioning/disposition, and this can be accomplished most effectively during the design process. Pollution prevention can also be incorporated into laboratory research and development activities.

The benefits of incorporating pollution prevention into the design of a project or facility include life cycle cost avoidance and improved perception by regulators and the public. In addition to those benefits, incorporating pollution prevention into design can offer simplified environmental management for a project, flexibility in design for change, and integrated project concepts. During the design phase there is much more flexibility for making major modifications to a project or facility. Changing drawings or project plans is far less expensive and time-consuming than retrofitting a facility or replacing a piece of equipment (Luper, 1996).

The challenges to incorporating pollution prevention into the design phase of a project tend to be the lack of available tools and information, cost and schedule impacts, lack of design data/drawings, lack of cost data, and addressing the needs of people working on this in the future. Lack of experience in the pollution prevention and the design fields was a primary obstacle for many years, but it has been diminishing over the years. Engineers are now expected to understand the concept of pollution prevention when they take the Engineer in Training exam, and more commercially available tools and publicly available literature on the topic of pollution prevention are available. The challenge now seems to be keeping up to date on all the information that is available.

As with any project, cost and schedule are primary drivers. Incorporating all the new information into the design while keeping down the design costs and impacts on schedule is a major challenge on some projects. At the same time, the cost savings that pollution prevention offers are often not realized by the designers, so it can be difficult to justify added design costs. Design is also a time of many unknowns that can make investigating pollution prevention ideas a challenge. The waste types and quantities are forecasts and the design of the process or facility may not be complete when pollution prevention is being considered. Even with these challenges, it has been estimated that while only 20 percent of life-cycle costs are incurred during design, decisions made during design determine up to 80 percent of the project's total life cycle costs and therefore is a perfect time for incorporating pollution prevention (U.S. DOE, 1996 and 2000).

12.2.1   Methods for Identifying Pollution Prevention Opportunities in Design

Many techniques are available for identifying and evaluating pollution prevention opportunities during the design of a process, product, or facility. For example, process simulation tools can be an effective method for identifying pollution prevention opportunities in chemical process plants (Hilaly and Skidar, 1996).

The method that will be discussed in detail in this chapter is a pollution prevention design assessment, also known as a design charrette when occurring during facility design. A *pollution prevention design assessment* is a systematic, documented approach to determining where pollution prevention opportunities may exist. Once opportunities are identified, the assessment can also reveal the best means of implementing the opportunity. As part of the design process, the assessment identifies items that can be incorporated into the current design and provides suggestions on areas for continuous improvement as the work continues.

A pollution prevention design assessment also helps estimate the cost of the future waste streams, the cost of changing to a cleaner process, the cost avoidance of a cleaner system, and the time needed to repay an investment. Determining the cost impacts of waste streams is an essential part of the pollution prevention process. The assessment can point toward innovative or emerging technologies that will be cleaner, safer, and more cost-effective, and will sometimes yield better results.

The steps to completing an assessment are basic but effective (see Fig. 12.2):

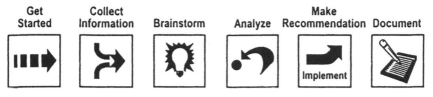

FIGURE 12.2   Steps in a pollution prevention design assessment. *(Engel-Cox and Fowler, 1999; graphic reprinted with author's permission.)*

1. Choose a diverse team to perform the assessment.
2. Gather data about the process, product, or facility design.
3. Brainstorm pollution prevention opportunities.
4. Research and analyze pollution prevention opportunities for waste reduction, cost avoidance, and payback or return on investment (see Sec. 12.6 for details on how to calculate payback and return on investment).
5. Make recommendations on which pollution prevention opportunities to incorporate into the design, on the basis of the waste and cost analysis.
6. Document the work and incorporate the opportunities into the design.

To initiate a pollution prevention design assessment, the scope of the project needs to be defined, management support and financial resources to perform the assessment need to be secured, and a technically diverse team and team leader need to be identified to perform the assessment. The next major step is to gather information about the design requirements, customer/user needs, expected waste generation qualities, quantities, and sources (where feasible), and investigate previous, similar activities that could provide design insights. Once the data have been gathered and reviewed, the team brainstorms design opportunities that may be applicable to its design project and selects specific design options to investigate further. Further investigation may involve both qualitative and quantitative analysis of the proposed costs and benefits. It can be difficult to identify cost avoidance for items incorporated into design for lack of a baseline to compare against. Therefore it is recommended that part of the data collection effort include identifying a comparable process/project/facility for baseline comparisons to assist in the analysis of pollution prevention opportunities. The final steps are to identify the opportunities worthy of incorporation into design, document the efforts taken throughout the design process to incorporate environmental considerations into the design, and share the results with management and peers. It is easier and less expensive to replicate successful modifications already used in a design than it is to evaluate every opportunity during every design.

The level of detail investigated in the design assessment may vary, depending on the stage of design it was performed in and length of the design process. If the design process is lengthy, involves many people, and/or is very complicated, multiple iterations of a design assessment may offer an effective approach to address the pollution prevention opportunities. Opportunities identified in the early stages of design may not be applicable in later stages, and, as the design changes, new opportunities may arise. Considering pollution prevention opportunities throughout the design process is the most effective way to ensure pollution prevention is incorporated into the product, process, or facility.

The different types of pollution prevention opportunities that can be considered during design include: (1) opportunities that minimize waste generated during the design (comparatively minimal) and (2) opportunities that minimize the impact of the process, product, or facility. The focus of a design assessment is on minimizing the environmental impact of the thing being designed.

12.2.2 Possible Pollution Prevention Opportunities in Design

Engineering design is the act of translating ideas into things. Those things may include processes, products, equipment, systems, and facilities. Below, opportunities are identified as either for products or facilities, where product design is assumed to include everything but facility design, and where it is all of the above it is referred to as *things*.

Solid Waste. The first step a design team should undertake is to evaluate whether a new thing is truly necessary. It may be that the upgrade of an old thing may be appropriate and cost-effective. Once it is determined that a new thing must be designed, it is the design team's responsibility to practice pollution prevention in how they do business while they are incorporating it into the design. Items that the design team can include:

- Perform the design using computer technology and only printing copies of the design when necessary
- Understand the purpose of the process, product, or facility being designed, know how it is expected to be marketed, used, misused, and its likely route of disposition. This understanding will become the basis for the some of the design decisions. Use this information to develop clear, detailed design specifications that are anchored on needs rather than assumptions based on previous designs.

The bulk of the waste generated by a process, product, or facility occurs during its operation. How a thing is designed can affect its overall impact on the environment during its operation. For solid waste, the layout or siting of the thing, material selection, and designing for flexibility offer many pollution prevention opportunities.

Layout and siting consider both the size of the thing and the location where it is placed. Understanding the needs of the users, the thing's function, and how the thing will be used, are key to knowing how large or small the thing being designed should be. To minimize the space that it takes up and to minimize the resources used to make it, designing something to the smallest feasible size is likely to minimize its environmental impact. When considering location, where appropriate, use space that has been used before, rather than undisturbed areas, and space near related processes or facilities. Areas that have already been used are more likely to be ready for immediate connection to energy and water sources, thereby avoiding costs and wastes and saving time.

Material selection is a primary component of design and the decisions made on materials can affect many other aspects of a design. Options include:

- Selecting materials because they meet performance-based design specifications, such as a cementlike product with a specific strength requirement rather than a named product.

- Selecting nonhazardous materials that do not generate hazardous waste or use hazardous materials during extraction or manufacturing, and will not require the use of hazardous materials during their useful life. For example, lead soldering of different metals would make a product hazardous.

- Selecting materials made from renewables, such as plastics made from corn-based feedstock rather than petroleum-based feedstock or wallboard made from wheat straw rather than gypsum.

- Selecting materials that are manufactured locally or raw materials that are available locally to minimize transportation-related environmental considerations and costs.

- Selecting materials made from recycled content to reduce consumption of raw materials, such as cloth made from plastic pop bottles or concrete made with fly ash.

- Minimizing the different kinds of materials that are selected (use only one type of plastic throughout the whole product or one type of metal, etc.) so that the product can be easily recycled or disassembled.

- Selecting materials that have an appropriate useful life for the product or facility, such as quality glassware and a dishwasher for research activities that allow for reuse of materials and plastic vials for research activities that cannot allow for reuse of materials.

- Selecting materials so that they do not negatively affect the people occupying the surrounding space, such as furniture without formaldehyde resin and low biocide paint.

Designing for flexibility involves considering the potential future uses of the process, product, or facility so that it can be easily adapted with minimal impact to the environment. The renovation of a thing can result in considerable wastes. If a process, product, or facility could be designed initially so that it is flexible or adaptable, the long-term environmental impact of the thing may be minimized.

Liquid Waste. Many industries use considerable quantities of water at their manufacturing facilities. The traditional approach to water use was to use the water once and then dispose of it as permitted wastewater. During the design process, water use can be eliminated, minimized, recirculated in a closed-loop system, or reused. The key to water-related pollution prevention opportunities is having an understanding of the water quality requirements/standards for the different water uses, of why the approach/technique needs water, and of what technological changes or research would need to be done to eliminate water use. For a water supply to be managed effectively, it is likely the plant's engineering team will have to design a custom set of technologies and process modifications that address the specific process flow of the plant (Jessen and Kemp, 1996).

One technique that can be used to eliminate water usage is keep the work environment clean so that scrubbers are not needed. For example, dry dust collectors can be put in place to replace or minimize the need for water-based scrubbers, thus eliminating the need for scrubber water (Jessen and Kemp, 1996). If water cleaning steps are still needed, see if multiple washing steps can be combined into one step. Consider using carbon dioxide for cleaning rather than water, where feasible.

Opportunities to minimize the need for new water into a system include recirculating, reusing, and treating wastewater and other liquid wastes. Examples include (Jessen and Kemp, 1996):

• Closed-loop cooling water systems, eliminating once-through cooling systems
• Noncontact cooling water for landscape irrigation or other nonpotable water uses
• Reduce the need for freshwater for scrubbing by installing spray dryer rotoclones to recirculate water
• High-pressure cleaning equipment
• Regenerate aqueous cleaners using ceramic ultra- or microfilters
• Regenerate caustic-etch systems by removing dissolved aluminum
• Recover aqueous cleaners, caustic, oil, and solvents with ceramic membrane systems
• Recover usable constituents from wastewater by chemical reclamation
• Purify water by a continuous deionization system with ion exchange resins, ion exchange membrane, and an electrical current
• Electrodialysis systems segregate contaminated wastewaters into two recyclable products (purified filtrate and metal-rich concentrate)
• Capture and reuse heat or energy from steam
• Recycle unused paint
• Recover solvents using distillation equipment

Other Waste.   In addition to solid and liquid wastes, the energy resources used to operate processes, products, and facilities can have a major impact on the environment in the form of carbon dioxide emissions. Energy conservation should be the goal during design. The equipment and materials chosen, along with the layout of the process, product, or facility, are key to being able to accomplish energy conservation. While purchasing equipment, evaluate its expected energy use while in operation and select the most efficient product for the intended use. High-efficiency equipment sized appropriately for the expected use is the ideal selection. At a minimum, select equipment that has been designated by the U.S. Environmental Protection Agency as Energy Star equipment to help minimize energy consumption.

Materials selection is important from three energy-use perspectives: (1) the energy intensity of the material during its manufacture, (2) the transportation distance of the materials, and (3) the use of appropriate materials to minimize heat loss/gain. Where possible, select materials that are less energy intensive in their production and that are locally available or manufactured to minimize fuel use related to transportation. Selecting materials that minimize heat loss/gain could include decisions related to appropriate insulation, windows, and conduction of the heat where appropriate.

Energy use as it relates to layout probably has the most impact during design. When designing a new facility, keep in mind how much heat would be gained or lost through the windows, doors, walls, lights, equipment, etc. Consider the flow of the air in the facility and how that may affect the energy requirements. Consider the number of people in the facility and how that may affect the energy requirements. When designing a new piece of equipment, optimize the quantity of energy needed to operate the equipment and the heat given off by the equipment. Consider whether there are ways to capture lost heat. Consider whether the source of energy could be changed to be more environmentally friendly or whether it could be changed in the future if new sources of energy become available.

The concept of hybrid lighting is the combination of artificial and natural light to meet the lighting needs of building occupants. Hybrid light fixtures use natural light as it is available, supplementing it with artificial light as needed to meet the selected illumination level. Lighting is 25 percent of the United States electricity usage, costing more than $100 million a day. Incorporating concepts such as hybrid lighting or daylighting techniques would not only conserve energy and avoid costs, but would also improve the quality of lighting for the building occupants (Cates, 2000).

12.2.3   Case Histories of Pollution Prevention in Design

Ciba's Toms River, New Jersey, plant cut its raw water consumption by 97 percent and recovered reusable dyestuffs from its on-site wastewater through a combination of water conservation projects. Ciba invested $6 million in a wastewater minimization and recovery system through five separate projects that lowered the total hydraulic load. Those five projects included installing chillers, recharging process steam condensate to the ground, installing spray dryer rotoclones, installing dry dust collectors, and using high-pressure cleaning equipment to clean blending equipment and dryers. Once those projects were in place, the plant's wastewater volume was greatly reduced, so Ciba installed two ceramic membrane-type ultrafiltration units and one reverse-osmosis unit. These units separated and recovered nearly all of the dye from the wastewater, resulting in a wastewater stream clean enough to be reused in some of the plant's cleaning operations (Jessen and Kemp, 1996).

Toyota Motor Corporation has embarked on a vision to provide "clean, safe products," dedicated to making the "planet more comfortable to live on." Toyota set pollution prevention goals that required designing new high-performance, low-price vehicles while optimizing fuel economy (Iwai, 1995). The result of this and other pollution prevention and performance goals resulted in the design of the Prius model, a hybrid electric–gas engine vehicle, which is advertised at achieving 48 miles per gallon on average. Toyota has also developed an electric RAV4 sport utility vehicle, an electric small vehicle called e-com, and a Camry that runs on compressed natural gas (Toyota, 2000). Most of the major automakers have also made significant design investments in hybrid and alternative fuel vehicles and are expected to have the vehicles ready for sale in the next few years.

12.3   POLLUTION PREVENTION/WASTE MINIMIZATION IN INDUSTRIAL PROCESSES

Gouchoe et al. (1996) give an organized approach for determining planning requirements for pollution prevention in process industries. There are a number of federal regulatory triggers, including the following, which may apply in addition to state and local requirements:

- RCRA requires a Waste Minimization and Contingency Plan if more than 1000 kg/mo of hazardous waste is generated or if the operation is a permitted hazardous waste treatment, storage, or disposal facility.
- The Clean Water Act requires the following:
  1. Stormwater pollution prevention plan if there is a need to file for a Stormwater Pollution Prevention permit
  2. Spill prevention control and countermeasures (SPCC) plan if there is underground or aboveground oil storage and certain volumes are exceeded
  3. Oil spill response plan if there is insufficient secondary containment, or an accidental discharge could affect environmentally sensitive areas, or a reportable spill over 10,000 gal occurred in the past 5 years
  4. Toxic organic management plan and a slug discharge control plan if there is a National Pollutant Discharge Elimination System (NPDES) permit or a Significant Industrial User Industrial Pretreatment permit
- The Clean Air Act requires a risk management program (RMP) if operations involve regulated substances in quantities exceeding specified thresholds.

Gouchoe et al. (1996) suggest that pollution prevention plans can be integrated with some of these other plans and programs and give a matrix that describes for each federal regulation listed above the elements of the pollution prevention plan.

12.3.1 Methods for Identifying Pollution Prevention Opportunities for Industrial Processes

Quinn (1996) lists World Wide Web sites that give information on a variety of pollution prevention opportunities. Examples include substitutes for solvents, metal parts cleaners, and materials; targeted processes such as specific chemicals manufacturing; construction/demolition waste reduction; and environmental processes for manufacturing operations.

Dyer and Mulholland (1998) describe an approach for selecting industrial plant waste streams and process components that are candidates for pollution prevention opportunities. The waste stream components that are of concern (e.g., hazardous RCRA compounds, hazardous air pollutants, carcinogens) and volumetric flow rates are identified. Components that have not been minimized and volume reduction become the focus. Then, process components are listed in two ways:

List 1 consists of raw materials that can become salable products, are intermediates, and are salable products.

List 2 consists of all other materials in the process, such as nonsalable by-products, solvents, water, air, nitrogen, acids, and bases.

Then the materials in list 1 are studied to determine if any can do the same function of the compounds in list 2, and modifications to the process are considered to eliminate the need for materials in list 2. Also, process modifications are considered that will minimize the materials in list 2 that are the result of producing nonsalable products. The core assessment team that accomplishes these studies is augmented with process separation and reactions specialists, an environmental specialist, a chemist, and a business expert. The extended team brainstorms possible improvements.

The brainstorming ideas are screened, in order to minimize the number of feasible alternatives to be evaluated, in two steps:

1. A first cut is made by using simple methods, such as judge yes or no, rank high/medium/low against specific criteria, and vote using a points award system.
2. A second cut is made by using a yes/no rating method and the criteria of technical feasibility, economic viability, and waste reduction potential.

The alternatives remaining after screening are evaluated, starting with development of revised mass and energy balances, process flow diagrams, and operating requirements. The investment, operating costs, and net present value for each option are then determined and compared against the base case.

Goldman (2000) describes systems that flag or prevent purchase of materials or sales of products that potentially could involve regulatory action or use of inappropriate substances. The system employs "watch lists" of environmental concerns. A match between an ingredient and a material on a watch list raises a flag that alerts the user or prevents proceeding. Lists may be related to regulations, or company-specific lists of chemicals targeted for reduction, or substances that are banned by company policy or banned from sales in certain countries.

An example is related by Goldman (2000) for a process plant in Massachusetts that uses these four watch lists:

• Substances for which reporting is required under the Emergency Planning and Community Right-to-Know act (EPCRA)
• Chemicals specified in the Massachusetts Contingency Plan, which includes chemicals similar to those in EPCRA
• Chemicals in the company's toxics use reduction plan
• Banned substances that the company does not want on site

12.3.2 Possible Pollution Prevention Opportunities for Industrial Processes

Wastewater.   A systematic approach to wastewater minimization starts with preparing a mass balance or volumetric balance of water inputs, uses, and disposal/losses for a processing facility. A complex process may involve steam generation, cooling, rinsing, incorporation of water into products, air pollution control scrubbing, etc. Some water reuse opportunities are obvious, e.g., recovery of steam condensate for boiler feedwater, and recirculation of cooling water through a cooling tower.

The development of a water balance begins with a block flow diagram, such as in the simplified example in Fig. 12.3.

Usually the balance of water in feed streams does not exactly match the discharges and losses that are accounted for. This situation provides an opportunity for finding leaks, checking flow meters, and establishing corrections to quantities of feeds, products, and effluent discharges. The exercise of developing a water balance can also trigger ideas for water conservation in general, and recycling and reuse specifics.

Recycling of Water in Industrial Processes.   Recirculation of cooling water and of steam condensate are the most common industrial reuse applications. Where wet scrubbing is used for control of acid gas and of particulate emissions, recirculation of

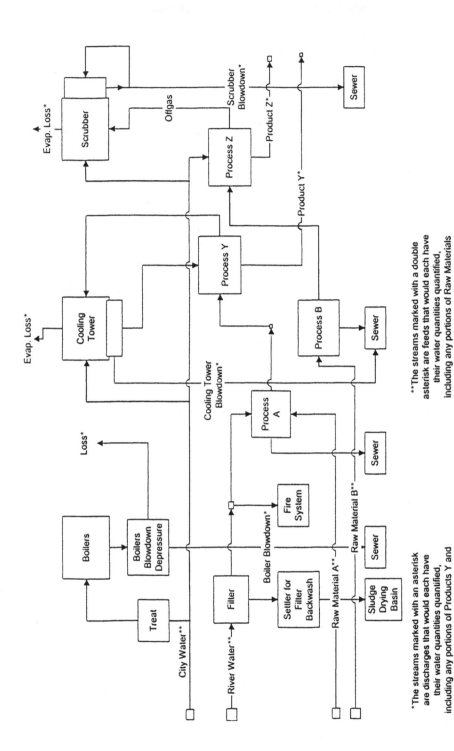

*The streams marked with an asterisk are discharges that would each have their water quantities quantified, including any portions of Products Y and Z that are water

**The streams marked with a double asterisk are feeds that would each have their water quantities quantified, including any portions of Raw Materials A and B that are water

FIGURE 12.3 Water balance flow diagram.

12.16

scrubber water is also common. All such reuse applications involve some water blowdown, with makeup to balance the blowdown quantity and losses and to prevent contaminant buildup beyond tolerable concentrations. It should be noted that cooling towers not only have evaporative losses that are required to produce the cooling effect, but also incur drift of water droplets that fall as rain and may ultimately follow the path of stormwater or be evaporated. For developing a cooling tower water balance, the drift is sometimes treated as an additional blowdown stream.

Recycling of steam condensate to a boiler feedwater system is not always direct. If the steam contacts process fluid or is not indirectly condensed with a heat exchanger, it becomes contaminated with volatile substances that get into the steam condensate. In this event, the condensate is usually treated before returning it to the boiler. For example, if organics contaminate the condensate, it is passed through a bed of granular activated carbon before recycling it.

Any process that involves drying or evaporation may present an opportunity for recycle of the water. Capital investment in and operating expenses for condensing equipment are needed. The benefit of recycling the water is mainly reduced makeup and makeup treatment needs. Treatment of fresh makeup water used in some industrial processes can be involved and expensive. Examples of makeup treatment include sterilization, demineralization (deionization), filtration, and softening. It should be noted that water condensed from drying processes is soft and generally inherently very pure. Consequently, it may require little or no treatment.

Reuse of Water in and from Industrial Processes. A major reuse concept is in cascading applications. The prime example is where multiple rinsing steps are employed in the process. A product is rinsed with water and then rerinsed one or more times. Reuse by cascade rinsing is done by using freshwater as the rinse fluid in the last step, and recovering the rinsate from the last step and using it as the rinse fluid in the previous step. Figure 12.4 shows this concept.

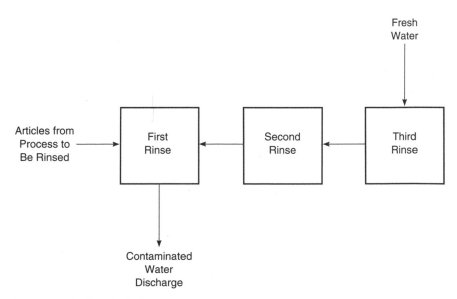

FIGURE 12.4 Cascade rinsing.

Reuse by application to land is done in two major ways:

- Dust control
- Irrigation

These uses may be within the industrial operations involved or they may be for a neighboring property. An example of reuse for on-site dust control is in the cleanup (remediation) of soil that has been contaminated with organics. The organics are removed by incineration or thermal desorption (volatilization). The exhaust gas is passed through a high-temperature afterburner for final organics destruction and cooled, and then particulates are removed with a fabric filter or a wet venturi scrubber. Acid gases may be removed in some instances with a low-energy wet scrubber. The scrubbers recirculate the water and have blowdown streams that can be used for dust control when excavating and handling the contaminated soil. The blowdown streams are filtered and passed through granular activated carbon before they are used for dust control.

The potential for using wastewater for irrigation depends on what is growing on the land needing irrigation and what contaminants are in the wastewater. If only lawns and ornamental shrubs and trees are involved, the purity of the irrigation water can be less than if the growth is for animal foraging or food crops.

Miscellaneous Industrial Water Reuse.   Consider the water effluent as a product if it is removed from the facility and used for another purpose. The water may or may not be treated before reuse, depending on the next user's water quality requirements. A common water reuse application is the deployment of wastewater for construction purposes, ranging from dust control to formulating concrete mixes. A potential use at both Air Force bases and civilian aircraft maintenance installations is for washing of aircraft.

Wastewater from one industrial process at one company may be shared with another company where the input water quality requirements are not as stringent. Wastewater could be reused by fish farms as it was at the Kalundborg ecoindustrial park. Wastewater can also be reused for cooling, fire protection, and flushing, in some instances without treatment (Ehrenfeld and Gertler, 1997).

Wastewater Effluent Volume Minimization.   The ultimate goal for wastewater effluent for some industrial processes is zero discharge. Short of this goal, minimization of the effluent volume is most desirable. Zero discharge and effluent minimization have a number of advantages, including these possibilities:

- Reduced effluent treatment costs
- Reduced space requirements for treatment equipment
- Smaller sewage disposal fees
- Less demand for freshwater makeup and less volume of freshwater to be treated
- No discharge permit needed if zero discharge is achieved.

To accomplish zero discharge of wastewater effluent streams, internal reuse and recycle are used to the greatest extent practical, and then the following practices are invoked:

- Incorporation of water into products
- Evaporation of blowdown and watery wastes
- Treatment of wastewater followed by return of treated water to groundwater aquifers, using infiltration galleries, percolation ponds, or injection wells, as may be permitted

This last practice may be required where the source of fresh makeup is from water wells and maintaining the groundwater table is mandated.

Methods of treating wastewater for recycling within a process plant and life cycle costs (accounting for capital investment, chemicals, and power) are reviewed by Zinkus, Byers, and Doerr (1998), including the following:

- Biological treatment, $40 to $500 per million gallons for streams with under 1000 mg/L biochemical oxygen demand
- Carbon adsorption for removal of organics, $70 to over $1000 per million gallons
- Centrifugal separation of suspended solids, $60 to $2000 per million gallons treated
- Chemical oxidation for removal of organics, ammonia, cyanide, sulfides, and mercaptans, $200 to $10,000 per million gallons
- Crystallization for removal of dissolved solids, more than $5000 per million gallons
- Electrodialysis for removal of dissolved solids, $50 to more than $1000 per million gallons
- Evaporation for removing dissolved salts, heavy metals, and nonvolatile organics, $20 per million gallons for ponds and more than $10,000 per million gallons for thermomechanical systems
- Filtration to remove suspended solids and/or oil and grease, $20 to more than $100 per million gallons
- Flotation to remove suspended solids and/or oil and grease, $20 to more than $100 per million gallons
- Gravity separation to remove suspended solids and/or oil and grease, $50 to $500 per million gallons
- Ion exchange to remove dissolved ions, $250 to more than $1000 per million gallons
- Membrane technologies to remove dissolved constituents (reverse osmosis, membrane electrolysis, diffusion dialysis) or suspended constituents (cross-flow microfiltration and ultrafiltration), $30 to more than $2000 per million gallons
- Chemical precipitation to remove dissolved compounds, $50 to more than $2000 per million gallons
- Solvent extraction to remove certain dissolved inorganics and organics, $1000 to $10,000 per million gallons minus the value of recovered constituents
- Stripping to remove for dissolved hydrogen sulfide, ammonia, carbon dioxide, hydrogen cyanide, and volatile organics, $40 to $250 per million gallons
- Incineration for destruction of concentrated organics, more than $1 per gallon

Completing a plant water balance and updating it once reuse and recycling are maximized helps quantify how much wastewater should be considered as a candidate for final evaporation or treatment and return to the subsurface.

Industrial Solid Wastes.   As with water use, the first step in addressing industrial solid waste is to prepare a mass balance of the materials purchased, the products manufactured, and the remaining solid wastes. In complex manufacturing processes, the mass balance can be performed on smaller subsystems. As with water, it can be difficult to identify and document a 100 percent balanced system. In those situations, focus on the largest volume and most expensive input materials and waste products. Once that part of the system has been optimized for waste reduction, the other components of the system can be reviewed.

Source Reduction of Solid Waste in Industrial Processes. Modifications to industrial processes that result in reducing the wastes at the source fit into three categories: major process modifications, procedure modifications, and administrative controls. All three can be useful tools for pollution prevention and yet they may have implementation challenges. Process modifications may require major cash outlays, and, depending on the industry, that may be difficult to afford. Procedure modification could involve asking experienced production workers to change what they have been doing for the last 20 years. Administrative controls have to be written so that they accomplish the goal without hindering the ability to work efficiently.

Process modifications include concepts such as reducing the scale of the work being done, updating equipment, and redesigning the manufacturing process. Procedure modifications include substituting chemicals, changing maintenance activities, and changing operating practices. Administrative controls include purchasing in bulk, returning all packaging to the vendors, purchasing controls/oversight, inventory management, identifying and evaluating the material balance, and supply chain management.

Designing new processes and/or new process equipment can be very expensive. However, depending on the process efficiency improvements associated with the design, it can increase a company's ability to compete. When the costs of the new process or equipment are being evaluated, it is important to compare the current materials and waste costs to those costs of the new process or equipment. If the new process completely eliminates a waste stream, the potential cost savings associated with the environmental compliance, safety, and liability should also be considered.

Examples of procedures modification include waste segregation and full utilization of materials. Segregation can involve separating waste, recyclables, and reusable items from each other following use. It can also include keeping hazardous and nonhazardous materials separate during production to minimize the number of materials exposed to the hazardous materials. Training staff to segregate solid waste materials on the production floor is one way to reduce solid waste and to increase the quality of the recycling program. Providing procedures that encourage/require the full utilization of materials can also reduce solid waste by having products used, repaired, and reused until their useful life has expired. Administrative controls often involve methods of managing purchases and inventories. Purchasing products appropriate for their expected life is also an important cost and waste reduction strategy. If a material will have to be disposed of after one use, purchase only the quanity needed for the one use. Purchase a material that is recyclable for the one use if possible. If the life of the material is intended to be long, purchase a quality material that can be reused and easily maintained.

The careful management of inventories helps companies know the quantity of materials purchased, quantity of materials stored, and quantity of materials used. It helps identify which materials might benefit from purchasing in bulk, and whether or not certain materials need to be purchased in smaller quantities. Optimizing inventories allows for minimizing storage space and purchasing costs.

When materials are purchased, an arrangement with the suppliers can be made so that all packaging is returned to them for reuse. For example, packing boxes, packing peanuts, and uniquely shaped packing material can be sent back to a repeat supplier as they deliver each shipment. The return of the packaging can also include an agreement that the packaging will be reused until the quality no longer meets shipping requirements. In addition to controlling the packaging received by the suppliers, guidance for suppliers can be written so that it strongly encourages incorporation of pollution prevention/environmental considerations into the supplies they are providing.

Reuse of Solid Waste in Industrial Processes.   The simplest reuse alternative for industrial process solid waste is to purchase components that can be frequently reused and easily maintained and cleaned. Industrial wastes can be exchanged with other private and public organizations that are in need of those supplies. For example, outdated computers and other equipment can be provided to local schools for use and or disassembly for vocational training courses.

Industrial Ecoparks are created through a network of companies that have shared input and output material needs. For example, company A generates waste toner ink, company B needs black dye; company B generates waste sludge rich in nitrogen, company C can use the sludge to fertilize its tree farm; company C generates steam, company A needs steam to process its toner chemicals.

Recycle of Solid Waste in Industrial Processes.   The most widespread recycling activity of industrial solid waste involves common recyclables such as white paper, newspaper, cardboard, magazines, glass, plastic, wood, metals, mixed paper, software, toner cartridges, fly ash, and concrete. Major challenges include establishing appropriately placed collection points, establishing the habit of recycling, and selecting the recycling vendor or vendors to work with. The next step in recycling is to work with recycling vendors to identify other solid wastes that may be recyclable but are not currently considered as such. For example, waste toner ink that is manufactured for printers has high quality standards. When those quality standards are not met, the toner becomes waste. Toner manufacturers are working with others, including the textiles industry, to identify potential uses for the toner. Solid sludge waste may be recycled by being used a fertilizer for landscaping or agricultural purposes, depending on its nutrient content.

In addition to recycling the industrial solid wastes that are recyclable, designing the product that is being manufactured so that it is recyclable is a form of pollution prevention. One way to accomplish a recyclable end product is to use only one type of recyclable material (e.g., plastic type 1, plastic type 2, aluminum only, etc.) to make the product. If creating a product with only one material is not a possibility, then designing the product so that it can be easily disassembled into recyclable parts or into parts that can be salvaged for reuse is another option.

Other Industrial Wastes.   Examples here include solvents, oils, and chemicals. Some solvents and oils can be combusted as fuels or recycled via distillation (re-refining) or activated carbon adsorption/recovery. Herbst and Fitzgerald (2000) list these types of used fluids and oils that can be recycled: synthetic oils, engine oil, transmission fluid, refrigeration oil, compressor oils, metalworking fluids and oils, laminating oils, industrial hydraulic fluid, copper and aluminum wire-drawing solution, electric insulating oil, industrial process oils, and oils used as buoyants. They point out that filtering a portion of some oils results in recyclable oil, up to 90 percent less waste and up to 85 percent reduced chemical usage because unconsumed active ingredients can be reused many times over.

Gouchoe et al. (1996) suggest that each chemical used in each process be accounted for in how it is received, handled, stored, used, reused, and released as follows:

- Quantity received
- Release rate in by-products (accounting for amounts recycled, treated, or disposed of on-site)

- Release rate in emissions (accounting for amounts released to air, water, or land; amounts transferred off site to be recycled, disposed, or treated; or amounts disposed of on-site
- Release rate in products

Mulholland and Dyer (2000) give 50 pollution prevention strategies applicable to process industries. Included are the following:

- Consider buying purer raw materials or removing the impurities before they enter the process. The supplier may best accomplish this.
- Use oxygen instead of air for oxidation reactions. This minimizes introduction of noncondensible gases, which must be purged from the process and may require treatment. The cost to treat each incremental 100 standard cubic feet per minute (scfm) is $1000/year or more.
- Make packaging materials out of the end product so the customer can grind them and use them as a feedstock.
- Strip wastewater that contains volatile organics and recover the organics.
- Use improved piping and vessel cleaning technology, such as pipe-cleaning pigs, rotating spray heads, high-pressure jets, antistick coatings, better draining equipment, mechanical cleaning and sweeping, and multiple small rinses instead of filling and draining. Minimize or eliminate washdowns. (More details are given by Dyer, Mulholland, and Keller, 1999, and in Mulholland and Dyer, 1999.) Each gallon per minute of water costs $1000/year to get to the process and then discharge.
- Keep process-contaminated air separate from ventilation air. Stage ventilation from cleaner air to a smaller volume of dirtier air requiring abatement.
- Segregate process wastewater from storm runoff by covering areas and diking, rather than building larger treatment facilities to handle the additional flow.
- Use pressure or gravity feed instead of pumps that may have seal leaks.
- Turn off surplus aerators in wastewater treatment, which saves $27,000/year for each 100 hp.
- To reduce acid makeup requirements, use waste acid or by-product acid from another part of the plant. This strategy saves money for acid and for base to neutralize the acid in the effluent.
- Reduce caustic scrubber pH to 8 for controlling chlorine, bromine, sulfur dioxide, and hydrochloric acid. At pH 10, caustic scrubbing removes carbon dioxide from oxidizer offgases, at an incremental cost of $100 to $700/year per scfm.
- Minimize the number and size of samples, especially those containing hazardous substances; minimize dead volumes in sample lines; and recycle samples to the process.
- Minimize start-ups and shutdowns—consider the waste generated.
- Return impurities, unused raw materials, and packaging to suppliers.

12.3.3 Case Histories of Pollution Prevention in Industrial Processes

Matthews (1996) describes a process plant producing ion exchange resins and resin fiber mixtures that used up to 10,000 gal/day of deionized water that achieved zero discharge. Its old wastewater treatment system primarily reduced suspended solids and adjusted pH for discharge to a municipal sewer system. A membrane filter system is now used to produce permeate that is 100 percent recycled within the plant.

Wastewater Engineers (1999) relate an example where all water-based processed fluids are recycled (zero discharge) and of 90 percent of a chemical (soap cleaner) is recycled, with a payback period of only 3 months. The facility uses vibratory metal finishing. Treatment includes screening, filtering, and chemical precipitation. Savings by this treatment versus hauling wastes as an alternative amount to 60 percent of estimated hauling costs.

Dyer, Mulholland, and Keller (1999) relate two case histories for reducing organics emissions in batch processes:

- Leaks from atmospheric mixing tanks were controlled by changing lids to bolted, gasketed types and adding pressure/vacuum conservation vents. The site reduced air emissions by 40 percent and saved $426,000/year in methylene chloride costs.

- A batch herbicide intermediate production facility was converted to a continuous operation, which increased throughput and reduced methanol emissions by 29 percent.

Mulholland and Dyer (1999) recount these case histories for operations related to equipment and parts cleaning:

- A facility manufacturing multiple types of chlorinated aromatic compounds eliminated the practice of using solvent washes between campaigns. The washing operation generated a waste solvent stream containing some valuable product and contributed to lengthening turnaround times between campaigns. Drain valves were placed at low spots and residual product collected at the end of each campaign, and stored for reuse in the next campaign of that product. When flushing is necessary, the flush is retained, and small quantities are periodically recycled back into the process. The capital investment of $10,000 resulted in net present value of more than $2 million and a 100% reduction in waste generation.

- At a chemical manufacturing site, a series of distillation columns and tanks were used to purify different product crudes in separate campaigns. A portion of product was used to wash out equipment, until it became so contaminated it was destroyed in a hazardous-waste incinerator. For one tank, the wash procedure was analyzed and wash volume reduced by a factor of ten. A dedicated pipeline for each product was installed, thus eliminating the need to flush the line between campaigns. An improved drainage procedure was developed for one of the distillation columns. Product specifications were studied, and relaxing certain standards resulted in fewer washes being required to maintain product specifications. The capital investment of $700,000 had a net present value of more than $3 million. Waste generation was reduced by 78 percent.

- At a chemical plant, kettles were cleaned manually with solvent. Changing to a high-pressure rotating spray head reduced labor time, improved safety, and reduced solvent usage. Capital costs were $69,000 and savings are $61,500/yr.

12.4 POLLUTION PREVENTION/WASTE MINIMIZATION FOR MAINTENANCE AND OPERATIONS

Pollution prevention related to maintenance and operations is primarily focused on good housekeeping, optimizing systems, addressing opportunities for continuous improvement, and minimizing the impact of the materials being used. As the term *continuous improvement* implies, pollution prevention opportunities can be identified and implemented at any time during regular maintenance and operations activities. It is recommended that, in addition to the everyday incorporation of pollution prevention into maintenance and operations, any major change also be reviewed for new pollution prevention opportunities. For the consideration of the materials that are used during operations, see the suggestions provided in Sec. 12.2.2.

12.4.1 Possible Pollution Prevention Opportunities for Maintenance and Operations

Good housekeeping pollution prevention techniques include inventory management to ensure all hazardous materials are appropriately tracked, minimizing spills and cleanup volume by keeping work areas clear of clutter, and using nonhazardous cleaning products for the janitorial services.

Optimizing existing systems includes keeping the equipment clean and operating in the manner it is expected to. For example, when heating, ventilation, and air conditioning equipment is installed in a facility, it is typically set at the manufacturers' recommended settings. If the initial settings are not immediately pleasing to the occupants, the settings are likely to be changed. After the initial tweaking of the system, it is likely to remain at these settings throughout its operation. The pollution prevention opportunity is to re-evaluate the settings for the equipment to optimize its use for energy efficiency as well as occupancy comfort. A good start for a pollution prevention opportunity would be to evaluate all of the sources of energy use in the facility and inform the occupants of the steps that they can take to reduce energy usage.

Other pollution prevention opportunities include:

- Procure environmentally friendly products
- Minimize use of herbicides
- Compost kithcen and yard-related wastes
- Landscape using xeriscape techniques, i.e., no water use or low water use techniques
- Review previously identified pollution prevention opportunities to see if they are currently feasible for implementation

12.5 POLLUTION PREVENTION/WASTE MINIMIZATION IN ENVIRONMENTAL RESTORATION

Pollution prevention in the environmental restoration stage of a project may seem unfeasible, but in fact there are many ways to reduce waste during a cleanup project.

To maximize pollution prevention impacts during an environmental restoration project

1. Incorporate pollution prevention into the planning stage of the project and revisit it at appropriate times throughout the life of the project.
2. Offer technical exchange of successful techniques to key project staff.
3. Train all staff on the concept of pollution prevention.

Environmental restoration wastes are typically separated into two categories: (1) primary wastes (wastes that are the focus of the cleanup activity) and (2) secondary wastes (wastes generated during the cleanup process). Both potential waste streams can be reduced by source reduction, reuse, recycling, and treatment techniques. Reuse/recycle is the most effective technique for primary waste streams. Segregation is the most cost-effective technique for primary waste because it reduces the volume of hazardous and radioactive waste streams while increasing the value of recyclables. Source reduction is the most effective for secondary waste streams (Doe, 1996b).

One of the first things to do is consider the future use of the property. This helps define "how clean" the environmental restoration project will need to be to return the site. There is no need to return a site to pristine condition if a new industrial facility will be located there. Next, identify whether or not something has to be identified as waste in the first place—characterize it. For those items that don't have to be considered waste, segregate them from the waste streams, determine whether the item has to be removed from the site, and, if it does, whether there is a reuse or recycle option for the item. If something is categorized as waste, identify whether it can be reused in its existing form, recycled, or treated. Having pollution prevention as part of the planning process is key; working with the project managers and on their schedule will help make pollution prevention activities a success.

As in the design stage, one of the major challenges for the environmental restoration projects is lack of waste and cost data, since none of the wastes have been generated.

12.5.1 Methods for Identifying Pollution Prevention Opportunities for Environmental Restoration

The Minnesota Pollution Control Agency has developed a toolkit that assists in the identification of pollution prevention opportunities during an environmental restoration project. The toolkit includes a decision tree to select potential opportunities, remediation scenarios that are linked to pollution prevention opportunities, a checklist of issues to consider before pursuing opportunities, and points of implementation like performance measures, progress tracking, and recognition of successes. The Minnesota Pollution Control Agency breaks the toolkit into the following four steps (MPCA, 2000):

1. Select the most promising options for your site
2. Things to keep in mind before pursuing an option
3. Implement the options, track the progress, and document and promote lessons learned and success stories
4. Awards and recognition for implementation efforts

The U.S. Department of Energy suggests that pollution prevention be an integrating factor for environmental restoration projects rather than a separate activity

(U.S. DOE, 1999a). The U.S. Department of Energy suggests the following pollution prevention evaluation steps (U.S. DOE, 1999b):

1. Forecast and characterize project wastes
2. Identify potential pollution prevention opportunities
3. Evaluate opportunities for life cycle cost, schedule impact, environmental impact, economic impact, worker safety, public health, and social preferences
4. Implement cost-effective opportunities
5. Document and track implemented opportunities

12.5.2 Possible Pollution Prevention Opportunities for Environmental Restoration

The first pollution prevention opportunity that should be considered on an environmental restoration project is ensuring that the site has been well characterized and that an appropriate level of cleanup has been determined. Reducing the volumes of waste that must be removed, treated, and/or disposed results in avoidance of waste, cost, and risk.

Solid Waste. The U.S. Department of Energy has catalogued 198 specific pollution prevention opportunities, some of which are listed below (U.S. DOE, 1999a):

- Consider the potential value of the waste product for reuse or recycle.
- Consider whether a facility can be reused rather than demolished.
- Consider whether deconstructing a facility, rather than decommissioning or demolishing it, can result in the reuse of the building materials rather than disposal. When the building materials cannot be reused, they can often be recycled to be used in future building materials.
- Minimize the quantity of equipment being used during an environmental restoration project if there is the potential of it becoming contaminated and designated as waste.
- Reuse equipment and furniture rather than dispose of it, where feasible.
- Use in situ treatment to reduce the wastes associated with removing and transporting the wastes from the environmental restoration site.
- Restore habitat following environmental cleanup.
- Segregate during deconstruction or decommissioning to avoid generating mixed waste.
- Recycle clean concrete and scrap metal.
- Field-screen soils or other contaminated media and use clean soils as backfill.
- Use the "direct push" sampling method for subsurface soil and groundwater investigations rather than the conventional auger drill method.
- Replace plastic sheeting with launderable tarps.
- Replace tape used to secure personnel protective equipment with launderable ankle and wrist gauntlets (cuffs).

Liquid Waste. Some pollution prevention opportunities for liquids include:

- Using high-pressure water flushing with in situ inspection instrumentation to clean underground piping and to verify the quality of the cleanup
- Developing sampling plans so that sampling frequency can be reduced for water sources that do not show a significant change in contaminant concentration (U.S. DOE, 1999a)
- Reusing treated water for irrigation
- Allowing for natural attenuation of chlorinated hydrocarbons in groundwater, which eliminates the possibility of cross-media transfer and reduces energy usage (MPCA, 2000)

12.5.3 Case Histories for Environmental Restoration

The Princeton Plasma Physics Laboratory (PPPL) has been using a micropurge technique rather than the traditional method of purging a groundwater well 3 to 5 well volumes or until field parameters such as temperature, pH, and specific conductance have stabilized. The micropurge technique uses dedicated, pneumatic, in-well pumps to transfer very small purge and sample volumes to a portable analyzer. PPPL is using a flow-through type analyzer that returns the purge and sample volumes to the well after analysis. In addition to minimizing the volume of waste that is being generated, the micropurge technique increases the sample quality and decreases the turbidity in the sample because of its low flow velocities and minimal hydraulic disturbances. It also provides an increase in sampling efficiency because of the reduced need for equipment decontamination, reduced setup and breakdown time, and the reduced time needed for purging (PPPL, 2000).

Airborne particles generated from remediation activities are controlled by a high-efficiency particulate air (HEPA) filtering system. To reduce the loading on the HEPA filter and extend operating life, disposable prefilters can be installed upstream of the HEPA filter to capture the larger particulates. Disposable filters perform well; however, the cost of frequent shutdowns for prefilter changeouts and the increased radiologic waste volume provide an opportunity for pollution prevention.

The Energy Technology Engineering Center developed and completed the initial demonstration of a self-cleaning filter that can be used instead of the disposable prefilter. The new filter system is configured with six cartridge filter elements inside a collection hopper. Pulsed air is used to dislodge the particulates coating the elements into a collection drum. The self-cleaning capability eliminates the cost and increased volumes of radioactive waste previously incurred with the prefilter changeouts (U.S. DOE 1999a).

12.6 POLLUTION PREVENTION/WASTE MINIMIZATION COST ANALYSIS TECHNIQUES

Pollution prevention opportunities can be evaluated for cost avoidance by using quantitative and/or qualitative measures. There are many techniques for calculating the quantitative values. A unit-of-product technique uses production-adjusted measurements to provide waste generation data that relates to the volume of production over a given time period (Malkin et al., 1997). First a unit-of-product ratio is calculated and then that value is used to identify what an equivalent or expected level of waste generation would be in the current year. That value can be used to compare actual waste generation with the expected value. If production increases and waste generation remains constant, the actual waste generation would be lower than the expected waste generation, quantitatively demonstrating a waste reduction. Without using the unit-of-product technique, it would have appeared that waste generation had not decreased. The unit-of-product technique described in Malkin et al. (1997) uses the following equations:

$$\text{Unit-of-product ratio} = \frac{\text{quantity of product in year } n}{\text{quantity of product in year } (n-1)}$$

Expected waste generation in year n = unit-of-product ratio

$$\times \text{waste generation in year } (n-1)$$

Profitability factors include net present value and internal rate of return. These calculations depend on the period of evaluation. Net present value is the value of avoided costs after the payback of investment costs. The internal rate of return is the discount rate that produces a zero net present value at the end of the period of evaluation. With these techniques, the longer the time the project realizes cost avoidance, the more valuable the proposed change (Meenaham and Martz, 1999).

Qualitative measures may include reduced liability, increased productivity, improved product quality, improved safety and health, reduced worker exposure, and improved company image. The recommendation for qualitative measures is to document and present them as part of the overall benefit or cost of implementing a specific pollution prevention opportunity. Two methods, payback and return on investment, for calculating quantitative measures are described below.

The quantitative analysis evaluates the potential waste reduction and cost avoidance and answers four essential questions:

1. How much waste will the opportunity reduce or eliminate?
2. What financial costs will the opportunity avoid?
3. What will it cost to implement the pollution prevention opportunity?
4. Is the pollution prevention opportunity worth investing in?

On the basis of the data gathered above, calculate either the payback or return on investment and document qualitative benefits to assist in the implementation recommendation decision.

There are two kinds of waste reduction: annual and one-time reductions. Annual reductions come from changing a multiyear process so that waste reduction and cost avoidance will be yielded every year after the new technology or technique has been implemented. An example of an annual reduction is using a closed-loop cooling system for routinely used equipment to prevent once-through cooling water from being discharged. One-time savings come from making changes to a single action, such as sending a unique batch of excess materials to another organization for reuse, instead of disposal.

The following steps can be used to determine the amount of waste that will be reduced for each pollution prevention opportunity:

1. Review the quantity of previously generated waste, from a routine process, or estimate the amount of waste to be generated, from a one-time activity.
2. Estimate quantity of waste expected to be reduced from this pollution prevention opportunity, based on the information gathered when developing the pollu-

tion prevention idea. The reduction may be based on the difference between the waste generated by the previous approach to that of the opportunity, efficiency of new equipment, percent reduction in the amount of materials required, or amount of materials recycled or reused.

3. Determine what type of waste will be reduced, e.g., sanitary, hazardous, or radioactive. In this context, *hazardous* refers to waste sent to a qualified hazardous waste disposal company. *Sanitary,* or *nonhazardous,* refers to waste sent to a sanitary landfill or sanitary sewer. *Radioactive* waste is any material that spontaneously emits ionizing radiation. Hazardous and radioactive wastes are much more expensive and time-consuming to dispose of than nonhazardous waste. The waste form may also be relevant: solid waste (includes liquids that have been solidified prior to disposal), liquid waste going to a process or sanitary sewer, and gaseous air emissions.

4. Identify which wastes will be eliminated or reduced and/or if reductions in the toxicity of a waste (such as reducing hazardous waste to nonhazardous waste) will occur.

Total cost avoidance involves comparing the operating costs of the current practice with those of the proposed action (pollution prevention opportunity). Costs to be considered include the following, and any other costs for which reliable data can be gathered:

- Waste disposal costs
- Material costs
- Labor costs

Waste disposal costs involve the quantity of waste reduced multiplied by the unit cost of disposing of those wastes. Waste disposal costs include the cost to collect, log, handle, package, ship, and store the waste.

Material costs include the quantity of materials purchased multiplied by unit costs (including energy costs).

Labor costs involve the quantity of labor time needed to perform the operation multiplied by the appropriate unit labor rate. Labor costs should be documented when process efficiencies from new equipment or methods are expected to change the labor investment, time spent ordering materials, or maintaining equipment, and when reduced compliance and safety requirements would affect labor costs.

The total costs are calculated for the current practice, or way of doing business, and for the proposed pollution prevention opportunity. The total cost avoidance for a pollution prevention opportunity is calculated as the difference between the current practice and proposed action:

Subtotal of current practice costs = waste disposal costs

+ material costs + labor costs

Subtotal of proposed action costs = waste disposal costs

+ material costs + labor costs

Total cost avoidance = current practice − proposed action

As mentioned above, the total cost avoidance can be annual or one-time savings. It is important to note the type of costs involved so that the investment payback can be calculated.

Implementation cost is the *one-time expense* involved in putting a pollution prevention opportunity into practice. Typically, implementation costs come from two types of expenses:

- Equipment and initial material investment purchasing costs
- Labor costs for installation, testing, and training

Equipment costs include the costs paid to vendors to purchase new equipment and installation materials, plus any software or ancillary equipment, taxes, purchase adders, or additional direct costs needed to set up the pollution prevention opportunity.

Labor costs typically involve the number of hours required to implement this new activity, multiplied by a labor rate. Labor costs for implementation include costs for research of a new method; changing permits or other regulatory permissions; selecting, installing, and testing equipment; training; writing new procedures; and providing any reports required for start-up.

$$\text{Implementation cost} = \text{equipment costs} + \text{labor costs}$$

Payback.   A simple payback calculation is one method that is used to determine if a pollution prevention opportunity is worthy of investment. The payback calculation may provide additional data about a pollution prevention opportunity that can be used to assist decision makers in investment decisions:

$$\text{Payback (years)} = \frac{\text{implementation cost (\$)}}{\text{total cost avoidance (\$/year)}}$$

The payback is the number of years or the fraction of the year needed to repay the initial investment in implementing the opportunity. A 3-year payback is typically considered "worth the investment." However, some businesses consider 1-year paybacks, while others accept longer-term paybacks. Payback is usually expressed in years as it is above. For one-time cost avoidance projects, payback can be expressed in dollars. If the result is positive (greater than zero), it is worthwhile to implement.

$$\text{Payback (\$)} = \text{total cost avoidance (\$)} - \text{implementation costs (\$)}$$

Return on Investment (ROI).   A return-on-investment calculation is the second method described in this chapter that can be used to determine if a pollution prevention opportunity is worthy of investment. Like payback, the return-on-investment calculation provides additional data about pollution prevention opportunities for assisting decision makers in investment decisions. Businesses that use a percent return on investment method often require at least a 33 percent ROI, though some require as much as a 100 percent return and others as low as 10 percent.

To calculate ROI, inverse the payback calculation and multiply by 100 to get a percentage:

$$\%\,\text{ROI} = \frac{\text{total cost avoidance (\$/year)}}{\text{implementation costs (\$)}} \times 100$$

One technique that considers the life of the pollution prevention project follows:

$$\%\,\text{ROI} = \frac{\text{TCA} - \text{IC/LP}}{\text{IC}} \times 100$$

where   TCA = total cost avoidance, $/year
         IC = implementation cost, $
         LP = life of project, years

The *life of project* is the length of time the new equipment or new process will be used, typically 5 to 20 years.

The cost analysis data are used to help decide which pollution prevention opportunities to implement. Accepted payback times and return on investment percentages vary with each company.

Depending on the detail of the analysis desired and the quantity of data available, a qualitative analysis of a pollution prevention opportunity may be appropriate. Additionally, even if quantitative data are available, qualitative information should also be documented to ensure critical issues are discussed.

Items to consider in a qualitative analysis include: increased productivity, increased safety, reduced worker exposure, improved capacity, higher quality product, and marketability of an innovative technique.

REFERENCES

Anderson, R. C. (1998), *Mid-Course Correction*, ISBN 0-9645953-5-4, Peregrinzilla Press, Atlanta.

Cannon, S. D. (1999), Pollution Prevention Successes Web site, http://p2.pnl.gov:2080/p2/scs9906.htm

Cates, M. R. (2000), "A New Vision for Lighting," *Environmental Design + Construction*, pp. 58–60, September/October.

Dyer, J. A., and K. L. Mulholland (1998), "Follow This Path to Pollution Prevention," *Chemical Engineering Progress*, pp. 34–42, Jan.

Dyer, J. A., K. L. Mulholland, and R. A. Keller (1999), "Prevent Pollution in Batch Processes," *Chemical Engineering Progress*, pp. 24–29, May.

Ehrenfeld, J., and N. Gertler (1997), "Industrial Ecology in Practice: The Evolution of Interdependence at Kalundborg," *Journal of Industrial Ecology*, vol. 1, no. 1, pp. 67–79, ISSN 1088-1980, School of Forestry and Environmental Studies, Yale University, MIT Press, Cambridge, Mass., http://www.indigodev.com/Kal.htm.

Engel-Cox, J. A., and K. M. Fowler (1999), "*Pollution Prevention Opportunity Assessments for Research and Development Laboratories,*" ISBN 1-57477-070-5, Battelle Press, Columbus, Ohio.

Goldman, Matthew (2000), "Integrate Environmental Management into Business Functions," *Chemical Engineering Progress*, pp. 27–33, March.

Gouchoe, Susan, et al. (1996), "Integrate Your Plant's Pollution Prevention Plans," *Chemical Engineering Progress*, pp. 30–43, November.

Hawkin, P. (1993), *The Ecology of Commerce*, ISBN 0-88730-655-1, Harper Collins Publishers, New York.

Herbst, S., J. Fitzgerald (2000), "Reaping the Benefits of Waste Recycling," *Pollution Engineering*, pp. 46–49, April.

Hilaly, A. K., and S. K. Sikdar (1996), "Process Simulation Tools for Pollution Prevention: New Methods Reduce the Magnitude of Waste Streams," *Chemical Engineering*, pp. 98–104, February.

Iwai, T. (1995), "Toyota's Activities for the Environment," *International Journal of Environmentally Conscious Design & Manufacturing*, vol. 4, no. 1, pp. 29–42, ISSN 1062-6832, ECM Press, Albuquerque, N.M.

Jessen and Kemp. (1996). "Zero Discharge," *Environmental Engineering World*, November-December, pp. 15–18.

Luper, Deborah (1996), "Integrate Waste Minimization into R&D and Design," *Chemical Engineering Progress,* pp. 58–60, June.

Malkin, M., J. Baskir, and T. J. Greiner (1997), "Developing and Using Production-Adjusted Measurements of Pollution Prevention," EPA/600/SR-97/048, U.S. Environmental Protection Agency, Cincinnati, Ohio, November.

Matthews, Gary A. (1996), "Achieving Zero Wastewater Discharge," *Environmental Technology,* pp. 42ff, July/August.

Meenaham, J., and D. Martz (1999), "Pollution Prevention Comes of Age: Accounting for the True Cost of Waste," *Pollution Engineering,* February.

Minnesota Pollution Control Agency MPCA (2000), Web site: http://www.pca.state.mn.us/programs/p2-s/remediation/index.html.

Mulholland, K. L., and J. A. Dyer (1999), "Prevent Pollution in Equipment and Parts Cleaning Operations," *Chemical Engineering Progress,* pp. 30–34, May.

Mulholland, K. L., J. A. Dyer (2000), "Reduce Waste and Make Money," *Chemical Engineering Progress,* pp. 59–62, January.

Princeton Plasma Physics Laboratory (2000), Success Story Web site: http://epic.er.doe.gov/epic/docs/mcroprge.txt.

Quinn, Barbara (1996), "Catch These Pollution Prevention Sources on the Web," *Pollution Engineering,* pp. 21–22, April.

Shahinpoor, M., and J. Lantz (1996), "An Environmentally Conscious Robotically Operated Fluorescent Lamp Recycling Plant," *International Journal of Environmentally Conscious Design & Manufacturing,* vol. 5, no. 2, pp. 37–39, ISSN: 1062-6832, ECM Press, Albuquerque, N.M.

Toyota Motor Sales, U.S.A. (2000), "New Beginnings," PMO-099, Web site: http://prius.toyota.com/.

U.S.C. Title 42, "Pollution Prevention Act of 1990," The Public Health and Welfare, Chap. 133–Pollution Prevention, 13101, Findings and Policy.

U.S. Department of Energy (1996), "Life Cycle Asset Management: Waste Minimization/Pollution Prevention," Good Practice Guide, GPG-FM-25.

U.S. Department of Energy (1999a), "Cleanup, Stabilization, and Decommissioning Activities Waste Reduction Goal Guidance Document: Case Studies Appendix," Web site address: http://www.em.doe.gov/p2/index.html.

U.S. Department of Energy (1999b), "Guidance for Applying Pollution Prevention and Waste Minimization During Environmental Restoration Activities," Web site: http://www.em.doe.gov/p2/index.html.

U.S. Department of Energy (2000), "Roadmap for Integrating Sustainable Design into Site-Level Operations," PNNL-13183, Pacific Northwest National Laboratory, Richland, Wash.

U.S. Executive Order (E.O.) 13101 (1999), "Greening the Government through Waste Prevention, Recycling, and Federal Acquisition."

Wastewater Engineers, Inc. (1999), "Case Histories in Wastewater Recycling," *Environmental Technology,* pp. 24–26, November/December.

Zinkus, G. A., W. D. Byers, and W. W. Doerr (1998). "Identify Appropriate Water Reclamation Technologies," *Chemical Engineering Progress,* pp. 19–31, May.

INDEX